Gas Turbine Propulsion

Gas Turbine Propulsion

D.P. Mishra

MV Learning
London • New Delhi

Reprinted 2024

MV Learning

3, Henrietta Street
London WC2E 8LU
UK

4737/23, Ansari Road,
Daryaganj, New Delhi 110 002
India

A Viva Books imprint

ISBN: 978-81-309-2752-7

Printed and bound in India.

SUB-0002/SSUB-0008

Dedicated to

My parents and students
who are my teachers in disguise

Preface

This textbook is designed for undergraduate and postgraduate engineering students interested in learning the fundamental aspects of aircraft propulsion. The basic theory of the gas turbine engine based on the principle of thermodynamics and aerodynamics is covered extensively, as most of the recent commercial and military aircraft are powered by the gas turbine engine. Emphasis has been placed on precise and logical presentation of the basic concepts and principles, which are essential for the better understanding of aircraft propulsion. Care has been taken to present the material in a gradual manner to instil confidence in the mind of the reader. Attempts have also been made to keep the derivation of fundamental equations and principles at an uncomplicated mathematical level. Several examples are provided in each chapter for the reader to assimilate and reinforce the ideas developed in this book. Adequate emphasis has been laid on problem solving techniques. The rudiments of the design, operation and behaviour of individual components of the gas turbine engine are included. Typical values of design parameters, experimental data and examples of actual engines are also included in several places to help the reader appreciate the subject matter. However, the details of the practical aspects such as mechanical design, control systems and manufacturing procedures etc., have not been included here deliberately, because these are not considered to be part of fundamental theory, as it undergoes frequent changes. The book starts with Chapter 1 that gives an introduction to various aspects of propulsion with a brief historical account. Chapter 2 is devoted to the review of the fundamental aspects of thermodynamics, combustion and fluid mechanics that are very essential for the understanding of the gas turbine engine. In Chapter 3, the reader is exposed to the basic definition of certain performance parameters such as thrust, thrust specific fuel consumption, propulsive, thermal and overall efficiencies etc. An expression for uninstalled thrust has been derived from the basic concept of momentum conservation. The performance of aircraft engines is presented in Chapters 4 and 5, which cover both ideal and real thermodynamic cycle analysis. The importance of the parametric cycle analysis for the design of gas turbine engine for an aircraft is highlighted in these chapters. The steps involved in parametric cycle analysis are also enumerated in detail. The cycle analysis is illustrated first for a simple system like ramjet and is gradually extended to the complex mixed flow turbofan engines. In Chapter 5, the engines are analysed more rigorously by taking account of losses incurred in various components of an engine. A comparison of real and ideal cycle analysis for the same set of parameters is made for all types of aircraft engines. The parametric analysis will help the reader to appreciate the intricacies of design problems. The rudiments of turbomachines are discussed in Chapter 6 in detail to familiarise the reader with their aerothermodynamic aspects. The Euler's Equation for turbomachine is derived from the Newton's Second Law of Motion. Chapters 7 and 8 are devoted to the basics of axial and centrifugal compressor – the crux of the gas turbine engine. The mean line analysis of the compressor is discussed with suitable examples to illustrate how to carry out preliminary analysis of a compressor. The performance parameters are discussed in detail. The phenomena of surging and rotating stall as well as their associated problems and their control are also covered appropriately. The basic theory and mean line analysis aspects of axial turbine are discussed in detail in Chapter 9. The performance of a turbine is dealt with while taking account of losses incurred in the turbine. A separate section is devoted to turbine blade cooling, including simple analysis of heat transfer of the turbine blades. Chapter 10 is devoted to the combustion system of the engine. Both the main combustor and afterburner are discussed in detail covering most of the important components of the combustion system. A separate section is provided for the fuel atomiser, whose performance dictates the performance of the main combustor. The performance parameters of the combustion systems are enumerated in detail. A full section is included on the emission of the gas turbine engine and its control. The operation and performance characteristics of both subsonic and supersonic air intake and exhaust nozzles are discussed in Chapters 11 and 12, respectively. Simple aerothermodynamic analysis is presented to illustrate the processes involved during deceleration of the flow to lower velocity. The starting problem of supersonic air intake is covered in Chapter 11. The basic principles of controlling its operation are illustrated by considering practical examples. In Chapter 12, the aerothermodynamic analysis of the exhaust nozzle is covered with a brief section on its performance. A separate section is devoted to various aspects of thrust vector control of the exhaust nozzle. Certain useful data, particularly required for a full appreciation of the material covered in this book, are given in the Appendices. This book will be both interesting and rewarding for the students and practicing engineers.

I am indebted to many individuals who have contributed directly or indirectly towards the preparation of this manuscript. First of all, I am thankful to several authors and researchers in this specialised field who have dedicated their life for unraveling knowledge which has been presented in this book for the benefit of students. I am indebted to my graduate students, Ugandhar, Swarup, Mahesh, Ezhil and Manisha for suggesting corrections in the manuscript. I am also thankful to my student P.K. Ezhil who has helped me in preparing graphs used in cycle analysis chapters. Besides this I appreciate the efforts of Ranjeet, Pinki and Sunil who have helped me in preparing figures presented in this book. Several students, namely, Bharthi, Sridhar, Ashok Kumar, Rajini, Subhranshu, Sanat, Gauravendra, Bichitra, Ankit, Jitendra, Asuthosh, Somprakash, Sivam, Yash, and many more students and my project staff members, namely, Ashok, Alok and Ankit have helped me in several ways. The typing support provided by Sunil and Pinki is highly appreciated. The financial support provided by the Continuing Education Centre of the Indian Institute of Technology Kanpur, Kanpur, for preparation of the manuscript is gratefully acknowledged. Lastly, the persistent support of my spouse, Itishree and son, Rutamvar for this time consuming project is highly appreciated.

D.P. MISHRA

Contents

1. Introduction

I have discovered that a screw-shaped device such as this, if it is well made from starched linen, will rise in the air if turned quickly.

Leonardo Da Vinci, 1480

I confess that in 1901, I said to my brother Orville that man would not fly for fifty years... Ever since, I have distrusted myself and avoided all predictions.

Wilbur Wright, in a speech to the Aero Club of France, 5 November 1908

Man's desire to fly has culminated into the development of exotic modern aircraft. Most of us might have been fascinated during our childhood by the flight of an aircraft in the sky. All early attempts of human flight were based on man's desire to imitate the flying of birds. Several attempts were made by people during the medieval time by leaping off the mountains, or tree tops or towers using flapping mechanisms attached to their limbs to simulate a bird's wing. Most of these earlier attempts to fly were unsuccessful and some even had disastrous consequences. During childhood my grandmother used to recount an incident how my great grandfather had met with a serious accident when he attempted to fly by strapping a pair of wings made out of bamboos to his arms. Interestingly, this kind of concept was described even in the early Greek mythology of *Daedalus* and his son *Icarus* who had attempted to escape from the labyrinth of *Crete* King *Minos* by mimicking a bird's flight with the help of feathers made of wax. The young and enthusiastic *Icarus* did not pay any heed to his father's advice not to fly close to the sun. As he flew closer to the sun, the wax melted due to the extreme heat and his wings disintegrated. As a result, *Icarus* plunged into the Mediterranean Sea. In due course of time, man abandoned the idea of strapping a pair of wings to his arms, realising the fact that human beings do not have the muscles and the bone structure of birds, which enables the latter to fly. However, gliders were built by taking cue from kite flying – an activity that most of us might have enjoyed during our childhood, and even adulthood. Sir George Cayle was the first person to build a glider in 1804. After around 100 years, the Wright Brothers – Orville and Wilbur – were the first humans to succeed in developing and demonstrating the flight of the first powered aircraft in 1903. This event was an epoch making starting point in the glorious development of modern aircraft.

1.1 Principle of Jet Propulsion

Propulsion is a method by which an object is propelled in a particular direction. The verb 'propel' means to drive or cause to move an object in a specified direction. The study of propulsion is concerned with this propelling force, the motion caused and the bodies involved. This study is not only concerned with an aircraft but also with other vehicles such as spacecraft, missiles, automobiles, trains, ships etc. The theory of jet propulsion is based on the principles of Newton's Laws of motion. Jet propulsion can be explained mainly by Newton's Second and Third Laws of motion. Let us consider an aircraft, flying at uniform speed in a straight flight path. Now the resultant force in the horizontal direction must be zero to satisfy Newton's Second Law of motion, which states that "*an unbalanced force acting on the body tends to produce an acceleration in the direction of the resultant force, which is proportional to the product of mass and acceleration of the body*". In other words, the aircraft must produce an amount of thrust, which must be equal to the drag force caused due to the fluid motion over the aircraft body. For accelerating an aircraft, one needs to supply higher thrust than that of the drag forces acting on it.

According to Newton's Third Law of motion "*for every acting force, there is an equal and opposite reacting force*". The acting force is the force exerted by one body on another, while the reacting force is the force acted by the second body on the first. Although these forces have equal magnitude and occur in opposite directions, they never cancel each other, because these forces always act on two different objects. Several examples of this phenomena are encountered in our day-to-day life such as recoil of a pistol, backward movement of a boat when a man jumps off it etc. For self-propelled vehicles such as automobiles, the second body is the earth itself and the process of propulsion involves the transfer of momentum between the automobile and the earth. The principle of jet thrust can be better understood by considering an inflated balloon (Fig. 1.1).

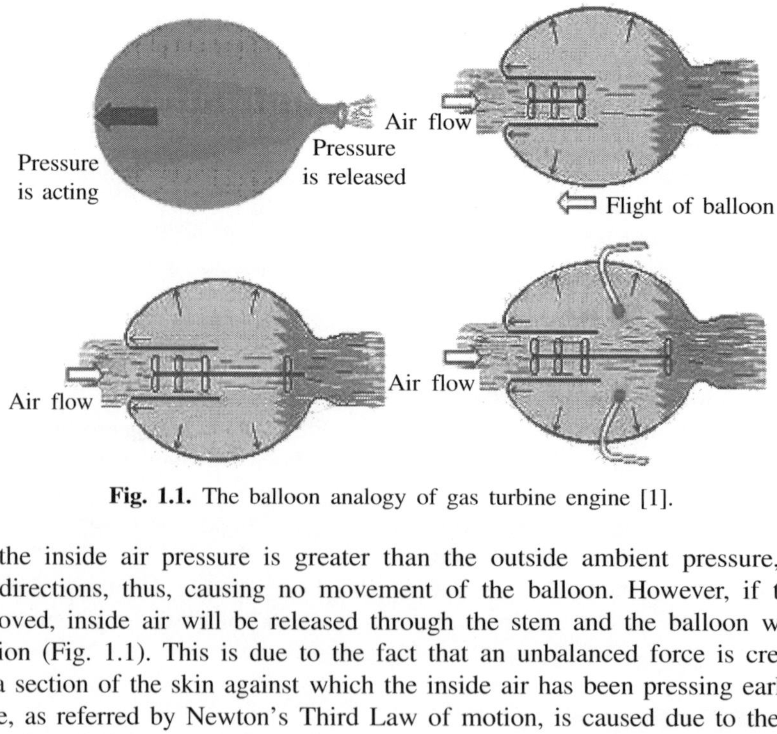

Fig. 1.1. The balloon analogy of gas turbine engine [1].

Although the inside air pressure is greater than the outside ambient pressure, equal force is exerted in all directions, thus, causing no movement of the balloon. However, if the stem of the balloon is removed, inside air will be released through the stem and the balloon will move in the opposite direction (Fig. 1.1). This is due to the fact that an unbalanced force is created as a result of removal of a section of the skin against which the inside air has been pressing earlier. In this case the acting force, as referred by Newton's Third Law of motion, is caused due to the acceleration of the air escaping through the opening of the stem. As the force occurs in pair, the reaction to this force is known as thrust, which acts in opposite direction. Hence, this definition clearly states that the study of propulsion must cover: (i) the propelling force; (ii) the motion caused due to the force on the body in a preferred direction; and (iii) the bodies involved. Thus, the process of propulsion always involves an object (airplane) to be propelled and one or more additional bodies, known as the *propellant* (fuel and air). It can be appreciated that this flight of the balloon would not last long, as the inside pressure gets quickly reduced to the ambient pressure. However, the flight of the balloon can be sustained if we can manage to supply high pressure air externally for maintaining a certain amount of air flow rate at the exit. This task can be accomplished by the use of a compressor (Fig. 1.1). In order to make it a self-sustained device, one needs to supply shaft power to run the compressor. The shaft power can be obtained by placing a turbine in the path of air stream. The energy must be supplied to the air by burning of fuel in a combustor (Fig. 1.1). After power is extracted by the turbine, the remaining energy is extracted by expanding the hot gas through the stem of the balloon, which resembles a nozzle in an actual gas turbine engine. Hence, the gas turbine engine consists of three main components – compressor, combustor and turbine. Besides these, an air-intake and nozzle are essential for aircraft jet engine. The balloon analogy of the jet engine must be kept in mind for better understanding of jet propulsion. In this book, we would restrict our discussion to the aircraft propulsion only.

1.2 Aircraft Propulsion: A Brief History

One can find several examples of devices related to aircraft propulsion, utilising the reaction principle, during the early part of recorded history. All early attempts of mankind to develop a flying device were not successful due to lack of sufficient technical knowledge in the field of engineering, manufacturing and metallurgy, although the principle of jet propulsion had been demonstrated around 2,000 years ago (150 BC) by Hero, an Egyptian scientist living in Alexandria. The schematic of Hero's *aeropile* (Fig. 1.2) shows that the heat from the burning fuel was used to convert water into steam. Two tubular uprights were attached to the head of the boiler, which discharged steam at right angles to the axis of rotation. This is really an interesting device for demonstrating the principle of reactive thrust. It is believed that the Chinese had developed windmills based on the principle of reactive thrust around the same period. Leonardo da Vinci, the famous artist, had created several sketches of flying machines during 1425-1519 AD. The diagram of a chimney jack made by da Vinci (Fig. 1.3) illustrates the application of the reaction principle. Hot air from the fire is passed through the blades, which helps in turning the roasting spits. Later on, this device was widely used for turning roasting pits. In contrast, it is believed that the Chinese developed rockets around 1232 following the invention of gunpowder whose main constituents are charcoal, sulphur and potassium nitrate. The Chinese scholar, Wan Hu developed a rocket sled (Fig. 1.4) which comprised of a series of rockets

Fig. 1.2. Aeropile by Hero of Alexandria [1].

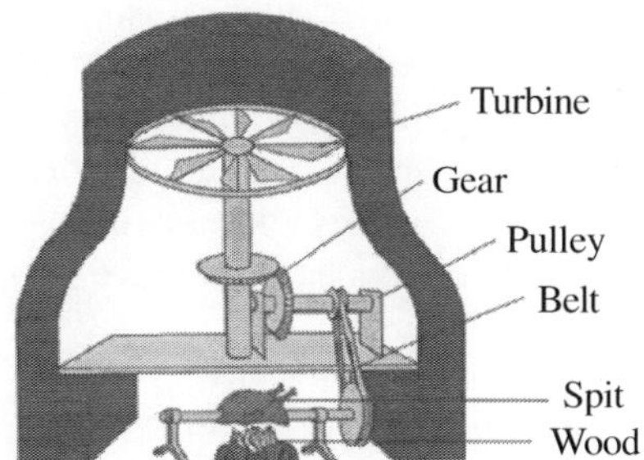

Fig. 1.3. Leonardo da Vinci's chimney jack [1].

attached to a chair. Unfortunately, he died while operating this device. We find that the reaction thrust principle was used by several people long before Sir Isaac Newton could formulate the laws of motions in the 17th century based on which modern jet engines are designed and developed. But the first gas turbine engine was designed by John Barber, an Englishman in 1791. This engine comprised of a reciprocating compressor, a combustion system, and a turbine. It is interesting to note that Barber had used water cooled turbine blades even at that time. Later on, several attempts were made to devise aircraft engines. A few of them are discussed in this section. Sir George Cayley invented in 1807, a reciprocating hot air engine that operated on the principles of modern closed cycle gas turbine engine. In 1851, the first aircraft was flown with average speed of 11 km/h from Paris to Trapper by Henerery Gifford who used a steam engine. It is believed that the aircraft was kept afloat by the balloon attached to it. This idea of using a steam engine was abandoned due to its very heavy weight. Later on, Dr. F. Stoltz designed and developed an engine in 1872 that never ran under its own power as the component efficiencies were quite low. Although the basic principles of jet propulsion were known at that time, but engineering know-how, which is dependent on availability of materials, skills, machine tools, were not matured enough to demonstrate this concept successfully. Hence, it must be kept in mind that although the scientific knowledge is essential but it may not be sufficient enough for the successful development of a device. However, the first successful aircraft was flown in 1903 by the Wright Brothers. The aircraft was powered by a piston engine driving a propeller and it marked the beginning of a glorious era of air flight.

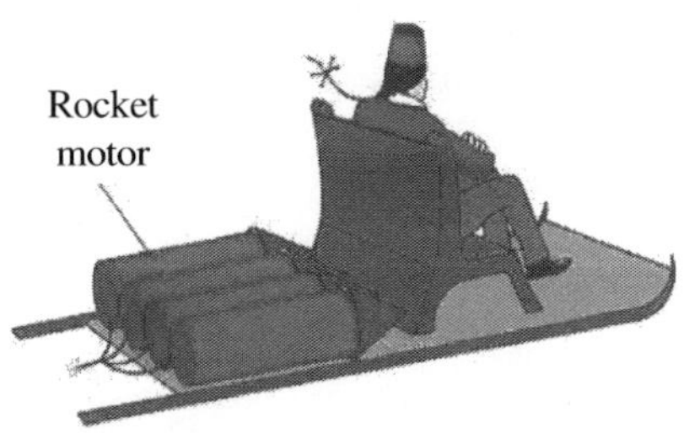

Fig. 1.4. Wan Hu's rocket sled [1].

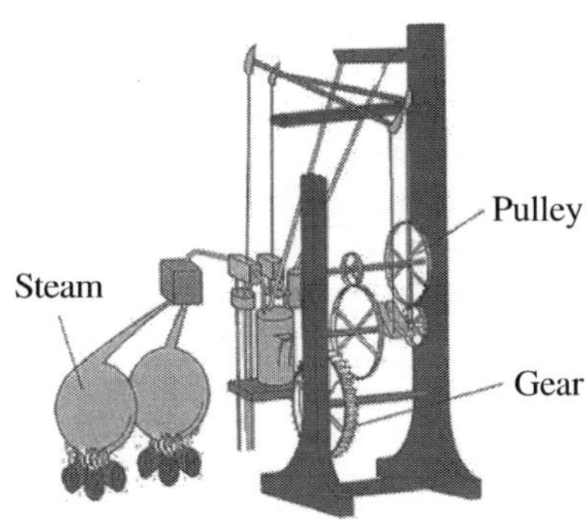

Fig. 1.5. Lorin's air breathing engine [1].

The development of jet engines followed soon. Lorin was the first to patent a jet engine based on the piston engine (Fig. 1.5) as early as 1908. Later on, in 1913, he coined another interesting and important idea of ramjet engine using ram pressure. This engine was a very simple and elegant propellant device from the design point of view. In 1921, M. Guillaume patented a jet engine based on turbomachinery. In this engine, air was compressed in an axial compressor and gas was expanded by an axial flow turbine. Although these concepts managed to demonstrate the concept of air breathing jet engines, they did not find any immediate application in aircraft, because the propulsive efficiency of these engines were quite low, particularly, at low speed. At that time, the flight speed of an aircraft was limited to a very low value (Mach number $M < 0.3$). In 1939, a German inventor, Hans von Ohain became the first person to fly a jet engine powered by a gas turbine engine. In 1941, working independently, Sir Frank Whittle was also instrumental in developing gas turbine based jet engines (Fig. 1.6). But it was only in 1991 that von Ohain was recognised as a co-inventor of the gas turbine based jet engine along with Sir Whittle. In 1950, the first supersonic aircraft, the Concord

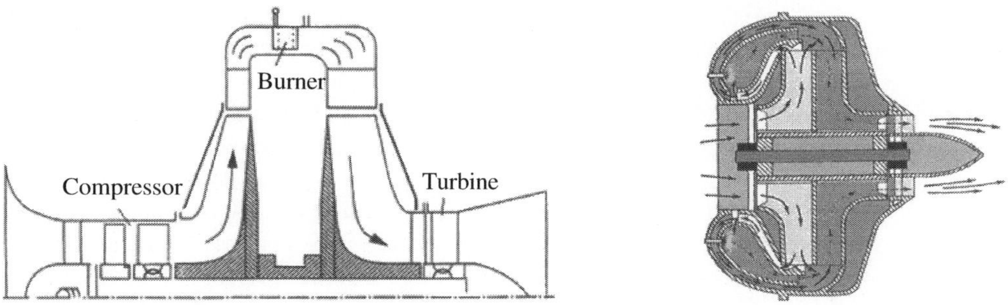

(a) (b)

Fig. 1.6. Gas turbine engine developed by (a) von Ohain and (b) Whittle [2].

was developed successfully. After an illustrious, long service record this aircraft was grounded in 2002.

1.3 Classification of Propulsive Engines

In the 20th century, several propulsive devices were designed and developed, which are used in modern aircraft. These devices can be broadly divided into two categories viz. airbreathing and non-airbreathing engines (Fig. 1.7). The energy of the fuel carried by aircraft is extracted by burning the fuel with an oxidiser in the combustion chamber of the engine. In case of air breathing engines, the oxidiser is taken from the ambient air for burning of fuels, whereas in non-airbreathing engines the oxidisers are carried on board the vehicle. Such kind of engines are known as *rockets*. This pertains to chemical rocket engines only. The term *rocket* is more broad-based, and its definition can be conceived as "*the device in which mass contained in the vehicle is ejected rearward by which thrust*

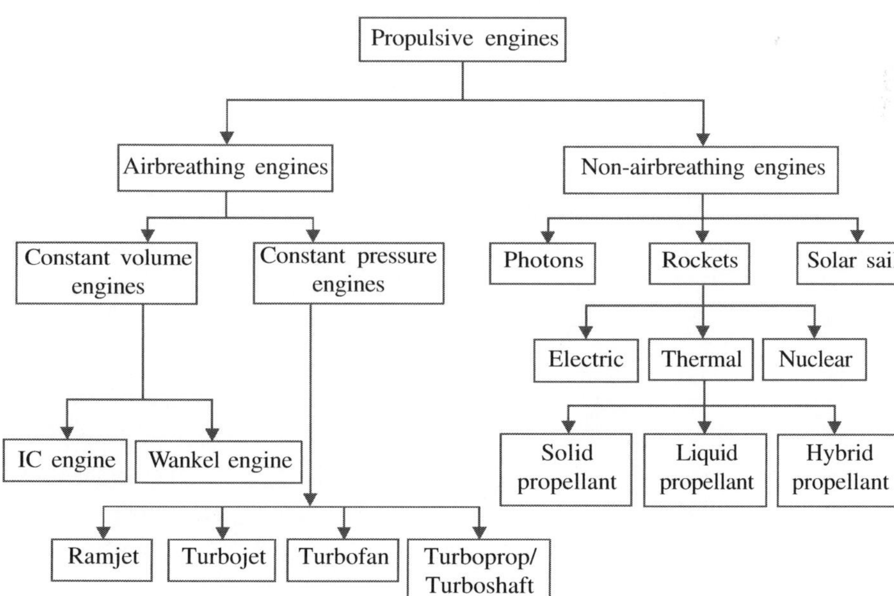

Fig. 1.7. Types of propulsive engines.

is produced for its propulsion". For rocket engines, chemical, solar, electrical and nuclear energy can be utilised for producing propulsive power. Chemical rockets, based on the types of propellants used, are divided into three categories (Fig. 1.7) viz. solid, liquid, and hybrid.

Airbreathing engines can be broadly divided into two categories viz. constant pressure combustion and constant volume combustion. The first category is the most important and is used in most medium and large aircraft engines. The engines are based on the Brayton cycle, discussed in subsequent sections. This kind of engine can be further divided into two categories viz. gas turbine and ramjet. Of course, the use of ramjet engines is restricted to missile applications. There is another exotic jet engine known as the *scramjet* engine. In recent times, considerable amount of efforts have been made to develop the scramjet engine, which is believed to be the future power plant for hypersonic air transport. Besides this, in recent times, efforts are underway, particularly in developed countries, to design and develop pulse detonation engine, which is likely to enhance the engine efficiency. Note that the gas turbine engine can be classified broadly into three categories viz. turbojet, turbofan, and turboprop/turboshaft. An extensive introductory discussion on gas turbine engines will be given in the subsequent sections.

1.4 Aircraft Engines

Both piston and gas turbine engines are used to propel aircrafts. However, the present discussion is restricted to only the gas turbine operated aircraft engine. The gas generator is the genesis of any gas turbine engine. The schematic of a gas generator (Fig. 1.8) shows that it consists of three main components viz. compressor, combustor, and turbine. The main function of the gas generator is to harness power to run the compressor while producing high temperature and pressure gas for generating thrust or shaft power as per requirement. This engine works on the principle of Brayton cycle, discussed in detail in subsequent chapters. Air is ingested into air intake and, subsequently, it gets compressed in the compressor. Then liquid hydrocarbon fuel is injected into the high pressure gas in the combustor where burning of fuels occurs, liberating large amount of heat. Hot gas at high pressure is then expanded in the turbine to produce mechanical power by converting thermal energy of fuel into kinetic energy. Note that the gas generator is common to all types of gas turbine based aircrafts. Besides this, gas turbines are used for power generation and surface transport.

A turbojet engine can be devised by adding an air intake and a nozzle (Fig. 1.9) to the basic gas generator configuration. When air enters into the intake of the turbojet engine, it gets compressed, resulting in an increase in pressure. This compression is known as *ram compression*. The major portion of compression is generally carried out mechanically by either axial or centrifugal compressor. Note

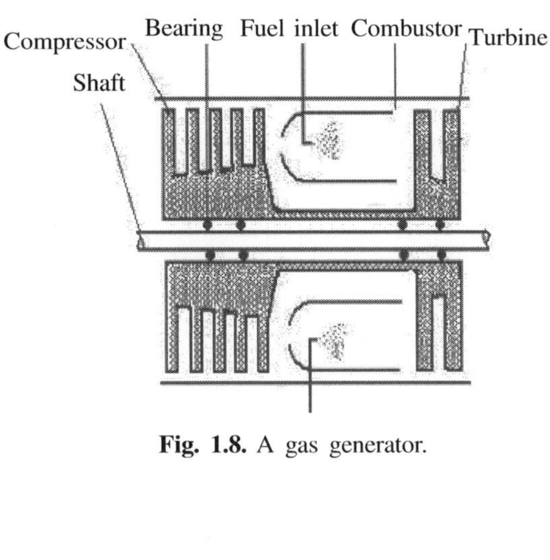

Fig. 1.8. A gas generator.

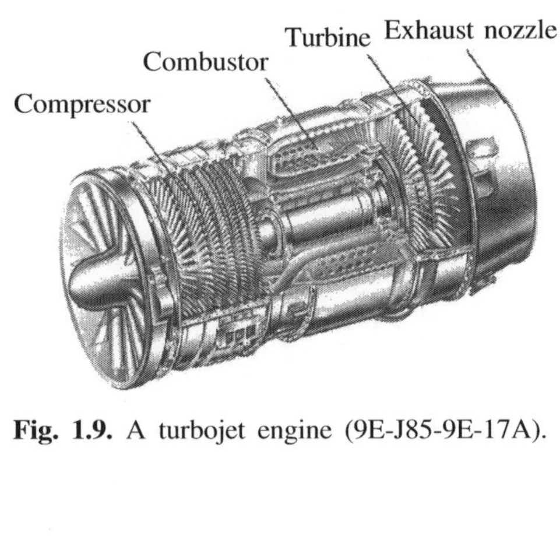

Fig. 1.9. A turbojet engine (9E-J85-9E-17A).

that the ram compression is predominant only when the aircraft speed is quite high. Generally, in modern aircraft engines pressure ratio across the compressor in the range of 10 to 40 is used. After compression of air, the temperature of air is raised to around 1400-1800 K by combustion of liquid hydrocarbon fuel. Thereafter, gases pass through the turbine in which they are expanded to produce mechanical work. In case of turbojet engine, the function of the turbine is to supply power to run the compressor. Then gas is further expanded in the nozzle, which provides the requisite thrust to the aircraft. Note that the net thrust produced by the engine is due to conversion of internal energy into kinetic energy. In case of turbojet engine, loss of kinetic energy in the exhaust gas is quite exorbitant, resulting in lower propulsive efficiency.

The variations of pressure, temperature and velocity of gas while passing through a typical turbojet engine are shown in Fig. 1.10. Note that an air intake acts like a diffuser, resulting in an increase in pressure and temperature. As the air flows through the compressor, both pressure and temperature increase along its length as the work is done on the air. The decrease in air velocity is very minimal. The diffuser portion of the combustor further reduces the velocity. As air passes through the main combustor, its temperature is increased by considerable amount due to the burning of fuel. The velocity is increased in the main combustor portion due to release of heat (gas expansion). Pressure remains almost constant across the combustor. These hot gases pass through the stages of turbine which extract work resulting in a drop in temperature and pressure at the turbine exit. Finally, the hot products are expanded in the nozzle and leave its exit at higher velocity. In the process, the requisite thrust is generated by the engine. The hot gases at the end of the turbine are at higher pressure than the ambient and still contain a large amount of oxygen. These gases containing enough amount of oxygen can be utilised to burn more fuel in the afterburner before being expanded in the nozzle. The burning of additional fuel in this afterburner is preferred in comparison to the main combustor as it is quite simpler to augment the thrust. In this way, higher thrust can be obtained easily. Such kind of engine is known as *afterburning turbojet engine*.

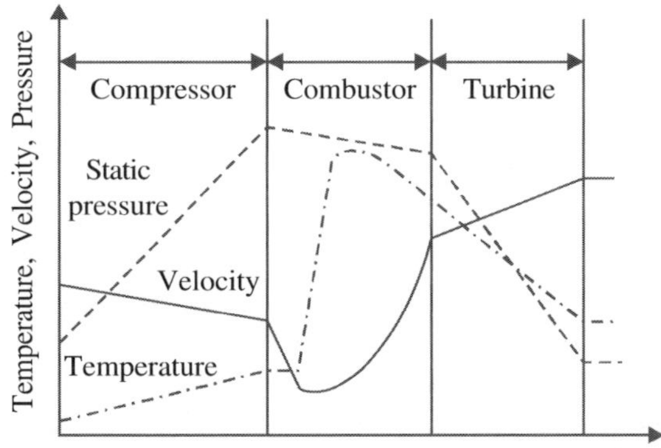

Fig. 1.10. Variation of pressure, temperature and velocity across various components of a typical turbojet engine.

In some configurations, a second turbine is used to run a fan. Such an engine is known as *turbofan engine* (Fig. 1.11) in which an air intake, fan and nozzle are used along with the generic gas generator. This fan is similar to a propeller but smaller in size and with multiple blades. Besides this, multi-bladed rotors are enclosed in a duct (Fig. 1.11) that suck large amounts of air at a lower velocity. Mostly, the fan is placed at the front, but in another configuration the fan is placed at the rear. The discharge from the fan can be either mixed with the main jet or expanded in a separate nozzle. Note that the frontal area of a turbofan is quite large as compared to a turbojet, which incurs more drag and, also, higher weight. However, this finds several applications, particularly, in high subsonic transport aircrafts.

In another configuration, the gas is expanded further in a multistage turbine, which provides power not only to the compressor but also to the propeller. Such an engine is known as *turboprop engine* (Fig. 1.12) which finds applications in helicopters. It has performance advantage over other engine configurations, particularly, for low speed flight and short takeoff. In some other design, mechanical power is also available at the shaft rather than at the propeller. This configuration is known as *turbo-shaft engine*, which finds applications in vertical/short takeoff and landing (V/STOL) vehicles. In contrast to a helicopter, turboshaft powered aircraft can attain comparable speed of winged aircraft while retaining V/STOL capabilities. The turboshaft engine can be used to start the main engines without external assistance. It can also provide air-conditioning, electrical, and hydraulic power, particularly, when the main engine is stopped temporarily.

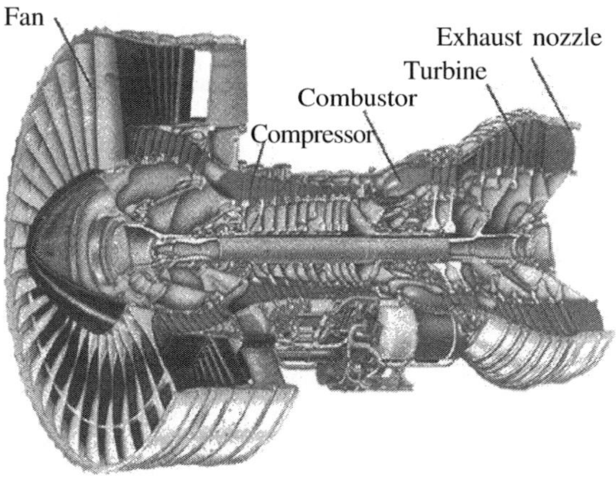

Fig. 1.11. A turbofan engine (Pratt and Whitney PW4000).

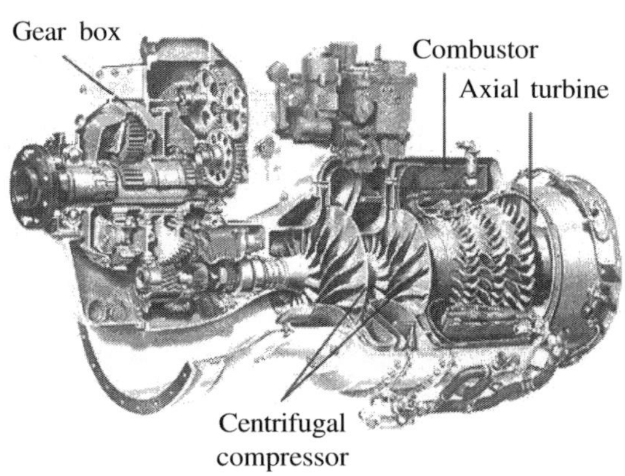

Fig. 1.12. A turboprop engine (Garretd Ai Research TPE 331).

1.5 Other Applications of Gas Turbine Engines

Gas turbine engines are commonly used for aircrafts, but they are also used for surface transportation (rail and road), power generation and several other applications. Besides this, gas turbine engines also find applications in military trucks, tanks, armoured vehicles, landing craft, hovercraft etc. In case of hovercraft, gas turbine is used to drive the lifting fan and air propeller. The modern naval ships are also powered by gas turbine engines as these engines are well suited for marine applications. The power shaft from the gas turbine engine drives the propeller or water jet units. In case of a hovercraft, the turboshaft engine is used to drive the lifting fan and air propeller. Gas turbines are also finding application in liquefied natural gas (LNG) plants. The gas turbine engine is quite suitable for driving an alternator as it is possible to maintain a constant operating speed easily, which is essential for accurate frequency control of alternative current. Small captive power plants generally employ such kind of engines. These engines also find application in gas turbine based thermal power plants. Currently, an attempt is being made to maximise the efficiency of a power plant by cogeneration. i.e. both gas and steam turbines are used to produce power to achieve maximum plant efficiency. Gas turbine engines are extensively used, due to their low weight and size, by the oil and gas industry, particularly, for off-shore drilling. These engines are also employed for pumping oil and gas through long pipelines. Gas turbine engines can be employed in the air cycle refrigeration system, which offers a portable source of cold air for blast freezing. It can be made available at an instant without any need for freon or ammonia. In this section, an attempt is made to provide a brief overview of the applications of gas turbine engines. For an exhaustive treatment, readers may refer to other books on gas turbine engines [1-4].

1.6 Chapter Summary

The present chapter gives a brief introduction to man's earnest desires to fly like a bird and the subsequent efforts to fulfil the same to create an interest in this exciting and exotic subject of aircraft propulsion. The basic principles of jet propulsion invoking Newton's laws of motions are discussed briefly. The concept of gas turbine engine is introduced using a balloon's analogy. A brief account of historical development of aircraft propulsion is presented starting from the concept of Hero's reaction machine to supersonic aircraft engine. A classification of propulsion engines based on the operating principle is presented. Various kinds of propulsion engine are classified in diagrammatic form based on operating principles. Three types of aircraft gas turbine engines – turbojet, turbofan, and turboprop/turboshaft – are introduced briefly. The qualitative variation of pressure, temperature and velocity across the aero gas turbine engine is discussed to explain the processes involved in each component. In the end, a brief description of other applications of gas turbine engines is given.

References

1. Treager, I.E., *Aircraft Gas Turbine Engine Technology*, 3rd Edn., Tata McGraw-Hill Publishing Company Ltd., New Delhi, 1996.
2. Mattingly, J.D., *Elements of Gas Turbine Propulsion*, 1st Edn., McGraw-Hill Publishing Company Ltd., Singapore, 1996.
3. Kerrebrock, J.L., *Aircraft Engines and Gas Turbines*, 1st Edn., MIT Press, UK, 1992.
4. Mukunda, H.S., *Understanding Aerospace Chemical Propulsion*, Interline Publishing, 2004.

Review Questions

1. Recount the story of Icarus's attempt to escape from the prison of Crete.
2. Who was the first person to illustrate the concept of jet propulsion? Describe the various features with the help of a diagram.
3. What does the drawing of Leonardo da Vinci illustrate?
4. What was the progress made in the development of aircraft engines during 1700-1900 AD?
5. What were the main contributions of the German engineer, Hans von Ohain to gas turbine engine? When was he recognised as the co-inventor of gas turbine powered jet engine?
6. What were the main contributions of the British engineer, Sir Frank Whittle to the gas turbine engine?
7. When did the first supersonic engine fly?
8. What is the difference between a rocket and a gas turbine engine?
9. How is a piston engine different from a gas turbine engine?
10. How is a ramjet engine different from a gas turbine engine?
11. What is meant by pulse detonation engine? How is it different from a jet engine?
12. How is a turbojet engine different from a turbofan engine?
13. How is a turboprop engine different from a turbofan engine?
14. Enumerate the relative advantages and disadvantages among turbojet, turbofan and turboprop engines?
15. What are the different applications of gas turbine engines?

2. Review of Basic Principles

The information needed by design engineers of either aircraft or flow machinery is the pressure, the shear stress, the temperature and heat flux vector imposed by the moving fluid over the surface of a specified solid body or bodies in a fluid stream of specified conditions. To supply this information is the main purpose of the discipline of gas dynamics.

H.S. Tsien, 1953

2.1 Introduction

The knowledge of thermodynamics and gas dynamics is essential for understanding aircraft propulsion. You may have been exposed to these topics in your previous studies. However, it is essential to recapitulate these topics for better understanding of propulsion. Hence, this chapter briefly reviews the rudiments of thermodynamics and gas dynamics by discussing the basic principles of mass, momentum, energy and entropy. To begin, we review the basics of classical thermodynamics starting with its definition and various terminologies.

2.2 Basic Principles of Thermodynamics

The term *thermodynamics* is a combination of two Greek words – *therme* and *dynamics*. *Therme* means heat while *dynamics* means work. Thus, thermodynamics can be conceived as a science that deals with work and heat and their interaction between the system and surrounding. Recall that thermodynamics is based on mainly three concepts viz., *energy*, *equilibrium* and *entropy* [1].

2.2.1 Basic Definitions

A thermodynamic *system* is a collection of materials of fixed identity in a region of space with certain boundaries. The boundary of a system is not necessarily fixed; it may be flexible and across which energy in the form of heat or work can transfer. For instance, let us consider an elastic balloon that contains gas. The physical boundary of the balloon can be treated as the system boundary. Let us assume the initial shape of the balloon to be spherical. But it undergoes some changes and takes cylindrical shape. We find that the system boundary has changed its shape and size. Thus, the system boundary may change its size, shape or position when the system undergoes a change. The boundary need not necessarily be real; it may even be an imaginary one. Now a question may arise: how does one choose the system boundary? Truly speaking, there is no basic principle, which can be used to fix the system boundary. The choice may vary from person to person depending on the nature of the problem. In other words, the system and its boundary can be chosen as per convenience and not by any principle. A system can be very simple, just containing a certain amount of matter; or it can be very complex, such as a complete thermal power plant. Everything external to the system is known as *surrounding*. Usually, surrounding means those things outside the system boundary, which interact in some way with the system. A system and its surrounding are known as a *Universe*.

The system can be conveniently classified into three categories – closed, open and isolated. If a system consists of a particular quantity of matter such that no matter will cross its boundary, it is called a closed system. For example, consider a piston cylinder arrangement (Fig. 2.1(a)), which contains fixed quantity of a gas. The inner surfaces of the piston and the cylinder constitute the system boundary. Everything external to the gas, including the piston and the cylinder, can be called as the surrounding. Since no matter (gas) but heat energy is transferred across the system boundary, we can label it as a closed system.

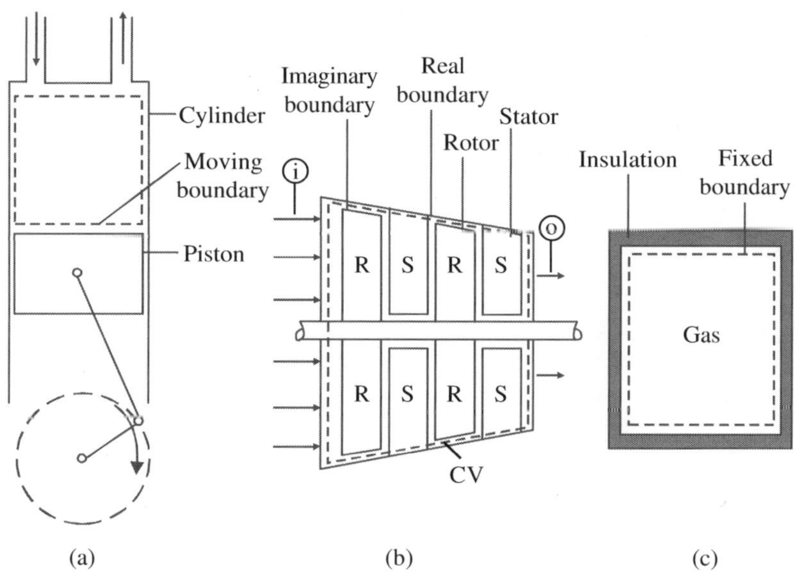

Fig. 2.1. Types of thermodynamic systems: (a) Control mass system, (b) Control volume system, and (c) Isolated system [1].

The jet engine can be considered as an open system in which air and fuel enter into the system and leave as exhaust gas. Let us say, we want to analyse the compressor. Hence, we can choose the outline of the compressor as our open system. In this example, attention is focused on certain volume corresponding to the outline of the compressor (Fig. 2.1(b)), which is known as the *control volume* (CV). The system boundary is known as the *control surface*. In addition to the air, work crosses through the system boundary by the driving shaft and a small amount of heat crosses through the control surface of the compressor. If neither matter nor energy crosses a closed system boundary, then it is known as an *isolated system*, i.e. it does not interact with its surroundings. At first instant, we may think that such a system would be of no relevance. But if two or more systems interact with each other only, the combination of these systems can be surrounded by a boundary to define an isolated system. Then an isolated system can be used for analysis of this kind of problem. In this book, the control volume approach is used very often rather than the control mass approach while analysing the various components of gas turbine engine viz., compressor, turbine, nozzle, etc.

We will be dealing with mostly gases for analysis of gas turbine engine. While carrying out thermodynamic analysis of any component of a gas turbine engine, we will have to deal with the macroscopic properties of the gas, such as pressure, temperature, volume, etc. In other words, the state of gas can be specified by the macroscopic properties, which can represent, basically, space averaged properties of gas. If this macroscopic approach is employed for thermodynamic analysis, then it is known as *classical thermodynamics*. Note that the concept of classical thermodynamics is used throughout the book. In this approach, the thermodynamic system attains an equilibrium state whenever a system undergoes a change.

Let us now understand the concept of equilibrium, which is very much used while dealing with the subject of thermodynamics. Equilibrium, according to Webster's dictionary, is a state of balance or even adjustment between opposing influences, interests, etc. In other words, the properties do not change across certain space and time. To illustrate this concept, let us take an example. At the end of the daily news bulletin on a particular day, the day time temperature is shown as 35°C in New Delhi. This does not mean that the temperature across entire New Delhi is 35°C. Rather, the temperature can vary as much as several degrees in such a big city. But the reported temperature can be assumed as the same throughout the city. Thus, one can assume the temperature of New Delhi is at the equilibrium over several kilometres around the meteorological stations.

A system is said to be in the state of thermodynamic equilibrium only when the mechanical, thermal and chemical equilibrium are satisfied. Let us define all three equilibrium briefly:

- *Mechanical Equilibrium*: It can be maintained when there are no unbalance forces either in the system or between the system and its surrounding.
- *Thermal Equilibrium*: It is possible only when the same temperature is maintained throughout the system and its surrounding.
- *Chemical Equilibrium*: It can be maintained only when the system does not undergo any chemical composition change, however slow it may be.

2.3 Thermodynamic Properties

In analysing the various components of gas turbine engines, we will consider the gas prevailing in the various components to be perfect because the pressure of the gas is not quite high and high temperature prevails from the compressor exit to the nozzle exit. Let us now discuss the concept of a perfect gas.

2.3.1 Perfect Gas

One may ask: what is a perfect gas? We know that a gas is a collection of particles consisting of molecules, atoms, ions, electrons, etc. These particles are in random motion. Around any particle, a force field prevails because of its electronic structure. When the distances among these particles are not much, then there will be interactions among the particles, thus, giving rise to intermolecular forces. However, if the particles are far apart from one another, then these intermolecular forces are too small and can be neglected. Moreover, the molecules of gas at high temperature and ambient pressure occupy negligibly small space in comparison to the volume of the system. Therefore, a gas can be termed as perfect gas only when its individual molecules occupy no space and do not experience intermolecular forces.

From experimental observations, we know that the thermodynamic state of a system containing a single pure gas can be depicted by the equation of state, i.e. a relationship between the pressure P, volume V and temperature T

$$PV = NR_uT \tag{2.1}$$

where N is the number of moles, R_u the universal gas constant (R_u = 8.314 kJ/kmol K). These thermodynamic properties do not depend on the process, rather on the state of the gas.

2.3.2 Internal Energy and Enthalpy

We know that gas consists of molecules, atoms etc., which are in random motion. Let us consider a diatomic molecule N_2 in air. The molecule will move randomly in space and, hence, give rise to translation energy. Besides this, a single molecule consists of atoms that are chemically bonded together. This molecule can be subjected to rotation about two orthogonal axes. Two nuclei may also vibrate as a spring (Fig. 2.2), which is manifested into vibrational energy. Moreover, electrons move around the nuclei of each molecule and, hence, contribute electronic energy. Thus, the N_2 molecule possesses translational, rotational, vibrational and electronic energy.

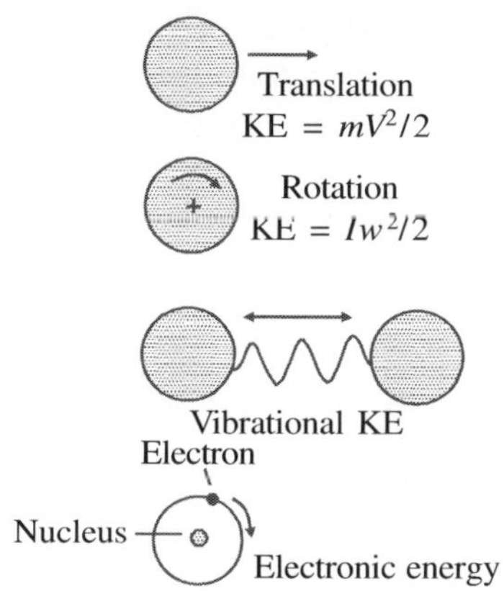

Fig. 2.2. Examples of microscopic modes of energy.

If we consider a system consisting of a large number of molecules, then the sum total of all the energies (translational, vibrational, rotational and electronic) of all the molecules is known as the *internal energy* of the system. The internal energy per unit mass is termed as *specific internal energy*, denoted by u. Another quantity, very often used for an open system, is the *specific enthalpy* and it is defined as

$$h = u + Pv \tag{2.2}$$

Generally, the internal energy u is a function of temperature and volume. Similarly, the enthalpy h is a function of temperature and pressure. These can be expressed as

$$u = u(T, v) \tag{2.3}$$

$$h = h(T, P) \tag{2.4}$$

The general differential expression for both u and h can be obtained by differentiating Eqs. (2.3) and (2.4), respectively

$$du = \left(\frac{\partial u}{\partial T}\right)_V dT + \left(\frac{\partial u}{\partial v}\right)_T dv \tag{2.5}$$

$$dh = \left(\frac{\partial h}{\partial T}\right)_P dT + \left(\frac{\partial h}{\partial P}\right)_T dP \tag{2.6}$$

For a perfect gas, Eqs. (2.5) and (2.6) will be

$$du = \left(\frac{\partial u}{\partial T}\right)_v dT = C_v dT \tag{2.7}$$

as $u = u(T)$ for a perfect gas.

$$dh = \left(\frac{\partial h}{\partial T}\right)_P dT = C_p dT \tag{2.8}$$

Now the specific heat of a substance at constant volume C_v can be defined as the rate of change of specific internal energy with respect to temperature when the volume is kept constant. Similarly, the specific heat of a substance at constant pressure C_p can be defined as the rate of change of specific enthalpy with respect to temperature when the pressure is kept constant. Remember that u and h are thermodynamic variables as they depend on the state of the gas and are independent of any process. In other words, Eqs. (2.7) and (2.8) are relations for the thermodynamic state variables and have nothing to do with the processes that may be taking place. The C_v and C_p are expressed as

$$C_v = \left(\frac{\partial u}{\partial T}\right)_v; \quad C_p = \left(\frac{\partial h}{\partial T}\right)_P \tag{2.9}$$

From Eq. (2.9), it is obvious that C_v and C_p are functions of temperature. The variations of specific heat C_p and C_v with temperature are given in Table 2.1 for dry air. It can be noted that the variations

Table 2.1. Average specific heat ratio at 1200 K along with heat of reactions of chemical species for CH_4 air system [2]

Species	$h^0_{f,i}$ (kJ/kmol) at 298 K	C_{pi} (kJ/mol K)
CH_4	– 74,830	-
CO_2	– 393,546	56.2
H_2O	– 241,845	43.87
N_2	0	33.7
O_2	0	-

of C_p and C_v are not significant when the temperature is below 1000 K and, hence, can be assumed to be constant within this range. A perfect gas with constant C_p and C_v is known as *calorically perfect gas*. For a specific ideal gas, let us derive the relation between C_p and C_v as

$$h = u + Pv = u + RT \tag{2.10}$$

By differentiating Eq. (2.10), we have

$$dh = du + RdT \tag{2.11}$$

$$C_p - C_v = R \tag{2.12}$$

where R is the specific gas constant. By defining $\gamma = C_p/C_v$, the C_p and C_v for a perfect gas can be expressed in terms of γ and R as

$$C_p = \frac{\gamma R}{\gamma - 1} \tag{2.13}$$

$$C_v = \frac{R}{\gamma - 1} \tag{2.14}$$

These equations will be useful for our subsequent discussions on gas turbine engines.

2.4 Thermodynamic Laws

The various components of gas turbine engines can be analysed using the laws of thermodynamics. These laws are stated below. For the details of these laws and derivation, one can refer some standard thermodynamic books [1, 3, 4].

2.4.1 First Law of Thermodynamics

The principle of conservation of energy is of paramount importance in engineering science. The First Law of Thermodynamics is basically a statement of this principle. It states that energy can neither be created nor destroyed but only transforms from one form to another. This statement was put forth by Nelson Obert. In other words, it is not possible to extract heat continuously without any other change in the system. To illustrate this let us consider a fixed mass system (Fig. 2.1a).

Let δq be the heat added to the system and δw the work done by the system on the surrounding by the displacement of boundary. For example, the system boundary in an Internal Combustion (IC) engine moves with the displacement of the piston. The δq and δw cause a change in energy in the system. According to the First Law of Thermodynamics, it is given by

$$de_t = \delta q - \delta w \tag{2.15}$$

where de_t is the change in total energy content of the system. But $de_t = du + dke + dpe$, whereas, e_t = Internal energy u + Kinetic energy ke + Potential energy pe. Note that de_t is the exact differential that exists for the state function and its value depends on the initial and final states of the system. In contrast, δq and δw depend on the path followed by the process. Remember that the First Law of Thermodynamics is an empirical relation between heat, work and internal energy. It was first put forth by James Prescott Joule in 1851 by conducting a series of experiments. Of course, the law has been verified thoroughly by subsequent experimental results. Till date, nobody has managed to disprove this law.

2.4.1.1 First Law for Control Volume (CV)

Now we will consider the CV (Fig. 2.1b). The fluid enters the CV at inlet i and leaves at outlet o. This form of First Law of Thermodynamics is very useful for propulsion problem. The reader is urged to get familiarised with this important thermodynamics law, by referring to some standard thermodynamics books [1, 3, 4]. Before writing down the expression for the First Law of Thermodynamics, let us look at the following assumptions made for deriving this law:

- The CV is fixed with respect to the coordinate system. As a result, the work interactions associated with the moving body are not considered.
- The fluid properties are uniform over the inlet and the outlet flow areas.

With these assumptions, the First Law of Thermodynamics for CV is given by

$$\frac{dE}{dt} = \left[\dot{m}_{\text{i}}\left(h_{\text{i}} + V_{\text{i}}^2/2 + gZ_{\text{i}}\right) - \dot{m}_{\text{o}}\left(h_{\text{o}} + V_{\text{o}}^2/2 + gZ_{\text{o}}\right)\right] + \dot{Q} - \dot{W}_{\text{sh}} \tag{2.16}$$

Equation (2.16) will be employed for the analysis of various components of a gas turbine engine. The reader is urged to keep this expression in mind and learn how to use it properly.

2.4.2 Second Law of Thermodynamics

The First Law of Thermodynamics does not say anything about the feasibility of the direction in which the process may proceed. Let us consider an example of water falling from the rooftop of a tall building. In the process, the potential energy of water is converted into kinetic energy as per First Law of Thermodynamics. Let us imagine that the reverse process can occur whereby water goes up to the rooftop of the building by converting kinetic energy to equivalent potential energy. Is it possible? Of course, it does not violate the First Law of Thermodynamics. Certainly, the same cannot happen in real life. This example clearly indicates the existence of the direction in which a process can proceed, which is dictated by the Second Law of Thermodynamics. Let us consider another example of a hot cup of tea kept on a table. After certain time has elapsed, we find that the hot cup of tea has already been cooled by virtue of heat transfer to the surrounding according to First Law of Thermodynamics. Is it possible for the heat to be transferred back from its surrounding to the tea cup as per First Law of Thermodynamics? We know from experience that it cannot happen. Now comes the Second Law of Thermodynamics to stipulate the direction of the process. According to the famous scientist Kelvin Plank, the Second Law of Thermodynamics states that *it is impossible to construct a cyclically operating device such that it produces no other effects than absorbing energy as heat from a single thermal reservoir and performs an equivalent amount of work.* The only option for the engine is then to convert a part of the energy as heat it receives from the thermal reservoir of higher temperature into work and reject the rest of heat energy to another thermal reservoir of lower temperature. In other words, it is impossible to have a heat engine with thermal efficiency of one. Although, we may not use this statement directly, but we should keep the same in mind. Further, the Second Law of Thermodynamics defines an important property of a system, known as *entropy*, which stipulates the feasibility of a certain process that occurs only in a particular direction. The entropy can be defined as

$$ds = \frac{dq_{\text{rev}}}{T} \quad \text{(reversible process)} \tag{2.17}$$

where ds is the change of entropy of the system during an incremental reversible heat exchange dq_{rev} when the system is at temperature T. The entropy being a state variable can be used either for a reversible or irreversible process. An alternative, more general, relation for entropy is

$$ds = \frac{dq_{\text{rev}}}{T} + ds_{\text{irrev}} \quad \text{(irreversible process)} \tag{2.18}$$

where dq is the actual amount of heat added to the system during which entropy change ds_{irrev} is generated due to the irreversible, dissipative phenomena. The processes involving friction, heat transfer with finite temperature gradient, and mass transfer with finite concentration gradient are some of the examples of irreversible processes. These dissipative phenomena always result in increase of the entropy as

$$ds_{\text{irrev}} \geq 0 \tag{2.19}$$

Combining Eqs. (2.18) and (2.19), we have

$$ds \geq \frac{dq}{T} \tag{2.20}$$

For adiabatic process ($dq = 0$), Eq. (2.20) becomes

$$ds \geq 0 \tag{2.21}$$

This form of Second Law of Thermodynamics (Eqs. (2.20) and (2.21)) provides the direction in which the process can proceed. A process can only proceed in such direction where the entropy of the system and its surrounding can be increased or, at least, remains unchanged. When the entropy of the system decreases, the process will not be feasible. Generally, irreversibility reduces the engineering usefulness of any given process. Therefore, care must be taken to reduce the irreversibility to a greater extent though it is impossible to eliminate absolutely while designing practical devices. The calculation of entropy can be carried out by assuming the heat to be added to the system reversibly by

$$\delta q = Tds \tag{2.22}$$

Substituting Eq. (2.22) in the energy equation (Eq. (2.15)), we can get

$$Tds = du + Pdv \tag{2.23}$$

From the definition of enthalpy, we have

$$dh = du + Pdv + vdP \tag{2.24}$$

For thermally perfect gas, we can have $dh = C_p dT$. Substituting this relation in Eq. (2.24) and using Eq. (2.23), we get

$$ds = C_p \frac{dT}{T} - \frac{vdP}{T} \tag{2.25}$$

Substituting $v/T = R/P$ from the equation of state into the last term of Eq. (2.25), we have

$$ds = C_p \frac{dT}{T} - R\frac{dP}{P} \tag{2.26}$$

For an isentropic (reversible and adiabatic) process $ds = 0$, we can integrate Eq. (2.26)

$$0 = C_p \ln \frac{T_2}{T_1} - R \ln \frac{P_2}{P_1} \tag{2.27}$$

$$\frac{P_2}{P_1} = \left(\frac{T_2}{T_1}\right)^{C_p/R} \tag{2.28}$$

However, for a calorically perfect gas, specific heat can be expressed in terms of specific heat ratio γ as

$$\frac{C_P}{R} = \frac{\gamma}{\gamma - 1} \tag{2.29}$$

Using Eq. (2.29), we may express Eq. (2.28) in the form

$$\frac{P_2}{P_1} = \left(\frac{T_2}{T_1}\right)^{\frac{\gamma}{\gamma-1}} \tag{2.30}$$

By employing the perfect gas law, we can transform Eq. (2.30) to

$$\frac{\rho_2}{\rho_1} = \left(\frac{T_2}{T_1}\right)^{\frac{1}{\gamma-1}} \tag{2.31}$$

$$\frac{P_2}{P_1} = \left(\frac{\rho_2}{\rho_1}\right)^{\gamma} \tag{2.32}$$

Combining Eqs. (2.30) to (2.32), we can summarise the isentropic relation as

$$\frac{P_2}{P_1} = \left(\frac{\rho_2}{\rho_1}\right)^{\gamma} = \left(\frac{T_2}{T_1}\right)^{\frac{\gamma}{\gamma-1}} \tag{2.33}$$

Equation (2.33) is very important as it relates the pressure, density and temperature for an isentropic process. It will be used very often while dealing with the gas turbine engine. Now questions may arise in one's mind about how the isentropic relation can be used so profusely for analysis of practical problems? It can be used so because the process must be adiabatic and reversible for applying the isentropic relation. This may seem very restrictive, but the fact is that isentropic process can be assumed for a large number of practical systems. For example, consider a flow through a nozzle. A viscous boundary layer is formed in the region next to the solid wall of the plate where the dissipative mechanism of viscosity is quite predominant. Of course, one cannot assume the flow to be isentropic in this region. But, the flow outside the boundary layer can be safely assumed to be reversible. The flow in this region can also be assumed to be adiabatic as no heat is being transferred. Since the thickness of boundary layer is very thin in comparison to the entire flow domain, one can safely analyse the problem assuming the flow to be isentropic. Thus, this isentropic relation (Eq. (2.33)) can be used for analysis of a wide range of practical problems.

2.4.3 Reacting System

We know that for the analysis of a combustion system, one has to carry out mass and energy balance. The mass balance in a chemical reaction, describing exactly how much oxidiser has to be supplied for complete combustion of certain amount of fuel, is termed as *stoichiometry* that basically originated from the Greek word 'stoicheion', which means element. Hence, an elemental balance is to be carried out to estimate the stoichiometric ratio for a chemical reaction. In other words, stoichiometry is the relationship between the mass of the reactants and the products of a chemical reaction. Let us consider an example in which a certain amount of methane (CH_4) is burnt in the presence of a certain amount of oxygen (O_2) to liberate certain amount of heat energy. The equation representing such a reaction is

$$CH_4\ (16\ g) + 2O_2\ (64\ g) \rightarrow CO_2\ (44\ g) + 2H_2O\ (36\ g) + \Delta H^0_{R,298} \tag{2.34}$$

where $\Delta H^0_{R,298}$ is the heat of reaction evolved during this exothermic reaction. From the above reaction, we find that one mole of CH_4 is reacted with two moles of O_2 to produce one mole of CO_2 and two moles of H_2O. Note that this is a balanced equation in which the numbers of chemical elements are balanced. It is interesting to note that 16 g of CH_4 reacts with 64 g of O_2 to produce 44 g of carbon dioxide (CO_2) and 36 g of water (H_2O). Thus, the total mass on the left hand side of the reaction is same as that on the right hand side, which indicates that mass is conserved. It is the underlying principle of chemical reaction that no mass is created or destroyed in a chemical reaction. In contrast, the number of moles need not be conserved. This type of reaction is known as a *stoichiometric reaction*. In this case, the quantity of oxidiser (air), sufficient to burn certain quantity of fuel completely in a chemical reaction, is known as *stoichiometric*. Specifically, it is the ratio of oxidiser to fuel that is sufficient to burn certain amount of fuel forming complete combustion products. In this example, the stoichiometric ratio m_{ox}/m_{fuel} is 4. However, oxidiser/fuel ratio always need not be in stoichiometric proportions in practical combustion devices. On several occasions, excess oxidisers are supplied to ensure complete combustion of fuel in practical devices. When the quantity of oxidiser exceeds the stoichiometric proportion, the mixture is termed as *fuel lean* or *lean mixture*. In contrast, if less quantity of oxidiser than the stoichiometric quantity is present, then the mixture is known as *fuel rich* or *rich mixture*. On several occasions, hydrocarbon fuels are burnt in the presence of air. For a hydrocarbon fuel represented by C_xH_y, the stoichiometric relation [1] is

$$C_xH_y + a(O_2 + 3.76N_2) \rightarrow xCO_2 + (y/2)H_2O + 3.76aN_2 \tag{2.35}$$

In the above reaction and throughout this book, for simplicity, air is assumed to contain 21% O_2 and 79% N_2 by volume. When the above reaction is balanced, a turns out to be $x + y/4$. Then the stoichiometric air-fuel (A/F) ratio would be

$$(A/F)_{st} = (m_{ox}/m_{fuel}) = 4.76a(MW_{air}/MW_{fuel}) \tag{2.36}$$

where MW_{air} and MW_{fuel} are the molecular weights of air and fuel, respectively. By using Eq. (2.36), the stoichiometric ratio can be obtained very easily for most of hydrocarbon fuels, such as methane, propane, butane, etc., when they are reacted with air. We find that stoichiometric air-fuel ratio for methane is 17.11 while for higher hydrocarbon fuels, such as propane and butane, it is around 15. Hence, it must be kept in mind that stoichiometric air-fuel ratio for higher hydrocarbons (kerosene, diesel, gasoline, aviation turbine fuel (ATF)) up to hexadecane ($C_{16}H_{34}$) is around 15. Also, it must be appreciated that, in practice, one has to supply the oxidiser in large proportion than the stoichiometric ratio to ensure complete combustion. Another useful quantity known as *equivalence ratio* ϕ is used very often in combustion literature to describe air-fuel mixture, which is defined as

$$\phi = \frac{(A/F)_{st}}{(A/F)} = \frac{(F/A)}{(F/A)_{st}} \tag{2.37}$$

Note that equivalence ratio is a non-dimensional number in which fuel-air ratio is expressed in terms of mass. This ratio describes quantitatively whether a fuel-oxidiser mixture is rich, lean or stoichiometric. From Eq. (2.37), it is very clear that the stoichiometric mixture has equivalence ratio of unity. For fuel rich mixture, equivalence ratio is greater than unity ($\phi > 1$) while fuel lean mixture has equivalence ratio less than unity ($\phi < 1$). By lean and rich mixture, we basically mean the extent of fuel in the mixture. In several combustion applications, the equivalence ratio is one of the most important parameter that dictates the performance of the system. Another parameter, used very often to describe the proportion of mixture is *percent stoichiometric air*, which can be related to equivalence ratio as

$$\%\ \text{stoichiometric air} = \frac{100\%}{\phi} \tag{2.38}$$

Another useful parameter to define relative stoichiometry of mixture is *percent excess air*, given by

$$\%\ \text{excess air} = \frac{(1-\phi)}{\phi}100\% \tag{2.39}$$

Example 2.1 Hydrocarbon fuel (C_6H_{14}) is burnt with dry air. The volumetric analysis of the products on dry basis is CO_2 = 10%, O_2 = 5.5%, CO = 1% and N_2 = 83.5%. Determine: (a) A/F ratio, and (b) Equivalence ratio ϕ.

Solution Let us consider the following reaction for Octane-air mixture

$$xC_6H_{14} + a(O_2 + 3.76N_2) \rightarrow 10CO_2 + 1.0CO + 5.5O_2 + 83.5N_2 + bH_2O$$

By mass balance, we have

N $\quad 3.76a = 83.5; \Rightarrow a = 22.207$

C $\quad 6x = 10 + 1.0; \Rightarrow x = 1.833$

H $\quad 14x = 2b; \Rightarrow b = 12.831$

O $\quad 2a = 10 \times 2 + 1.0 + (2 \times 5.5) + b$

Now the above balanced equation becomes

$$1.833C_6H_{14} + 22.207(O_2 + 3.76N_2) \rightarrow 10CO_2 + 1.0CO + 5.5O_2 + 83.5N_2 + 12.831H_2O$$

Let us recast the above chemical reaction in terms of 1 mole of fuel as

$$C_6H_{14} + 12.115(O_2 + 3.76N_2) \rightarrow 5.455CO_2 + 0.545CO + 3O_2 + 45.628N_2 + 7H_2O$$

Then the air-fuel ratio by mass becomes

$$(A/F)_{\text{actual}} = m_{\text{air}}/m_{\text{fuel}} = \frac{12.115(32+(3.76\times28))}{(6\times12)+14(1)} = 19.338$$

In order to find out the equivalence ratio, we will have to estimate stoichiometric fuel-air ratio by considering a balance equation as

$$C_6H_{14} + 9.5(O_2 + 3.76N_2) \rightarrow 6CO_2 + 7H_2O + 35.72N_2$$

Then the stoichiometric fuel-air ratio becomes

$$\left(\frac{F}{A}\right)_{\text{st}} = \frac{m_{\text{fuel}}}{m_{\text{air}}} = \frac{86}{9.5(32+3.76\times28)} = 0.0659$$

Therefore $$\left(\frac{A}{F}\right)_{\text{st}} = 15.174$$

Note that the stoichiometric air-fuel ratio is 15.164. Then the equivalence ratio becomes

$$\phi = \frac{(F/A)}{(F/A)_{\text{st}}} = \frac{0.0517}{0.0659} = 0.784$$

The above estimated equivalence ratio indicates that the C_6H_{14} and air mixture is fuel lean.

Thermochemistry: Several chemical reactions take place simultaneously in the combustion process. In some reactions, heat will be evolved and in certain others, heat will be absorbed. Hence, it is important to evaluate the heat liberated or absorbed in a chemical reaction. Let us take an example of a burner (Fig. 2.3) in which 1 kmol of CH_4 reacts with 2 mole of O_2 to produce to CO_2 and H_2O as per the following chemical reaction

$$CH_4 + 2O_2 \Rightarrow CO_2 + 2H_2O \tag{2.40}$$

Let us assume that this reaction proceeds in a steady flow reactor (Fig. 2.3) with negligible change in kinetic energy and potential energy and it does not involve shaft work. The heat liberated during the chemical reaction can be obtained by applying First Law of Thermodynamics for the CV. Then

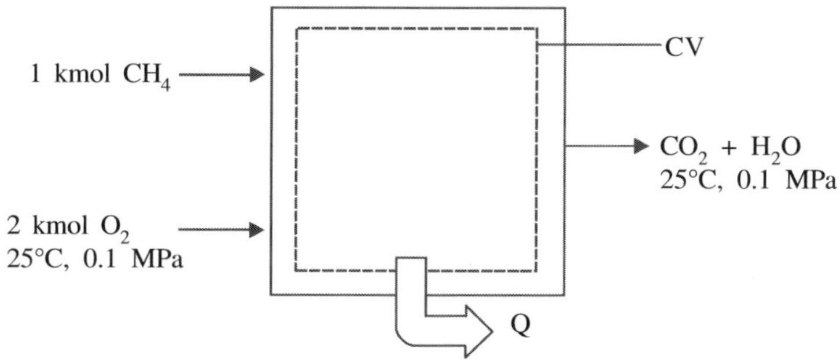

Fig. 2.3. Control volume for combustion chamber.

the First Law of Thermodynamics for this CV (Eq. (2.16)) becomes

$$Q = dH = H_P - H_R = \sum_{iP} n_{iP} h_{iP} - \sum_{iR} n_{iR} h_{iR} = \Delta H^0_{R,298} \tag{2.41}$$

where H_R is the total enthalpy of reactants, H_P the total enthalpy of products, n_{iR} the number of moles of ith reactant, n_{iP} the number of moles of ith product, h_{iR} the heat of formation per unit mole of ith reactant, h_{iP} the heat of formation per unit mole of ith product, and $\Delta H^0_{R,298}$ the heat of reaction at standard states.

The enthalpy change due to chemical reactions $\Delta H^0_{R,298}$ is defined as the difference between the enthalpy of products and enthalpy of reactants at the specified states, which is known as the *heat of reaction*. Generally, the heat of reaction is stated at the standard condition. As per international convention, one atmospheric pressure and 298.15 K is considered to be the standard state for gas. However, for pure liquids and solids, the standard reference state at a given temperature is the real state of substance at one atmospheric pressure. We know that for an ideal gas, enthalpy is a function of temperature. Therefore, the heat of reaction for gaseous reaction can be represented as shown in Fig. 2.4, which clearly illustrates that heat of reaction is dependent on temperature. If one of the reactants happens to be the fuel and the other an oxidiser, then combustion takes place, liberating a certain amount of heat. The heat of reaction is then known as the *heat of combustion*. We will consider the previous example of CH_4 fuel, reacting with O_2 for which heat of reaction can be estimated by

$$\Delta H^0_{R,298} = \left(n_{CO_2} h^0_{f,CO_2} + n_{H_2O} h^0_{f,H_2O}\right) - \left(n_{CH_4} h^0_{f,CH_4} + n_{O_2} h^0_{f,O_2}\right) \tag{2.42}$$

Note that the heat of formation of each species must be known to evaluate the heat of reaction. The heat of formation of a particular species can be defined as the heat of reaction per mole of that species formed isothermally from elements in their standard states. It must be kept in mind that the heat of formation of an element in its standard state is assigned a value of zero as per international norms. For example, graphite is the most stable in solid state and its heat of formation at standard state is zero. Similarly, the heat of formation of gaseous nitrogen at standard state is zero since this state is its natural state. The heat of formation of certain important pure substances is given in Appendix A1. The main advantage of tabulating heat of formation is that one has to keep track of heat of formation of only a few species to estimate heat of reaction of several reactions.

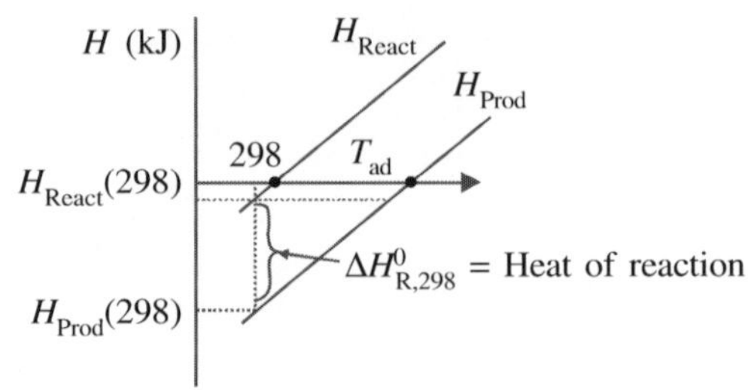

Fig. 2.4. The variation of enthalpy with temperature for typical fuel-air system to illustrate heat of reaction.

2.4.4 Adiabatic Flame Temperature

Let us consider a reactant mixture containing a certain amount of fuel and air in a certain ratio passing through a combustor. If the chemical reactions are initiated by producing a spark, the reactions will proceed to completion (equilibrium state) under constant pressure. Then the final temperature attained under adiabatic condition is known as *adiabatic flame temperature* (T_{ad}).

The T_{ad} depends on initial pressure, initial temperature and composition of the reactants. Of course, if the final compositions of the products are known, then the First Law of Thermodynamics is sufficient to determine T_{ad}. Let us consider the physical boundary of combustor as the CV (Fig. 2.4). We know that change in both kinetic energy and potential energy is zero. There would not be any shaft work involved in this process, which is assumed to be adiabatic. Under this condition, the First Law of Thermodynamics (Eq. (2.16)) turns out to be

$$H_P(T_{ad}, P) = H_R(T_u, P) \tag{2.43}$$

$$H_R = \sum_{i=R} n_i h_i = \sum_{i=R} n_i \left[h^0_{f,i} + \int_{T_i}^{T_u} C_{pi}(T)dT \right] \tag{2.44}$$

$$H_P = \sum_{i=P} n_i h_i = \sum_{i=P} n_i \left[h^0_{f,i} + \int_{T_u}^{T_{ad}} C_{pi}(T)dT \right] \tag{2.45}$$

where $h^0_{f,i}$ is the heat of formation of ith species (see Appendix A1), n_i the number of moles of ith species and $C_{p,i}$ the specific heat of ith species as a function of temperature. But we can use some average values to estimate an approximate value of adiabatic temperature. However, care must be taken to choose appropriate value of specific heats for products to predict the accurate adiabatic temperature.

At 298 K, we know that the heat of reaction is expressed as

$$\Delta H^0_{R,298} = \sum_{i=P} n_i h_{f,i} - \sum_{i=R} n_i h_{f,i} \tag{2.46}$$

In Eq. (2.46), if we know the final composition of the product then we can easily evaluate T_{ad}. But the problem is that final equilibrium composition is dependent on final T_{ad}. Hence, one has to resort to an iterative technique. However, if the final equilibrium composition are known, the estimation of T_{ad} is straightforward, as illustrated in the following example:

Example 2.2 In a burner, CH_4 and air at T_u = 298 K and P_u = 0.1 MPa undergo constant pressure combustion process to produce CO_2, H_2O and N. Estimate the adiabatic temperature T_{ad} of the stoichiometric mixture.

Given T_u = 298 K and P_u = 0.1 MPa.

Solution The hydrogen will react with air at 298 K and 0.1 MPa by the following reaction

$$CH_4 + 2(O_2 + 3.76N_2) \rightarrow CO_2 + 2H_2O + 7.52N_2$$

Applying First Law of Thermodynamics for CV system for this combustion process, we get energy balance as

$$H_R = \sum_{i=R} n_i h_i = 1(-74{,}830) + 2(0) + 7.52(0) = -74830 \text{ kJ}$$

$$H_P = \sum_{i=P} n_i h_i = \sum_{i=P} n_i \left[h^0_{f,T_u} + \bar{C}_{pi}(T_{ad} - T_u) \right]$$

Note that the specific heat of species is dependent on temperature. We will have to choose proper average values, as the values depend on temperature. Let us take the approximate value in the temperature range around 1200 K in this example.

Substituting the above values in the above equation for H_P along with the heat of reaction as above, we have

$$H_R = \sum_{i=R} n_i h_i = 1(-74{,}830) + 2(0) + 7.52(0) = -74830 \text{ kJ}$$

$$H_P = \sum_{i=P} n_i \left[h^0_{f,T_u} + \bar{C}_{pi}(T_{ad} - T_u) \right] = 1\left[-393{,}546 + 56.2(T_{ad} - 298)\right]$$

$$+ 2\left[-241{,}845 + 43.87(T_{ad} - 298)\right] + 2\left[0 + 33.7(T_{ad} - 298)\right]$$

$$H_P(T_{ad}, P) = H_R(T_u, P)$$

By solving the above equations, T_{ad} can be estimated as

$$T_{ad} = 2315 \text{ K}$$

Thus, T_{ad} is estimated to be 2315 K. It is interesting to note that this estimated value of T_{ad} is about as accurate as that can be obtained from detailed calculation.

2.5 Basic Principles of Compressible Flow

We know that fluid mechanics is a science that deals with analysis of the motion of a fluid (liquid or gas) and the associated forces. But we are more interested in the flow of gases, i.e. the gas dynamics aspect of the complete flow field. Gas dynamics is highly a developed subject, which cannot be covered in the present chapter. The purpose of this chapter is to review some aspects of gas dynamics that will be helpful for analysing various components of the propulsive system viz., nozzle, turbine, compressor, diffuser, etc. The interested readers may refer some standard textbooks [5, 6] for exhaustive treatment on this subject.

2.6 Fundamental Equations

We know that compressible flow is governed by the equations of continuity, momentum, energy and, of course, with the equation of state. While dealing with propulsive devices, the working fluid can be safely assumed to be perfect in nature. For algebraic convenience, we can assume the specific heat of the gas to be constant. This section states the fundamental equations of flow for a calorically perfect gas.

2.6.1 Definition of Stagnation State

Let us consider that a fluid element with local velocity V, pressure P, temperature T and density ρ is passing through a point in a steady flow. These properties are known as static properties, which can be felt one glides along with the fluid element with the same local velocity. If the fluid element is brought to rest adiabatically, one may expect these properties of the fluid to undergo some changes.

When the fluid is brought to rest adiabatically and without any work, the energy equation (Eq. (2.16)) becomes

$$(dh + VdV) = 0 \tag{2.47}$$

By integrating Eq. (2.47), we obtain

$$h_t = h + \frac{V^2}{2} \tag{2.48}$$

where h_t is the total enthalpy, h is the static enthalpy and V is the flow velocity. For constant C_p, we can rewrite Eq. (2.48) in terms of temperature as

$$T_t = T + \frac{V^2}{2C_p} \tag{2.49}$$

where T_t is the total enthalpy, T is the static enthalpy and $\frac{V^2}{2C_p}$ is the temperature due to dynamic head. The above expression will be used very often in this book.

2.6.2 Definition of Mach Number

We will have to analyse the compressible flow in the components of gas turbine engines. We know that the compressible flow can be characterised by Mach number M, which is defined as the ratio of the local flow velocity to the local speed of sound as

$$M = \frac{V}{a} = \frac{V}{\sqrt{\gamma RT}} \tag{2.50}$$

But the speed of sound a can be defined as

$$a^2 = \left(\frac{\partial P}{\partial \rho}\right)_{isen} = \gamma RT \tag{2.51}$$

Note that for a perfect gas, $a = \sqrt{\gamma RT}$. Generally, both V and a are dependent on spatial and temporal locations. This non-dimensional number is very useful in delineating the regimes of flow. If $M < 1$, the flow is termed as *subsonic flow* ($0 < M < 0.8$). But the flow is termed as *supersonic* if $M > 1$ ($1.2 < M < 5.0$). The *transonic flow* regime occurs around unity M ($0.8 < M < 1.2$). If $M > 1$ of the flow, then it is termed as *hypersonic flow*.

Example 2.3 Air is flowing through a duct with inlet velocity of 200 m/s, pressure 0.1 MPa and temperature 298 K. Determine (i) stagnation pressure and temperature, and (ii) inlet Mach number.

Given V = 200 m/s, P = 0.1 MPa, T = 298 K.

Solution

$$T_t = T + \frac{V^2}{2C_p}$$

We need to estimate specific heat C_P assuming the gas as an ideal gas

$$C_p = \frac{\gamma R}{\gamma - 1} = \frac{1.4 \times 287}{0.4} = 1005 \text{ J/kg K}$$

Then total temperature becomes

$$T_t = 298 + \frac{200^2}{2 \times 1005} = 318 \text{ K}$$

Assuming the flow to be isentropic, we can estimate pressure ratio as

$$P_t = \left(\frac{P_t}{P}\right) \times P = \left(\frac{T_t}{T}\right)^{\frac{\gamma}{\gamma-1}} = \left(\frac{318}{298}\right)^{3.5} \times 0.1 = 0.126 \text{ MPa}$$

Therefore, the inlet Mach number is estimated as

$$M = \frac{V}{a} = \frac{V}{\sqrt{\gamma R T}} = \frac{200}{\sqrt{1.4 \times 287 \times 298}} = 0.57$$

2.7 Steady Quasi-One Dimensional Flow

We will have to analyse the flow in various components of the gas turbine engine. Note the flow across the components are inherently three dimensional and unsteady in nature. But it is quite difficult to handle three dimensional characteristics of fluid flow for flow analysis of propulsive devices. However, some meaningful study can be made by simplifying the actual problems with certain assumptions. The first assumption is to consider flow to be quasi-one dimensional steady flow, which we will be using profusely for the analysis of propulsive devices. Although, we have adopted quasi-one dimensional flow in this text for simplicity reason, it does not mean that such analysis will be of no value. We will realise that many real life problems, e.g. nozzle, diffuser, etc., have varying area duct, which can be well illustrated by one dimensional steady flow analysis. This analysis provides average properties at a particular cross-section. Let us consider the flow across a diffuser (Fig. 2.5) whose cross-sectional area increases gradually along its length. If this rate of increase of area for subsonic flow is quite small, then we can assume flow properties to be a function of x-direction only, which remains uniform across any given stream tube. Such kind of flow is known as *quasi-one dimensional flow*.

Let us apply the equations of mass, momentum and energy along with the equation of state for perfect gas to a quasi-one dimensional differential CV (Fig. 2.5) with the following assumptions:

- Steady quasi-one dimensional flow
- Inviscid flow
- No body force
- Ideal gas with constant thermo-physical properties

Note that the heat transfer rate to the CV is dQ and work done by the CV is dW (Fig. 2.6). The integral form of mass conservation equation for steady flow in the CV is given by

$$\iint \rho(V.n)\, dA = 0 \tag{2.52}$$

For steady one-dimensional flow in the CV (Fig. 2.6), we can reduce Eq. (2.52) to

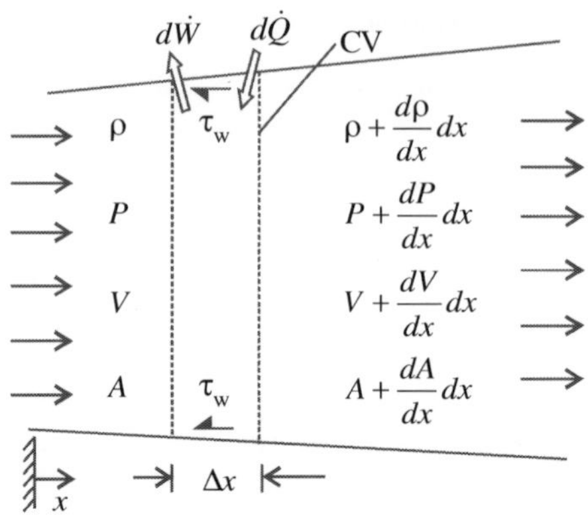

Fig. 2.5. CV for one dimensional steady flow.

$$\frac{d(\rho VA)}{dx} = 0 \tag{2.53}$$

$$\frac{d\rho}{\rho} + \frac{dA}{A} + \frac{dV}{V} = 0 \tag{2.54}$$

Similarly, the integral form of momentum equation for steady flow one dimensional in CV can be expressed as

$$\iint \rho V(V.n)dA = \sum F \tag{2.55}$$

where $\sum F$ is the sum of forces due to pressure, body force, wall shear etc. From Eq. (2.55), the differential form of momentum equation for quasi-one dimensional flow can be derived as

$$\rho V\, AdV = -\,(AdP + \tau_w Cdx) \tag{2.56}$$

$$\rho V \frac{dV}{dx} = -\left(\frac{dP}{dx} + \frac{\tau_w C}{A}\right) \tag{2.57}$$

where τ_w is the wall shear stress and C the duct circumference. Similarly, the integral form of energy equation can be deduced for the present case

$$\rho VA(dh + VdV) = dQ - dW \tag{2.58}$$

$$(dh + VdV) = dq - dw \tag{2.59}$$

where q and w are the heat and work interaction per unit mass.

Assuming the gas to be perfect, the equation of state (Eq. (2.1)) will be

$$P = \rho RT \tag{2.60}$$

With knowledge of the flow area, heat transfer and shear stress, we can use these equations to obtain the changes across the fluid stream. In general, these equations cannot be expressed in simple algebraic form. However, the closed form solution to these equations can be obtained for some special cases. We will discuss these cases in detail as follows

2.8 Isentropic Flow through Variable Area Duct

We now analyse the isentropic flow through a variable area duct. The mass conservation expression (Eq. (2.54)) for quasi-one dimensional steady flow in a varying area duct is

$$\frac{d\rho}{\rho}+\frac{dA}{A}+\frac{dV}{V}=0 \tag{2.61}$$

Similarly, by using momentum equation (Eq. (2.57)) we can derive the expression for quasi-one dimensional steady flow as

$$\rho V dV + dP = 0 \tag{2.62}$$

Combining the continuity equation (Eq. (2.54)) and momentum equation (Eq. (2.62)), we can get

$$dP+\rho V^2\left(-\frac{dP}{\rho a^2}-\frac{dA}{A}\right)=0 \tag{2.63}$$

As per the definition of speed of sound, we know that

$$d\rho=\frac{dP}{a^2} \tag{2.64}$$

By rearranging Eq. (2.63), we can have

$$dP(1-M^2)=\rho V^2\frac{dA}{A} \tag{2.65}$$

where M is the Mach number. Note that Eq. (2.65) depicts the effect of Mach number on compressible flow through a duct of varying cross-section. For subsonic flow ($M < 1$), the term on left hand side of Eq. (2.65), $(1 - M^2)$ is positive. Hence, the pressure will increase when there will be an increase in cross-sectional area (dA = + ve). From Eq. (2.65), we learn that there will be an increase in the velocity with decrease in area for subsonic flow (Fig. 2.6). We can also derive an expression relating area with velocity using the momentum equation (Eq. (2.63)) as

$$\frac{dA}{A}=(M^2-1)\frac{dV}{V} \tag{2.66}$$

This is a very important result, known as *area-velocity relation*. Let us appreciate the efficacy of this simple but important relation, which can help us to understand various flow regimes. If the Mach number is quite low ($M \to 0$), then Eq. (2.66) indicates that AV = constant, indicating the continuity equation for incompressible flow. When the flow is subsonic ($0 < M < 1$), the velocity decreases in divergent duct and increases in convergent duct (Fig. 2.6). For supersonic flow ($M > 1$) the velocity increases in divergent duct and decreases in convergent duct (Fig. 2.6), which is opposite to subsonic flow. For sonic flow ($M = 1$), Eq. (2.66) indicates that change in area becomes zero ($dA = 0$). Of course, change in velocity ($dV = 0$) is finite. Mathematically, this corresponds to local maximum or minimum area. Interestingly, the area must have a minimum value to have a physically realistic solution. Let us see why is it so? Suppose we want to bring the fluid at rest to supersonic speed. How do we achieve it? Obviously, we will have to accelerate the flow to high subsonic speed using the convergent nozzle. When it reaches sonic speed, the convergent duct will act as a diffuser leading to decrease in velocity. However, we will have to provide a divergent duct for accelerating the flow to a supersonic speed. Hence, the flow has to pass through a minimum area. Such kind of convergent-divergent duct is

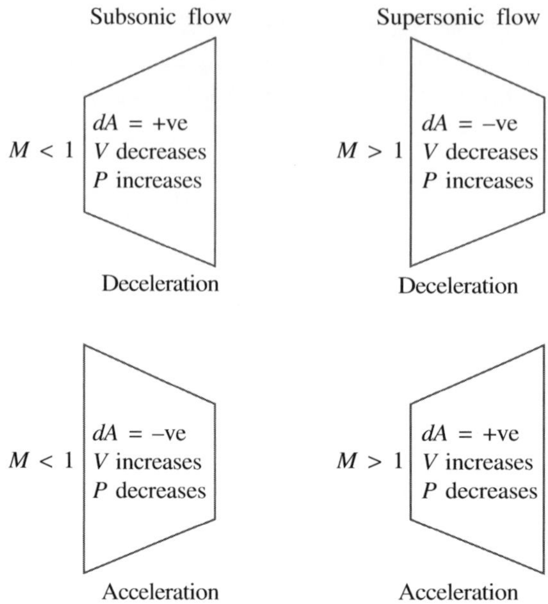

Fig. 2.6. Flow in convergent and divergent ducts.

known as *de Laval nozzle* (Fig. 2.7), which is used profusely in gas turbine engine based aircraft, rocket engines, wind tunnel etc. The de Laval nozzle was designed and developed for the first time by a Swedish engineer Carl G.P. de Laval for a single stage steam engine used in marine application. The minimum area of this nozzle is known as *throat.* It must be appreciated that flow can also be decelerated from supersonic to subsonic using CD diffuser (Fig. 2.7). Hence, we can infer that flow cannot be continuously accelerated from subsonic to supersonic flow and vice versa without pass through the minimum cross-sectional area, known as throat.

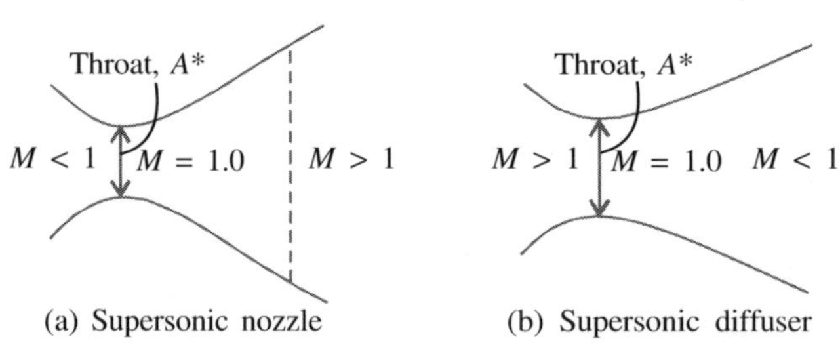

Fig. 2.7. Schematic of de Laval (a) nozzle and (b) diffuser.

2.8.1 Area-Mach Number Relation

In the last section we learnt that compression in a diffuser and expansion in a nozzle can be achieved by using varying area duct. For designing these components, it will be useful to find a relation between the ratio of duct area to the sonic throat area and Mach number.

Let us consider a varying area duct (Fig. 2.8). We can assume the flow at the throat with cross-sectional area A^* to be sonic condition. Note that the asteric mark (*) corresponds to sonic flow

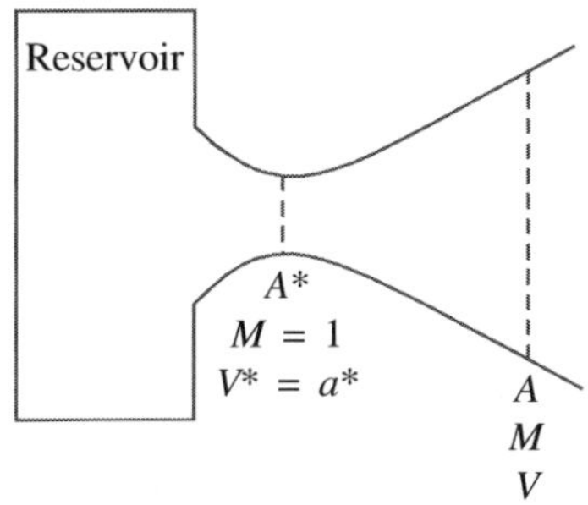

Fig. 2.8. Schematic of varying area duct.

condition. Considering isentropic quasi-steady one dimensional flow, we can derive the Area-Mach number relation by using continuity equation as

$$\dot{m} = \rho VA = \rho^* V^* A^* = \text{constant} \tag{2.67}$$

where ρ, V and A are density, velocity and area, respectively. We first attempt to derive an expression for mass flow parameter. Then using Eq. (2.67), we can derive an expression for mass flux as

$$\frac{\dot{m}}{A} = \rho V = \rho M\sqrt{\gamma RT} = \frac{\rho_t}{\left(1+\frac{\gamma-1}{2}M^2\right)^{\frac{1}{\gamma-1}}} M \sqrt{\frac{\gamma RT_t}{1+\frac{\gamma-1}{2}M^2}} \tag{2.68}$$

Here we have assumed the flow to obey calorically perfect gas law. By simplifying Eq. (2.68) and expressing density in terms of pressure and temperature, we can have

$$\frac{\dot{m}\sqrt{T_t}}{AP_t} = \frac{\sqrt{\gamma}}{\sqrt{R}} M \left(\frac{1}{1+\frac{\gamma-1}{2}M^2}\right)^{\frac{\gamma+1}{2(\gamma-1)}} = f(M) \tag{2.69}$$

Note that $\frac{\dot{m}\sqrt{T_t}}{AP_t}$ is known as *mass flow parameter* (MFP), which is a unique function of Mach number and specific heat ratio γ for calorically perfect gas. The variation of MFP with M for air (Fig. 2.9) shows that maximum MFP occurs at sonic condition for given stagnation condition. So MFP that occurs at sonic condition is expressed as

$$\text{MFP}_{\max} = \frac{\dot{m}_{\max}\sqrt{T_t}}{A^* P_t} = \frac{\sqrt{\gamma}}{\sqrt{R}}\left(\frac{2}{\gamma+1}\right)^{\frac{\gamma+1}{2(\gamma-1)}} \tag{2.70}$$

Dividing Eq. (2.69) by (2.70) we can get

$$\frac{A}{A^*} = \frac{1}{M}\left[\frac{2}{\gamma+1}\left(1+\frac{\gamma-1}{2}M^2\right)\right]^{\frac{\gamma+1}{2(\gamma-1)}} \tag{2.71}$$

with Mach number for air (γ = 1.4 and 1.33).

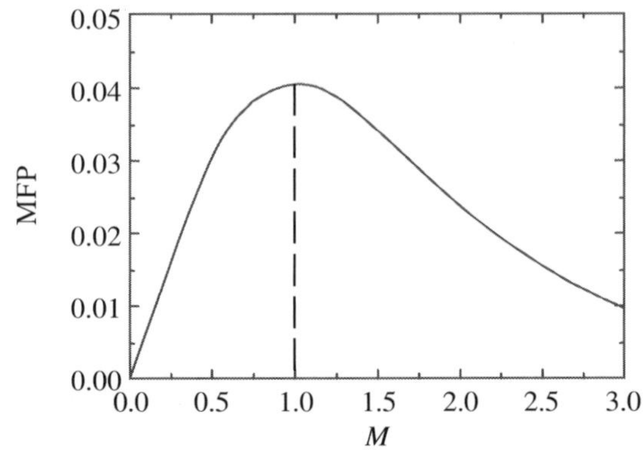

Fig. 2.9. Variation of MFP with *M*.

Equation (2.71) is known as *Area-Mach number relation*. It can be appreciated that *M* at any location of the duct is a function of the ratio of local cross-sectional area to sonic throat area and γ. As discussed earlier, local duct area must be greater than sonic throat area. Interestingly, for a particular area ratio ($A/A^* > 1$), two values of *M* viz., subsonic and supersonic, can be obtained from Eq. (2.71), which will be dictated by the pressure at the inlet and exit of the duct. The variation of A/A^* with *M* (Fig. 2.10) for air ($\gamma = 1.4$) shows that for subsonic flow, A/A^* decreases with increase in *M*, indicating that the duct is convergent. Similarly, for supersonic flow, A/A^* increases with *M*, indicating that the duct is divergent. Of course, at $M = 1$, A/A^* attains unity value indicating it to be a throat. By knowing the local *M*, pressure ratio P/P_t and temperature ratio T/T_t can be easily estimated (Fig. 2.10).

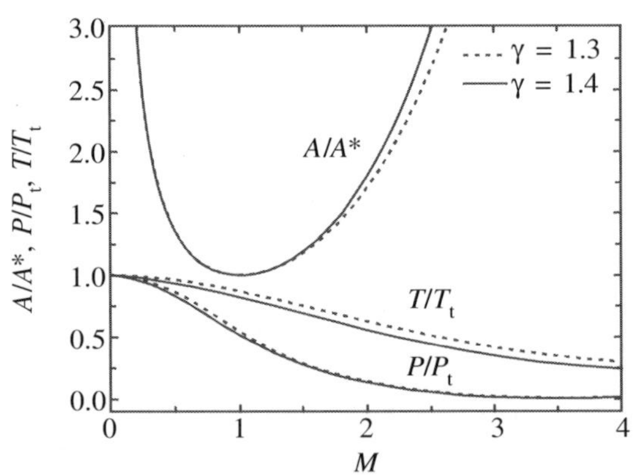

Fig. 2.10. Variation of A/A^*, T/T_t and P/P_t with *M* for air (γ = 1.4 and 1.33).

Example 2.4 A CD nozzle is to be designed for exit Mach number of 2.0. Air is entering from a tank at pressure 0.1 MPa and temperature 298 K. If the exit diameter of the nozzle is 100 mm, determine (i) throat and exit area ratio, and (ii) maximum mass flow rate.

Given $M = 2.0$, $d = 100$ mm $= 0.1$ m, $P_t = 0.1$ MPa, $T_t = 298$ K.

Solution We can estimate area throat and exit area ratio as

$$\frac{A}{A*}=\frac{1}{M}\left[\frac{2}{\gamma+1}\left(1+\frac{\gamma-1}{2}M^2\right)\right]^{\frac{\gamma+1}{2(\gamma-1)}}=\frac{1}{2}\left[\frac{2}{1.4+1}(1+0.2\times 2^2)\right]^{\frac{1.4+1}{2(1.4-1)}}=1.69$$

The exit area of nozzle is estimated as

$$A=\frac{\pi}{4}d^2=\frac{3.14}{4}0.1^2=7.85\times 10^{-3}m^2;\text{ Then, }A*=\frac{A}{1.69}=\frac{7.85\times 10^{-3}}{1.69}=4.64\times 10^{-3}m^2$$

We can estimate maximum mass flow rate through the nozzle as

$$\dot{m}_{\max}=\rho * A * V*=\frac{0.6847}{\sqrt{RT_t}}P_tA*=\frac{0.6847}{\sqrt{287\times 298}}0.1\times 10^6\times\frac{7.85\times 10^{-3}}{1.69}=1.088\text{ kg/s}$$

2.9 Normal Shocks

A *shock wave* is a compression wave caused by supersonic motion of a body in a medium. In other words, the shock can be conceived as irreversible discontinuities in the flow that can occur either in internal or external supersonic flow field. Proper understanding of shock behaviour is very important in designing supersonic diffusers either for high performance aircraft or wind tunnels. Normal shock can occur in front of an air intake and inside a nozzle in supersonic flow (Fig. 2.11). Hence, the next section is devoted to analysing the flow containing normal shock. Before analysing the normal shocks, it is essential to get a clear physical picture of the shock itself. Let us consider a steady one dimensional shock standing in a passage of varying area (Fig. 2.12).

The thickness of the shock is very thin, around 0.3 μm (roughly five times the mean free path of gas molecules). We know that there will be finite changes taking place inside the shock. As a result, the finite changes in fluid properties such as temperature, pressure and velocity lead to large gradients across the small distance of the shock. Hence, there will be viscous and heat conduction dissipation inside the shock wave, which render the shock process internally irreversible. Besides this, the local fluid deceleration attains around tens million of gravity. Due to these sudden changes, we can treat normal shock as an abrupt discontinuity, although it is physically impossible to have discontinuities

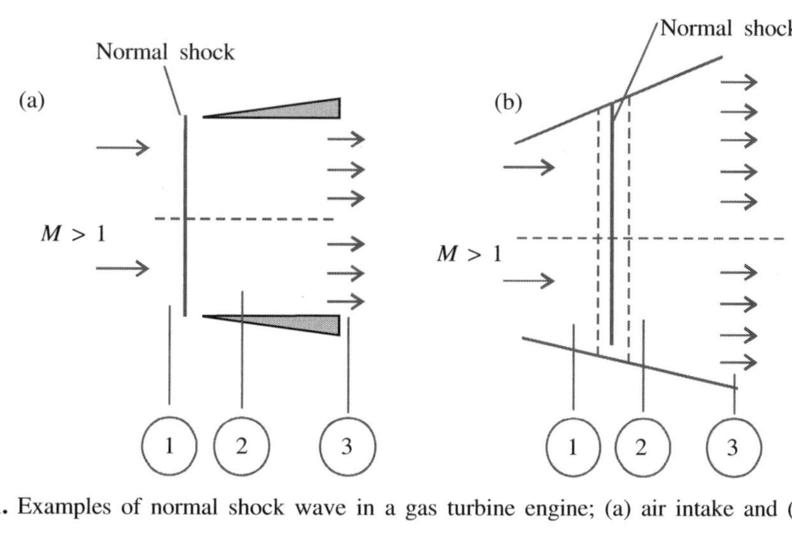

Fig. 2.11. Examples of normal shock wave in a gas turbine engine; (a) air intake and (b) nozzle.

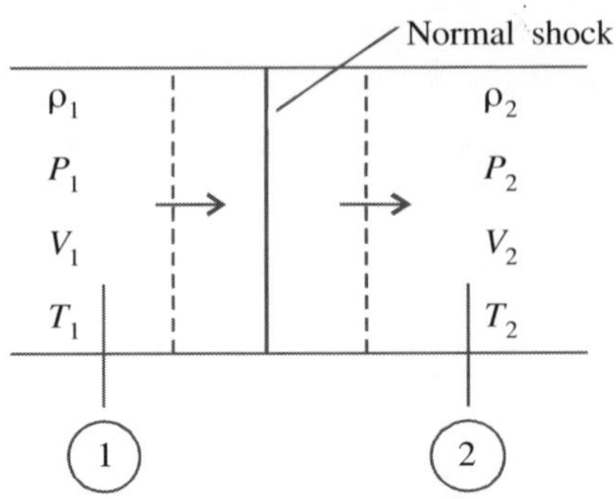

Fig. 2.12. Schematic of CV for analysis of normal shock wave.

in fluid flow. Note that we are interested in an analysis to predict the changes that occur across the shock rather than its detailed structure.

Analysis

Let us apply the basic governing equation to the thin CV (Fig. 2.12) and make the following assumptions before proceeding further:

(i) Steady flow
(ii) Uniform flow at each section
(iii) $A_1 = A_2 = A$ (as the shock is quite thin)
(iv) Negligible friction at channel wall as the shock is quite thin
(v) Negligible gravity force
(vi) Adiabatic flow
(vii) No shaft power
(viii) No work due to shear
(ix) Ideal gas with constant thermodynamic properties

Continuity Equation

The integral form of mass conservation is given by

$$\frac{\partial}{\partial t}\int \rho d\forall + \int_A \rho V.dA = 0 \tag{2.72}$$

where $\forall$ is the volume of fluid element.

Integrating Eq. (2.72) for steady flow and taking $A_1 = A_2 = A$ (thin shock), between two stations (1) and (2), we can have

$$\rho_1 V_1 = \rho_2 V_2 \tag{2.73}$$

Momentum Equation

Let us write down the integral momentum conservation equation as

$$\frac{\partial}{\partial t}\int (\rho V)\, d\forall + \int V(\rho V.dA) = -\int PdA + F_s + \int \rho_f d\forall \tag{2.74}$$

By integrating Eq. (2.74) for steady flow and taking $F_s = 0$, $F_g = 0$, $A_2 = A_1 = A$, between two stations (1) and (2), we can have

$$P_1 + \rho_1 V_1^2 = P_2 + \rho_2 V_2^2 \tag{2.75}$$

Energy Equation

The integral form of the energy equation can be expressed as

$$\frac{\partial}{\partial t}\int \rho\left(e + \frac{V^2}{2}\right)d\forall + \int \rho V\left(e + \frac{V^2}{2}\right)dA = -\int PV.dA + \int \rho(f.V)d\forall + \int \rho\dot{Q}d\forall \tag{2.76}$$

By considering assumptions (v) to (viii), we can integrate Eq. (2.76) between two stations (1) and (2) as

$$-\rho_1\left(e_1 + \frac{V^2}{2}\right)V_1A_1 + \rho_2\left(e_2 + \frac{V^2}{2}\right)V_2A_2 = P_1V_1A_1 - P_2V_2A_2 \tag{2.77}$$

Using Eq. (2.73), we can simplify Eq. (2.77) to get

$$\frac{P_1}{\rho_1} + e_1 + \frac{V_1^2}{2} = \frac{P_2}{\rho_2} + e_2 + \frac{V_2^2}{2} \tag{2.78}$$

As $h = e + \dfrac{P}{\rho}$, Eq. (2.78) becomes

$$h_1 + \frac{V_1^2}{2} = h_2 + \frac{V_2^2}{2} \tag{2.79}$$

We can rewrite Eq. (2.79) in terms of total enthalpy as

$$\Rightarrow h_{t1} = h_{t2} \tag{2.80}$$

Hence, the total energy of the flow remains constant as there is no energy addition. For an ideal gas, the equation of state is given by

$$P = \rho RT \tag{2.81}$$

Let us first obtain property ratios in terms of M_1 and M_2. The temperature ratio can be expressed as

$$\frac{T_2}{T_1} = \frac{T_2}{T_{t2}}\cdot\frac{T_{t2}}{T_{t1}}\cdot\frac{T_{t1}}{T_1} \tag{2.82}$$

For a calorically perfect gas, the total enthalpy equation becomes

$$h_{t1} = h_{t2} \Rightarrow T_{t1} = T_{t2} \text{ (for an ideal gas)} \tag{2.83}$$

Note that stagnation temperature remains constant across the shock. We can express static temperature ratio in terms of Mach number using isentropic relationship between total and static temperature as

$$\frac{T_2}{T_1} = \frac{1 + \frac{\gamma - 1}{2}M_1^2}{1 + \frac{\gamma - 1}{2}M_2^2} \tag{2.84}$$

We can find a relationship for velocity ratio as

$$\frac{V_2}{V_1}=\frac{M_2 a_2}{M_1 a_1}=\frac{M_2\sqrt{\gamma RT_2}}{M_1\sqrt{\gamma RT_1}}=\frac{M_2}{M_1}\sqrt{\frac{T_2}{T_1}} \tag{2.85}$$

From the above equations, we can have

$$\frac{V_2}{V_1}=\frac{M_2}{M_1}\left[\frac{1+\frac{\gamma-1}{2}M_1^2}{1+\frac{\gamma-1}{2}M_2^2}\right]^{\frac{1}{2}} \tag{2.86}$$

From the continuity equation, a relationship between the density ratio can be obtained as

$$\frac{\rho_2}{\rho_1}=\frac{V_1}{V_2}=\frac{M_1}{M_2}\left[\frac{1+\frac{\gamma-1}{2}M_2^2}{1+\frac{\gamma-1}{2}M_1^2}\right]^{\frac{1}{2}} \tag{2.87}$$

Similarly, from the momentum equation, we can obtain a relationship between pressure ratio as

$$P_1\left(1+\frac{V_1^2}{RT_1}\right)=P_2\left(1+\frac{V_2^2}{RT_2}\right) \tag{2.88}$$

As $V^2=\gamma RTM^2$, we can have

$$\frac{P_2}{P_1}=\frac{1+\gamma M_1^2}{1+\gamma M_2^2} \tag{2.89}$$

To find relationship between M_1 and M_2, let us use equation of state and express temperature ratio as

$$\frac{T_2}{T_1}=\frac{P_2}{P_1}\frac{\rho_1}{\rho_2}=\left[\frac{1+\gamma M_1^2}{1+\gamma M_2^2}\right]\frac{M_2}{M_1}\left[\frac{1+\frac{\gamma-1}{2}M_1^2}{1+\frac{\gamma-1}{2}M_2^2}\right]^{\frac{1}{2}} \tag{2.90}$$

Using Eq. (2.90) and the expression for static temperature ratio, we can get

$$\left[\frac{1+\frac{\gamma-1}{2}M_1^2}{1+\frac{\gamma-1}{2}M_2^2}\right]^{\frac{1}{2}}=\left[\frac{1+\gamma M_1^2}{1+\gamma M_2^2}\right]\frac{M_2}{M_1} \tag{2.91}$$

Squaring Eq. (2.91), we can get

$$\frac{1+\frac{\gamma-1}{2}M_1^2}{1+\frac{\gamma-1}{2}M_2^2}=\frac{M_2^2}{M_1^2}=\left[\frac{1+\gamma M_1^2}{1+\gamma M_2^2}\right]^2 \tag{2.92}$$

By solving Eq. (2.92), the following two solutions can be obtained

$$M_2 = M_1 \text{ (trivial solution)} \tag{2.93}$$

$$M_2^2 = \frac{M_1^2 + \dfrac{2}{\gamma-1}}{\dfrac{2\gamma}{\gamma-1}M_1^2 - 1} \tag{2.94}$$

If we take this result for $\gamma = 1.4$ (Fig. 2.13), it is found that when inlet Mach number is supersonic ($M_1 > 1$), the exit Mach number M_2 will be subsonic ($M < 1$) and a compression shock will be produced. But when inlet flow is subsonic ($M_1 < 1$), an expansion wave will be formed ($M_1 > 1$).

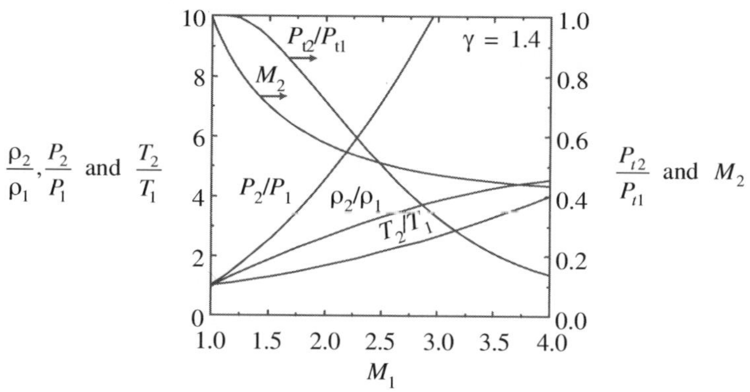

Fig. 2.13. Variation of properties across a normal shock wave with upstream Mach number.

Let us now find the relationship between stagnation pressure across the shock wave as

$$\frac{P_{t2}}{P_{t1}} = \frac{P_{t2}}{P_2}\cdot\frac{P_2}{P_1}\cdot\frac{P_1}{P_{t1}} = \frac{P_2}{P_1}\left[\frac{1+\dfrac{\gamma-1}{2}M_2^2}{1+\dfrac{\gamma-1}{2}M_1^2}\right]^{\frac{\gamma}{\gamma-1}} \tag{2.95}$$

Combining Eqs. (2.89) and (2.94), we can express static pressure ratio in terms of inlet Mach number M_1 as

$$\frac{P_2}{P_1} = \frac{1+\gamma M_1^2}{1+\gamma M_2^2} = \frac{1+\gamma M_1^2}{1+\gamma\left(\dfrac{M_1^2 + \dfrac{2}{\gamma-1}}{\dfrac{2\gamma}{\gamma-1}M_1^2 - 1}\right)} \tag{2.96}$$

By simplifying Eq. (2.96), we can get density and temperature ratio as

$$\frac{P_2}{P_1} = \frac{2\gamma}{\gamma+1}M_1^2 - \frac{\gamma-1}{\gamma+1} = \frac{2\gamma}{\gamma+1}\left(M_1^2 - 1\right) + 1 \tag{2.97}$$

Similarly, we can get the expression

$$\frac{\rho_2}{\rho_1} = \frac{(\gamma+1)M_1^2}{(\gamma-1)M_1^2+2} \tag{2.98}$$

$$\frac{T_2}{T_1} = \frac{P_2}{P_1}.\frac{\rho_1}{\rho_2} = \left[\frac{2\gamma}{\gamma+1}\left(M_1^2-1\right)+1\right]\left[\frac{2+(\gamma-1)M_1^2}{(\gamma+1)M_1^2}\right] \tag{2.99}$$

Entropy Across the Shock

$$s_2 - s_1 = C_P \ln\left[\left(\frac{2\gamma}{\gamma+1}\left(M_1^2-1\right)+1\right)\frac{2+(\gamma-1)M_1^2}{(\gamma+1)M_1^2}\right] - R\ln\left[\frac{2\gamma}{\gamma+1}\left(M_1^2-1\right)+1\right] \tag{2.100}$$

Equation (2.100) indicates that entropy change across the normal shock wave is only a function of M_1 and γ. We find that when flow is supersonic ($M_1 > 1$), the change in entropy across the normal shock ($s_2 - s_1$) becomes positive. But if the flow is subsonic, the change in entropy across the normal shock ($s_2 - s_1$) becomes negative, which violates the Second Law of Thermodynamics. Hence, it is not possible to accelerate the flow across a shock (expansion) wave, i.e. the shock can appear only when the flow is supersonic.

The property ratios P_{t2}/P_{t1}, T_2/T_1, P_2/P_1 and V_2/V_1 are given in Appendix B in terms of M_1 for a flow of an ideal gas through a normal shock (γ = 1.4). These data can be used easily for solving the problem. The following example illustrates the use of these relationships and normal shock waves.

Example 2.5 A normal shock occurs at an air intake area ratio of 3 for a gas turbine engine powered aircraft travelling at Mach number 2.0 at an altitude of 10 km. Determine the Mach number, static pressure and stagnation pressure of the flow leaving the diffuser.

Given Al = 10 km, P_1 = 26.48 kPa.

Solution Assuming isentropic flow in the diffuser, let us first find out flow properties just downstream of the shock. We can use normal shock wave table of Appendix B for γ = 1.4 to get the flow properties for M_1 = 2.

$$M_1 = 2.0;\ M_2 = 0.5774 \simeq 0.58;\ \frac{P_2}{P_1} = 4.5;\text{ and }\frac{\rho_2}{\rho_1} = 2.667$$

$$\frac{T_2}{T_1} = 1.668;\text{ and }\frac{P_{t2}}{P_{t1}} = 0.7209$$

From isentropic flow table given in Appendix B2 we can get flow properties for M_1 = 2.0

$$\Rightarrow \qquad \frac{P_1}{P_{t1}} = 0.1278;\ \frac{T_1}{T_{t1}} = 0.5556;\text{ and }\frac{A}{A^*} = 1.688$$

Hence, P_{t1} = 2.07 × 10^5 Pa.
Now the total pressure at downstream of the shock is

$$P_{t2} = 0.7209 \times 207 \times 10^3 = 149.4 \times 10^3 \text{ Pa}$$

As the flow downstream of the shock is isentropic, we have $P_{t3} = P_{t2}$ = 149.4 kPa.
For isentropic flow from (2) to (3), with A^*_3, we have

$$\frac{A_3}{A_3^*} = \frac{A_3}{A_2}\times\frac{A_2}{A_3^*}$$

From isentropic table, for $M_2 = 0.58$, $\frac{A}{A^*} = 1.21$.

Therefore, we can evaluate $\frac{A_3}{A_3^*}$ as

$$\frac{A_3}{A_3^*} = 3 \times 1.21 = 3.63 \text{ as } \frac{A_3}{A_2} = 3$$

For corresponding $\frac{A_3}{A_3^*}$ from isentropic table in Appendix B, we get

$$M_3 = 0.165; \ \frac{P_3}{P_{t3}} = 0.981$$

Hence, $P_3 = 0.981 \times 149.4 = 146.56$ kPa.
It can be observed from the above example that shock relations can be used to design supersonic air intake.

2.10 Oblique Shocks

In Section 2.9, we have discussed about normal shock wave. This is a special case of an oblique shock wave that occurs in supersonic flow. We know that normal shock wave is more likely to occur in one dimensional flow in which a compression wave occurs normal to the flow direction. If the supersonic flow is two/three dimensional, then a compression wave is likely to occur inclined at an angle to the flow, which is known as *oblique shock*. There are several physical situations, where an oblique shock can be formed, e.g. if a wedge-shaped object is placed in a supersonic flow, an oblique shock can be formed attached to the nose of the body (Fig. 2.14(a)).

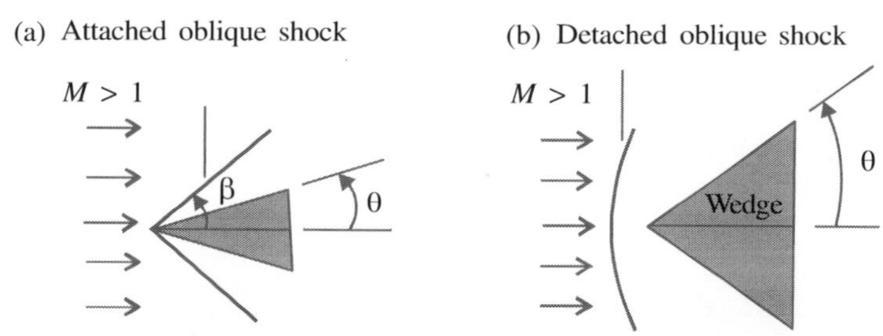

Fig. 2.14. Types of oblique shock waves (a) attached shock and (b) detached shock.

If the wedge is thick, then the oblique shock wave will be detached (Fig. 2.14(b)) but a major portion of the shock can be considered as planar wave. However, for a blunt wedge, a curved detached shock can be formed. Hence, we find that free stream Mach number wedge angle can determine the types of oblique shocks that are likely to be formed. Oblique shocks can occur at air intake and exit of the turbine blades under supersonic flow condition. The drag due to shock as an aerodynamic body increases with increase in wave angle. Hence, it is important to study oblique shocks so that drag or losses due to shock formation can be minimised by proper design.

Let us analyse oblique shock inclined at an angle to the flow as shown in Fig. 2.15. It can be noted that as the flow passes through the shock wave, it is deflected through an angle θ. The oblique shock makes a wave angle β with respect to upstream velocity V_1. The velocity behind the shock wave is V_2, which is parallel to the wedge surface with an angle θ. Let us consider a CV as indicated by the dashed lines (Fig. 2.15). The upper and lower sides of this CV coinciding with the flow stream lines on the left and right hand sides are parallel to the oblique shock. The velocity component perpendicular and parallel to the shock wave are V_n and V_t, respectively (Fig. 2.15). We then apply integral conservation equations as applied to a normal shock wave with the following assumptions:

(i) Steady inviscid flow
(ii) No body force
(iii) Adiabatic flow
(iv) Ideal gas with constant properties
(v) Negligible gravity force

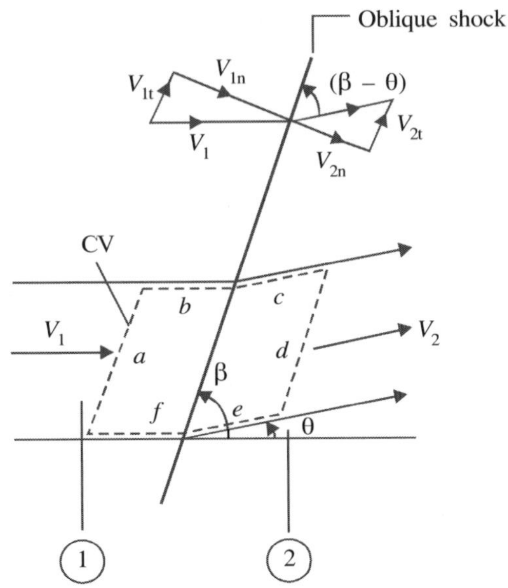

Fig. 2.15. CV for an oblique shock wave.

Mass Conservation Equation

By integrating the mass conservation equation across the CV between stations (1) and (2), we have

$$\rho_1 V_{1n} A_1 = \rho_2 V_{2n} A_2 \tag{2.101}$$

The mass contribution due to surface integral ($V.dA = 0$) is zero as the velocity in the normal and tangential directions are parallel to their respective faces (Fig. 2.16). Assuming $A_1 = A_2$, we have

$$\rho_1 V_{1n} = \rho_2 V_{2n} \tag{2.102}$$

Momentum Equation

Let us first apply the momentum equation in normal and tangential direction. Invoking the momentum balance along the tangential direction, we have

$$\int (\rho V \cdot dA) V = -\int P dA \tag{2.103}$$

For CV shown in Fig. 2.15, Eq. (2.103) becomes

$$-\rho_1 V_{1n}^2 + \rho_2 V_{2n}^2 = -(-P_1 + P_2) \tag{2.104}$$

The tangential component of the momentum equation would be

$$-\rho_1 V_{1n} V_{1t} + \rho_2 V_{2n} V_{2t} = 0 \tag{2.105}$$

Note that tangential component of $P \cdot dA$ is zero as the momentum due to pressure component on *b* and *c* cancel that on faces *f* and *e*, respectively. Using Eq. (2.105) and continuity equation, we have

$$V_{1t} = V_{2t} \tag{2.106}$$

The result indicates that the tangential component of flow velocity remains constant across an oblique shock. Hence, an oblique shock can be easily considered as a normal shock relative to the coordinate system moving with velocity V_{1t}. As a result, we can easily use the normal shock relations but in terms of normal component of upstream Mach number M_{1n}. The normal Mach number is given by

$$M_{1n} = M_1 \sin\beta \tag{2.107}$$

where β is the angle between the wedge centerline and the shock, which is generally known as *shock angle*. Hence, the properties ratios across oblique shock waves are obtained easily by using the following normal shock relationships

$$\frac{P_2}{P_1} = 1 + \frac{2\gamma}{\gamma+1}\left(M_{1n}^2 - 1\right) \tag{2.108}$$

$$\frac{\rho_2}{\rho_1} = \frac{(\gamma+1)M_{1n}^2}{(\gamma-1)M_{1n}^2 + 2} \tag{2.109}$$

$$\frac{T_2}{T_1} = \frac{P_2}{P_1} \cdot \frac{\rho_1}{\rho_2} \tag{2.110}$$

$$M_{2n}^2 = \frac{M_{1n}^2 + [2/(\gamma-1)]}{[2\gamma/(\gamma-1)]M_{1n}^2 - 1} \tag{2.111}$$

The Mach number at downstream of the shock can be found out by using geometry (Fig. 2.15) as

$$M_2 = \frac{M_{2n}}{\sin(\beta-\theta)} \tag{2.112}$$

From Eq. (2.112), it can be observed that we need to know the flow deflection angle θ, to find out M_2. But θ is uniquely related to M_1 and β. Let us derive a relation between θ, β and M_1 by considering the velocity triangle (Fig. 2.15), as

$$\tan\beta = \frac{V_{1n}}{V_{1t}} \tag{2.113}$$

$$\tan(\beta-\theta) = \frac{V_{2n}}{V_{2t}} \tag{2.114}$$

By using Eqs. (2.113) and (2.114) and noting that $V_{1t} = V_{2t}$, we have

$$\frac{\tan(\beta-\theta)}{\tan\beta} = \frac{V_{2n}}{V_{1n}} \tag{2.115}$$

But from the continuity, we know that

$$\frac{V_{2n}}{V_{1n}} = \frac{\rho_1}{\rho_2} \tag{2.116}$$

Now, using Eqs. (2.107), (2.109), and (2.114) and simplifying further, Eq. (2.116) can be expressed in terms of inlet Mach number as

$$\frac{\tan(\beta-\theta)}{\tan\beta} = \frac{(\gamma-1)M_1^2\sin^2\beta + 2}{(\gamma+1)M_1^2\sin^2\beta} \tag{2.117}$$

Equation (2.117) is an implicit relation between θ, β for given M_1. By certain trigonometric manipulation, the above expression can be written as

$$\tan\theta = 2\cot\beta\left[\frac{M_1^2\sin^2\beta - 1}{M_1^2(\gamma+\cos 2\beta)+2}\right] \tag{2.118}$$

It can be observed that this is an explicit relation for θ in terms of M_1 and β, which is known as β-θ-M relation. The β-θ-M relation is very important for the analysis of oblique shock. For wide range of θ and β, the results are shown in Fig. 2.16 for certain Mach numbers. From Fig. 2.16 it can be observed that for a given upstream Mach number M_1, the deflection angle has a maximum value ($\theta > \theta_{max}$) beyond which no solution exists for a straight shock wave. Rather, the shock gets detached and curved as shown in Fig. 2.16. If for a given upstream Mach number M, the deflection angle is less than θ_{max} ($\theta < \theta_{max}$), two values of β can be obtained from the above β-θ-M relation (Fig. 2.16). For example, for given $\theta = 20°$, $M_1 = 3.0$, β can either be 37.76° (week solution) or 82.15° (strong solution). It must be kept in mind that for strong shock solution, the flow becomes subsonic behind the shock wave. In contrast, the flow becomes supersonic except for a small range of θ values around the θ_{max} behind the shock wave. Note that weak shock is more likely to occur in nature. However, strong shock can also occur very rarely in special situations, where the back pressure is increased by some independent mechanisms. Such situations may occur in supersonic air intake, particularly, as the back pressure downstream of air intake can be changed by the engine. If the deflection angle becomes zero, then the shock wave angle can be either 90°, leading to the formation of a normal shock wave, or β decreases to a limiting value ($1/M_1$) giving rise to a Mach wave (disappearance of shock). It can also be observed for a particular deflection angle θ, the wave angle β increases for the weak solution with decrease in free stream Mach number from high to low supersonic value. Besides these observations, we also learn from oblique shock relations along with Fig. 2.16 that for a fixed value of M_1, shock wave angle β, P_2, T_2 and ρ_2 increases and M_2 decreases with increase in deflection angle θ. But if the value of θ increases beyond θ_{max}, the shock wave gets detached. Let us now see what happens when the upstream Mach number increases beyond unity for a fixed deflection angle. Note

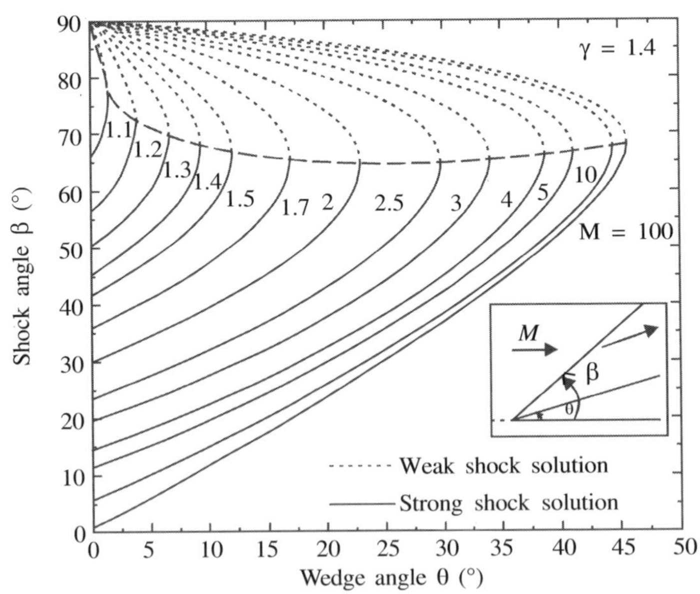

Fig. 2.16. β-θ-M diagram for air (γ = 1.4) for an oblique shock wave.

that when M_1 increases beyond unity for a fixed value of θ, the shock wave gets attached (Fig. 2.16). With further increase in M_1, the shock remains attached but moves closer to the wedge leading to decrease in β and increase in P_2 T_2, ρ_2 and M_2. Remember that the above discussions are meant only for weak shock, which is very common in nature.

Example 2.6 A steady supersonic flow with M_1 = 3.0, P_1 = 100 kPa and T_1 = 298 K enters an air intake with a wedge of included angle of 40°. Determine (i) Shock angle β, (ii) Exit Mach number M_2, (iii) Exit total temperature T_{t2}, and (iv) Exit total pressure P_{t2}.

Given M_1 = 3.0, P_1 = 100 kPa, T_1 = 298 K and θ = 20°.

Solution From β-θ-*M* diagram (Fig. 2.16), θ = 20° and shock angle β = 37.5°. Now we can estimate normal component of inlet Mach number as

$$M_{1n} = M_1 \sin\beta = 3 \times \sin 37.5° = 1.826$$

From normal shock Table (Appendix B2) for M_{1n} = 1.826, we can get

$$\frac{T_2}{T_1} = 1.551;\ \frac{P_2}{P_1} = 3.723;\ M_{2n} = 0.6108 \text{ and } \frac{P_{t2}}{P_{t1}} = 0.801$$

Hence, we can easily evaluate properties ratio across oblique shock wave as

$$P_2 = \frac{P_2}{P_1} \times P_1 = 3.723 \times 100 = 372.3 \text{ kPa}$$

$$T_2 = \frac{T_2}{T_1} \times T_1 = 1.551 \times 298 = 462.2 \text{ K}$$

$$M_2 = \frac{M_{2n}}{\sin(\beta - \theta)} = \frac{0.61}{\sin 17.5°} = 2.03$$

From isentropic flow table (Appendix B1), for M_1 = 3.0 we can get $\frac{T_{t1}}{T}$ = 2.8 and $\frac{P_{t1}}{P}$ = 36.73. Using these values we can estimate total properties at its exit as

$$P_{t2} = \frac{P_{t2}}{P_{t1}}\frac{P_{t1}}{P_1} \times P_1 = 0.8 \times 36.73 \times 100 = 2942 \text{ kPa}$$

$$T_{t2} = T_{t1} = \frac{T_{t2}}{T_1} \times T_1 = 2.8 \times 298 = 834.4 \text{ K}$$

Interestingly, the Mach number at the downstream of oblique shock is less than inlet Mach number M_1 but remains within the supersonic flow regime.

2.11 Constant Area Flow with Friction

In propulsive devices such as air intake, compressor, turbine, nozzle etc., we can encounter friction due to gas flow over the surface of wall and blades, which cannot be neglected. As a result, the fluid properties can change due to wall friction in the duct. Of course, we can assume the flow to be adiabatic. Let us now consider constant area duct with friction (Fig. 2.17) for which we will have to carry out compressible flow analysis. This kind of flow is known as *Fanno flow*. The following assumptions can be made for analysing this problem:

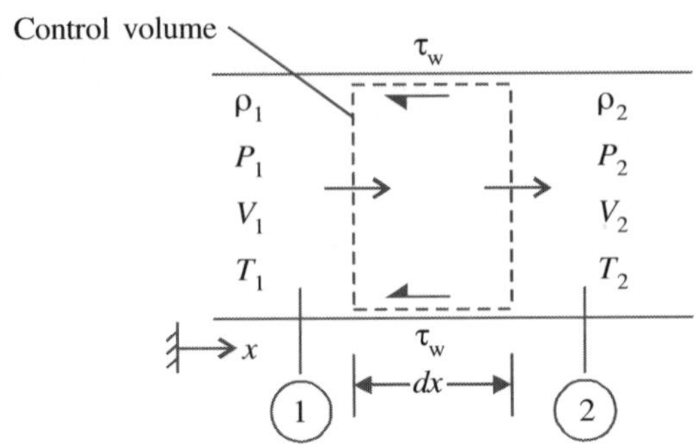

Fig. 2.17. CV for flow in a constant area duct with friction.

(i) One dimensional steady flow
(ii) Adiabatic flow
(iii) No gravitational force
(iv) Constant thermodynamic properties
(v) Ideal gas
(vi) No work interaction across system boundaries
(vii) No work done by shear force

Continuity Equation

The integral form of mass conservation for this problem becomes

$$\frac{\partial}{\partial t}\int \rho \, d\forall + \int \rho V \cdot dA = 0 \tag{2.119}$$

Equation (2.119) for the CV shown in Fig. 2.18 becomes

$$\Rightarrow \rho_1 V_1 = \rho_2 V_2 = \frac{\dot{m}}{A} (As\ A_1 = A_2 = A) \tag{2.120}$$

Momentum Equation

We know that the integral form of the momentum equation along x direction can be written as

$$\frac{\partial}{\partial t}\int V_x \rho \, d\forall + \int V_x(\rho V \cdot dA) = \int P\, dA + \int \rho f \, d\forall + \int \rho f \, d\forall + F_w \tag{2.121}$$

where F_w is the frictional force on the wall. For CV shown in Fig. 2.18 and considering the aforementioned assumptions, the above momentum equation becomes

$$\rho_2 V_2^2 A - \rho_1 V_1^2 A = (P_1 - P_2)\, A + F_w \tag{2.122}$$

Equation (2.122) can be written in terms of mass flow rate as

$$\rho_2 V_2^2 - \rho_1 V_1^2 + (P_2 - P_1) = -\frac{4}{D}\int_0^L \tau_w \, dx \tag{2.123}$$

where τ_w is frictional shear stress acting on the cylindrical CV, and D the hydraulic diameter of the duct.

Energy Equation

We know that the integral form of the energy equation can be written as

$$\frac{d}{dt}\int \rho\left(e+\frac{V^2}{2}\right)d\forall+\int \rho V\left(e+\frac{V^2}{2}\right)dA \tag{2.124}$$

$$=-\int PVdA+\int \rho(f\cdot V)\,d\forall+\int \rho\dot{q}d\forall \tag{2.125}$$

By using the above assumptions, we have

$$h_1+\frac{V_1^2}{2}=h_2+\frac{V_2^2}{2} \tag{2.126}$$

The equation of state

$$P = \rho RT \tag{2.127}$$

The differential form of continuity, momentum and energy equations can be written as

$$\frac{d\rho}{\rho}+\frac{dV}{V}=0 \tag{2.128}$$

$$\rho VdV+dP=-\frac{1}{2}\rho V^2\frac{4f\,dx}{D} \tag{2.129}$$

$$dh + V\,dV = 0 \tag{2.130}$$

where f = coefficient of friction = $\dfrac{\tau_w}{\left(\frac{1}{2}\rho V^2\right)}$.

For a calorically perfect gas, we can rewrite Eq. (2.129) using Eq. (2.127) and Eq. (2.130) in terms of Mach number as

$$\frac{2(1-M^2)}{\gamma M^2\left[1+\frac{(\gamma-1)}{2}M^2\right]}\frac{dM}{M}=\frac{4f}{D}dx \tag{2.131}$$

Integrating Eq. (2.131) between two stations (1) and (2), we have

$$\left[-\frac{1}{\gamma M^2}-\frac{\gamma+1}{2\gamma}\ln\left(\frac{M^2}{1+\frac{\gamma-1}{2}M^2}\right)\right]_{M_1}^{M_2}=\int_{x_1}^{x_2}\frac{4f}{D}dx \tag{2.132}$$

The ratios of static pressure, density and total pressure between the two stations can be obtained as derived earlier by considering the flow to be adiabatic. These properties relationships are

$$\frac{P_2}{P_1}=\frac{M_1}{M_2}\left[\frac{1+\frac{\gamma-1}{2}M_1^2}{1+\frac{\gamma-1}{2}M_2^2}\right]^{\frac{1}{2}} \tag{2.133}$$

$$\frac{\rho_2}{\rho_1} = \frac{M_1}{M_2}\left[\frac{1+\frac{\gamma-1}{2}M_1^2}{1+\frac{\gamma-1}{2}M_2^2}\right]^{-\frac{1}{2}} \tag{2.134}$$

$$\frac{P_{t2}}{P_{t1}} = \frac{M_1}{M_2}\left[\frac{1+\frac{\gamma-1}{2}M_2^2}{1+\frac{\gamma-1}{2}M_1^2}\right]^{\frac{\gamma+1}{2(\gamma-1)}} \tag{2.135}$$

2.12 One Dimensional Flow with Heat Addition

In the case of gas turbine or rocket engine, heat is released in the combustion chamber due to burning of fuel. For simplicity, we can assume the flow in the combustion chamber to be one dimensional and inviscid flow. In real flow, however, friction effects will be present, particularly, near the solid surfaces. Besides this, both wall friction and heat transfer are interlinked. However, such kind of ideal flow without friction is also known as *Rayleigh flow*. In order to analyse constant area flow with heat addition, we will have to use the governing equations of mass, momentum and energy with the following assumptions:

(i) One dimensional steady flow
(ii) Inviscid flow
(iii) No gravitational force
(iv) Ideal gas
(v) No work interaction
(vi) No work done by shear force
(vii) Constant thermodynamic properties

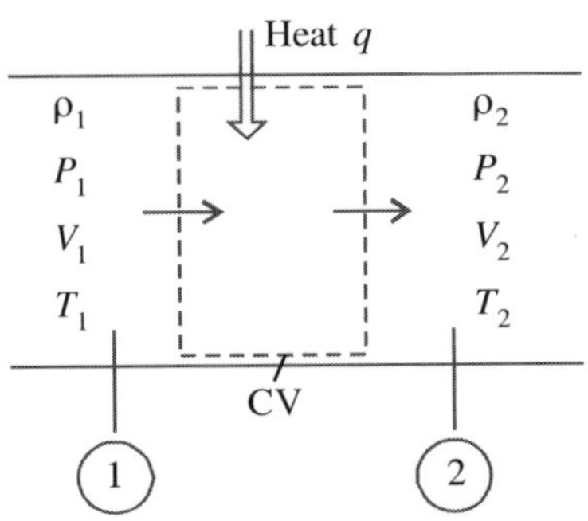

Fig. 2.18. CV for flow in a constant area duct with heat addition.

Let us consider a CV with heat addition (Fig. 2.18). By invoking the integral conservation equation of mass, momentum and energy between stations (1) and (2), we can have the following equations

Mass Conservation $$\rho_1 V_1 = \rho_2 V_2 \tag{2.136}$$

Momentum Conservation $$\rho_1 V_1^2 + P_1 = \rho_2 V_2^2 + P_2 \tag{2.137}$$

Energy Conservation
$$h_1 + \frac{V_1^2}{2} + q = h_2 + \frac{V_2^2}{2} \tag{2.138}$$

where q is the amount of heat added per unit mass (kJ/kg). The flow properties: P_2, ρ_2, T_2 and V_2 can be obtained by solving the above equations for known upstream flow properties and amount of heat addition $\dot{q}$.

The momentum equation can be rewritten as

$$P_2 - P_1 = \rho_1 V_1^2 - \rho_2 V_2^2 = \gamma P_1 M_1^2 - \gamma P_2 M_2^2 \tag{2.139}$$

We can now find out pressure ratios between stations (1) and (2) as

$$\frac{P_2}{P_1} = \frac{1+\gamma M_1^2}{1+\gamma M_2^2} \tag{2.140}$$

Using equation of state and mass conservation equation, we get expression for temperature ratio as

$$\frac{T_2}{T_1} = \frac{P_2}{P_1}\frac{\rho_1}{\rho_2} = \frac{P_2}{P_1}\frac{V_2}{V_1} = \frac{P_2}{P_1}\frac{M_2 a_2}{M_1 a_1} = \frac{P_2 M_2}{P_1 M_1}\sqrt{\frac{T_2}{T_1}} \tag{2.141}$$

Using Eq. (2.141), we have

$$\frac{T_2}{T_1} = \left[\frac{1+\gamma M_1^2}{1+\gamma M_2^2}\left(\frac{M_2}{M_1}\right)\right]^2 \tag{2.142}$$

We can also find out a relationship for ρ_2/ρ_1 as

$$\frac{\rho_2}{\rho_1} = \frac{P_2}{P_1}\frac{T_1}{T_2} = \frac{1+\gamma M_2^2}{1+\gamma M_1^2}\left(\frac{M_1}{M_2}\right)^2 \tag{2.143}$$

We will be more interested to find out an expression for total pressure ratio, which can be obtained by invoking the isentropic relationship as

$$\frac{P_{t2}}{P_{t1}} = \frac{P_{t2}}{P_2}\cdot\frac{P_2}{P_1}\cdot\frac{P_1}{P_{t1}} = \frac{1+\gamma M_1^2}{1+\gamma M_2^2}\left[\frac{1+\frac{\gamma-1}{2}M_2^2}{1+\frac{\gamma-1}{2}M_1^2}\right]^{\frac{\gamma}{\gamma-1}} \tag{2.144}$$

Similarly, the total temperature ratio relationship can be obtained as

$$\frac{T_{t2}}{T_{t1}} = \left[\frac{1+\gamma M_1^2}{1+\gamma M_2^2}\left(\frac{M_2}{M_1}\right)\right]^2\left[\frac{1+\frac{\gamma-1}{2}M_2^2}{1+\frac{\gamma-1}{2}M_1^2}\right] \tag{2.145}$$

For given heat addition rate per unit mass q, the energy equation can be used to get an expression for T_{t2} as

$$T_{t2} = T_{t1} + \frac{q}{C_P} \tag{2.146}$$

It can be noted that total temperature of the flow can be enhanced by the heat addition. By knowing the inlet Mach number M_1, T_{t2} and q, we can find out M_2 from Eq. (2.145). But Eq. (2.145) has to be solved by hit and trial method and $\frac{P_2}{P_1}, \frac{T_2}{T_1}, \frac{\rho_2}{\rho_1}$ and $\frac{P_{t2}}{P_{t1}}$ can be evaluated by using Eqs. (2.140), (2.142), (2.143) and (2.144), respectively. However, a more direct method can be devised by considering the sonic flow as a reference state. The properties ratio expressions corresponding to sonic condition ($M = 1$) are rewritten as

$$\frac{P}{P^*} = \frac{1+\gamma}{1+\gamma M^2} \tag{2.147}$$

$$\frac{T}{T^*} = M^2\left[\frac{(1+\gamma)}{1+\gamma M^2}\right]^2 \tag{2.148}$$

$$\frac{\rho}{\rho^*} = \frac{1}{M^2}\left(\frac{1+\gamma M^2}{1+\gamma}\right) \tag{2.149}$$

$$\frac{P_t}{P_t^*} = \frac{1+\gamma}{1+\gamma M^2}\left[\frac{2+(\gamma-1)M^2}{\gamma+1}\right]^{\frac{\gamma}{\gamma-1}} \tag{2.150}$$

$$\frac{T_t}{T_t^*} = \frac{(\gamma+1)M^2}{(1+\gamma M^2)^2}\left[2+(\gamma-1)M^2\right] \tag{2.151}$$

It is interesting to note that the fluid properties can be presented in terms of local Mach number. The data for properties ratios for $\gamma = 1.4$ in terms of local Mach number are given in Appendix B4 for Rayleigh flow, which can be used for solving problems. It can be observed that the flow approaches toward the sonic condition ($M = 1$) when the flow is either subsonic or supersonic. Once the flow attains a sonic condition, any further heat addition cannot be possible, unless the initial condition is changed. Then the flow is said to be thermally chocked. It must also be noted that the stagnation pressure drops significantly whenever heat energy is added to the flow. Hence, heat addition will incur significant stagnation pressure loss for both subsonic and supersonic flow. In contrast, heat extraction from the flow (cooling) causes a decrease in the stagnation pressure. We now consider an example to illustrate how the expressions corresponding to sonic condition can be used to analyse the constant flow with heat addition.

Example 2.7 The fuel-air mixture enters a gas turbine combustor with a velocity of 78 m/s at static pressure of 50 kPa and temperature of 400 K. During combustion, heat per unit mass of 1500 kJ/kg is released. Assuming the combustion chamber to be a constant area duct with negligible friction, estimate the properties such as T_2, P_2, and M_2 at its downstream. Take $\gamma = 1.33$, $C_p = 1.281$ kJ/kg K.

Given $V_1 = 78$ m/s, $T_1 = 400$ K, $P_1 = 50$ kPa and $q = 1500$ kJ/kg, $\gamma = 1.33$.

Solution The inlet Mach number can be estimated as

$$M_1 = \frac{V_1}{a_1} = \frac{78}{\sqrt{1.33\times 287\times 400}} = 0.2$$

From isentropic relation we can estimate total temperature and pressure as

$$T_{t1} = T_1\left(1+\frac{\gamma-1}{2}M_1^2\right) = 400\times\left(1+0.165\times(0.2)^2\right) = 402.6\text{ K}$$

$$P_{t1} = P_1\times\left(\frac{T_{t1}}{T_1}\right)^{\frac{\gamma}{\gamma-1}} = 50\times\left(\frac{402.6}{400}\right)^{4.0} = 51.33\text{ kPa}$$

Now we can estimate total temperature at the exit as

$$T_{t2} = T_{t1} + \frac{q}{C_P} = 402.6+\frac{1500\times10^3}{1281} = 1573.6\text{ K}$$

For M_1 = 0.2, by using Eqs. (2.148) and (2.147), we can estimate T_1/T^*, T_{t1}/T^* and P_1/P^*, P_{t1}/P^* as

$$\frac{T_1}{T^*} = M^2\left[\frac{(1+\gamma)}{1+\gamma M^2}\right]^2 = 0.2^2\left[\frac{(1+1.33)}{1+1.33\times0.2^2}\right]^2 = 0.196$$

$$\frac{P_1}{P^*} = \frac{1+\gamma}{1+\gamma M^2} = \frac{1+1.33}{1+1.33\times0.2^2} = 2.21$$

$$\frac{P_{t1}}{P_t^*} = \frac{1+\gamma}{1+\gamma M^2}\left[\frac{2+(\gamma-1)M^2}{\gamma+1}\right]^{\frac{\gamma}{\gamma-1}} = \frac{1+1.33}{1+1.33\times0.2^2}\left[\frac{2+(1.33-1)0.2^2}{1.33+1}\right]^{\frac{1.33}{1.33-1}} = 1.23$$

$$\frac{T_{t1}}{T_t^*} = \frac{(1.33+1)0.2^2}{(1+1.33\times0.2^2)^2}\left[2+(1.33-1)0.2^2\right] = 0.17$$

Now we can estimate exit temperature ratio as

$$\frac{T_{t2}}{T_t^*} = \frac{T_{t2}}{T_{t1}}\frac{T_{t1}}{T_t^*} = \frac{1573.6}{402.6}\times0.17 = 0.66$$

From the table in Appendix B4, we can estimate exit Mach number, M_2 = 0.48. Also, for M_2 = 0.48 T_2/T^* = 0.7587, and P_2/P^* = 1.815, P_{t2}/P^* = 1.122. We can determine properties at the exit of the combustor as

$$T_2 = \frac{T_2}{T^*}\frac{T^*}{T_1}T_1 = 0.7587\frac{1}{0.17}\times400 = 1785.2\text{ K}$$

$$P_2 = \frac{P_2}{P^*}\frac{P^*}{P_1}P_1 = 1.815\frac{1}{2.21}\times50 = 41.06\text{ kPa}$$

2.13 Boundary Layer Flow

In an earlier section, we have discussed inviscid flow. Of course, the viscous effect can be negligibly small as compared to the inertial effect only at very high Reynolds number. In the real sense, the viscous effect cannot be neglected near the wall region, particularly, when it is desirable to estimate drag due to friction, because the viscosity of fluid can cause the layer of fluid to stick to the wall surface. This layer of fluid near the wall has zero velocity with respect to the surface, which is known as "no slip" condition. To illustrate this point, let us consider the flow of a viscous fluid over a flat plate (Fig. 2.19). Due to viscous effect, the fluid adheres to the flat surface from the leading edge of the plate causing a layer of fluid stick to this plate. Figure 2.19 shows that velocity of the fluid increases rapidly along the vertical direction, to free stream velocity. As a result, the flow field can

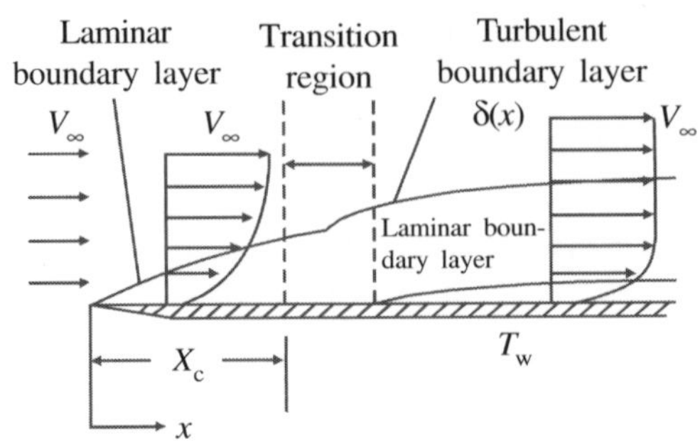

Fig. 2.19. Flow over a flat plate.

be divided into two parts (Fig. 2.19). The effect of viscosity is confined to a very thin layer, known as *boundary layer*, in the vicinity of the surface. Everywhere outside the boundary layer the velocity V_∞ and pressure P_∞ prevail. In this boundary layer, the velocity varies from zero at surface to V_∞ at the outer edge of the boundary layer. The flow can be assumed to be inviscid beyond this boundary layer. This concept of boundary layer was first discovered by the great scientist Ludwig Prandtl in 1904. Note that the thickness of the boundary layer increases along the length as more fluid slows down due to viscous effect. Although the thickness of the boundary layer is quite less, it is very important to consider the boundary layer flow, particularly, when we wish to estimate drag coefficient or heat transfer coefficient. We will briefly review the boundary layer flow, which the reader may have learnt from fluid mechanics [7, 8].

2.13.1 Laminar Boundary Layers

Let us consider a simple situation of boundary layer flow, which is formed over an infinitely flat plate (Fig. 2.19). It can be noted that the flow in the boundary layer near the leading edge is in smooth layers in which fluid particles remain in the same layer. Hence, the flow in this region is laminar in nature. With increase in the distance from the leading edge, the *laminar boundary layer* breaks down and becomes unstable. Hence, the flow becomes random in nature. This portion of boundary between the laminar and turbulent boundary layers is known as *transition region.* The turbulent boundary layer region witnesses random movement of slower moving particles into faster moving layers and vice versa due to mixing action. As a result, a more uniform velocity profile is produced in the turbulent boundary layer (Fig. 2.19). In the turbulent boundary layer, a thin layer known as *laminar sublayer* also prevails (Fig. 2.19). The shear stress due to viscous forces on its surface is dependent on the type of flow – laminar or turbulent – prevailing in the boundary layer. For example, for laminar flow boundary layer, the shear stress is given by

$$\tau_l = \mu \left[\frac{dv}{dy} \right]_{y=0} \tag{2.152}$$

But in the case of turbulent flow, momentum exchange is caused due to random motion of fluid particles from one layer to another. This leads to turbulent shear stress, which can be expressed in terms of eddy viscosity as

$$\tau_t = \rho\, \varepsilon \left[\frac{dv}{dy} \right]_{y=0} \tag{2.153}$$

Note that the eddy viscosity ε is much greater (100 times) than the kinematic viscosity ν. It must also be kept in mind that ε is not a physical property of fluids, rather it is dependent on the nature of the flow. Hence, it is not really possible to evaluate ε directly. Therefore, we will use approximate method while handling turbulent boundary layer.

In order to ascertain whether the flow is laminar or turbulent, we will have to evaluate the critical Reynolds number based on the distance along x-direction on which the boundary layer is being developed. The value of the critical Reynolds number depends on the roughness of surface, pressure gradient, turbulence level, etc. Generally, the critical Reynolds number for a flow over a flat plate is considered to be around 5×10^5. For example, if air is passing over a flat plate with velocity of 50 m/s, the transition from laminar to turbulent flow occurring at a distance from the leading edge is estimated as

$$x = \frac{\mathrm{Re}_c\,\nu}{V_\infty} = \frac{5\times10^5(1.6\times10^{-5})}{50} = 0.16\text{ m} \tag{2.154}$$

Keep in mind that we will be mostly dealing with the flow through a duct. In most cases, flow is likely to be under developed. Hence, the boundary layer analysis over a flat plate can be used approximately for ducts in propulsion devices.

Laminar Boundary Layer

For two dimensional steady flow, we can write conservation of mass and momentum equation as

$$\frac{\partial V_x}{\partial x} + \frac{\partial V_y}{\partial y} = 0 \tag{2.155}$$

$$V_x\frac{\partial V_x}{\partial x} + V_y\frac{\partial V_x}{\partial y} = -\frac{1}{\rho}\frac{\partial P}{\partial x} + \nu\left(\frac{\partial^2 V_x}{\partial x^2} + \frac{\partial^2 V_x}{\partial y^2}\right) \tag{2.156}$$

$$V_x\frac{\partial V_y}{\partial x} + V_y\frac{\partial V_y}{\partial y} = -\frac{1}{\rho}\frac{\partial P}{\partial y} + \nu\left(\frac{\partial^2 V_y}{\partial x^2} + \frac{\partial^2 V_y}{\partial y^2}\right) \tag{2.157}$$

In the vicinity of the boundary layer, we can consider that velocity along normal direction V_y is much smaller than velocity parallel to the plate. The rate of change of any parameter along the flow $\partial/\partial x$ must be much smaller than that in the normal to the flow direction as

$$V_y << V_x \text{ and } \frac{\partial}{\partial x} << \frac{\partial}{\partial y} \tag{2.158}$$

As a result, y-momentum equation turns out to be $\dfrac{\partial P}{\partial y} = 0$, which means that the pressure remains invariant along the y-direction. As a result, there is no pressure variation throughout the entire flow field when we consider uniform flow over a flat plate along the plate. Hence, the governing equations for the boundary layer are

$$\frac{\partial V_x}{\partial x} + \frac{\partial V_y}{\partial y} = 0 \tag{2.159}$$

$$V_x \frac{\partial V_x}{\partial x} + V_y \frac{\partial V_x}{\partial y} = \nu \frac{\partial^2 V_x}{\partial y^2} \tag{2.160}$$

The boundary condition for the above governing boundary layer equation are

$$y = 0;\ V_x = V_y = 0 \text{ (no slip condition)};\ 0 < y < \infty \quad V_x \to V_\infty \tag{2.161}$$

Mathematically, the velocity in x-direction in the boundary layer approaches to upstream velocity asymptotically. Practically, we can consider V_x to attain 0.99 V_∞ at distance δ, known as *boundary layer thickness*, from the plate. Note that the above boundary layer equations are non-linear in nature, which are extremely difficult to solve. However, H. Blasius converted the above partial differential equations by transforming the coordinate by $\eta = y\sqrt{\frac{V_\infty}{\nu x}}$. This was based on the argument that the velocity profile in a flat plate is self-similar in nature. Note that it is a single interdependent variable. Hence, V_x/V_∞ can be expressed in terms of unknown $f'(\eta)$, which needs to be determined. By using this transformation the boundary layer equations become

$$2f''' + f f'' = 0 \tag{2.162}$$

The boundary conditions given in Eq. (2.161) can be rewritten as

$$f = f' = 0 \text{ at } \eta = 0 \text{ and } f' \to 1 \text{ at } \eta \to \infty \tag{2.163}$$

Note that the original partial differential equation and boundary conditions are reduced to a third order non-linear ordinary differential equation by using similarity variable η. The three boundary conditions given in Eq. (2.161) are sufficient to solve the Blasius equation numerically. Hence, the dimensionless velocity profile of the boundary layer flow, V_x/V_∞ can be presented as $\eta = y\sqrt{\frac{V_\infty}{\nu x}}$ in a single curve (Fig. 2.20). It can be noted that $V = 0.99\ V_\infty$ approaches 1.0 as $\eta \to \infty$. We know that boundary layer thickness δ is the distance from wall to point where V_x becomes 0.99 V_∞. Hence, we can write

$$\delta\sqrt{\frac{\rho V_\infty}{\mu x}} = 5.0;\ \Rightarrow \frac{\delta}{x} = \frac{5.0}{\sqrt{\mathrm{Re}_x}} \tag{2.164}$$

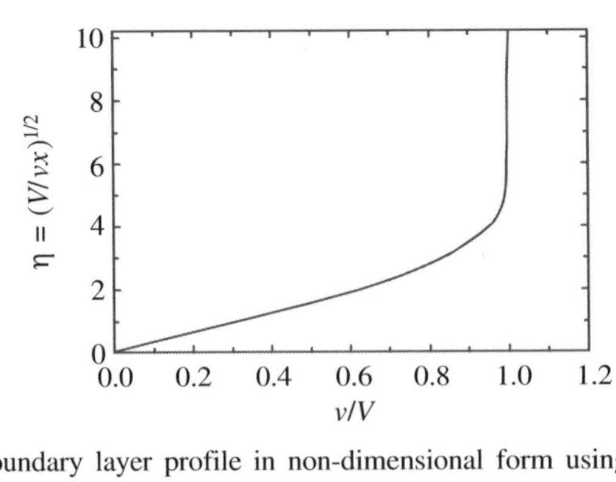

Fig. 2.20. Blasius boundary layer profile in non-dimensional form using similarity variable η.

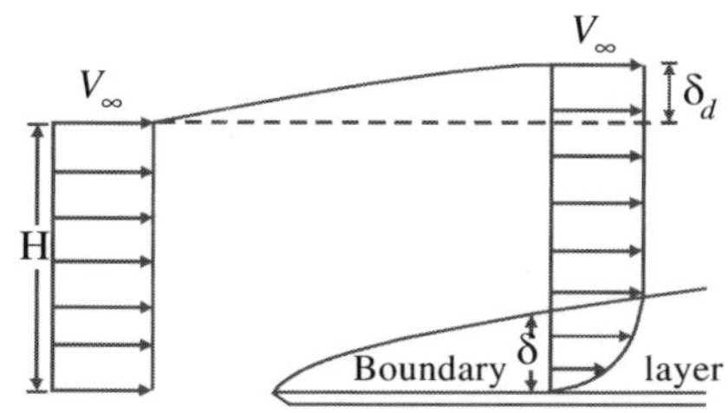

Fig. 2.21. Boundary layer flow defining momentum thickness δ_d.

This definition of boundary layer thickness is quite arbitrary. We can define another thickness that has more physical meaning. Therefore, let us see how displacement of fluid takes place by considering again uniform flow with V_∞ passing over a flat plate (Fig. 2.21). It can be noted from Fig. 2.21 that the flow initially contained in the height H is displaced upward due to decrease in velocity in the boundary layer by a distance δ_d to satisfy the mass conservation. Mathematically, we can write

$$V_\infty H = V_\infty \delta_d + V_\infty (H - \delta) + \int_0^\delta V_x dy \tag{2.165}$$

By simplifying, we get

$$V_\infty \delta_d = V_\infty \delta - \int_0^\delta V_x dy; \Rightarrow \delta_d = \int_0^\delta \left(1 - \frac{V_x}{V_\infty}\right) dy \tag{2.166}$$

From $\delta < y < \infty$, $V = V_\infty$, we can express Eq. (2.166) as

$$\delta_d = \int_0^\infty \left(1 - \frac{V_x}{V_\infty}\right) dy \tag{2.167}$$

By using Blasius solution for $\frac{V_x}{V_\infty}$ we can show that

$$\delta_d = \frac{1.73\, x}{\sqrt{\text{Re}_x}} \tag{2.168}$$

In analogy to the displacement thickness, another thickness known as *momentum thickness* δ_m can be defined on the basis of momentum conservation

$$\delta_m = \int_0^\infty \frac{V_x}{V_\infty}\left(1 - \frac{V_x}{V_\infty}\right) dy \tag{2.169}$$

By using the Blasius solution for $\frac{V_x}{V_\infty}$ we can get

$$\delta_m = \frac{0.664}{\sqrt{\text{Re}_x}} \tag{2.170}$$

In order to find out the propulsive thrust required for an aircraft, it is important to find the drag force due to viscous effects. We know that it is easy to determine the shear stress over a flat plate from its basic definition by knowing the velocity profile as

$$\tau_w = \mu\left(\frac{\partial V_x}{\partial y}\right)_{y=0} = 0.332\mu\frac{V_\infty}{x}\sqrt{\text{Re}_x} \tag{2.171}$$

From Eq. (2.171) we find that the shear stress at the wall decreases with increase in x. This is due to decrease of velocity gradient with increase in x, with increase in boundary layer thickness. We can express shear stress at the wall in terms of local skin friction coefficient C_{fx} as

$$C_{fx} = \frac{\tau_w}{\frac{1}{2}\rho V_\infty^2} = \frac{0.664}{\sqrt{\text{Re}_x}} \quad \text{for flat plate} \tag{2.172}$$

The total drag force F_D over a plate length L can be obtained as

$$F_D = \int_0^L 0.332\,\mu\frac{V_\infty}{x}\sqrt{\frac{\rho V_\infty x}{\mu}}\,b\,dx = \frac{0.664\,b\rho V_\infty^2 L}{\sqrt{\text{Re}_L}} \tag{2.173}$$

where b is the plate width. Then the skin friction drag coefficient C_{Df} can be obtained as

$$C_D = \frac{F_D}{\frac{1}{2}\rho V_\infty^2\,b\,L} = \frac{1.328}{\sqrt{\text{Re}_L}} \tag{2.174}$$

Note that C_D is the drag coefficient integrated over length L of the flat plate.

Example 2.8 A model aircraft is flying at an altitude of 2 km (T_1 = 288 K, P_1 = 79 kPa) and velocity of 3 m/s. It has wings having L = 20 and b = 25 cm. Assuming it as a flat plate, determine (i) δ, (ii) δ_d at trailing edge, and (iii) Aerodynamic drag force F_D on the wing.

Given V_∞ = 3.0 m/s, T_1 = 288 K, P_1 = 79 kPa, L = 20 cm, and b = 25 cm.

Solution Let us estimate Reynolds number at the trailing edge as

$$\text{Re}_L = \frac{V_\infty L}{\nu} = \frac{3.0\times 0.2}{1.5\times 10^{-5}} = 4\times 10^4$$

where the kinematic viscosity ν for air at (T_1 = 288 K, P_1 = 79 kPa) is considered to be 1.5 × 10^{-5} m^2/s.

Note that this is less than critical Reynolds number for transition to turbulent flow, i.e. 5 × 10^5. Hence, the boundary layer over the entire wing can be considered to be laminar in nature. Thus, the equations developed earlier can be utilised to estimate the requisite quantities. The boundary thickness δ at V_x/V_∞ = 0.99 is given by

$$\delta_d = \frac{5.0x}{\sqrt{\text{Re}_x}} = \frac{5.0\times 0.2}{\sqrt{4\times 10^4}} = 0.5\text{ cm}$$

Similarly, displacement thickness δ_d can be estimated as

$$\delta_d = \frac{1.721x}{\sqrt{\text{Re}_x}} = \frac{1.73\times 0.2}{\sqrt{4\times 10^4}} = 0.172\text{ cm}$$

We can easily estimate drag D as

$$F_D = 2\left(\frac{1}{2}\rho V_\infty^2 bL\right)\times C_D = 2\left(\frac{1}{2}\rho V_\infty^2 bL\right)\frac{1.328}{\sqrt{\text{Re}_L}} = 2\left(1.2\times 3^2\times 0.2\times 0.25\right)\frac{1.328}{\sqrt{4\times 10^4}} = 3.58\times 10^{-3}\ N$$

2.13.2 Turbulent Boundary Layer Flow

We know that the turbulent boundary layer is very complex in nature. Hence, the fluid variables such as velocity V, temperature T, etc., are unsteady and random in nature. The various fluid quantities are converted across the boundary layer in the direction perpendicular to the plate by random motion due to transport of eddies. As a result, there is a considerable net transfer of x momentum perpendicular to the plate. Thus, the shear force for turbulent boundary layer is considered to be larger as compared to that of the laminar boundary layer flow.

To evaluate the shear stress in turbulent boundary layer flow, we need to obtain velocity profile by solving boundary layer equations. Unfortunately, no exact solution for turbulent boundary layer flow is possible till date. Turbulent boundary layer flow can be solved numerically, but it is beyond the scope of the present section. However, we will discuss momentum integral method developed by von Karman. In this method, we will write down continuity and momentum equations for a CV extending from the wall surface to the outer edge of the boundary layer (Fig. 2.22).

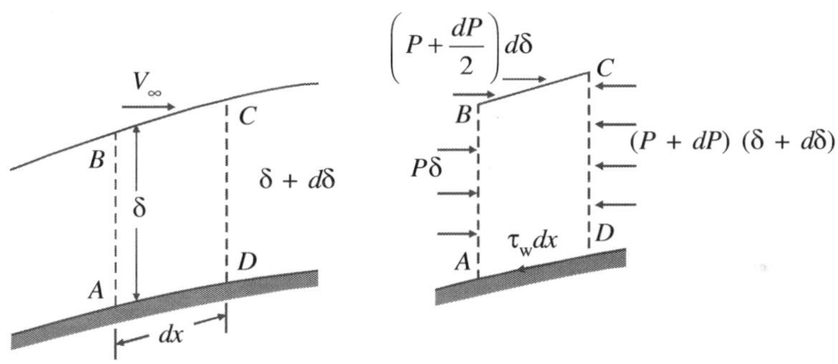

Fig. 2.22. CV for turbulent boundary layer flow illustrating the momentum integral method.

Assuming the flow to be two dimensional and steady, let us apply integral momentum equation for CV (ABCD) (Fig. 2.22) in the x-direction

$$\int_{ABCD} V(\rho V\cdot dA) = \Sigma F_x \tag{2.175}$$

The rate of momentum flux through AB is given by

$$\int_0^\delta \rho V_x^2\, dy \tag{2.176}$$

The rate of momentum across CD is given by

$$\int_0^\delta \rho V_x^2 dy + \frac{\partial}{\partial x}\int_0^\delta \left(\rho V_x^2 dy\right) dx \tag{2.177}$$

From the continuity equation the mass flow rate through BC is given by

$$\frac{d}{dx}\int_0^\delta \left(\rho V_x dy\right) dx \tag{2.178}$$

Keep in mind that the flow into CV through *BC* enters with velocity V_∞. Hence, the rate of momentum across *CD* is given by

$$V_\infty \rho \frac{d}{dx}\int_0^\delta (V_x dy)\, dx \tag{2.179}$$

Let us now evaluate the forces, mainly contributed due to pressure and wall shear stress, which can be expressed as

$$\Sigma F_x = P\,\delta + \left(P + \frac{dP}{2}\right) d\delta - (P + dP)(\delta + d\delta) - \tau_w dx \tag{2.180}$$

By dropping the second order terms and simplifying, Eq. (2.180) becomes

$$\Sigma F_x = -\,\delta\, dP - \tau_w dx \tag{2.181}$$

Substituting the above terms in the momentum equation (Eq. (2.175)) and simplifying, we can get

$$\rho \frac{d}{dx}\int_0^\delta V_x^2 dy - \rho V_x \frac{d}{dx}\int_0^\delta V_x dy = -\delta \frac{dP}{dx} - \tau_w \tag{2.182}$$

Equation (2.182) is known as *Momentum Integral equation*, which was first proposed by T. von Karman, a German engineer, in 1921. This equation is valid both for laminar and turbulent boundary layer flows. To evaluate the shear stress on the wall, we will have to assume the velocity profile along the y-direction.

Let us assume a velocity profile for turbulent boundary layer flow from experimental data as

$$\frac{V_x}{V_\infty} = \left(\frac{y}{\delta}\right)^{\frac{1}{7}} \tag{2.183}$$

This velocity profile is known as *power law*. It is not really valid very near to the wall as this laminar sub-layer exists near the wall. Hence, we can use a different empirical expression for τ_w obtained from experimental data as

$$\frac{\tau_w}{\rho V_\infty^2} = 0.0225\left(\frac{\nu}{V_\infty \delta}\right)^{\frac{1}{4}} \tag{2.184}$$

This above expression can be applied for Re in the range from 5×10^5 to 5×10^7. The momentum integral equation becomes

$$\rho V_\infty^2 \frac{d}{dx}\int_0^\delta \left(\frac{y}{\delta}\right)^{\frac{2}{7}} dy - \rho V_\infty^2 \frac{d}{dx}\int_0^\delta \left(\frac{y}{\delta}\right)^{\frac{1}{7}} dy = -\,\rho V_\infty^2 (0.0225)\left(\frac{\nu}{V_\infty \delta}\right)^{\frac{1}{4}} \tag{2.185}$$

By simplifying Eq. (2.185), we can get

$$\delta^{\frac{1}{4}}\, d\delta = 0.231\left(\frac{\nu}{V_\infty}\right)^{\frac{1}{4}} dx \tag{2.186}$$

Integrating Eq. (2.186), we have

$$\frac{4}{5}\delta^{\frac{5}{4}} = 0.231\left(\frac{\nu}{V_\infty}\right)^{\frac{1}{4}} x + C \tag{2.187}$$

Let us assume that the turbulent boundary layer starts at the leading edge, i.e. $x = 0$; $\delta = 0$; hence, $C = 0$. Then Eq. (2.187) becomes

$$\delta = 0.371\left(\frac{V_\infty x}{\nu}\right)^{-\frac{1}{5}} x \tag{2.188}$$

But we know that wall shear stress can be expressed as a function of x-distance as

$$\tau_w = \rho V_\infty^2\, 0.0225\left[\frac{\nu}{V_\infty}\frac{1}{\delta}\right]^{\frac{1}{4}} \tag{2.189}$$

By using the expression for δ in Eq. (2.189), we get

$$\tau_w = \frac{0.029\,\rho V_\infty^2}{\mathrm{Re}_x^{\frac{1}{5}}} \tag{2.190}$$

The drag force over the plate length L can be obtained as

$$F_D = \int_0^L \tau_w dx = 0.036\ \rho V_\infty^2 \frac{bL}{\mathrm{Re}_L} \tag{2.191}$$

Now the skin friction drag coefficient C_D can be obtained as

$$C_D = \frac{F_D}{\frac{1}{2}\rho V_\infty^2 bL} = \frac{0.072}{\mathrm{Re}_L^{\frac{1}{5}}} \tag{2.192}$$

Although Eq. (2.192) for drag coefficient is obtained by approximate integral method, it predicts experimental data reasonably well. Keep in mind that the friction drag coefficient is valid over Reynolds number from $\mathrm{Re}_L = 5 \times 10^5$ to 5×10^7.

2.14 Chapter Summary

In this chapter we revisited the basic concepts of thermodynamics and gas dynamics, which are essential for the analysis of each component of a gas turbine engine. The basic principles of thermodynamics are discussed briefly, introducing the concepts of both control mass and control volume systems. In addition, thermochemical properties such as heat capacity, heat of reaction and heat of combustion for individual species and their mixtures are reviewed. All three laws of thermodynamics are restated for completeness. A brief discussion on reacting system, introducing certain important parameters such as equivalence ratio and stoichiometric proportion, heat of combustion, etc. are also covered. Another important parameter is adiabatic flame temperature, which indicates maximum possible temperature in a combustion process. Subsequently, basic principles of compressible flow are discussed briefly in order to highlight basic definitions and concepts, which are very useful in analysing the various components of gas turbine engines. Steady quasi-one dimensional flow problems are analysed deriving area-Mach number relationship, which is very important for analysis of both air intake and nozzle. Both normal and oblique shocks are analysed under certain ideal conditions. Besides this, some other constant area flows such as Fanno (frictional) and Rayleigh (heat addition) have been covered adequately. The concept of boundary layer flow is introduced, which may be helpful for understating complex flows in gas turbine engines. This chapter also touches briefly upon the concept of turbulent boundary layer.

Nomenclature

A	Area of cross-section
B	Width of the plate
C	Duct circumference
C_v	Specific heat of constant volume
C_p	Specific heat of constant pressure
C_{fx}	Local skin friction coefficient
C_D	Drag coefficient
E	Total energy of the system
e_t	Total energy constant of the system
F_D	Drag force
F_w	Frictional force at the wall
F_x	Force along x-direction
H	Enthalpy
$\Delta H_{R,298}$	Heat of reaction at standard states
H_{ip}	Heat of formation per unit mole of ith product
H_{iR}	Heat of formation per unit mole of ith reactant
H_P	Total enthalpy of product
H_R	Total enthalpy of reactant
L	Length of the flat plate
M	Mach number
m_{ox}	Mass of oxygen
m_{fuel}	Mass of fuel
MW_{air}	Molecular weight of air
MW_{fuel}	Molecular weight of fuel
$\dot{m}$	Mass flow rate
N	Total number of moles
n_{ip}	Number of moles of ith product
n_{Ir}	Number of moles of ith reactant
P	Pressure
P_t	Total pressure
Q	Heat
q_{rev}	Reversible heat exchange
R_u	Universal gas constant
Re	Reynolds number
S	Entropy
S_{irrev}	Entropy due to irreversible process
T	Temperature
U	Internal energy
V	Local flow velocity
W	Work done

Greek Symbols

γ	Specific heat ratio
ρ	Density
ϕ	Equivalence ratio
ε	Eddy viscosity
ν	Kinematics viscosity
δ_m	Momentum thickness
δ	Boundary thickness

Subscripts/Superscripts

ad	Adiabatic
D	Drag
f	Local skin friction
irrev	Irreversible
ip	i^{th} product
iR	i^{th} reactant
ox	Oxygen
P	Pressure
p	Product
R	Reactant
rev	Reversible
t	Total
u	Universal
V	Volume
W	Wall

Abbreviations

CV	Control volume
KE	Kinetic energy
PE	Potential energy

References and Suggested Readings

1. Mishra, D.P., *Engineering Thermodynamics*, Cengage Learning India Pvt. Ltd., 2011.
2. Mishra, D.P., *Fundamentals of Combustion*, PHI Learning Pvt. Ltd., 2008.
3. Sonntag, R.E., Borgnakke and Van Wylen, *Fundamentals of Thermodynamics*, John Wiley & Sons, Inc., 1999.
4. Cengel, Y.A. and Boles, M.A., *Thermodynamics – An Engineering Approach*, Tata McGraw-Hill Publishing Company Ltd., 2008.
5. Rathakrishnan, E., *Gas Dynamics*, PHI Learning Pvt. Ltd., 2008.
6. Anderson, J.D., Jr., *Modern Compressible Flow*, McGraw-Hill, New York, 1982.
7. White, F.M., *Fluid Mechanics*, 6th Edition, McGraw-Hill Inc., 2008.
8. Shames, H., *Mechanics of Fluids*, McGraw-Hill, New York, 2005.

Review Questions

1. What do you mean by calorically perfect gas? Explain it invoking kinetic theory of gas. Can it be employed for analysis of components of a gas turbine engine?

2. Draw the variation of specific heats of mono-, di- and tri-atomic species with temperature and explain the difference among the three types of species.
3. State First Law of Thermodynamics. What are the limitations of this law? Derive an expression for First Law of Thermodynamics for a control volume system by stating the assumptions.
4. Why is Second Law of Thermodynamics important for analysis of various component of a gas turbine engine?
5. What do you mean by Clasius Equation? How can it be utilised for analysis of a gas turbine engine?
6. What do you mean by equivalence ratio? Why is it used very often while dealing with a gas turbine engine?
7. What do you mean by adiabatic flame temperature? How can it be estimated? Explain using an example.
8. Draw the variation of adiabatic flame temperature with equivalence ratio for C_3H_8-air mixture at ambient temperature and pressure.
9. How does the adiabatic flame temperature vary with increase in initial temperature? Why is it so? Explain it by providing proper argument.
10. Enumerate the procedure to estimate adiabatic flame temperature under constant volume condition. How will it be different from the adiabatic flame temperature under constant pressure condition?
11. What do you mean by compressible flow? How does it differ from incompressible flow?
12. Explain why is it desirable to locate normal shock at the throat of a CD nozzle.
13. What is difference between compression and expansion wave in a compressible flow?
14. When does a compression wave change into a shock wave?
15. What do you mean by shock strength? Explain it physically. Derive an expression for shock strength.
16. Derive the Rankine-Hugoniot relation for oblique shock.
17. What is choking? Why does it occur? When a nozzle is choked, is there any way to change its mass flow rate? Elaborate.
18. In order to reduce pressure losses in supersonic air intake, which kind of shock is preferred.
19. If the shock appears over two dimensional wedge and axi-symmetric cone with same cone angle for same flow Mach number, then in which case would shock losses be minimum? Explain why is it so?
20. What is thermal chocking?
21. What is Rayleigh flow? When will it occur? Provide two practical examples. Derive an expression for exit Mach number for one dimensional Rayleigh flow.
22. Draw a Rayleigh line on T-s plane describing the processes involved.
23. What is Fanno flow? When will it occur? Provide two practical examples. Derive an expression for exit Mach number for one dimensional Rayleigh flow.
24. Show that upper and lower curves of Fanno flow represent subsonic and supersonic flows, respectively. Prove that flow becomes sonic at maximum entropy point.
25. What do you mean by laminar flow? How is it different from turbulent flow?
26. Why does the boundary layer develop whenever flow takes place over solid surface? Define boundary layer thickness for laminar flow.
27. Draw the boundary layer growth over a flat plate covering all three regimes of flow.

Problems

Q1. Air enters a nozzle at $T_1 = 300$ K and $P_1 = 0.25$ MPa. Assuming the flow to be isentropic, estimate density and temperature at point when pressure attains 0.21 MPa.

Q2. In a nozzle, air is expanded from initial state ($P_1 = 0.3$ MPa, $T_1 = 450$ K) to final state ($P_2 = 0.1$ MPa, $T_2 = 300$ K). Determine specific enthalpy change between these two states. If the process takes place between two states isobaric and isochoric process, respectively, show that entropy change between the two states is same.

Q3. When air ($T = 1000$ K) is expanded isentropically in a cylinder piston arrangement the original volume increases by five times. Determine (i) Pressure ratio, (ii) Final temperature, and (iii) Work done by the piston.

Q4. An air tank is at pressure of 0.5 MPa and temperature 300 K. If a 5 mm hole is made in this tank, determine the velocity of air flow rate and mass flow rate.

Q5. For an air intake the pressure coefficient is to be determined by $C_p = \dfrac{P - P_\infty}{0.5\rho_\infty V_\infty^2}$ where P_∞, ρ_∞ and V_∞ are the pressure, density and velocity, respectively, in the free stream. Derive an expression for C_p in terms of velocity ratio, Mach number and γ.

Q6. A conical diffuser ingests 10 kg/s air at $M = 0.8$, 0.1 MPa, and 30°C. If air is decelerated to 0.2, find total and static pressure, as well as temperature at inlet and exit of the diffuser.

Q7. The hot gas from a turbine at 0.1 MPa, 1000 K and $M = 0.8$ enters a CD nozzle with a throat area of 25 cm^2. If its exit Mach number is 1.5, then determine (a) Throat condition, (b) Exit pressure temperature, (c) Exit area, and (d) Mass flow rate through the nozzle. ($\gamma = 1.33$)

Q8. When a pitot probe is introduced into an air stream flowing at 400 m/s, a detached bow shock is formed ahead of it. Determine the pressure ratio across the bow shock assuming the central portion to be a normal shock.

Q9. Air is expanded through a CD nozzle from a large reservoir at 0.5 MPa and 30°C, respectively. If the back pressure is maintained at 200 kPa, determine (i) A_e/A_{th}, (ii) V_e, and (iii) At what back pressure will a normal shock exist at its exit plane.

Q10. Air is accelerated from $M = 0.8$ to sonic speed in a tube of 50 mm diameter. Determine the length of the tube. Determine the length of the tube if friction factor $f = 0.006$.

Q11. Air flows in an insulated pipe with $V = 200$ m/s, $T = 300$°C and $P = 0.5$ MPa. Determine (i) Exit velocity if pressure is dropped to 0.4 MPa, due to friction, and (ii) Distance between two sections if the pipe diameter is 100 mm and friction factor $f = 0.006$.

Q12. Air enters a combustion chamber at $V = 150$ m/s, 0.5 MPa and 400°C. During combustion 500 kJ/kg of heat is added. Determine pressure, temperature and Mach number at its exit. Assume flow to be one dimensional inviscid flow.

Q13. A gas turbine combustor receives air at 0.4 MPa, 300°C and 100 m/s. If the air-fuel ratio by mass is 40, determine (i) Mach number at exit, (ii) Maximum total temperature, and (iii) Pressure loss (%). ($\gamma = 1.33$)

Q14. Gas at $P_1 = 0.2$ MPa, $T_1 = 400$ K and $V_1 = 100$ m/s enters a constant area duct with diameter of 50 mm and length 300 mm. How much heat is required to accelerate the gas to sonic velocity at its exit. Determine the pressure and temperature of gas at sonic condition.

Q15. An aircraft is flying at an altitude of 10 km with velocity of 50 m/s. It has wings with $L = 2$ and $b = 15$ m. Assuming it to a flat plate, determine (i) δ, (ii) δ_d at trailing edge, and (iii) Aerodynamic drag D on the wing.

3. Aircraft Engine

My interest in aircraft propulsion began in the fall of 1933 at the age of 21 years while I was a student at Georgia Augusta University of Gottingen in Physics under Prof. R. Pohl with a minor in Applied Mechanics under Prof. Ludwig Prandtl. The strong vibrations and noise of the propeller piston engines triggered my interest in aircraft propulsion. It appeared to me that a steady thermodynamic flow process is needed. Such a process would not produce vibrations. Also, an engine based upon such a process could probably be lighter and more powerful than reciprocating engine with a propeller.

Hans von Ohain, inventor of the gas turbine jet engine

3.1 Introduction

This chapter is devoted to the fundamental aspects of gas turbine engine powered aircraft. As thrust is the most important characteristic of a jet engine, the equation for uninstalled thrust is derived invoking the conservation of mass and momentum. In order to characterise the performance of jet engine quantitatively, propulsive, thermal and overall efficiencies are defined, particularly, for a single stream turbojet engine. In addition, takeoff thrust, static thrust, specific thrust and thrust specific fuel consumption (TSFC) – used for carrying out cycle analysis in Chapters 4 and 5 – are defined and discussed extensively. It is important to know the range covered by an aircraft for certain amount of fuel. Hence, a relation for flight range known as *Breguet Range Equation* is derived from the first principle, which is also related to overall efficiency of the engine. As a jet engine operates on the Brayton cycle, the fundamental thermodynamic principles are invoked to analyse it considering air standard assumptions. It must be kept in mind that several engine configurations such as turbojet, turbofan turboprop, ramjet and combinations thereof are developed based on the Brayton cycle. A brief comparative description of these engines is discussed along with their respective applications.

3.2 Thrust Equation

A generalised equation for the thrust of an airbreathing jet engine can be derived by invoking the conservation equation of mass and momentum without considering the internal details of a particular engine. For this purpose, a CV enclosing the whole jet engine is considered (Fig. 3.1). Figure 3.1 can be understood as if an observer is moving alongwith the jet engine. The control surface (shown as a dashed line) is chosen at station (1) that is far away upstream from the inlet of the engine. The downstream control surface is fixed at station (2), which passes through the exhaust jet plane of the nozzle. Note that the control surfaces, on upstream and downstream sides, are selected such that the flow can be assumed to be one dimensional in nature. Thus, the analysis can be simplified, particularly, for estimating the momentum flux both at upstream and downstream control surfaces. The side surfaces of the CV are chosen parallel to the flight velocity V and are far away from the engine such

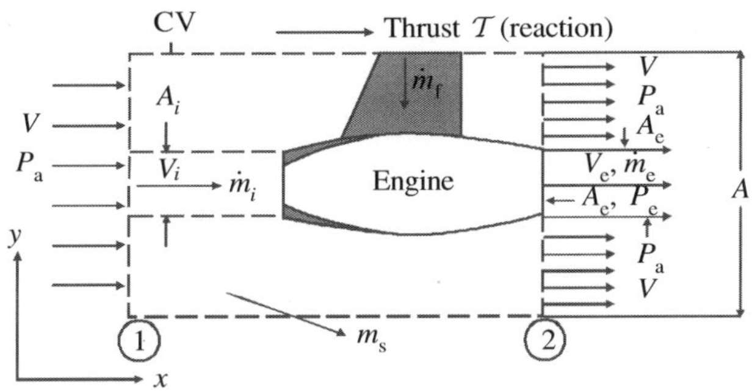

Fig. 3.1. CV of a gas turbine engine.

that one dimensional flow conditions can prevail with negligibly small velocity in y-direction. We can assume that the thrust and all other conditions at all points within the CV remain same with respect to time. The thrust $\mathcal{T}$ produced by the engine is transmitted through the structural support (e.g. pod, in case of pod mounted turbojet). Thus, the engine thrust can be defined as the vector summation of all forces on the external and internal surfaces of the nacelle. However, we can assume that the flow external to the engine is reversible. It means that there would not be any drag acting on the external surface such as nacelle or pod/strut of the engine.

Air enters into the CV through the control surface indicated as station (1) in Fig. 3.1 with flight velocity V and ambient density ρ. Let us say that air that is drawn into the engine at a rate of $\dot{m}_a$, which is captured by an area A_i (Fig. 3.1). Keep in mind that the fuel flow rate $\dot{m}_f$ is added from outside the CV (Fig. 3.1). Air leaves the exit plane of the CV with the same flight velocity V and density ρ except at the engine exist. However, the gas leaves the engine exit with a velocity V_e and density ρ_e (Fig. 3.1). The mass flux $\dot{m}_e$ crossing the engine exit area A_e is equal to $\rho_e V_e A_e$. Considering steady flow process, the mass conservation equation becomes

$$\iint \rho V \cdot n dA = 0 \tag{3.1}$$

Equation (3.1) for mass flow rate through the engine turns out to be

$$\rho_e V_e A_e - \rho V A_i - \dot{m}_f = 0 \tag{3.2}$$

By expressing Eq. (3.2) in terms of mass flow rate, we have

$$\dot{m}_e = \dot{m}_i + \dot{m}_f = \dot{m}_a + \dot{m}_f \tag{3.3}$$

Now by invoking the mass conservation equation for the whole CV, we can get

$$\rho_e V_e A_e + \rho V(A - A_e) + \dot{m}_s - \rho V A_i - \dot{m}_f = 0 \tag{3.4}$$

where $\dot{m}_s$ is the mass flow through the side surface of the CV and A the cross-sectional area of the CV normal to velocity V. Combining Eqs. (3.2) and (3.4) and simplifying, we get

$$\dot{m}_s = \rho V(A_e - A_i) \tag{3.5}$$

The upper and lower surfaces of the CV are chosen at sufficient distance from the engine such that the mass flow rate $\dot{m}_s$ crosses the CV (Fig. 3.1) with velocity component in x-direction.

Now we will derive an expression for the reaction to the thrust $\mathcal{T}$, which is transferred through the structural support (Fig. 3.1). Keep in mind that the thrust is the sum of all forces on the external and internal surfaces of the nacelle and the engine, respectively. Considering the flow to be steady and neglecting body forces, the momentum equation in x-direction becomes

$$\iint V(\rho V\cdot n)\, dA = \sum F_x \tag{3.6}$$

In the case of reversible external flow, both velocity and pressure can be considered to be constant over the entire control surface except over the engine's exit area A_e. The forces acting on control surfaces are contributed due to pressure and the reaction to the thrust, which is expressed as

$$\sum F_x = \mathcal{T} + P_a A - P_a (A - A_e) - P_e A_e \tag{3.7}$$

$$\sum F_x = \mathcal{T} + A_e (P_a - P_e)$$

In Eq. (3.7), the exit pressure P_e may be different from the ambient pressure P_a, particularly, when the exhaust velocity V_e attains supersonic speed. Now let us evaluate the momentum flux part given on the left hand side of Eq. (3.7) by

$$\iint V(\rho V\cdot n)\, dA = \dot{m}_e V_e + \dot{m}_s V + \rho V^2 (A - A_e) - \dot{m}_a V - \rho V^2 (A - A_i) \tag{3.8}$$

Substituting the expression for $\dot{m}_s$ (Eq. (3.5)) in Eq. (3.8) and simplifying, we get

$$\iint V(\rho V\cdot n)\, dA = \dot{m}_e V_e - \dot{m}_a V \tag{3.9}$$

Substituting Eq. (3.9) in the momentum equation (Eq. (3.7)) the expression for thrust becomes

$$\mathcal{T} = \dot{m}_e V_e - \dot{m}_a V + A_e (P_e - P_a) \tag{3.10}$$

Of course, we know that $\dot{m}_e$ is equal to $(\dot{m}_a + \dot{m}_f)$. By defining the fuel-air ratio, $f = \dot{m}_f / \dot{m}_a$, Eq. (3.10) for the thrust becomes

$$\mathcal{T} = \dot{m}_a \left[(1+f)V_e - V\right] + A_e (P_e - P_a) \tag{3.11}$$

Note that this expression for the thrust consists of two terms viz., momentum thrust and pressure thrust. Again, the momentum thrust consists of two terms viz., gross momentum thrust, $\dot{m}_a (1+f) V_e$ and the intake momentum drag $\dot{m}_a V$. The thrust must be equal in magnitude and opposite in direction to the drag of the aircraft when the aircraft is flying at uniform velocity V in level flight.

So far, we have only considered the thrust expression for a single stream engine. But there are some jet engines, e.g. turbofan and turboprop engines, where two exhaust streams have to be considered for estimation of thrust. This will be discussed in Chapter 4.

3.3 Engine Performance Parameters

The performance of a gas turbine engine can be characterised by several parameters. The two most important performance parameters are specific thrust $\mathcal{T}_s$ and thrust specific fuel consumption (TSFC). These parameters are related to the more fundamental efficiencies – propulsive, thermal and overall. The definitions of these efficiencies are given below.

3.3.1 Propulsive Efficiency

The propulsive efficiency η_p is the ratio of the thrust/propulsive power P_s to the rate of production of kinetic energy (sum of propulsive power and used kinetic energy of the jet). The propulsive efficiency would not provide effectiveness of the overall power plant in any sense. Rather, it indicates the effectiveness with which the engine is being used for propelling the aircraft. The kinetic energy of the jet is basically relative to the Earth. But the propulsive power is the product of the thrust and the flight speed, which is given by $P_s = \mathcal{T}V$. So the propulsive efficiency for a single stream engine is given by

$$\eta_p = \frac{\mathcal{T}V}{\mathcal{T}V + \dot{m}_a (V_e - V)^2/2} \tag{3.12}$$

By ignoring the pressure thrust and fuel flow rate $\dot{m}_f$ (as $\dot{m}_f << \dot{m}_a$) in case of air breathing engine, Eq. (3.12) becomes

$$\eta_p \approx \frac{\dot{m}_a V(V_e - V)}{\dot{m}_a \left[V(V_e - V) + (V_e - V)^2/2 \right]} = \frac{2}{1 + V_e/V} \tag{3.13}$$

The propulsive efficiency can be considered as a measure of the effectiveness with which the engine is being used for propelling an aircraft. It is also known as *Froude efficiency* in commemoration of William Froude (1810-1879) who was the first to define it for the characterisation of marine propulsive devices.

The variation of η_p vs. velocity ratio V_e/V (Fig. 3.2) shows that η_p would be maximum only when V_e/V will be unity. Unfortunately, under this condition, the thrust will be zero. However, when the thrust is maximum, particularly, for zero flight speed ($V = 0$), then η_p becomes zero. Hence, the designer has to choose a compromising value between the above two extreme cases, which means that the jet velocity V_e must be greater than the flight speed V, but the difference should not be too great. Due to these reasons, several kinds of propulsive devices such as turboprop, turbofan, turbojet and ramjet (Fig. 3.3) have been developed with optimum propulsive efficiency. The turbojet engine has high values of V_e/V and, thus, has a low propulsive efficiency. In contrast, the turboprop engine has lower value of V_e/V, which results in highest propulsive efficiency among these three jet engines. The choice of power plant is dependent on the mission requirement of a particular aircraft. More specifically, the choice of power plant (engine) depends on the flight velocity, altitude, maximum rate of climb, and desired range of flight, because the performance parameters such as thrust, fuel consumption, propulsive efficiency etc., are dependent on the operating conditions. These aspects will be discussed further not only in this chapter but also in subsequent chapters.

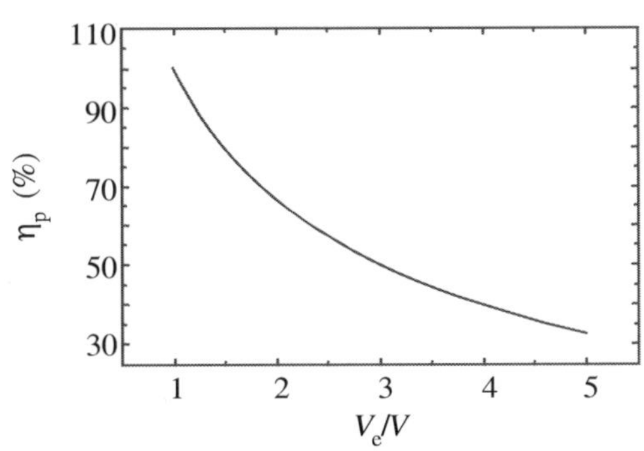

Fig. 3.2. Variation of propulsive efficiency η_p with velocity ratio V_e/V.

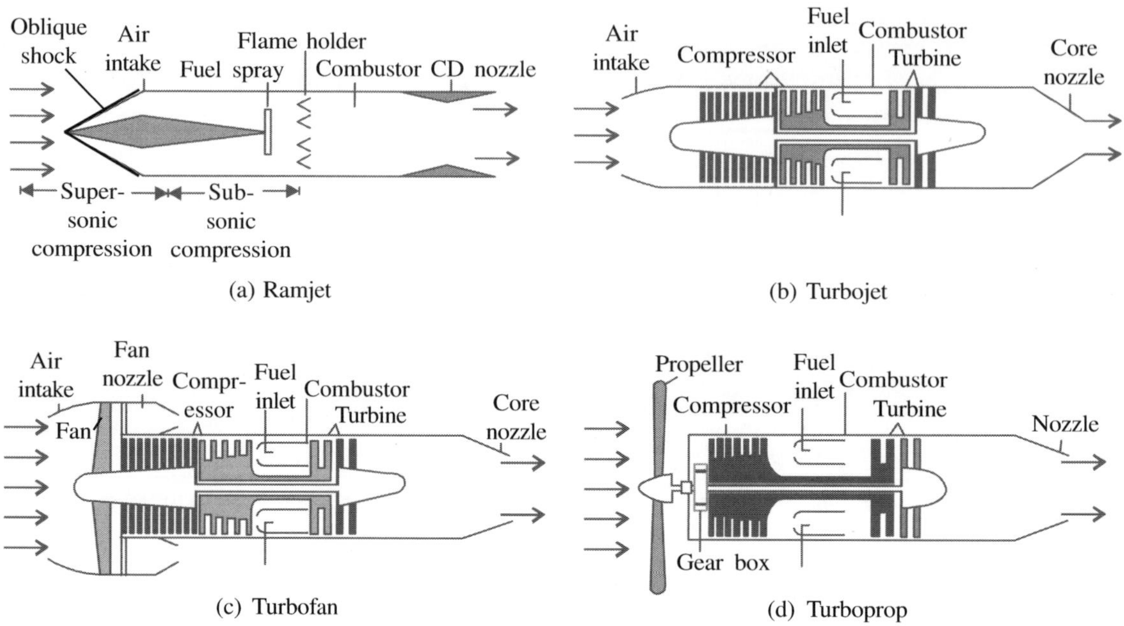

Fig. 3.3. Schematic of different types of turbojet engines.

3.3.2 Thermal Efficiency

Another useful performance parameter of the engine is its thermal efficiency η_{th}, generally, defined as the ratio of net useful kinetic energy (shaft power for turboshaft engine) available for propulsion to the rate of thermal energy available from the fuel in the engine. For single stream engine η_{th} is

$$\eta_{th} = \frac{\dot{m}_a\left[(V_e - V) + (V_e - V)^2\right]/2}{\dot{m}_f \Delta H_C} \approx \frac{\dot{m}_a\left[V_e^2 - V^2\right]/2}{\dot{m}_f \Delta H_C} \text{ (when } f \ll 1) \tag{3.14}$$

where ΔH_C is the heating value of fuel. Since the power output of a turboshaft or turboprop engine is in terms of shaft power $\dot{P}_{sh}$ the thermal efficiency can be expressed as

$$\eta_{th} = \frac{\dot{P}_{sh}}{\dot{m}_f \Delta H_C} \tag{3.15}$$

3.3.3 Overall Efficiency

The overall efficiency η_o of an engine can be defined as the ratio of thrust power to the total energy supplied by the fuel, and it is given by

$$\eta_o = \frac{TV}{\dot{m}_f \Delta H_C} = \eta_p . \eta_{th} = \frac{2}{1 + V_e / V} \eta_{th} \tag{3.16}$$

Interestingly, η_o is basically the product of η_p and η_{th}. In addition, η_o is dependent on V_e/V and η_{th}. The importance of η_o will be fully appreciated only when we discuss about aircraft range in a subsequent subsection. Assuming $\dot{m}_f$ is quite small as compared to $\dot{m}_a$, Eq. (3.16) for η_o becomes

$$\eta_o = \frac{\dot{m}_a (V_e - V)V}{\dot{m}_f \Delta H_C} = \frac{(V_e - V)V}{f \Delta H_C} \tag{3.17}$$

From Eq. (3.17) we find that for certain f, η_o will be dependent on V_e and V. For a certain type of fuel and fuel-air ratio f, an aircraft can attain maximum η_o only with certain V_e/V, which can be obtained by differentiating Eq. (3.17) with flight velocity V and equating to zero as

$$\frac{d\eta_o}{dV} = 0; \Rightarrow V = \frac{V_e}{2} \tag{3.18}$$

Hence, overall efficiency of an aircraft can attain a maximum value only when flight velocity is equal to half of the exhaust jet velocity of the engine.

Generally, the performance of different engines can be compared in terms of efficiencies, particularly, at the following two operating conditions:

(i) Sea level static performance at maximum power.
(ii) Cruise performance at optimum flight speed and altitude for which the engine is designed.

3.3.4 Static Thrust

A gas turbine engine must possess the ability to provide static and low speed thrust so that an aircraft can take off under its own power unlike ramjet engine. Hence, the static thrust $\mathcal{T}_{st}$ is a very important performance parameter of the engine and it is defined as

$$\mathcal{T}_{st} = \dot{m}_a V_e \tag{3.19}$$

Sometimes, this is also known as *takeoff thrust*. It is clear from Eq. (3.19) that $\mathcal{T}_{st}$ is directly proportional to V_e for certain air flow rate. Let us see how does $\mathcal{T}_{st}$ vary with V_e for a given fuel flow rate? To find the answer, let us recast Eq. (3.19) as

$$\mathcal{T}_{st} = \frac{2\dot{m}_f \Delta H_C \eta_{th}}{V_e} \tag{3.20}$$

Equation (3.20) clearly indicates that $\mathcal{T}_{st}$ is inversely proportional to V_e for a given fuel flow rate and η_{th}. Hence, we can enhance $\mathcal{T}_{st}$ for the same energy input to the engine by accelerating large air mass flow rate to the jet exit velocity. Based on this principle, turboprop, turbofan and turbojet engines are designed to provide $\mathcal{T}_{st}$ in decreasing order respectively. All these types of jet engines use the basic gas generator, which comprises of compressor, combustor and turbine. In case of the turbojet engine (Fig. 3.3), high pressure and high temperature gas is accelerated in the exhaust nozzle to produce thrust. The work produced due to expansion of hot gases in the turbine is used to run the compressor. But the turbofan engine (Fig. 3.3) consists of a ducted fan in addition to compressor. The gas must be expanded further to lower velocity as compared to a turbojet engine to run the ducted fan. As a result, this engine can produce higher $\mathcal{T}_{st}$ and η_p as compared to a turbojet jet engine under same operating condition. Similarly, a turboprop engine expands gases in the exhaust nozzle.

3.3.5 Aircraft Range

The aircraft range is an important performance parameter for an aircraft, which is dictated mainly by engine performance parameters. For estimation of range, let us consider the level flight of an aircraft by neglecting the distance covered during the climbing and descending phase. We do so because the distance covered during cruise flight, particularly, for long-range passenger aircraft is quite large as compared to the climb and descent phase. Let us assume that cruise flight is taking place at constant speed. Then the engine thrust becomes equal to the drag experienced by the aircraft while aircraft weight is equal to lift generated by the aircraft. Hence

$$\mathcal{T} = \mathcal{D} = \mathcal{L}\left(\frac{\mathcal{D}}{\mathcal{L}}\right) = \frac{mg}{\frac{\mathcal{L}}{\mathcal{D}}} \tag{3.21}$$

where $\mathcal{D}$ is the drag force, $\mathcal{L}$ is the lift generated by the aircraft, m is the instantaneous mass of the aircraft, g is the acceleration due to gravity and $\mathcal{L}/\mathcal{D}$ the lift-drag ratio. Therefore, the thrust power becomes

$$\mathcal{T}V = \frac{mgV}{\left(\frac{\mathcal{L}}{\mathcal{D}}\right)} \tag{3.22}$$

By using the equation for η_o (Eq. (3.16)) in Eq. (3.22), we have

$$\dot{m}_f = \frac{mgV}{\eta_o \Delta H_C \left(\frac{\mathcal{L}}{\mathcal{D}}\right)} \tag{3.23}$$

Note that the fuel gets consumed during its flight. As a result, the total mass of the aircraft decreases during its flight and it is given by

$$\dot{m}_f = -\frac{dm}{dt} = -\frac{dm}{dS}\frac{dS}{dt} = -\frac{dm}{dS}V \tag{3.24}$$

where S is the distance travelled by aircraft, also known as *flight range*. By substituting Eq. (3.23) in Eq. (3.24) and rearranging, we have

$$\frac{dm}{dS} = -\frac{mg}{\eta_o \Delta H_C \left(\frac{\mathcal{L}}{\mathcal{D}}\right)} \tag{3.25}$$

Assuming $\eta_o \Delta H_C$ and $\mathcal{L}/\mathcal{D}$ to be constant, Eq. (3.25) can be integrated to get an expression for range S as

$$S = -\eta_o \frac{\Delta H_C}{g}\left(\frac{\mathcal{L}}{\mathcal{D}}\right)\ln\left(\frac{m_i}{m_F}\right) \tag{3.26}$$

where m_i and m_F, respectively, are the initial and final mass of the aircraft. The consumption of fuel mass flow rate during flight causes changes in the initial and final mass of a vehicle. Equation (3.26) is known as *Breguet Range Equation*. Also, the range is directly proportional to the overall efficiency, heat of combustion and lift-drag ratio.

3.3.6 Specific Thrust

The specific thrust $\mathcal{T}_s$ is defined as the ratio of the thrust and total mass flow rate through the engine, which is given by

$$\mathcal{T}_s = \frac{\mathcal{T}}{\dot{m}_a} \tag{3.27}$$

This parameter indicates the relative size of engines producing the same amount of thrust as the size of engine is mainly dependent on the total mass flow rate passing through the engine. The size of the engine plays a very dominant role because it dictates not only the weight but also the frontal area of the engine. An increase in air mass flow rate will definitely increase the frontal area and, thus, result in an increase in drag. Hence, the designer must strive to achieve higher $\mathcal{T}_s$ such that the engine becomes more compact with minimum drag.

3.4 Thrust Specific Fuel Consumption

Thrust specific fuel consumption (TSFC) is the mass flow rate of fuel divided by the thrust produced by the engine and is given by

$$\text{TSFC} = \frac{\dot{m}_f}{\mathcal{T}} \tag{3.28}$$

Interestingly, both TSFC and $\mathcal{T}_s$ can be related by

$$\text{TSFC} = \frac{f}{\mathcal{T}_s} \tag{3.29}$$

where f is the fuel-air ratio. TSFC is widely used as an experimental indicator of the engine's quality. By ignoring the pressure term in the thrust expression for a turbojet engine, we obtain an expression for TSFC as

$$\text{TSFC} = \frac{\dot{m}_f}{\dot{m}_a\left[(1+f)V_e - V\right]} \tag{3.30}$$

From Eq. (3.30) we find that TSFC is strongly dependent on the flight speed V. In order to avoid ambiguity, TSFC, defined as $\dot{m}_f/\mathcal{T}$ is used, particularly, for turbojet and turbofan engines. However, for ramjet, one has to consider a particular flight Mach number. Typical values of TSFC for modern jet engines are given in Table 3.1. For turboshaft and turboprop engines, brake specific fuel consumption (BSFC) is used instead of TSFC, and it is defined as

$$\text{BSFC} = \frac{\dot{m}_f}{\dot{P}_{sh}} \tag{3.31}$$

where $\dot{P}_{sh}$ is the shaft power. In order to take account of the thrust due to hot gases, we can define equivalent brake specific fuel consumption (EBSFC) as

$$\text{EBSFC} = \frac{\dot{m}_f}{\dot{P}_{sh} + \mathcal{T}_h V} \tag{3.32}$$

where $\mathcal{T}_h$ is the thrust produced by the hot gas in the exhaust nozzle and V is the flight speed. Typical values of EBSFC for a modern turboprop engine lies between 0.27 and 0.36 kg/kWh.

Table 3.1. Typical values of TSFC for modern airbreathing engines [4]

Engine type	TSFC (kg/N.hr)
Turbojet (static)	0.075-0.11
Turbofan (static)	0.03-0.05
Ramjet (M = 2.0)	0.17-0.26

Example 3.1 A turbojet engine is flying at a velocity of 500 m/s at altitude of 10 km. The uninstalled specific thrust is 750 N.s/kg and TSFC is 50 g/(kN.s). Determine the engine performance parameters such as f, η_o, η_p, and η_{th} assuming that heat of combustion of fuel is 45,000 kJ/kg.

Given V = 500 m/s; Al = 10 km; $\mathcal{T}_s$ = 750 N.s/kg; TSFC = 50 g/(kN.s); ΔH_C = 43,000 kJ/kg.

Solution We know that fuel-air can be expressed in terms of specific thrust and TSFC using

$$f = \frac{\dot{m}_f}{\dot{m}_a} = \mathcal{T}_s \cdot \text{TSFC} = \frac{750 \times 50}{10^6} = 0.0375$$

By using Eqs. (3.16) and (3.28), we can express η_o in terms of V, ΔH_C and TSFC as

$$\eta_o = \frac{\mathcal{T}V}{\dot{m}_f \Delta H_C} = \frac{V}{\Delta H_C \cdot \text{TSFC}} = \frac{500 \times 10^3}{45{,}000 \times 50} \times 100 = 22.2\%$$

The η_p can be estimated as

$$\eta_p \approx \frac{\dot{m}_a V(V_e - V)}{\dot{m}_a \left[V(V_e - V) + (V_e - V)^2/2\right]} = \frac{2}{1 + V_e/V}$$

However, we will have to evaluate V_e/V as

$$\frac{V_e}{V} = \frac{1}{f+1}\left(\frac{\mathcal{T}}{\dot{m}_a V} + 1\right) = \frac{1}{0.0375+1}\left(\frac{750}{500} + 1\right) = 2.41$$

Now η_p is determined as

$$\eta_p = \frac{2}{1 + V_e/V} = \frac{2}{1 + 2.41} \times 100 = 58.65\%$$

By knowing η_o and η_p, the η_{th} can be determined easily by

$$\eta_{th} = \frac{\eta_o}{\eta_p} \times 100 = \frac{22.2}{58.65} \times 100 = 37.85\%$$

In this example, η_o turned out to be quite small due to lower value of η_p. By improving η_p, higher η_o can be achieved by using various engine configurations. In the next section, we will learn more about different types of gas turbine engines.

3.5 Comparison of Gas Turbine Engines

It would be interesting at this point to compare all three types of gas turbine engines – turbojet, turbofan and turboprop – considering the three performance parameters viz. thrust, TSFC and η_p. We have assumed that these engines having same compressor pressure ratio and turbine inlet temperature are installed in the same size aircraft. Let us carefully look at the variations of thrust and TSFC against flight speed (Fig. 3.4) at sea level and at an altitude of 12 km. From these plots we can observe the characteristics of each engine along with its applications, which are given below.

3.5.1 Turbojet Characteristics

- Low thrust at low forward speed, which results in relatively long takeoff.
- Relative higher TSFC at low altitude.
- Lowest η_p.
- Small frontal area results in reduced ground clearance problems.

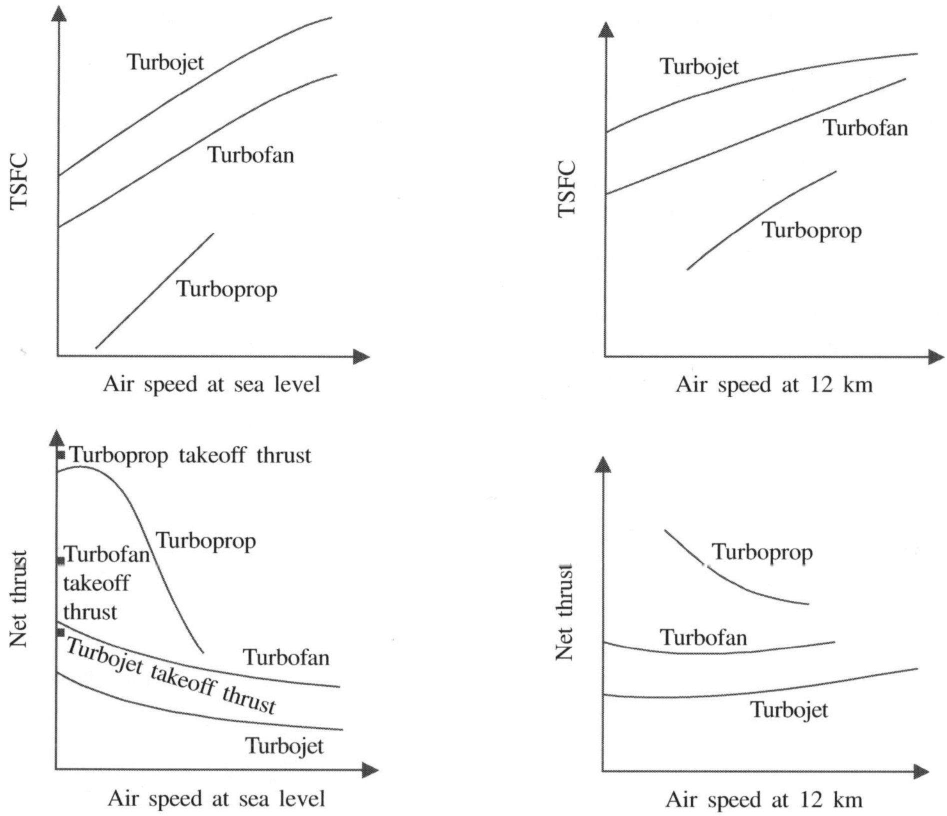

Fig. 3.4. Characteristics of typical gas turbine engines in terms of T_s and TSFC.

- Lightest specific weight (kg/N) of thrust produced.
- Ability to take advantage of high ram pressure. Thus, compressor having low pressure ratio is required.
- Higher jet noise level.

Application of Turbojet Engine

The turbojet engine is best suited for high speed, high altitude and medium distance aircraft. It is widely used for fighter aircraft and, of course, with afterburner system.

3.5.2 Turbofan Characteristics

- Increased thrust at low forward speed in comparison to turbojet, resulting in shorter takeoff.
- Relative lower TSFC in comparison to turbojet.
- The η_p is in between that of turbojet and turboprop.
- Larger frontal area as compared to turbojet.
- Ground clearance less than turboprop but higher than turbojet.
- Weight between that of turbojet and turboprop.
- Considerable reduction in noise level over turbojet.
- Superior to turbojet in *hot day* performance.

- Two thrust reversal system required if fan and core engine air are expanded in separate nozzles.

Application of Turbofan Engine

The turbofan engine is best suited for relatively high speed and long distance aircraft as compared to the turboprop engine.

3.5.3 Turboprop Characteristics

- Able to develop very high thrust at low forward speed, which results in shortest takeoff distance.
- Lowest TSFC.
- Highest η_p, particularly at low flight speed.
- Largest frontal area, which requires longer landing gears for a low wing airplane.
- Heavier and more complicated than turbojet engine.
- No jet noise.
- Efficient reverse thrust is possible.

Application of Turboprop Engine

The turboprop engine is best suited for lifting heavy aircraft on short and medium runways. But its flight speed is limited to about 500 km/h as its propeller efficiency falls rapidly beyond this value.

3.6 Brayton Cycle

We know that all gas turbine based aero engines contain a gas generator. As mentioned earlier, the gas generator consists of three components – compressor, combustor and turbine – which operate on an open cycle (Fig. 3.5a), in which fresh air is drawn into the compressor, where the air is compressed to higher pressure with increase in temperature. Then this high pressure gas enters the combustion chamber, where the fuel is burnt with air to increase the gas temperature. Subsequently, high pressure and high temperature gases are expanded in the turbine to produce shaft power. At the exit of the turbine, the exhaust gases are let out to the atmosphere. Hence, this cycle is termed as an open cycle. The open gas turbine cycle can also be modelled as a closed cycle (Fig. 3.5a), in which the exhaust gases are brought back to the original atmospheric condition by passing through a heat exchanger and, thus, a cycle is completed. George Brayton proposed this cycle around 1870 and, hence, it is known as *Brayton cycle*. Interestingly, although it is used widely for modelling of gas turbine engines in recent times, but it was originally developed for oil burning piston engine.

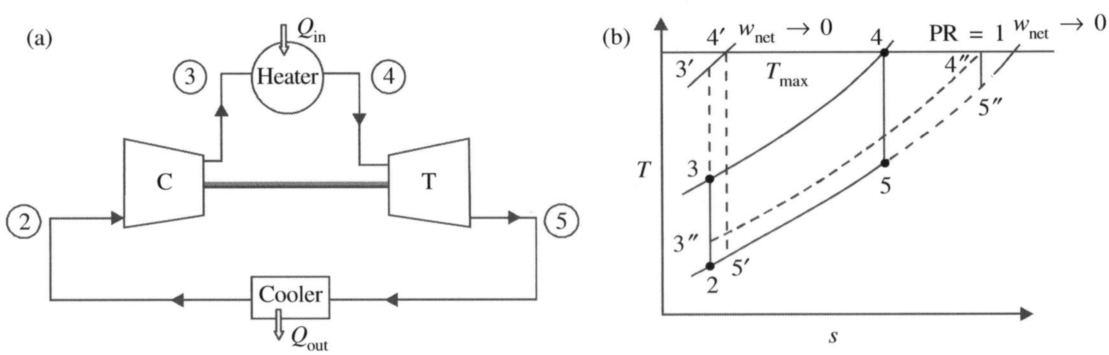

Fig. 3.5. Schematic of Brayton Cycle along with *T-s* diagram.

In the gas turbine engine, air is the primary working fluid, which mainly contains nitrogen. Of course, fuel is burnt in the combustion chamber, which changes the composition of the working fluid to combustion product, i.e. gas. Generally, an air-fuel ratio around 50:1 is used in a gas turbine engine. In addition, air contains large amount of nitrogen, which does not take part in combustion since it is an inert gas. Thus, for simplicity, the working fluid can be assumed as air. In order to simplify the analysis of this cycle, the following approximations – commonly known as *air standard assumptions* – are considered:

1. A fixed quantity of air as the working fluid is circulated throughout the entire cycle. Hence, the cycle is devoid of any inlet and exhaust process, resulting in a closed cycle.
2. Air behaves always as a perfect (ideal) gas with constant specific heats.
3. Compression and expansion processes are isentropic (reversible and adiabatic).
4. The change in kinetic energy and potential energy of the working fluid between inlet and outlet of each component can be neglected since it is quite a small value.
5. The combustion chamber (CC) is replaced by an equivalent reversible external heater.
6. The exhaust process is modelled as heat rejection in an equivalent reversible cooler, which rejects heat to the surrounding, restoring the working fluid to its initial state.

This air standard Brayton cycle is described in temperature-entropy (*T-s*) plane (Fig. 3.5b) by the following four processes:

- 2-3: Isentropic compression
- 3-4: Isobaric heat addition
- 4-5: Isentropic expansion
- 5-2: Isobaric heat rejection

All these processes are considered to have attained steady state. Thus, they can be analysed using the First Law of Thermodynamics for CV (Eq. (2.17)) for each component. Neglecting the change in kinetic energy and potential energy, Eq. (2.17) can be applied for the compressor and turbine as

$$w_c = w_{23} = -(h_3 - h_2) = -C_p(T_3 - T_2) \tag{3.33}$$

$$q_{cc} = q_{34} = (h_4 - h_3) = C_p(T_4 - T_3) \tag{3.34}$$

$$w_t = w_{45} = (h_4 - h_5) = C_p(T_4 - T_5) \tag{3.35}$$

$$w_{net} = w_t + w_c = (h_4 - h_5) - (h_3 - h_2) = C_p(T_4 - T_5) - C_p(T_3 - T_2) \tag{3.36}$$

We know that the cycle thermal efficiency η_{th} is a useful quantity to evaluate the overall performance of the system and is defined as

$$\eta_{th} = \frac{\text{Net work done}}{\text{Heat input in combustion chamber (CC)}} = \frac{w_{net}}{q_{CC}} = 1 - \frac{(T_5 - T_2)}{(T_4 - T_3)} \tag{3.37}$$

Note that the compression process (2-3) and expansion process (4-5) are isentropic and $P_3 = P_4$ and $P_2 = P_5$. Now we can have an expression for pressure ratio (PR) as

$$\text{PR} = \left(\frac{P_3}{P_2}\right)^{\frac{\gamma-1}{\gamma}} = \frac{T_3}{T_2} = \left(\frac{P_4}{P_5}\right)^{\frac{\gamma-1}{\gamma}} = \frac{T_4}{T_5} \tag{3.38}$$

where PR is the pressure ratio across the compressor and the turbine. Hence, Eq. (3.38) can be recast in terms of temperature

$$\frac{T_3}{T_2} = \frac{T_4}{T_5} \tag{3.39}$$

$$\frac{(T_5 - T_2)}{(T_4 - T_3)} = \frac{T_3}{T_2} = \left(\frac{1}{\text{PR}}\right)^{\frac{\gamma-1}{\gamma}} \tag{3.40}$$

Thus, the expression for η_{th} becomes

$$\eta_{th} = 1 - \frac{(T_5 - T_2)}{(T_4 - T_3)} = 1 - \left(\frac{1}{\text{PR}}\right)^{\frac{\gamma-1}{\gamma}} \tag{3.41}$$

From Eq. (3.41) we find that η_{th} of an air standard Brayton cycle depends only on the PR and specific heat ratio γ of the working fluid, provided it is different from that of air. The variation of η_{th} as a function of PR (Fig. 3.6(a)) for γ = 1.4 (air) indicates that η_{th} increases with increase in PR. This increase in efficiency can be explained physically with the help of the *T-s* process diagram (Fig. 3.5). It can be seen that the increase in PR changes the area from 2-3-4-5 to 2-3'-4'-5' on the *T-s* diagram. Obviously, the cycle 2-3'-4'-5' has a greater heat addition for the same heat rejection as that of original cycle 2-3-4-5. Therefore, from the definition of thermal efficiency, the cycle 2-3'-4'-5' has a higher efficiency. But it attains a maximum temperature, which is greater than the original cycle maximum temperature T_4. The maximum cycle temperature is limited by the metallurgical limitations of the turbine materials used in the gas turbine engine since it is quite difficult to design turbine blades that can withstand higher temperature under rotating condition. Let us now consider a new cycle 2-3"-4"-5" (Fig. 3.5) at lower PR, by keeping the same maximum cycle temperature T_4. Now the net work per unit mass will be reduced for this new cycle. This invariably limits the PR that can be used in the cycle. Hence, for a fixed turbine inlet temperature T_4, the work output increases with PR, attains a maximum value and then decreases (Fig. 3.6(b)). Therefore, we must find out the optimum PR that can provide the maximum work output. Let us now find an expression for specific work output w_s

$$w_s = \frac{w_{net}}{C_p T_2} = \frac{C_p(T_4 - T_5) - C_p(T_3 - T_2)}{C_p T_2} = \frac{T_4}{T_2}\left[1 - \left(\frac{1}{\text{PR}}\right)^{\frac{\gamma-1}{\gamma}}\right] - \left(\text{PR}^{\frac{\gamma-1}{\gamma}} - 1\right) \tag{3.42}$$

From Eq. (3.42) we find that w_s is dependent on PR, temperature ratio (TR = T_4/T_2), and γ. For a particular working fluid (γ = 1.4), the variation of w_s as a function of PR and TR (Fig. 3.6(b)) shows that the specific work increases initially and after attaining a maximum value it again decreases with increase in PR for a particular temperature ratio. Thus, there is an optimum PR corresponding to maximum work output. When PR is around 1, the work output will be quite small ($w_s \to 0$). However, when PR is around $(T_4/T_2)^{\frac{\gamma}{\gamma-1}}$, the work output tends toward zero (Fig. 3.6(b)). Interestingly, this optimum PR increases with increase in turbine inlet temperature for a fixed compressor inlet temperature. This optimum PR can be obtained for any given temperature ratio by differentiating Eq. (3.42) with respect to $\text{PR}^{\frac{\gamma-1}{\gamma}}$ and equating to zero as

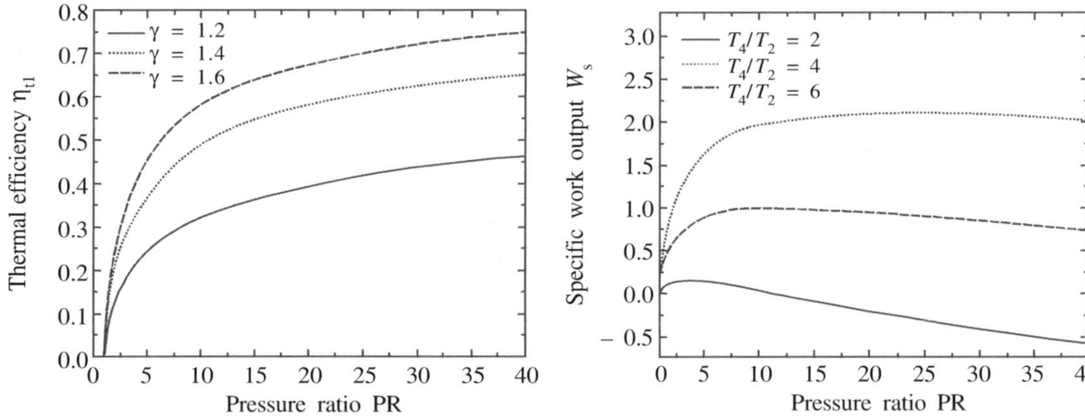

(a) (b)

Fig. 3.6. (a) Thermal efficiency vs PR and (b) specific work vs PR for three TR.

$$\frac{dw_s}{d(\mathrm{PR})^{\frac{\gamma-1}{\gamma}}} - 0 \tag{3.43}$$

By using Eq. (3.42) and manipulating it we get an expression for optimised pressure ratio $\mathrm{PR}_{\mathrm{opt}}$ as

$$\mathrm{PR}_{\mathrm{opt}} = \left(\frac{T_4}{T_2}\right)^{\frac{\gamma}{2(\gamma-1)}} \tag{3.44}$$

But we know that $\mathrm{PR}^{\frac{\gamma-1}{\gamma}}$ is equal to T_4/T_5 or T_3/T_2. Thus, Eq. (3.44) can be expressed as

$$\left(\mathrm{PR}^{\frac{\gamma-1}{\gamma}}\right)^2 = \mathrm{PR}^{\frac{\gamma-1}{\gamma}} \cdot \mathrm{PR}^{\frac{\gamma-1}{\gamma}} = \frac{T_4}{T_5} \cdot \frac{T_3}{T_2} = \frac{T_4}{T_2} \quad \text{for } T_5 = T_3 \tag{3.45}$$

From Eq. (3.45) we find that $\left(\mathrm{PR}^{\frac{\gamma-1}{\gamma}}\right)^2$ becomes $\frac{T_4}{T_2}$ only when $T_5 = T_3$. Hence, w_s attains maximum value when PR is such that the exit temperature of turbine is equal to that of the compressor.

3.7 Gas Turbine Engine Design Concept

This book intends to introduce the readers to the fundamental aspects of gas turbine propulsion. This section attempts to expose them to a gamut of gas turbine design concepts that will provide an ample chance to realise the importance of interactions among various specialists, which is essential for the successful design and development of a gas turbine engine. The schematic diagram of a typical design procedure for an aircraft gas turbine engine (Fig. 3.7) shows that gas turbine design involves broadly five disciplines: (i) Thermodynamics, (ii) Aerodynamics, (iii) Structure, (iv) Manufacturing, and (v) Control Systems. This book only deals with the aero-thermodynamics aspects of gas turbine engine.

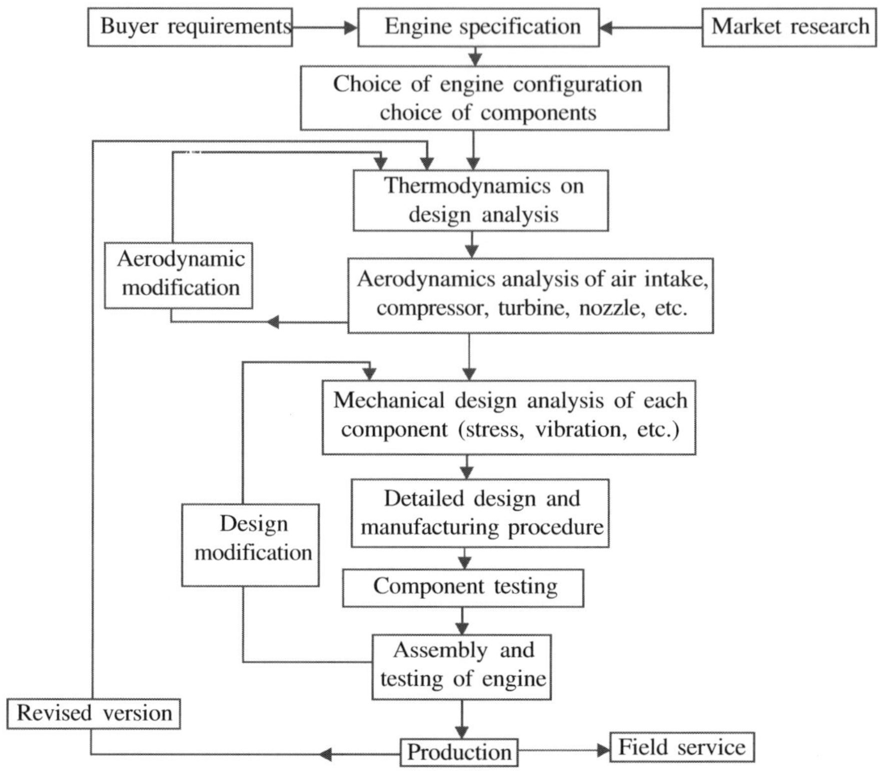

Fig. 3.7. Block diagram indicating gas turbine design procedure [1, 3].

The first important step is to identify the specification of the engine, which can be obtained either from market research or customer requirement. The design and development of a gas turbine engine is an extremely time consuming and expensive affair. Hence, gas turbine engines are being developed by multinational consortia, particularly, in recent times. It is important to design an engine with well planned specification so that it can cater to maximum number of applications. Power and efficiency are two important specifications that must be identified in the beginning. Other important specifications such as weight, volume, length, flight range, life, cost, noise level, emission level, etc. – mainly dictated by the type of applications – also need to be identified. These requirements may contradict each other. For example, higher efficiency will definitely incur higher cost. If the running hours are limited, then one can go for a cheaper engine with lower efficiency. The next important step is to decide the cycle to be adopted for the engine (discussed extensively in Chapters 4 and 5). Subsequently, one has to decide about the type of turbomachinery depending on the requirement of PR and thrust for a particular application. The layout of the engine must also be decided considering the mission requirement. The next major step is to carry out design point thermodynamic cycle analysis considering important factors such as variable fluid properties, pressure losses across each component, expected component efficiencies, etc. This analysis has to be carried out over a wide range of PR across the compressor, turbine inlet temperature, flight Mach number, altitude, etc., to determine $\mathcal{T}_s$, TSFC, engine efficiencies, etc. The choice of cycle parameters is dictated by air flow rate and engine size. After estimating the air flow rate, PR, and turbine inlet temperature, the aerodynamics analysis of various components of an aircraft engine has to be carried out to determine the dimensions of each component, number of stages and rotational speed of turbomachinery. It is important to consult other groups during this stage of design. The aerodynamic designer must take into account the manufacturing feasibility of components. The next step is to carry out structural analysis of these components in determining stress, strain, vibration level, etc., to ensure their structural integrity even under adverse situations. It is essential to consult aerodynamic and thermodynamic groups to arrive at a feasible design. Control is very essential for successful operation of any engine. Hence, control system design must be considered to ensure safe and automatic operation of the engine using predicted pressure and temperature across each engine component. It must be kept in mind that measured values of controlling parameters are to be used to operate the engine in controlled manner. After designing and manufacturing the engine components, testing of individual components needs to be carried out exhaustively. Subsequently, the engine is assembled and tested rigorously over a whole range of operating conditions. Modifications can be made to the engine design at each stage. During this process, several iterations have to be carried out to ensure successful design and development of the gas turbine engine. Thus, gas turbine design procedure is quite complex, which requires excellent coordination among various involved groups.

3.8 Chapter Summary

The fundamental aspects of gas turbine engine are discussed to familiarise the reader with various terminologies and important concepts. A generalised thrust equation for an airbreathing engine is derived from the first principle. Other engine performance parameters such as specific thrust, thrust specific fuel consumption, propulsive, thermal and overall efficiencies are defined in a lucid manner. General comparison of performance parameters among turbojet, turbofan and turboprop engines is carried out to illustrate the operating envelope of these engines. For fighter aircraft, turbojet engine is the best choice while turbofan engine is preferred for long range passenger aircraft applications. Turboprop/turboshaft engines can be employed to drive heavier aircraft for short-range application. As the Brayton cycle is the basis of all jet engines, the fundamental thermodynamic aspects of this cycle are discussed briefly to provide adequate background to the reader. In addition, a brief account of the design concepts for gas turbine engine is provided so that the reader can appreciate the intricacies involved during its design. Chapter 4 discusses on-design ideal cycle analysis for jet engines.

Nomenclature

A_e	Engine exit area
CV	Control volume
$\mathcal{D}$	Drag force
F	Force
G	Acceleration due to gravity
ΔH_c	Heating value of fuel
$\mathcal{L}$	Lift generated
m	Instantaneous mass
$\dot{m}$	Mass flow rate
P_e	Exit pressure
P_a	Ambient pressure
P_s	Shaft power
PR	Pressure ratio
S	Distance travelled

TSFC Thrust specific fuel consumption
V_e Exit velocity
V Flight velocity

Greek Symbols

$\mathcal{T}$ Thrust
$\mathcal{T}_s$ Specific thrust
$\mathcal{T}_h$ Thrust product by hot gases
ρ Ambient density
ρ_e Exit density
η_p Propulsive efficiency
η_{th} Thermal efficiency
η_o Overall efficiency
γ Specific heat ratio

Subscripts/Superscripts

a Ambient
e Exit
f Final
h Hot gases
i Initial
o Overall
p Propulsive
s Side surface
s Shaft
s Specific
th Thermal

Abbreviations

BSFC Break specific fuel consumption
CV Control volume
PR Pressure ratio
TSFC Thrust specific fuel consumption

References and Suggested Readings

1. Mattingly, J.D., *Elements of Gas Turbine Propulsion*, 1st Edn., McGraw-Hill Publishing Co. Ltd., Singapore, 1996.
2. Kerrebrcok, J.L., *Aircraft Engines and Gas Turbines*, 1st Edn., MIT Press, UK, 1992.
3. Cohen, H., Rogers, G.F.C. and Saravanamuttoo, H.I.H., *Gas Turbine Theory*, 4th Edn., Addison Wesley Longman Ltd., UK, 1998.
4. Hill, P.G. and Peterson, C.R., *Mechanics of Thermodynamics of Propulsion*, 2nd Edn., Addison-Wesley Publishing Company, Inc., Reading, MA, 1992.

Review Questions

1. What do you mean by specific thrust? Why is it used widely instead of thrust? How is it related to TSFC?
2. How is TSFC related to η_o for generic gas turbine engine?
3. How can static thrust be enhanced for a particular engine configuration?

4. Derive the expression for thrust for a turbofan engine.
5. Why is the propulsive efficiency of turboprop engine higher than that of turbojet engine?
6. Why is takeoff thrust for turboprop engine higher than that for turbojet engine?
7. What do you mean by BSFC? Why is it used for turboprop engine?
8. What are the differences between turbojet and turbofan engines?
9. Enumerate the differences between turbojet and turbojet engines.
10. Show the velocity of T_s, TSFC and η_p with flight speed for turbojet and turboprop engines.
11. For an ideal reheat gas turbine cycle, derive an expression for pressure ratio in terms of reheat temperature and inlet temperature at maximum specific work output.

Problems

Q1. An aircraft is accelerated from a static ground base through the atmosphere. Derive an expression for velocity increment in terms of mass ratio, MR (final mass/initial mass). State the assumptions made by you.

Q2. An aircraft designed to cover a distance of 3,000 km by using an airbreathing engine with an overall efficiency of 0.5 and $(1 - \psi)C_L/C_D = 4.5$, where ψ is the installation loss coefficient, C_L the lift coefficient and C_D the drag coefficient. Determine the fuel mass fraction for aviation turbine fuel (ATF) ($\Delta H_C = 45{,}000$ kJ/kg) and Hydrogen ($\Delta H_C = 105{,}000$ kJ/kg).

Q3. An aircraft with total mass of 10,000 kg consumes mass 1,000 kg of aviation turbine fuel (ATF) with caloric value of 45,000 kJ/kg, density of 950 kg/m^3 to cover 5,000 km distance. The combustor is changed, which can run on liquid natural gas fuel with 50,000 kJ/kg and density 500 kg/m^3. Find the volume of the tank required to cover the same distance.

Q4. Show that TSFC for an airbreathing engine is a function of specific thrust, flight velocity, thermal efficiency and heat of combustion. Suggest the methods of improving TSFC when flight velocity, thermal efficiency and heat of combustion remain constant.

Q5. A simple gas turbine powered gas generates 10 MW shaft power with compressor pressure ratio of 12. The turbine inlet temperature is found to be 1200 K. Determine air flow rate and TSFC if it sucks air from atmosphere at 0.1 MPa and 298 K.

Q6. An aircraft propelled by turbojet engine flies at $M = 0.7$ in level flight at an altitude of 10 km. Turbine inlet temperature is 1300 K. If the stagnation temperature increases across the air intakes and the compressor is at 300 K, determine (a) Fuel-air ratio, (b) Pressure ratio across compressor, turbine and nozzle, and (c) Exhaust nozzle exit velocity. ($\Delta H_C = 45{,}000$ kJ/kg)

Q7. An aircraft propelled by two turbojet engines of 2.5 MW with nozzle diameter of 0.2 m, while consuming 0.42 kg/kWh of fuel ($\Delta H_C = 43{,}000$ kJ/kg) flies at 10 km altitude with a speed of 350 m/s. If air-fuel ratio is 40, determine (a) Exhaust velocity, (b) Thrust, and (c) Propulsion, overall and thermal efficiencies, assuming.

Q8. A fighter aircraft is flying at $M = 0.9$ at an altitude of 10 km. Its engine produces a net thrust of 56 kN at level flight while consuming 50 kg/s of air and 2.0 kg/s of fuel. Estimate (i) T_s, (ii) TSFC, and (iii) η_p, η_{th} and η_o.

Q9. A turboprop engine powered aircraft flies at 400 km/h at an altitude of 7.5 km. The diameter of aircraft propeller is 3 m. If flight to exhaust jet velocity is 0.8, estimate (i) $\dot{m}_a$, (ii) T_s, and (iii) Thrust power.

Q10. Air at 0.1 MPa and 298 K with velocity of 100 m/s enters a conical diffuser with inlet and exit diameter of 10 and 20 cm, respectively. Estimate (i) Exit velocity, (ii) Pressure, (iii) Coefficient of pressure across diffuser, and (iv) Force exerted on the diffuser walls.

4. Ideal Cycle Analysis

I believe I have found a way to make a Machine lighter than air itself.... We live submerged at the bottom of an ocean of the element air, which by unquestioned experiments is known to have weight.

Francesco de Lana de Terzi of Brescia, the first person to propose a flying Machine based on sound scientific principles, 1670

The jet engine had, like the Phoenix, risen from its ashes.

Sir Frank Whittle, inventor of the gas turbine jet engine, 1935

4.1 Introduction

In the previous chapters, we have discussed in detail the basic principle of various kinds of gas turbine power plants. The performance parameters of these engines have been defined and discussed extensively. It may be recalled that the thermodynamic cycle analysis of the engine is a very essential tool for the design of various types of gas turbine engines. Hence, the present chapter is devoted to the basic concepts of cycle analysis for various kinds of jet engines. A detailed discussion about the basic principles of the Brayton cycle – already discussed in Chapter 3 – will be employed in this chapter for the analysis of jet engines.

The main objective of *cycle analysis* is to study the thermodynamic changes of the working fluid while it flows through the various components of the engine. The working fluid means the combination of air and combustion products due to burning of fuel in the combustion chamber (CC) and afterburner of jet engines. This cycle analysis helps us to seek answers to some pertinent questions: Whether turbine inlet temperature or the compressor pressure ratio is to be increased for obtaining higher thrust for a particular engine size or not? By how much thrust specific fuel consumption can be lowered when one uses a turbofan engine as compared to a turbojet engine? Is it easier to enhance the compressor efficiency or turbine efficiency? How does it affect the overall performance of the engine? The thermodynamic analysis of an engine cycle can be performed by adopting several levels of simplifications. If the losses incurred in each component of the engine are ignored assuming that all the components are operated under ideal conditions, then it is called *ideal cycle analysis*. In contrast, if the losses incurred in each component and change in properties of the fluid due to its composition are considered during the cycle analysis, then it is known as *real cycle analysis*. Depending on its purpose, cycle analysis can be broadly divided into two categories: (i) Parametric Cycle Analysis, and (ii) Engine Performance Analysis. The parametric cycle analysis determines the performance of an imaginary engine. In other words, it does not deal with the performance of any particular engine. Rather, it provides the performance analysis of the rubber engine as the effects of geometry of a particular engine are not included in this analysis. In contrast, engine performance analysis is employed to determine the performance of a specific engine under all flight conditions and throttle settings. We will restrict our discussion to parametric cycle analysis only.

Generally, designers initiate the design process of an aircraft engine by carrying out the thermodynamic cycle analysis with an open mind. During the preliminary design phase, there is a need to vary the parameters over a wide range of its operation. This parametric cycle analysis determines the performance of the hypothetical engine under different flight conditions such as altitude (*Al*), Mach number M etc., by varying the design choices and design limitations. For example, the compressor pressure ratio is one of the important design choices, whereas the inlet turbine temperature is one of the design limitations. As mentioned earlier, the parametric cycle analysis is not concerned with any particular engine configuration. Hence, it is also known as *on-design cycle analysis*. This analysis is intended to determine the performance of the engine with change in flight conditions (*Al* (T_0, P_0), M_0), design choices (e.g. compressor pressure ratio), design limitations (e.g. turbine inlet temperature), component performance (e.g. air intake efficiency), etc. The main objective of parametric cycle analysis is to relate the engine performance parameters such as thrust $\mathcal{T}$, specific thrust $\mathcal{T}_s$, thrust specific fuel consumption (TSFC), propulsion efficiency η_P, thermal efficiency η_{th}, overall efficiency η_o, etc., with its component parameters such as compressor pressure ratio, turbine temperature ratio, turbine inlet temperature (TIT), etc., to the flight environment (*Al* (T_0, P_0), M_0). By performing such parametric cycle analysis, the designer can easily find out the choice of parameters that can satisfy the requirement of a particular application. However, this would not provide the performance of a particular engine for which one has to carry out *off-design cycle analysis* while taking into account the effects of geometry of a particular engine. As mentioned earlier, this book mainly discusses parametric (on-design) cycle analysis. The readers may refer other advanced books for off-design analysis [1, 2].

4.1.1 General Notation

In this book, cycle analysis will be carried out by considering individual components with proper station numbers (Fig. 4.1) according to *Aerospace Recommended Practice (ARP) 755A* (1974). The station numbers are assigned from 0 to 9 for the core engine, whereas, for bypass stream (Fig. 4.1), station numbers from 13 to 19 are employed. In some cases, decimal numbers are used to indicate the intermediate stations. In this analysis, the ratio of total pressure and temperature across each component of the gas turbine engine is used extensively. Thus, let us denote total pressure ratio across a component by π with x as the subscript indicating a specific component

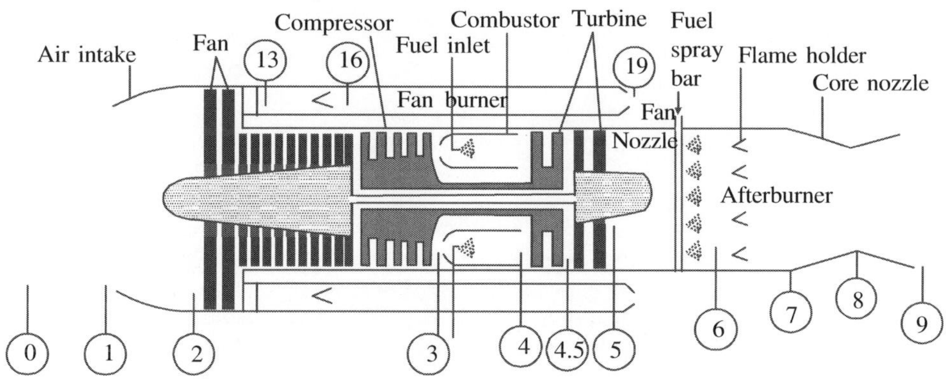

Fig. 4.1. Schematic of a gas turbine engine with station numbers.

$$\pi_x = \frac{\text{Total pressure at the exit of component } x}{\text{Total pressure at the inlet of component } x}$$

The subscripts used for various components of a gas turbine engine include diffuser/air intake/inlet (*d*), compressor (*c*), burner/combustor (*b*), turbine (*t*), nozzle (*n*), and fan (*f*).

Similarly, the total temperature ratio for a component is defined as

$$\tau_x = \frac{\text{Total temperature at the exit of component } x}{\text{Total temperature at the inlet of component } x}$$

We can also define the total/static temperature and pressure ratios of the free stream as

$$\tau_r = \frac{T_{t0}}{T_0} = \left[1 + \frac{(\gamma - 1)}{2} M_0^2\right] \tag{4.1}$$

$$\pi_r = \frac{P_{t0}}{P_0} = \left[1 + \frac{(\gamma - 1)}{2} M_0^2\right]^{\frac{\gamma}{\gamma - 1}} \tag{4.2}$$

However, the total enthalpy at the combustor exit with respect to ambient condition τ_λ is given by

$$\tau_\lambda = \frac{\text{Total enthalpy at burner exit}}{\text{Ambient enthalpy}} = \frac{(C_p T_t)_{exit}}{(C_p T_0)} \tag{4.3}$$

4.1.2 Assumptions of Ideal Cycle Analysis

The following assumptions are made while carrying out an ideal cycle analysis of a jet engine. A component can be considered to be operated under an ideal condition only when no loss is considered to have occurred during its operation. We can assume a compressor/turbine to be operated under ideal condition only if it undergoes an isentropic (reversible and adiabatic) process. For simplicity, the gas passing through the jet engine can be assumed to obey calorically perfect gas law. Besides these assumptions, pressure losses across any component are assumed to be negligibly small during its operation. The assumptions made for an ideal cycle analysis are as follows:

- The working fluid has the same composition throughout the engine. It behaves as a calorically perfect gas (constant specific heats).
- Fuel flow rate is negligibly small in comparison to air flow rate ($\dot{m}_f << \dot{m}_a$; hence, $\dot{m}_a + \dot{m}_f \simeq \dot{m}_a$).
- Isentropic compression in air intake, fan and compressor.
- Isentropic expansion in turbine and nozzle.
- Constant pressure combustion.
- No pressure losses in air intake, combustor and nozzle.
- The engine exhaust gas is expanded fully in the nozzle.

4.2 Ramjet

The ramjet is the simplest of all airbreathing jet engines, because it does not have any rotating elements (like compressor, turbine). The schematic of a typical ramjet (Fig. 4.2a) shows that it consists of an air intake, a combustion chamber, and an exhaust nozzle. Air enters the air intake at a high speed (generally supersonic) in which the velocity is reduced to a subsonic Mach number of around 0.3 with the help of a number of oblique and normal shocks, because combustion takes place efficiently

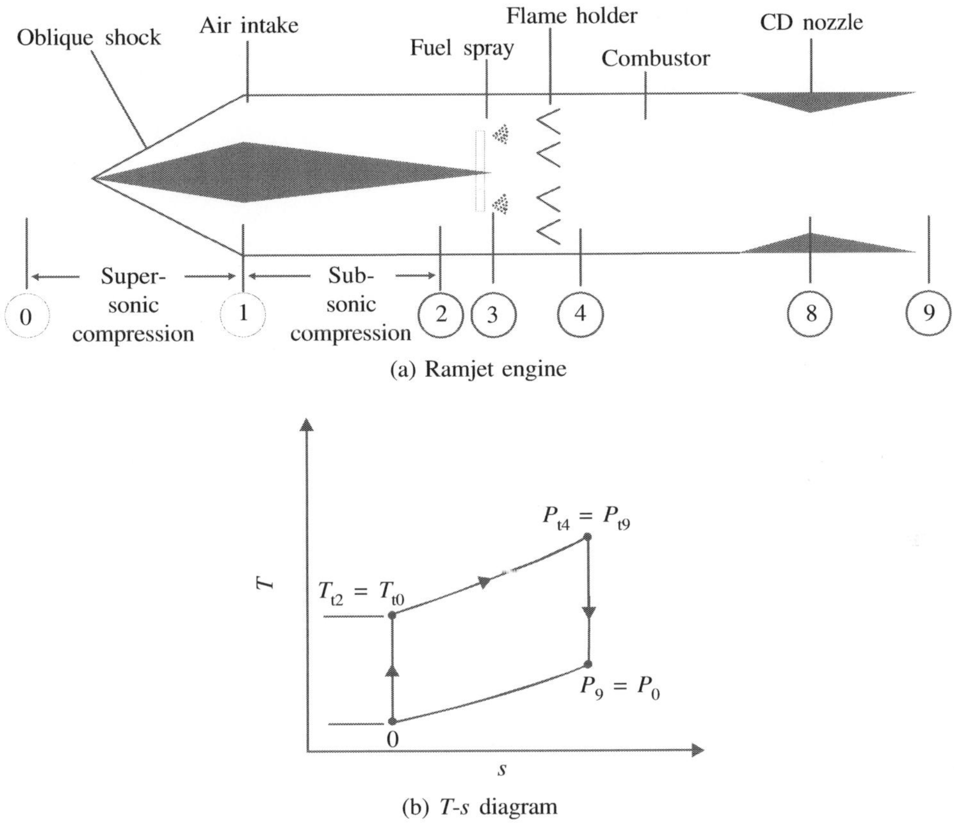

Fig. 4.2. (a) Schematic of (a) ramjet engine with station numbers and (b) *T-s* diagram of an ideal ramjet engine.

only at an inlet Mach number of around 0.3 or less. If the supersonic flow is decelerated in the air intake to low subsonic flow, the pressure rise will be substantial. For example, if the flight Mach number of 2.5 is to be decelerated isentropically to the Mach number of 0.3, the ideal static pressure ratio across the air intake will be around 17. Hence, there is no need to use a compressor to raise the pressure of ingested air into the air intake. The compressed gas passes through the combustion chamber in which liquid fuel is injected to form a fine spray and mixes with incoming air. The efficient burning of fuel is accomplished inside the combustor by stabilising the flame with the help of flame holders (Fig. 4.2a). As a result, the temperature of the working fluid rises to around 2800 K. Of course, it is not possible to go beyond 1600 K with the material available at present. However, the designers have managed to use combustion chamber temperature upto even 2000 K, with proper adjustment of fuel injection pattern such that the cooler fluid shields the wall of the combustion chamber to attain critical temperature of material erosion. However, the gas turbine engine is restricted to operate at turbine inlet temperature of 1600 K only, because the turbine blades cannot be cooled easily for it to be operated at a temperature similar to that of the ramjet engine. Moreover, turbine blades are subjected to centrifugal stress and, thus, cannot withstand higher thermal loads. As higher peak temperatures can be used in ramjet, one can operate the ramjet engine at high flight Mach number.

However, the disadvantage of the ramjet is that the pressure ratio is strictly limited by the flight Mach number and the performance of air intake. When we will learn more about air intake, we will appreciate that it is quite difficult to design an air intake for high efficiency, particularly, in supersonic flight regime. Note that the ramjet engine performance is governed mainly by air intake performance. Moreover, the most serious problem with the ramjet engine is that it cannot develop static thrust itself. With this in mind, let us carry out an ideal cycle analysis of a ramjet engine. We know that the thrust equation for the ramjet engine becomes

$$\mathcal{T} = \left(\dot{m}_9 V_9 - \dot{m}_0 V_0\right) + A_9\left(P_9 - P_0\right) \tag{4.4}$$

Assuming complete expansion in the exhaust nozzle ($P_9 = P_0$) and $\dot{m}_9 = \dot{m}_0$, Eq. (4.4) becomes

$$\mathcal{T} = \left(\dot{m}_0 V_9 - \dot{m}_0 V_0\right) = \dot{m}_9 a_0\left(\frac{V_9}{a_0} - M_0\right) \tag{4.5}$$

where a_0 is the speed of sound at station (0) $\left(a_0 = \sqrt{\gamma R T_0}\right)$. Now we can express V_9/a_0 in terms of Mach number and static temperature as

$$\left(\frac{V_9}{a_0}\right)^2 = \frac{a_9^2 M_9^2}{a_0^2} = \frac{\gamma_9 R_9 T_9 M_9^2}{\gamma_0 R_0 T_0} \tag{4.6}$$

However, $\gamma_9 = \gamma_0 = \gamma$ and $R_9 = R_0 = R$ for a calorically perfect gas. Hence, Eq. (4.6) becomes

$$\left(\frac{V_9}{a_0}\right)^2 = \frac{T_9}{T_0} M_9^2 \tag{4.7}$$

The exit Mach number M_9 can be expressed in terms of pressure ratio by

$$M_9^2 = \frac{2}{\gamma - 1}\left[\left(\frac{P_{t9}}{P_9}\right)^{\frac{\gamma-1}{\gamma}} - 1\right] \tag{4.8}$$

However, the pressure ratio at station (9), P_{t9}/P_9 can be expressed in terms of pressure ratio across each component as

$$\frac{P_{t9}}{P_9} = \frac{P_{t9}}{P_{t4}} \cdot \frac{P_{t4}}{P_{t2}} \cdot \frac{P_{t2}}{P_{t0}} \cdot \frac{P_{t0}}{P_0} \cdot \frac{P_0}{P_9} \tag{4.9}$$

By recasting Eq. (4.9) in terms of π of each component, we have

$$\frac{P_{t9}}{P_9} = \frac{P_{t9}}{P_0} = \pi_n \cdot \pi_b \cdot \pi_d \cdot \pi_r \quad \left(\text{as } P_9 = P_0\right) \tag{4.10}$$

But for an ideal engine, $\pi_n = \pi_b = \pi_d = 1$. Then Eq. (4.10) will be

$$\frac{P_{t9}}{P_9} = \pi_r \tag{4.11}$$

Now the expression for M_9 can be expressed as

$$M_9^2 = \frac{2}{\gamma-1}\left[\left(\frac{P_{t9}}{P_9}\right)^{\frac{\gamma-1}{\gamma}} - 1\right] = \frac{2}{\gamma-1}\left[\left(\pi_r\right)^{\frac{\gamma-1}{\gamma}} - 1\right] = \frac{2}{\gamma-1}\left[\left(\tau_r\right) - 1\right] \tag{4.12}$$

Since, $\pi_r^{\frac{\gamma-1}{\gamma}} = \tau_r = 1 + \frac{(\gamma-1)}{2} M_0^2$, Eq. (4.12) becomes

$$M_9^2 = \frac{2}{\gamma-1}\left[(\tau_r)-1\right] = M_0^2 \tag{4.13}$$

It is interesting to note that exit Mach number M_9 for an ideal ramjet engine is equal to flight Mach number. Now let us express T_9/T_0 in terms of temperature ratios, given by

$$\frac{T_9}{T_0} = \frac{T_{t9}/T_0}{T_{t9}/T_9} \tag{4.14}$$

We can express T_{t9}/T_0 in terms of τ across the various components as

$$\frac{T_{t9}}{T_0} = \frac{T_{t9}}{T_{t4}} \cdot \frac{T_{t4}}{T_{t2}} \cdot \frac{T_{t2}}{T_{t0}} \cdot \frac{T_{t0}}{T_0} = \tau_n \cdot \tau_b \cdot \tau_d \cdot \tau_r \tag{4.15}$$

But we know that for an ideal gas, $\tau_d = \tau_n = 1$, therefore, Eq. (4.15) becomes

$$\frac{T_{t9}}{T_0} = \tau_b \cdot \tau_r \tag{4.16}$$

Again, we know that $T_{t9}/T_9 = \left(P_{t9}/P_9\right)^{\frac{\gamma-1}{\gamma}}$. So the expression for T_9/T_0 can be rewritten as

$$\frac{T_9}{T_0} = \frac{T_{t9}/T_0}{T_{t9}/T_9} = \frac{\tau_b \tau_r}{\tau_r} = \tau_b \tag{4.17}$$

Now we will analyse the processes in the combustor by carrying out an energy balance across a CV (Fig. 4.3). Assuming the flow in the combustion chamber to be one dimensional steady flow, we can write down the energy equation across CV as

$$\dot{m}_2 h_{t3} + \dot{m}_f \Delta H_C = \dot{m}_4 h_{t4} \tag{4.18}$$

where ΔH_C is the heat of combustion. From the continuity equation, we know that

$$\dot{m}_4 = \dot{m}_f + \dot{m}_3 \tag{4.19}$$

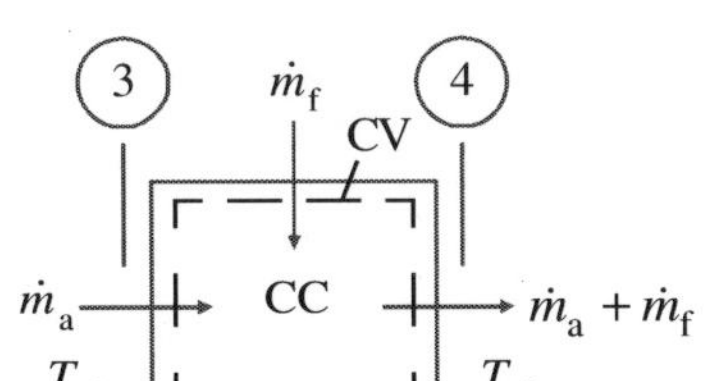

Fig. 4.3. Control volume for a combustor.

Since, $\dot{m}_f << \dot{m}_3$, therefore, $\dot{m}_4 \simeq \dot{m}_3 \simeq \dot{m}_a$. We also know that for ideal gas $C_{p3} = C_{p4} = C_p$, so the energy equation for the combustor becomes

$$\dot{m}_f \Delta H_C = \dot{m}_a C_p T_{t2}\left(\frac{T_{t4}}{T_{t2}} - 1\right) \tag{4.20}$$

Let us define f = fuel/air ratio = $\frac{\dot{m}_f}{\dot{m}_a}$, then Eq. (4.20) becomes

$$f = \frac{\dot{m}_f}{\dot{m}_a} = \frac{C_p T_{t2}}{\Delta H_C}\left(\frac{T_{t4}}{T_{t2}} - 1\right) = \frac{C_p T_{t0}}{\Delta H_C}(\tau_b - 1) \tag{4.21}$$

where $\tau_b = \frac{T_{t4}}{T_{t2}}$. For an ideal ramjet $T_{t0} = T_{t2} = T_0 \frac{T_{t0}}{T_0} = T_0 \tau_r$. Then the expression for f will be

$$f = \frac{\dot{m}_f}{\dot{m}_a} = \frac{C_p T_0 \tau_r}{\Delta H_C}(\tau_b - 1) \tag{4.22}$$

However, we can find out that $\tau_\lambda = \frac{T_{t4}}{T_0} = \frac{T_{t4}}{T_{t2}} \cdot \frac{T_{t2}}{T_0} = \tau_b \tau_r$. So the expression for f can be rewritten as

$$f = \frac{\dot{m}_f}{\dot{m}_a} = \frac{C_p T_0}{\Delta H_C}(\tau_r \tau_b - \tau_r) = \frac{C_p T_0}{\Delta H_C}(\tau_\lambda - \tau_r) \tag{4.23}$$

Now we can express $\left(\frac{V_9}{a_0}\right)^2$ in terms of T_9, T_0 and M_9 as

$$\left(\frac{V_9}{a_0}\right)^2 = \frac{T_9}{T_0} M_9^2 = \tau_b M_9^2 = \tau_b M_0^2 \tag{4.24}$$

As stated earlier $\tau_b = T_9/T_0$ and $M_9 = M_0$. Then the expression for the thrust will be

$$\mathcal{T} = \dot{m}_0 a_0 \left(\frac{V_9}{a_0} - M_0\right) = \dot{m}_0 a_0 M_0 \left(\sqrt{\tau_b} - 1\right) \tag{4.25}$$

We can recast Eq. (4.25) to get a relation for specific thrust as

$$\mathcal{T}_s = a_0 \left(\frac{V_9}{a_0} - M_0\right) = a_0 M_0 \left(\sqrt{\tau_b} - 1\right) \tag{4.26}$$

Now let us write an expression for TSFC as

$$\text{TSFC} = \frac{f}{\mathcal{T}/\dot{m}_0} = \frac{C_p T_0 \tau_r (\tau_b - 1)}{\Delta H_C a_0 M_0 \left(\sqrt{\tau_b} - 1\right)} \tag{4.27}$$

In order to characterise the ramjet engine performance, we must estimate the propulsion efficiency η_p, which is defined as the ratio of thrust power to the rate of kinetic energy of the incoming fluid into the engine. It is given by

$$\eta_p = \frac{2TV_0}{\dot{m}_a\left(V_9^2 - V_0^2\right)} \tag{4.28}$$

It can be shown that Eq. (4.28) for η_p can be expressed as

$$\eta_p = \frac{2}{\left(\sqrt{\tau_b}+1\right)} = \frac{2}{\left(\sqrt{\tau_\lambda/\tau_r}+1\right)} \tag{4.29}$$

Next, we can derive an equation for another important performance parameter, known as thermal efficiency η_{th}. Recall that it is defined as the ratio of rate of kinetic energy of the incoming fluid into the engine to the total thermal energy consumption rate, and it is given by

$$\eta_{th} = \frac{\dot{m}_a\left(V_9^2 - V_0^2\right)}{2\dot{m}_f \Delta H_C} \tag{4.30}$$

Using the above analysis, η_{th} of a ramjet engine can be written in a simplified form as

$$\eta_{th} = 1 - \frac{1}{\tau_r} \tag{4.31}$$

which the reader can derive. So the overall efficiency η_o, which is the product of η_p and η_{th}, is

$$\eta_o = \eta_p \eta_{th} = \frac{2\left(\tau_r - 1\right)}{\left(\sqrt{\tau_r \tau_\lambda} + \tau_r\right)} \tag{4.32}$$

The above equations are employed to carry out a parametric study of the performance of an ideal ramjet engine by varying the turbine inlet temperature and flight Mach number. Three representative combustor exit temperatures such as 1400, 1800 and 2200 K are considered for the present analysis. Higher combustor exit temperature is chosen for the ramjet engine as it does not have any rotating element (turbine) unlike a turbojet engine. Figure 4.4 shows the variation of T_s and TSFC with the flight Mach number for three values of combustor exit temperature at sea level condition. Figure 4.4

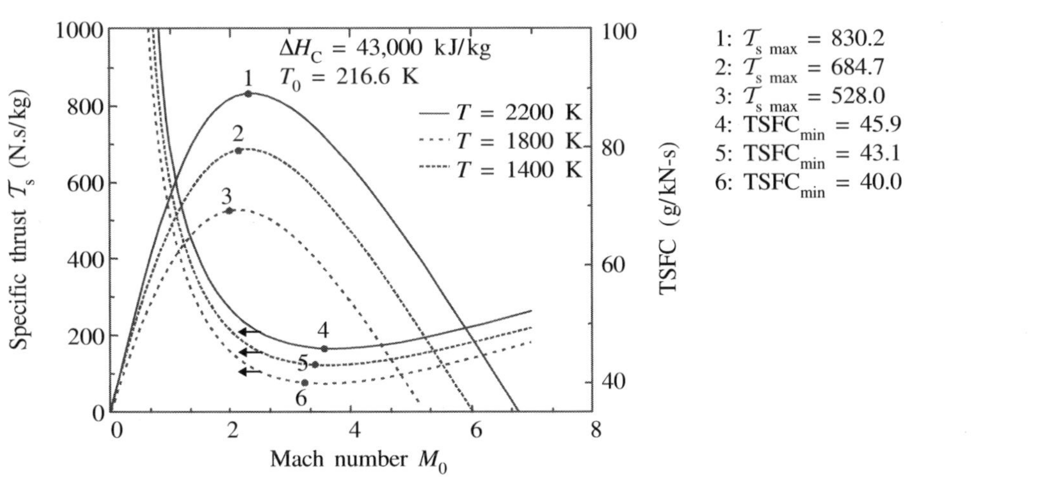

Fig. 4.4. Variation of T_s and TSFC with M_0 for T_{t4} = 1400, 1800 and 2200 K for ramjet engine.

shows that for a given combustor exit temperature, T_s increases from zero value at $M_0 = 0$ and attains a maximum value at certain flight Mach number and then drops down with further increase in M_0. It means there is an optimum flight Mach number corresponding to maximum T_s. In contrast, it can be observed from Fig. 4.4 that for a given combustor exit temperature, TSFC decreases from a large value at $M_0 = 0$ and attains a minimum value at certain flight Mach number and then goes up with further increase in M_0. It means there exists another optimum flight Mach number corresponding to minimum TSFC. It can be observed in Fig. 4.4 that the maximum specific thrust for all three cases as shown as 1, 2 and 3 occurs around flight Mach number of 2.2. In contrast minimum TSFC for all three cases (points 4, 5 and 6) occurs around flight Mach number of 3.6. Interestingly, this optimum Mach number does not coincide with the flight Mach number corresponding to the maximum T_s. Hence the designer has an option of choosing a particular flight Mach number corresponding to either maximum T_s or to the minimum TSFC. Of course, the mass flow rate ingested by air intake dictates the weight of the engine and, in turn, substantially affects the carrying capacity of the vehicle. However, higher fuel consumption will affect the operating cost of the flight. Hence, the designer has to select the best flight envelope for which both T_s and TSFC would attain their respective optimum value. Figure 4.4 also shows that with increase in combustor exit temperature, T_s increases substantially, whereas, the rate of increase in TSFC is quite less. It can also be observed that at lower flight Mach number (subsonic), the value of TSFC increases at a faster rate accompanied by decrease in T_s for any maximum combustor exit temperature. Hence, it can be concluded that for the ramjet, supersonic flight envelope must be chosen for obtaining adequate T_s with reasonable TSFC. It also demonstrates that the ramjet engine produces zero T and higher TSFC at static condition ($M_0 = 0$), which is a drawback of the ramjet due to which it cannot takeoff from the ground itself. In order to overcome zero T_s problem of the ramjet engine is it really possible to devise new methods to produce adequate static thrust for takeoff?

The variation of η_p with compressor pressure ratio (Fig. 4.5) for all three cases of combustor exit temperature shows an increase in η_p with increasing flight Mach number for a particular combustor exit temperature as there is an increase in propulsive thrust. Also, there is a marginal increase in η_p with combustor exit temperature a particular flight Mach number. The variation of η_{th} with compressor pressure ratio (Fig. 4.5) for all three cases of combustor exit temperature shows that η_{th} increases with

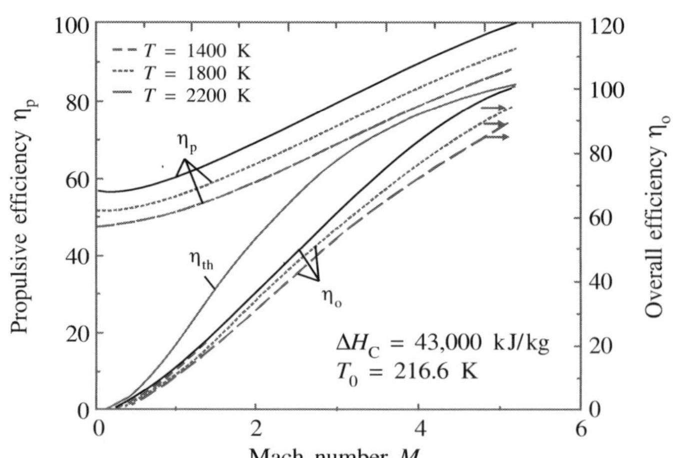

Fig. 4.5. Variation of η_p and η_{th}, respectively, with M_0 for ramjet engine.

flight Mach number. Interestingly, it is not affected by any change in combustor exit temperature. However, η_o exhibits similar trend as that of η_p, but with a lower value as indicated in Fig. 4.5.

Example 4.1 An ideal ramjet engine is flying with a flight Mach number of 2.5 at an altitude of 20 km. The maximum allowable temperature at the exit of the combustor is 2200 K. Determine $\mathcal{T}_s$, TSFC, η_p, η_{th}, and η_o if the heating value of fuel is 43,000 kJ/kg.

Given Al = 20 km, T_0 = 216.6, P_0 = 5.529 kPa, T_{t4} = 2200 K, M_0 = 2.5, C_P = 1.005 kJ/kgK, ΔH_C = 43,000 kJ/kg.

Solution The processes involved during the operation of ramjet engine are shown in Fig. 4.2 with station numbers.

Method I

We have already derived an expression for $\mathcal{T}_s$ as

$$\mathcal{T} = \dot{m}_0 a_0 M_0 \left(\sqrt{\tau_b} - 1\right)$$

The speed of sound a_0 can be evaluated using $a_0 = \sqrt{\gamma R T_0}$ = 295 m/s. Then the value of τ_b can be estimated as

$$\tau_b = \frac{T_{t4}}{T_{t2}} = \frac{2200}{487.35} = 4.51 \text{ as } T_{t2} = T_{t0} = T_0\left(1 + \frac{\gamma - 1}{2} M_0^2\right) = 216.6\left(1 + \frac{1.4 - 1}{2} 2.5^2\right) = 487.35 \text{ K}$$

Then $\mathcal{T}_s$ can be easily determined as

$$\mathcal{T}_s = \dot{m}_0 a_0 M_0 \left(\sqrt{\tau_b} - 1\right) = 295 \times 2.5\left(\sqrt{4.51} - 1\right) = 828.71 \text{ N} \cdot \text{s/kg}$$

The expression for TSFC can be used to evaluate its value as

$$\text{TSFC} = \frac{C_p T_0 \tau_r (\tau_b - 1)}{\Delta H_C a_0 M_0 \left(\sqrt{\tau_b} - 1\right)}$$

Substituting the respective values in the above equation, we get

$$\text{TSFC} = \frac{1.005 \times 216.6 \times 2.25(4.51 - 1)}{43000 \times 295 \times 2.5\left(\sqrt{4.51} - 1\right)} = 4.82 \times 10^{-5} \text{ kg/N} \cdot \text{s} = 48.2 \text{ mg/N} \cdot \text{s}$$

Now η_p can be determined as

$$\eta_p = \frac{2}{\left(\sqrt{\tau_b} + 1\right)} = \frac{2}{\left(\sqrt{4.51} + 1\right)} = 0.6403 = 64.03\%$$

Similarly, η_{th} can be evaluated as

$$\eta_{th} = 1 - \frac{1}{\tau_r} = 1 - \frac{1}{2.25} = 0.5555 = 55.55\%$$

Then η_o is estimated as

$$\eta_o = \eta_p \eta_{th} = 64.03 \times 55.55 = 35.57\%$$

Here we have directly used the expressions for various performance parameters and solved the given problem. But this method does not provide any feel for this problem. Let us explore another method of solving the problem.

Method II

$$\mathcal{T}_s = (V_9 - V_0) \text{ as } P_9 = P_0$$

$$V_9 = \sqrt{\gamma R T_9} \cdot \sqrt{\frac{2}{\gamma - 1}\left[\left(\frac{P_{t9}}{P_9}\right)^{\frac{\gamma-1}{\gamma}} - 1\right]}$$

But $V_0 = a_0 M_0 = \sqrt{\gamma R T_0} \cdot M_0 = \sqrt{1.4 \times 287 \times 216.6} \times 2.5 = 737.5$ m/s.

Let us now evaluate the pressure ratio P_{t9}/P_9 as

$$\frac{P_{t9}}{P_9} = \frac{P_{t9}}{P_{t4}} \cdot \frac{P_{t4}}{P_{t2}} \cdot \frac{P_{t2}}{P_{t0}} \cdot \frac{P_{t0}}{P_0} \cdot \frac{P_0}{P_9} = \pi_n \cdot \pi_b \cdot \pi_d \cdot \pi_r \quad (\text{as } P_9 = P_0)$$

But for an ideal engine $\pi_n = \pi_b = \pi_d = 1$. Therefore, the above equation will become

$$\frac{P_{t9}}{P_9} = \pi_r = \left(1 + \frac{\gamma - 1}{2} M_0^2\right)^{\frac{\gamma}{\gamma-1}} = \left(1 + \frac{0.4}{2} 2.5^2\right)^{3.5} = 17.09$$

Now we can evaluate the value of M_9 as

$$M_9 = \sqrt{\frac{2}{1.4 - 1}\left(17.09^{0.286} - 1\right)} = 2.5$$

It can be noted that the Mach number at the nozzle exit is same as that of the flight Mach number, which was discussed earlier. Now in order to evaluate the speed of sound a_9 at the nozzle exit, let us estimate T_9 as

$$T_9 = \frac{T_{t9}}{1 + \frac{\gamma - 1}{2} M_9^2} = \frac{2200}{1 + \frac{0.4}{2} 2.5^2} = 977.8 \text{ K}$$

Using the above two equations, we can evaluate V_9 as

$$V_9 = \sqrt{\gamma R T_9} \cdot M_0 = \sqrt{1.4 \times 287 \times 977.8} \times 2.5 = 1567 \text{ m/s}$$

Thus, $\mathcal{T}_s$ can be evaluated as

$$\mathcal{T}_s = (V_9 - V_0) = (1567 - 737.5) = 829.5 \text{ N} \cdot \text{s/kg}$$

In order to estimate TSFC, we will have to evaluate the fuel/air ratio as

$$f = \frac{\dot{m}_f}{\dot{m}_a} = \frac{C_p T_0}{\Delta H_C}(T_{t4} - T_{t2}) = \frac{1.005}{43{,}000}(2200 - 487.35) = 0.04$$

Then TSFC becomes

$$\text{TSFC} = \frac{f}{\mathcal{T}_s} = \frac{0.04}{829.9} = 4.82 \times 10^{-5} \text{ kg/N} \cdot \text{s}$$

The η_p can be evaluated as

$$\eta_p = \frac{2}{\frac{V_9}{V_0} + 1} = \frac{2}{\frac{1567}{737.5} + 1} = 0.64 = 64\%$$

The η_{th} can be evaluated as

$$\eta_{th} = \frac{\left(V_9^2 - V_0^2\right)}{2 f \cdot \Delta H_C} = \frac{1567^2 - 737.5^2}{2 \times 0.04 \times 43000 \times 10^3} = 0.5556 = 55.56$$

The η_o can be evaluated as

$$\eta_o = \eta_p \cdot \eta_{th} = 64.0 \times 55.56 = 35.56\%$$

It can be noted that the value of TSFC is quite closer to that of the turbojet engine. Besides this, η_o seems to be quite low. Method II provides the same answer to this problem as was obtained by Method I. Method II could manage to provide better physical feel for this problem, which is very important to develop an understanding of the subject. Hence, the reader is urged to use Method II for solving problems. However, Method I can be easily implemented in a computer program to carry out an on-design analysis, which is used for the analysis of aircraft cycles in aircraft industries.

4.3 Turbojet

In Section 4.2, we learnt that the ramjet engine cannot develop takeoff thrust and, hence, cannot fly itself. In fact, it cannot perform well, particularly, at high subsonic and low supersonic flight velocity as its pressure ratio is dictated by the flight Mach number. These problems of ramjet engine can be alleviated easily by using a compressor to enhance the pressure of the incoming air in addition to the air intake. Of course, a turbine is also to be employed to run the compressor. Such kind of engine is basically known as *turbojet engine*. Recall that turbojet engines were employed to propel aircraft during World War II with extensive research being pursued by both German and British researchers. At that time, turbojet engine was readily accepted by the aircraft manufacturers as it turned out to be a very light aeroengine with the capability of higher thrust for achieving higher flight speed as compared to piston engine driven aircraft. Even till date turbojet engine is preferred over other variants of gas turbine engines for fighter aircraft applications. Hence, it is important to understand the processes involved during its operation and to carry out the cycle analysis to characterise its performance over a wide range of operating parameters.

The layout of a typical turbojet engine is shown in Fig. 4.6a along with proper station numbers. The station number (0) corresponds to the freestream. Station number (2) represents air intake exit and compressor inlet. Station number (3) is the compressor exit and combustor inlet. The burner exit and turbine inlet is denoted by station number (5). The jet pipe is represented from (5) to (7). Finally, the nozzle exit is indicated as station number (9). Air enters the air intake with a flight velocity and gets decelerated by increasing its pressure at the air intake exit. Then air enters the compressor where its pressure gets increased while reducing its velocity. In the combustor, fuel is burnt to enhance the temperature of the high pressure air. Subsequently, air is expanded in a turbine to produce shaft power, which is used just to run the compressor. Then high pressure and temperature gas is expanded in an exhaust nozzle to produce thrust, which mainly propels the aircraft. All these processes involved during the operation of an ideal turbojet engine such as compression, heat addition and expansion are shown as a *T-s* diagram in Fig. 4.6b. It must be appreciated that this is a complex engine as compared to ramjet since it contains turbomachinery. The performance characteristics of the turbojet engine are likely to be quite different from that of the ramjet engine. The parametric cycle analysis is carried out below for an ideal turbojet non-afterburning engine. Subsequently, we will also carry out a similar analysis for a turbojet engine with afterburner configuration.

Let us write down an expression for the thrust produced by an ideal turbojet by assuming complete expansion in the exhaust nozzle ($P_9 = P_0$) and ($\dot{m}_9 = \dot{m}_0$) as

$$\mathcal{T} = \dot{m}_9 V_9 - \dot{m}_0 V_0 = \dot{m}_9 a_0 \left(\frac{V_9}{a_0} - M_0 \right) \tag{4.33}$$

where a_0 is the speed of sound and $M_0 = \sqrt{\gamma R T_0}$ is the flight Mach number at station number (0).

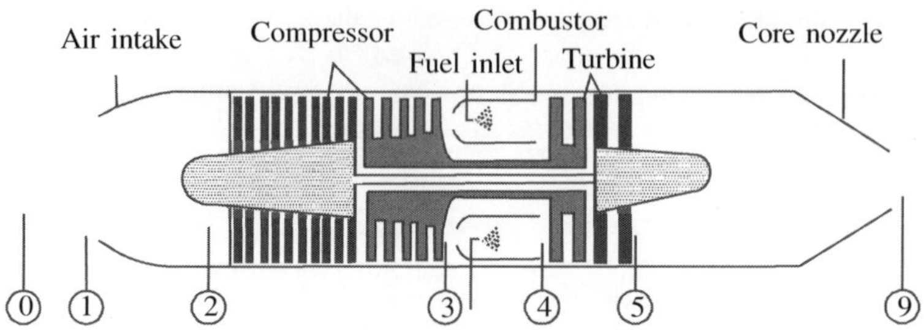

(a) Turbojet engine

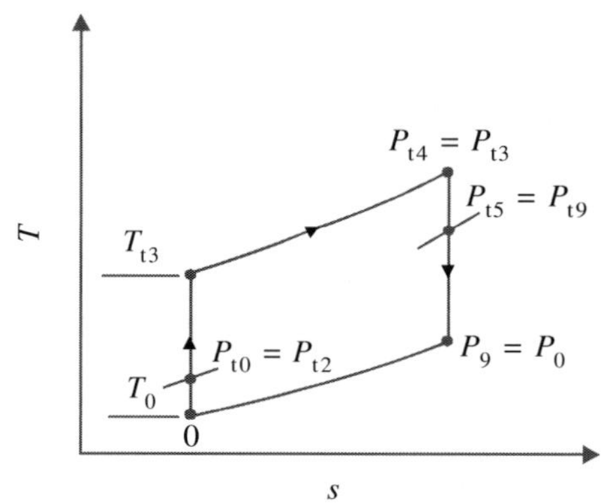

(b) *T-s* diagram

Fig. 4.6. (a) Schematic of a turbojet engine with station numbers and (b) *T-s* diagram of an ideal turbojet engine.

Specific Thrust

By using the definition, an expression for specific thrust $\mathcal{T}_s$ can be derived as

$$\mathcal{T}_s = \frac{T}{\dot{m}_0} = a_0\left(\frac{V_9}{a_0} - M_0\right) \tag{4.34}$$

But we know that

$$\left(\frac{V_9}{a_0}\right)^2 = \frac{a_9^2 M_9^2}{a_0^2} = \frac{T_9}{T_0} M_9^2 \tag{4.35}$$

The exit Mach number M_9 can be expressed in terms of pressure ratio by

$$M_9^2 = \frac{2}{\gamma - 1}\left[\left(\frac{P_{t9}}{P_9}\right)^{\frac{\gamma-1}{\gamma}} - 1\right] \tag{4.36}$$

Pressure ratio P_{t9}/P_9 can be expressed in terms of pressure ratio across individual components as

$$\frac{P_{t9}}{P_9} = \frac{P_{t9}}{P_{t5}} \cdot \frac{P_{t5}}{P_{t4}} \cdot \frac{P_{t4}}{P_{t3}} \cdot \frac{P_{t3}}{P_{t2}} \cdot \frac{P_{t2}}{P_{t0}} \cdot \frac{P_{t0}}{P_0} \cdot \frac{P_0}{P_9} \tag{4.37}$$

Assuming $P_9 = P_0$, Eq. (4.37) can be expressed in terms of pressure ratio across each component as

$$\frac{P_{t9}}{P_0} = \pi_n \cdot \pi_t \cdot \pi_b \cdot \pi_c \cdot \pi_d \cdot \pi_r \tag{4.38}$$

But for an ideal engine $\pi_n = \pi_b = \pi_d = 1$. Now Eq. (4.38) will be

$$\frac{P_{t9}}{P_9} = \pi_t \cdot \pi_c \cdot \pi_r \tag{4.39}$$

Using Eq. (4.39) the exit Mach number M_9^2 can be expressed as

$$M_9^2 = \frac{2}{\gamma-1}\left[\left(\frac{P_{t9}}{P_9}\right)^{\frac{\gamma-1}{\gamma}} - 1\right] = \frac{2}{\gamma-1}\left[(\pi_t\pi_c\pi_r)^{\frac{\gamma-1}{\gamma}} - 1\right] \tag{4.40}$$

Since $\pi_r = \tau_r^{\frac{\gamma}{\gamma-1}}$ and for an ideal turbojet $\pi_c = \tau_c^{\frac{\gamma}{\gamma-1}}$ and $\pi_t = \tau_t^{\frac{\gamma}{\gamma-1}}$, Eq. (4.40) becomes

$$M_9^2 = \frac{2}{\gamma-1}\left[\tau_t\tau_c\tau_r - 1\right] \tag{4.41}$$

But we can express T_9/T_0 in terms of temperature ratios as

$$\frac{T_9}{T_0} = \frac{T_{t9}/T_0}{T_{t9}/T_9} \tag{4.42}$$

Let us express T_{t9}/T_0 in terms of τ across the various components as

$$\frac{T_{t9}}{T_0} = \frac{T_{t9}}{T_{t5}} \cdot \frac{T_{t5}}{T_{t4}} \cdot \frac{T_{t4}}{T_{t3}} \cdot \frac{T_{t3}}{T_{t2}} \cdot \frac{T_{t2}}{T_{t0}} \cdot \frac{T_{t0}}{T_0} = \tau_n \cdot \tau_t \cdot \tau_b \cdot \tau_c \cdot \tau_d \cdot \tau_r \tag{4.43}$$

We know that for an isentropic process $\tau_d = \tau_n = 1$, so Eq. (4.43) becomes

$$\frac{T_{t9}}{T_9} = \tau_t \cdot \tau_b \cdot \tau_c \cdot \tau_r \tag{4.44}$$

Again, we know that $(T_{t9}/T_9) = (P_{t9}/P_9)^{\frac{\gamma-1}{\gamma}}$. So the expression for T_9/T_0 can be rewritten as

$$\frac{T_9}{T_0} = \frac{T_{t9}/T_0}{T_{t9}/T_9} = \frac{\tau_t\tau_b\tau_c\tau_r}{\tau_t\tau_c\tau_r} = \tau_b \tag{4.45}$$

An expression for $\mathcal{T}_s$ can be obtained by combining Eqs. (4.34), (4.35), (4.41) and (4.45) as

$$\mathcal{T}_s = \frac{\mathcal{T}}{\dot{m}_0} = a_0\left(\frac{V_9}{a_0} - M_0\right) = a_0\left[\sqrt{\left(\frac{2}{\gamma-1}\right)\frac{\tau_\lambda}{\tau_c\tau_r}(\tau_t\tau_c\tau_r - 1)} - M_0\right] \tag{4.46}$$

From Eq. (4.46) it can be noted that the turbine temperature ratio needs to be related to other variables. Since the power produced by the turbine is meant to run the compressor, it is clear that the turbine work output must be equal to the compressor input under ideal conditions. By applying one dimensional steady state energy equation to a CV containing the turbine, the work output $\dot{W}_t$ is given by

$$\dot{W}_t = \dot{m}_5 (h_{t4} - h_{t5}) = \dot{m}_0 C_p (T_{t4} - T_{t5}) \tag{4.47}$$

Similarly, the work input to the compressor $\dot{W}_c$ will be given by

$$\dot{W}_c = \dot{m}_3 (h_{t3} - h_{t2}) = \dot{m}_0 C_p (T_{t3} - T_{t2}) \tag{4.48}$$

But we know that $\dot{W}_t = \dot{W}_c$. Therefore, by equating Eqs. (4.47) and (4.48) we can have

$$\tau_t = 1 - \frac{\tau_r}{\tau_\lambda}(\tau_c - 1) \tag{4.49}$$

By carrying out one dimensional steady flow energy balance across the combustion chamber as done for the ramjet engine, the fuel/air ratio f can be expressed in terms of known variables as

$$f = \frac{\dot{m}_f}{\dot{m}_a} = \frac{C_p T_0}{\Delta H_C}(\tau_b \tau_r \tau_c - \tau_r \tau_c) \tag{4.50}$$

Thrust Specific Fuel Consumption (TSFC)

The expression for TSFC can be written as

$$\text{TSFC} = \frac{\dot{m}_f}{T} = \frac{f}{T_s} \tag{4.51}$$

Propulsive Efficiency

The relation for η_p from the basic definition can be expressed as

$$\eta_p = \frac{2TV_0}{\dot{m}_a (V_9^2 - V_0^2)} = \frac{2M_0}{V_9/a_0 + M_0} \tag{4.52}$$

Thermal Efficiency

The expression for η_{th} can be written as

$$\eta_{th} = \frac{\dot{m}_a (V_9^2 - V_0^2)}{2\dot{m}_f \Delta H_C} = 1 - \frac{1}{\tau_r \tau_c} \tag{4.53}$$

Overall Efficiency

We know that η_o is the product of η_p and η_{th} and it can be expressed as

$$\eta_o = \eta_p \eta_{th} \tag{4.54}$$

The above equations are employed to carry out parametric studies of the performance of an ideal turbojet engine by varying the compressor pressure ratio, turbine inlet temperature and flight Mach number. Three flight Mach numbers such as 0, 0.85 and 1.5 are considered, which represent the three flight regimes viz., sea level, high subsonic and supersonic motion, respectively. Each flight Mach number is assumed to be at different altitude with a different pair of ambient pressure P_0 and ambient temperature T_0 (Table 4.1).

Table 4.1. Flight conditions for an ideal parametric cycle analysis of a turbojet engine

Flight altitude (Flight Mach number)	Ambient pressure, P_0 (kPa)	Ambient temperature, T_0 (K)
Sea level (M = 0)	101.3	288.2
Al = 12 km (M = 0.85)	19.4	216.7
Al = 15 km (M = 1.5)	12.1	216.7

The plot of T_s and TSFC vs. compressor pressure ratio for TIT of 1600 K covering three Mach numbers of 0, 0.85 and 1.5 (Fig. 4.7) shows that T_s for $M_0 = 0$ increases, initially, to attain a maximum value at π_c of 20 and, subsequently, it decreases with further increase in pressure ratio across the compressor. For higher flight Mach number, the variation of T_s with π_c shows a similar trend. However, there is reduction in T_s value for a particular π_c when the flight Mach number is increased. Interestingly, the optimum pressure ratio for which T_s will be maximum is around 9 for supersonic flight (M = 1.5) and it is about 20 for high subsonic flight (M_0 = 0.85). This is due to the fact that ram compression contributes significantly to the overall compression and, thus, a compressor with smaller pressure ratio can provide the same level of thrust. If the flight Mach number increases beyond 3, then no compressor is required. Under this condition, the turbojet can operate like a ramjet engine. Figure 4.7 also shows that for a given flight Mach number, there is a pressure ratio that provides maximum T_s.

The variation of TSFC with π_c for all three Mach number (Fig. 4.7) shows that TSFC decreases with π_c as less fuel is consumed for the same thrust. The actual value of TSFC increases with flight Mach number for a particular π_c. However, a similar trend is witnessed with the pressure ratio across the compressor. The TSFC is likely to attain minimum value at exorbitant high pressure ratio (not shown in Fig. 4.7). For certain aircraft, it will be prudent to operate at minimum TSFC for which very high pressure ratio compressor has employed. Unfortunately, the optimum value of TSFC departs quite exorbitantly from the optimum T_s for a particular flight Mach number and altitude. The designer

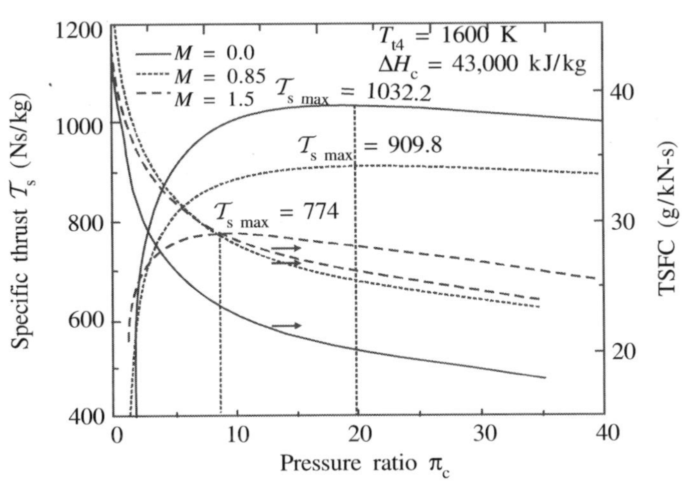

M	π_c	$T_{s,max}$ (Ns/kg)
0.0	20	1032.2
0.85	20	909.8
1.5	9	774

Fig. 4.7. Variation of T_s and TSFC with π_c for different values of M_0 for a turbojet engine at TIT of 1600 K.

has to choose proper pressure ratio for best cruising range while compromising on either optimum TSFC or T_s depending on the mission requirement of the aircraft. Of course, the weight of the engine is dependent strongly on mass flow rate of air and, in turn, it substantially affects the carrying capacity of an aircraft. However, higher fuel consumption affects the operating cost of the flight. Hence, the designer has to select the best compressor pressure ratio for optimum values of T_s and TSFC.

The variations of η_p and η_{th} with compressor pressure ratio for TIT of 1600 K are shown in Fig. 4.8. It can be observed that propulsion efficiency decreases marginally with π_c for $M = 0.85$ and 1.5. Of course, higher flight number case exhibits a lower propulsive efficiency as compared to lower Mach number ($M = 0.85$) because kinetic energy is being wasted at higher flight Mach number. In contrast the higher flight velocity exhibits higher value of thermal efficiency. But for all three flight Mach number the thermal efficiency increases with π_c.

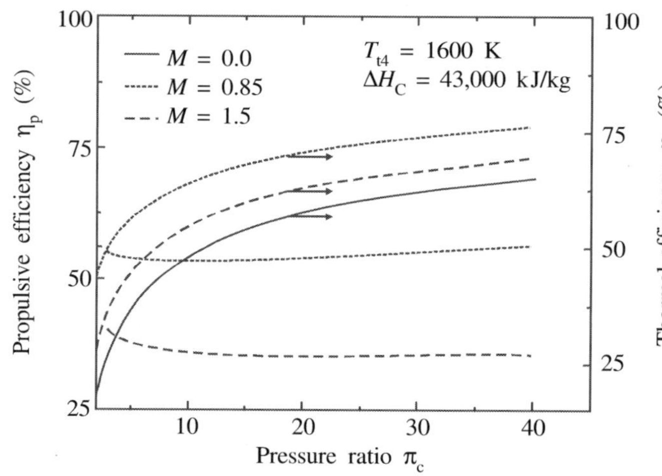

Fig. 4.8. Variation of η_p and η_{th} with π_c for a turbojet engine.

In order to study the effect of turbine inlet temperature, three representative temperatures viz., 1200, 1400 and 1800 K are considered for the present discussion. Figure 4.9 shows the variation of T_s and TSFC with the compressor pressure ratio for three values of TIT, respectively, for $M = 0$ at sea level condition. It can be observed (Fig. 4.9) that with increase in TIT, T_s increases substantially, whereas, there is a little increase in TSFC. A similar trend holds good for the other two Mach numbers. However, the maximum temperature of turbine inlet air must be limited to 1200 K for uncooled blades with currently available materials. With the inclusion of cooling system for turbine blades, TIT can be increased up to 1800 K with the available materials. The parametric studies similar to those discussed above can provide the designer ample data and wisdom to make choices about the engine operating parameters for a given mission requirement.

The variation of η_{th} along with f for three representative temperatures viz., 1200, 1400 and 1800 K (Fig. 4.10) show that η_{th} remains same for the three cases although fuel consumption increases with increase in TIT. In contrast, f increases with increase in TIT for the same compressor pressure ratio due to higher fuel consumption.

Example 4.2 An ideal turbojet is flying at flight Mach number $M_0 = 2.0$ at an altitude of 15 km. The pressure ratio across the compressor is 12. At the combustion chamber exit, temperature is 1600 K. Estimate the net thrust

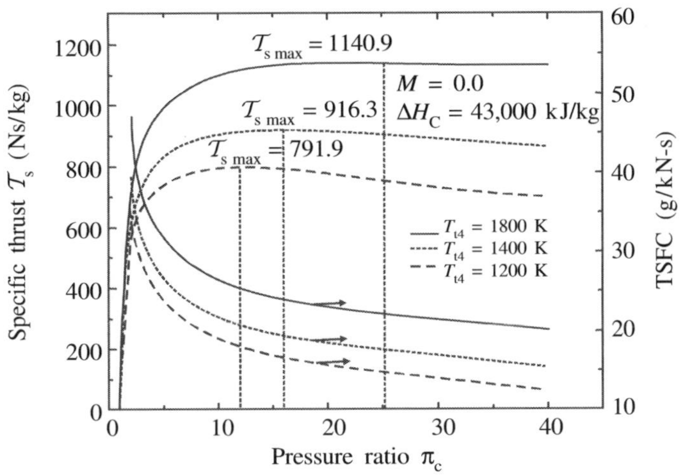

Fig. 4.9. Variation of $\mathcal{T}_s$ and TSFC with π_c for three TIT at sea level condition and $M_0 = 0$ for a turbojet engine.

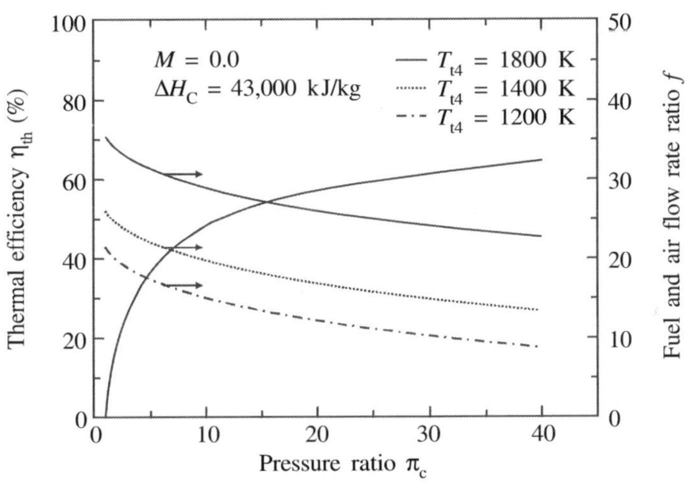

Fig. 4.10. Variation of η_p and η_{th} with π_c for three TIT at sea level condition and $M_0 = 0$ for a turbojet engine.

produced by the engine if it consumes 50 kg/s of air. Determine $\mathcal{T}_s$, TSFC, η_p, η_{th} and η_o if the heat of combustion is 43,000 kJ/kg.

Given Al =15 km, T_0 = 216.7, P_0 = 11.2 kPa, T_{t4} = 1600 K, M_0 = 2.0, $\dot{m}_0$ = 50 kg/s, C_p = 1.005 kJ/kgK and ΔH_C = 43,000 kJ/kg.

Solution The processes involved during the operation of a turbojet engine are shown in Fig. 4.6 with the station numbers. We know that $\mathcal{T}_s$ is expressed as

$$\mathcal{T}_s = (V_9 - V_0) \text{ as } P_9 = P_0$$

We know that flight velocity is given as

$$V_0 = \sqrt{\gamma R T_0} \cdot M_0 = \sqrt{1.4 \times 287 \times 216.7} \times 2 = 590.15 \text{ m/s}$$

But the nozzle exit velocity is to be determined by

$$V_9 = \sqrt{2C_p(T_{t5} - T_9)}$$

In order to evaluate T_{t5} and T_9, we need to estimate the temperature and pressure at each station numbers as follows:

Air Intake

The total pressure and temperature at station (2) can be evaluated as

$$T_{t2} = T_{t0} = T_0\left(1 + \frac{\gamma - 1}{2}M_0^2\right) = 216.7\left(1 + 0.2\times 2^2\right) = 390.06 \text{ K}$$

Let us now evaluate the pressure ratio P_{t2}/P_0 as

$$\frac{P_{t2}}{P_0} = \frac{P_{t0}}{P_0} = \left(\frac{T_{t0}}{T_0}\right)^{\frac{\gamma}{\gamma-1}} = \left(\frac{390.06}{261.7}\right)^{3.5} = 7.82 \Rightarrow P_{t2} = \frac{P_{t2}}{P_0}\times P_0 = 87.6 \text{ kPa}$$

Compressor

The total pressure across the compressor is given as

$$\frac{P_{t3}}{P_{t2}} = 12 \Rightarrow P_{t3} = \frac{P_{t3}}{P_{t2}}\times P_{t2} = 1051.2 \text{ kPa} = P_{t4}$$

Let us now evaluate the total temperature ratio T_{t3}/T_{t2} as

$$\frac{T_{t3}}{T_{t2}} = \left(\frac{P_{t3}}{P_{t2}}\right)^{\frac{\gamma-1}{\gamma}} = 12^{0.286} = 2.035 \Rightarrow T_{t3} = \frac{T_{t3}}{T_{t2}}\times T_{t2} = 794 \text{ K}$$

Turbine

The total temperature at the exit of the turbine is to be evaluated. We know that the function of a turbine is to supply the requisite power to the compressor. Hence, we have

$$\dot{W}_c = \dot{W}_t \Rightarrow C_p(T_{t3} - T_{t2}) = C_p(T_{t4} - T_{t5})$$

From the above equation, we can estimate the turbine exit temperature as

$$T_{t5} = T_{t4} - (T_{t3} - T_{t2}) = 1600 - (794 - 390.06) = 1196.1 \text{ K}$$

Then the total pressure, P_{t5} can be determined as

$$\frac{P_{t5}}{P_{t4}} = \left(\frac{T_{t5}}{T_{t4}}\right)^{\frac{\gamma}{\gamma-1}} = \left(\frac{1196.1}{1600}\right)^{3.5} = 0.361 \Rightarrow P_{t5} = \frac{P_{t5}}{P_{t4}}\times P_{t4} = 379.7 \text{ kPa}$$

Combustor

Let us evaluate the amount of fuel consumed in the combustor by striking an energy balance across the combustor

$$\dot{m}_f \Delta H_C = \dot{m}_0 C_p(T_{t4} - T_{t3}) \Rightarrow \dot{m}_f = \frac{50\times 1.005(1600 - 794)}{43000} = 0.94 \text{ kg/s}$$

Nozzle

By assuming the flow through the nozzle to be isentropic and the nozzle to be fully expanded to ambient pressure, nozzle exit temperature T_9 can be estimated as

$$\frac{T_{t5}}{T_9} = \frac{T_{t9}}{T_9} = \left(\frac{P_{t9}}{P_9}\right)^{\frac{\gamma-1}{\gamma}} = \left(\frac{379.7}{11.2}\right)^{0.286} = 2.739$$

Then we can evaluate T_9 as

$$T_9 = \frac{T_{t9}}{\left(T_{t9}/T_9\right)} = \frac{1196.1}{2.739} = 436.7 \text{ K}$$

Then using these two equations, we can evaluate V_9 as

$$V_9 = \sqrt{2 \times C_p \left(T_{t5} - T_9\right)} = \sqrt{2 \times 1005(1196.1 - 436.7)} = 1235.5 \text{ m/s}$$

Thus, $\mathcal{T}$ can be evaluated as

$$\mathcal{T} = \dot{m}_0 \left(V_9 - V_0\right) = 50(1235.5 - 590.15) = 32.27 \text{ kN}$$

TSFC can be evaluated as

$$\text{TSFC} = \frac{\dot{m}_f}{\mathcal{T}} = \frac{0.94}{32.27} = 0.029 \text{ kg/N} \cdot \text{s}$$

The η_p can be evaluated as

$$\eta_p = \frac{2}{\frac{V_9}{V_0} + 1} = \frac{2}{\frac{1235.5}{590.15} + 1} = 0.6465 = 64.65\%$$

The η_{th} can be evaluated as

$$\eta_{th} = \frac{\dot{m}_0 \left(V_9^2 - V_0^2\right)}{2\dot{m}_f \cdot \Delta H_C} = \frac{50\left(1235.5^2 - 590.15^2\right)}{2 \times 0.94 \times 43000 \times 10^3} = 0.7287 = 72.87\%$$

The η_o can be evaluated as

$$\eta_o = \eta_p \cdot \eta_{th} = 64.65 \times 72.87 = 47.11\%$$

It can be noted that η_p is comparable to that of ramjet engine while η_{th} of this turbojet is higher to that of the ramjet engine.

4.4 Thrust Augmentation

We know that certain types of aircraft such as interceptor and fighter need to have engines capable of providing higher thrust for limited periods of time of the order of a few minutes. During combat manoeuvres, a fighter aircraft requires temporary augmentation of thrust. Besides this, the initial $\mathcal{T}$ for certain aircraft is quite high, which is expected to be provided by the engine for initial takeoff period. For supersonic aircraft, the engine must also supply higher thrust for accelerating the aircraft from subsonic to supersonic speed. Hence, it is important to study available methodologies that can be used to augment the thrust produced by the engine without significantly increasing its weight, size etc. Let us now discuss how to increase the thrust of a gas turbine engine.

We know from our previous discussion that several methods can be devised to enhance the thrust of a gas turbine engine. One of them is to enhance thrust by increasing TIT, which enhances $\mathcal{T}_s$ and, in turn, the thrust for a given size. But the availability of materials that can withstand high thermal load under centrifugal load conditions poses serious problems. Another alternative to enhance the thrust is to increase the mass flow rate through the engine. Both methods provide significant increase in thrust, of course, with redesigning of the engine, which may not be the right choice. There are several other methods which can be devised to augment the thrust of an aircraft engine. However, we will restrict our discussion to two widely used methods viz., liquid injection and reheating (afterburning).

4.4.1 Liquid Injection

In this method, liquid such as water or mixture of water and methanol is injected into the compressor inlet in the form of fine spray. This liquid spray consists of fine droplets, which need to be evaporated quickly while passing through the entire compressor in a few milliseconds. The vaporisation of small droplets of spray entails heat extraction from the air resulting in reduction in its temperature. We know from our previous discussion that a reduction in compressor inlet temperature can cause an increase in compressor pressure ratio and mass flow rate due to increase in rotational speed, which results in an increase in thrust. This method is generally suitable for increasing takeoff thrust. Besides this, on a dry and hot day, the thrust in an aircraft can be easily enhanced by 50% using this method. The amount of water that can be injected is dictated by the volumetric constraints of the compressor. It must be noted that a large quantity of water will be required for such kind of thrust augmentation. However, if water is consumed during takeoff and initial climb, then it would not cause any significant weight penalty.

It is important to choose the injectant with higher latent heat of vaporisation. Also, it should not damage the engine parts. Hence, generally, distilled water is used as an injectant for thrust augmentation. But if the water is not consumed during the takeoff thrust, there is a chance that the leftover water may freeze at high altitudes. In order to overcome this problem, small quantity of alcohol is mixed with water. But the addition of methanol reduces the latent heat of vaporisation to some extent depending upon the percentage of alcohol mixed in water. Of course, the addition of alcohol can act as a fuel and generate heat while getting burnt in the combustion chamber. Besides this, the thrust can be augmented temporarily by injecting liquid (water or alcohol-water mixture) into the combustion chamber. It must be noted that the liquid must be injected at higher pressure well above the combustion chamber. Generally, a large quantity of water needs to be injected into the combustion chamber than that with compressor inlet. This liquid injection will cause a temporary blockage in the combustion chamber, which eventually compels operation of the compressor at a higher pressure ratio. Moreover, there will be cooling of gases and increase in mass flow rate through the turbine and nozzle. As a result, there will be substantial temporary augmentation of thrust. Although liquid injection is one of the effective methods of enhancing the thrust, it is not preferred in modern times over reheating system, which is described below in detail.

4.4.2 Reheat (Afterburner) System

We know that the thrust can be enhanced by increasing the exit velocity, which can be augmented by increasing the inlet temperature of the exhaust nozzle. In order to enhance the nozzle inlet temperature, additional amount of fuel must be burnt in the jet pipe located at the downstream of the turbine. This system is known as *reheating/afterburning*. It is very easy to achieve higher temperature even upto the maximum temperature (around 2200 K) for certain fuels as there is no highly stressed rotating components in the jet pipe. Interestingly, no additional air needs to be supplied for burning of additional fuel in the afterburner since the gas turbine engine uses a large amount of air to keep the TIT below the adiabatic temperature due to turbine blade material constraints. As a result, around three-quarter of the oxygen in the air will be available for the burning of additional fuel in reheat (afterburner) system. The burning of fuel in the afterburner can be accomplished in a much simpler system as compared to the main combustor. Liquid fuel is injected into the afterburner, forming sprays of fine droplets. The V-shaped gutter consisting of two concentric rings is employed to stabilise the flame inside the afterburner. The details about the combustion aspects of afterburner will be discussed in Chapter 10. In order to achieve maximum thrust, fuel must be burnt in stoichiometric fuel/air ratio raising the temperature of gas upto 2200 K in the afterburner. The fuel consumption increases significantly when the afterburner is operational, leading to substantial loss in overall efficiency. The afterburner system also causes loss in pressure even if it is not in operational mode, which is known as *dry loss*. In addition, an increase in nozzle area results in an enhancement in pressure thrust. Hence, a variable nozzle must be fitted with the afterburner jet engine to accommodate increased volume flow rate of gases during afterburner operation. This kind of variable area nozzle is quite difficult to design and manufacture, which leads to increased complexities as well as cost. It also introduces extra weight to the engine. Generally, the afterburner is activated only when the engine is operated at its peak unaugmented thrust. But we know that the maximum unaugmented thrust can occur only when the engine is operated at its maximum speed. Hence, generally, the afterburner is made operational at the maximum engine speed. As a result, the nozzle area increases significantly to allow the increased mass flow rate at lower density.

Since the afterburner is being widely used in modern aircraft, a parametric cycle analysis is presented in detail here. The schematic of a turbojet engine with afterburner system along with the station numbers is shown in Fig. 4.11(a). The station numbers are similar to that of a simple turbojet engine, except station number (7), which indicates the exit of the afterburner. The usual processes of compression, heat addition and expansion for a simple ideal turbojet engine are described in the *T-s* diagram (Fig. 4.11(b)). The heat added in the afterburner is shown as path 5-7 (Fig. 4.11), which can be assumed to be taking place at constant pressure. Then the gas is expanded in the exhaust

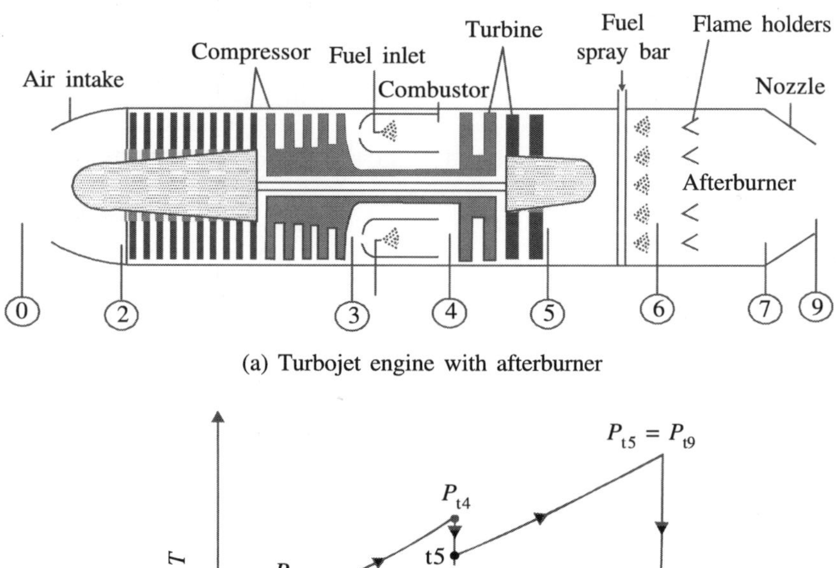

(a) Turbojet engine with afterburner

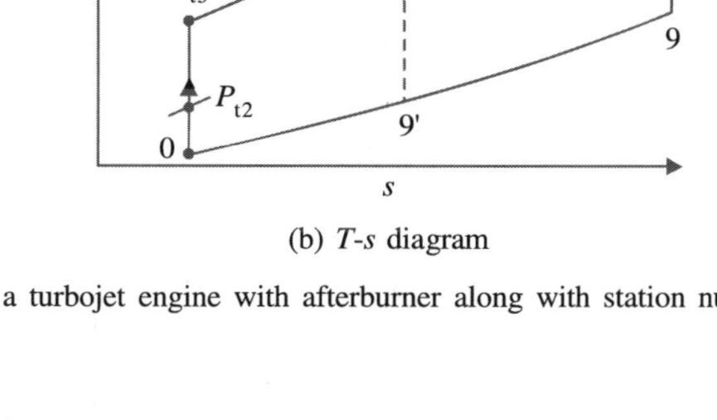

(b) *T-s* diagram

Fig. 4.11. (a) Schematic of a turbojet engine with afterburner along with station numbers and (b) *T-s* diagram.

nozzle, resulting in higher velocity as described by the path (7-9). When the afterburner is not operational, the gas is expanded through the exhaust nozzle to station number (9') (dashed line in Fig. 4.11).

The analysis of a turbojet with afterburner can be carried out in a similar way to that of a simple turbojet engine. However, we will follow a shortcut route to derive an expression for all the performance parameters using the results of a simple turbojet engine. We know that the equation for $\mathcal{T}_s$ is

$$\mathcal{T}_s = \left(\dot{m}_9 V_9 - \dot{m}_0 V_0\right) = \dot{m}_9 a_0 \left(\frac{V_9}{a_0} - M_0\right) \tag{4.55}$$

We can derive a relationship between the exhaust velocity $V_{9'}$ and exhaust velocity at the nozzle exit V_9 with the use of an afterburner. For this purpose, let us derive an expression for V_9 as

$$V_9 = \sqrt{2C_p\left(T_{t7} - T_9\right)} = \sqrt{2C_p T_{t9}\left[1 - \frac{T_9}{T_{t9}}\right]} = \sqrt{2C_p T_{t9}\left[1 - \left(\frac{P_9}{P_{t9}}\right)^{\frac{\gamma-1}{\gamma}}\right]} \tag{4.56}$$

Note that $T_{t7} = T_{t9}$ as no heat loss is likely to occur between stations (7) and (9) for an ideal cycle. When the afterburner is not operational, exhaust velocity at the nozzle exit with the afterburner $V_{9'}$ is given by

$$V_{9'} = \sqrt{2C_p\left(T_{t5} - T_{9'}\right)} = \sqrt{2C_p T_{t5}\left[1 - \frac{T_{9'}}{T_{t5}}\right]} = \sqrt{2C_p T_{t5}\left[1 - \left(\frac{P_9}{P_{t9}}\right)^{\frac{\gamma-1}{\gamma}}\right]} \tag{4.57}$$

By using Eqs. (4.56) and (4.57), we can relate $V_{9'}$ and V_9 as

$$\frac{V_9}{V_{9'}} = \frac{V_9/a_0}{V_{9'}/a_0} = \sqrt{\frac{T_{t9}}{T_{t5}}} \Rightarrow \frac{V_9}{a_0} = \frac{V_{9'}}{a_0} \cdot \sqrt{\frac{T_{t9}}{T_{t5}}} \tag{4.58}$$

Let us express the temperature ratio T_{t9}/T_{t5} in terms of respective total temperature ratio τ across various components as

$$\frac{T_{t9}}{T_{t5}} = \frac{\left(T_{t9}/T_{t7}\right)\cdot\left(T_{t7}/T_0\right)}{\left(T_{t5}/T_{t4}\right)\cdot\left(T_{t4}/T_0\right)} = \frac{\tau_{\lambda AB}}{\tau_t \cdot \tau_\lambda} \tag{4.59}$$

where temperature ratio in the afterburner $\tau_{\lambda AB}$ is defined as T_{t7}/T_0. Using the expression for V_9/a_0 from a simple jet analysis, we can find an expression for V_9/a_0 of a turbojet engine with afterburner as

$$\frac{V_9}{a_0} = \sqrt{\frac{2}{\gamma-1}\left(\frac{\tau_{\lambda AB}}{\tau_r \tau_c \tau_t}\right)\left(\tau_r \tau_c \tau_t - 1\right)} \tag{4.60}$$

However, we can relate τ_c and τ_t by assuming that work produced by the turbine $\dot{W}_t$ is consumed fully by the compressor work input $\dot{W}_c$ as derived for a simple turbojet engine

$$\tau_t = 1 - \frac{\tau_r}{\tau_\lambda}\left(\tau_c - 1\right) \tag{4.61}$$

Combustor

Let us evaluate the total amount of fuel consumed in both main combustor and afterburner by striking an energy balance across the combustor

$$\dot{m}_{f,tot}\Delta H_c = \dot{m}_0 C_p \left(T_{t9} - T_{t0}\right) \Rightarrow f_{tot} = \frac{C_p T_0}{\Delta H_c}\left(\tau_{\lambda AB} - \tau_r\right) \tag{4.62}$$

Thrust Specific Fuel Consumption (TSFC)

Similarly, the expression for TSFC for a turbojet engine with afterburner can be written as

$$\text{TSFC} = \frac{\dot{m}_f}{\mathcal{T}} = \frac{f_{tot}}{\mathcal{T}_s} \tag{4.63}$$

Propulsive efficiency

The relation for η_p from the basic definition can be expressed in a similar way

$$\eta_p = \frac{2\mathcal{T}V_0}{\dot{m}_a\left(V_9^2 - V_0^2\right)} = \frac{2M_0}{\dfrac{V_9}{a_0} + M_0} \tag{4.64}$$

Thermal efficiency

The expression for η_{th} for turbojet with afterburner can be recasted easily as

$$\eta_{th} = \frac{(\gamma - 1)C_p T_0\left[\left(\dfrac{V_9}{a_0}\right)^2 - M_0^2\right]}{2 f_{tot}\Delta H_C} \tag{4.65}$$

The above equations can be used to carry out parametric studies of the performance of an ideal turbojet engine with afterburner by varying the compressor pressure ratio, TIT and flight Mach number. However, we will now carry out comparative studies of the variation of $\mathcal{T}_s$ and TSFC with flight Mach number for $\pi_c = 12$, $T_{t4} = 1400$ and $T_{t9} = 2000$ K. The plot of $\mathcal{T}_s$ with flight Mach number (Fig. 4.12) indicates that the specific thrust gets enhanced with addition of afterburner as compared to a simple turbojet engine. It can be noted that afterburner augments $\mathcal{T}_s$ by as much as 125% over the whole range of flight Mach number ($M_0 = 0$ to 3). However, TSFC becomes significantly higher with afterburner indicating that thrust can be enhanced by paying heavy penalty in terms of fuel cost. Interestingly, TSFC of an afterburning turbojet engine attains optimum value at $\pi_c = 12$ and Mach number of 1.5. It is logical to choose a lower pressure ratio for an afterburning engine, particularly, at supersonic flight Mach number ($M_0 = 1.5$), since it provides a smaller engine with the same fuel consumption. Note that a General Electric GE4 turbojet afterburning engine with pressure ratio of 12 is employed for supersonic transport aircraft (e.g. Boeing). Besides this, the variation of η_p, η_{th} and η_o with flight Mach number for this case (Fig. 4.13) shows that all efficiencies get reduced with the use of afterburner. Of course, use of an afterburner leads to decrease in η_o and increase in fuel consumption while augmenting the thrust.

Example 4.3 An ideal turbojet with afterburner is flying at $M_0 = 2.0$ at an altitude of 15 km. The pressure ratio across the compressor is 12. The temperature at the exit of the main combustion chamber and afterburner is 1600

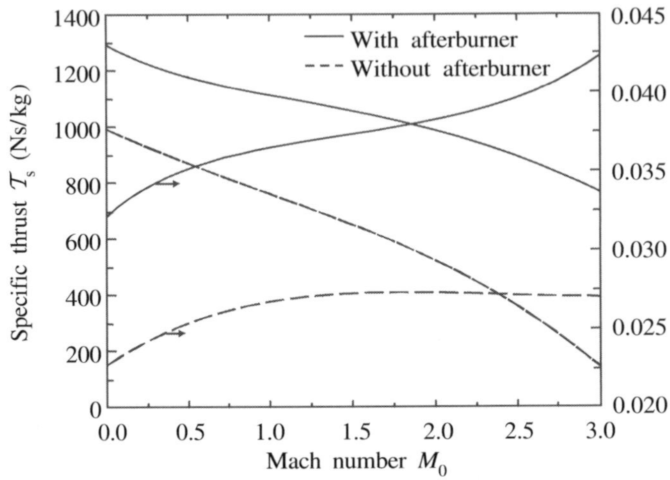

Fig. 4.12. Comparison of T_s and TSFC with π_c with flight Mach number for an ideal turbojet engine and turbojet with afterburner for π_c = 12, T_{t4} = 1400 K and T_{t9} = 2000 K.

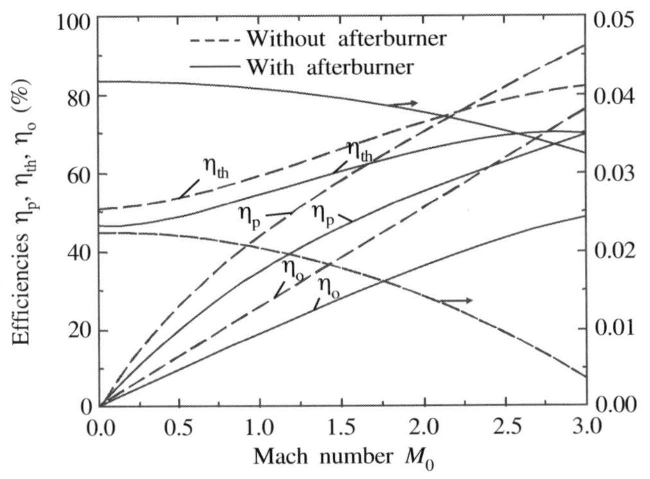

Fig. 4.13. Comparison of η_p, η_{th}, η_o and f with π_c with flight Mach number for an ideal turbojet engine and turbojet with afterburner for π_c = 12, T_{t4} = 1400 and T_{t9} = 2000 K.

and 2200 K, respectively. Estimate the net thrust produced by the engine if it consumes 50 kg/s of air. Determine T, TSFC, η_p, η_{th} and η_o with afterburner and compare them to that of a turbojet engine without afterburner as given in Example 4.2. Assume the heat of combustion to be 43,000 kJ/kg.

Given M_0 = 2.0, Al = 15 km, T_0 = 216.7 K, and P_0 = 11.2 kPa, π_c = 12, $\dot{m}_0$ = 50 kg/s, T_{t4} = 1600 K, T_{t7} = T_{t9} = 2200 K.

Solution The given example is similar to Example 4.2 with an afterburner. Hence, we can use the results of Example 4.2. It will now be illustrated how we can modify the equations such that we can use them to evaluate the performance of a turbojet engine with afterburner by using the results of a simple turbojet. Recall the expression for gas velocity at the nozzle exit is given by

$$V_9^2 = 2C_pT_{t9}\left[1-\left(\frac{P_9}{P_{t9}}\right)^{\frac{\gamma-1}{\gamma}}\right]$$

Assuming that $P_0 = P_e = P_9 = P_{9'}$ (Fig. 4.11), we can relate pressure ratio for dry and afterburning cases as

$$\frac{P_9}{P_{t9}} = \frac{P_{9'}}{P_{t5}} = \left(\frac{P_9}{P_{t9}}\right)_{AB}$$

Now we can easily relate velocity ratio between dry and afterburning cases in terms of total temperature as

$$\left(\frac{V_9}{V_{9'}}\right)^2 = \frac{T_{t9}}{T_{t5}} \Rightarrow \frac{V_9^2}{T_{t9}} = \frac{V_{9'}^2}{T_{t5}}$$

From nozzle relation, by using the values obtained previously we have

$$\frac{V_{9'}^2}{T_{t5}} = 2C_p\left[1-\left(\frac{P_9}{P_{t9}}\right)^{\frac{\gamma-1}{\gamma}}\right] = 2\times1005\left[1-\left(\frac{11.2}{379.7}\right)^{\frac{1.4-1}{1.4}}\right] = 1276.24\ \text{m}^2/\text{s}^2.\text{K}$$

Substituting the values of $V_{9'}^2/T_{t5}$ in the above equation, we have

$$V_9^2 = T_{t9}\cdot\frac{V_{9'}^2}{T_{t5}} = 2200\times1276.24 \Rightarrow V_9 = 1675.6\ \text{m/s}$$

It is given that $T_{t9} = T_{t7} = 2200$ K. By carrying out an energy balance between inlet (0) and exhaust nozzle outlet (9), we have

$$\dot{m}_{f,\,tot}\cdot\Delta H_C = \dot{m}_a\left(C_PT_{t9} - C_PT_{t0}\right)$$

$$\Rightarrow f_{tot} = \frac{C_P\left(T_{t9}-T_{t0}\right)}{\Delta H_c} = \frac{1005(2200-390.1)}{43000\times10^3} = 0.0423$$

Now we can evaluate the total $\mathcal{T}_s$ as

$$\mathcal{T}_s = V_9 - V_0 = 1675.6 - 590.15 = 1085.5\ \text{N/kg·s}$$

So the total thrust becomes

$$\mathcal{T} = \dot{m}_a\times\mathcal{T}_s = 50\times1085.5 = 54.27\ \text{kN}$$

The TSFC is evaluated as

$$\frac{f}{\mathcal{T}_s} = \frac{0.0423}{1085.5}\times10^6 = 38.97\ \text{g/kNs}$$

We can estimate η_p as

$$\eta_p = \frac{2}{\frac{1675.6}{590.15}+1}\times100 = 52.09\%$$

Then η_{th} is determined as

$$\eta_{th} = \frac{1675.6^2 - 590.15^2}{2 \times 0.0423 \times 43000 \times 10^3} \times 100 = 67.6\%$$

Thus, η_o is easily estimated as

$$\eta_o = \eta_p \cdot \eta_{th} = 35.2\%$$

It can be noted that thrust increased only by 68.2% but with about 12.5% increase in fuel consumption. Also, both η_p and η_o decrease with the use of an afterburner.

4.5 Turbofan

The schematic of a turbofan engine is shown in Fig. 4.14a along with the station numbers. The processes involved during compression, heat addition and expansion for an ideal engine are described in two *T-s* diagrams (Fig. 4.14b) for core and fan streams, respectively. The total thrust of an ideal turbofan, contributed due to both primary and secondary streams can be expressed as

$$\mathcal{T} = \dot{m}_c \left(V_9 - V_0\right) + \dot{m}_F \left(V_{19} - V_0\right) \tag{4.66}$$

where $\dot{m}_c$ is the mass flow rate of the primary stream through the core engine, and $\dot{m}_F$ the mass flow rate of the secondary stream through the fan. Complete expansion is assumed to occur in the exhaust nozzles of both primary ($P_9 = P_0$) and secondary ($P_{19} = P_0$) streams. Then $\mathcal{T}_s$ can be expressed as

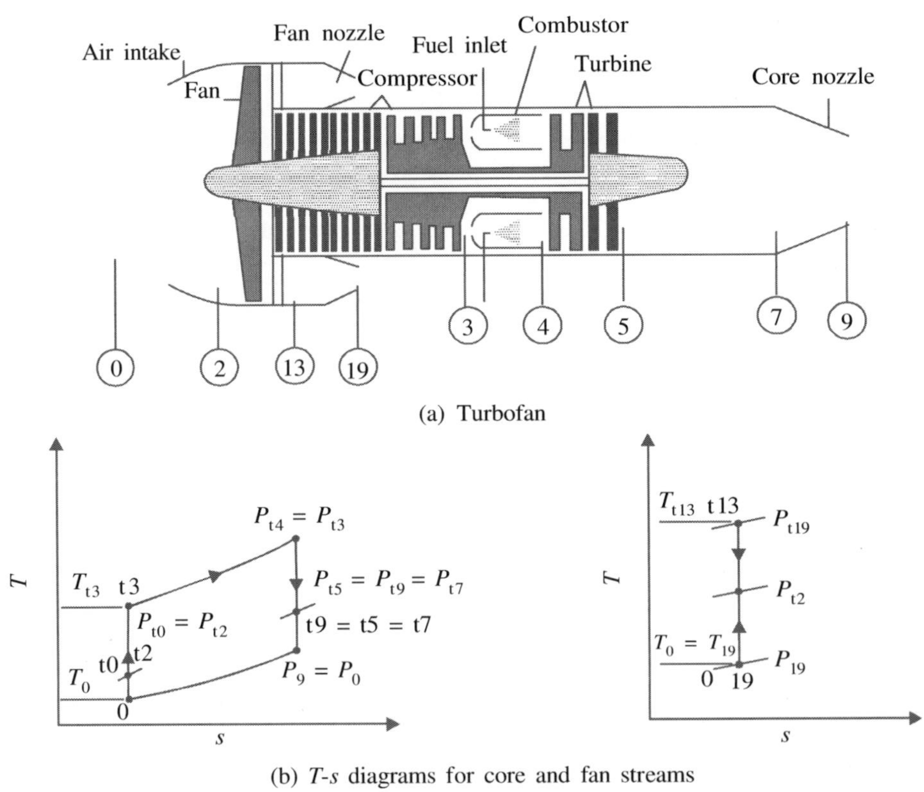

Fig. 4.14. (a) An ideal turbofan engine with station numbers and (b) *T-s* diagrams for core and fan streams.

$$\mathcal{T}_s = \frac{\mathcal{T}}{\dot{m}_c + \dot{m}_F} = \frac{a_0}{1+\alpha}\left[\left(\frac{V_9}{a_0} - M_0\right) + \alpha\left(\frac{V_{19}}{V_0} - M_0\right)\right] \tag{4.67}$$

where $\alpha = \dot{m}_F/\dot{m}_c$ is the bypass ratio. By carrying out similar analysis as was done for turbojet engine, we can derive an expression for $\mathcal{T}_s$ as

$$\mathcal{T}_s = \frac{a_0}{1+\alpha}\left[\left(\sqrt{\frac{2}{\gamma-1}\frac{\tau_\lambda}{\tau_c\tau_r}(\tau_t\tau_c\tau_r - 1)} - M_0\right) + \alpha\left(\sqrt{\frac{2}{\gamma-1}(\tau_F\tau_r - 1)} - M_0\right)\right] \tag{4.68}$$

We know that the power produced by the turbine is used to run a compressor and fan. Let us now strike a power balance for these components in an ideal cycle as

$$\dot{W}_t = \dot{W}_c + \dot{W}_F \tag{4.69}$$

By using First Law of Thermodynamics, we can express Eq. (4.69) in terms of temperature as

$$\dot{m}_c C_p (T_{t4} - T_{t5}) = \dot{m}_c C_p (T_{t3} - T_{t2}) + \dot{m}_F C_p (T_{t13} - T_{t2}) \tag{4.70}$$

By simplifying Eq. (4.70) and expressing it by temperature ratio, we have

$$\tau_t = 1 - \frac{\tau_r}{\tau_\lambda}\left[(\tau_c - 1) + \alpha(\tau_F - 1)\right] \tag{4.71}$$

Thrust Specific Fuel Consumption (TSFC)

The fuel/air ratio f for a turbofan can be expressed in terms of variables by carrying out the energy balance across the combustion chamber similar to that carried out for a turbojet engine as

$$f = \frac{\dot{m}_f}{\dot{m}_c} = \frac{C_p T_0}{\Delta H_C}(\tau_\lambda - \tau_r\tau_c) \tag{4.72}$$

Hence, the expression for TSFC can be written as

$$\text{TSFC} = \frac{\dot{m}_f}{\mathcal{T}} = \frac{f}{\mathcal{T}/\dot{m}_c} = \frac{f}{(1+\alpha)\mathcal{T}_s} \tag{4.73}$$

Propulsive Efficiency

The expression for η_p from the basic definition can be expressed as

$$\eta_p = \frac{2V_0\left[\dot{m}_c(V_9 - V_0) + \dot{m}_F(V_{19} - V_0)\right]}{\left[\dot{m}_c\left(V_9^2 - V_0^2\right) + \dot{m}_F\left(V_{19}^2 - V_0^2\right)\right]} = \frac{2\left[V_9/V_0 - 1 + \alpha(V_{19}/V_0 - 1)\right]}{\left[V_9^2/V_0^2 - 1 + \alpha\left(V_{19}^2/V_0^2 - 1\right)\right]} \tag{4.74}$$

Thermal Efficiency

The expression for η_{th} for the turbofan is the same as that derived for the turbojet engine

$$\eta_{th} = \frac{V_0^2\left[V_9^2/V_0^2 - 1 + \alpha\left(V_{19}^2/V_0^2 - 1\right)\right]}{2f\Delta H_C} = 1 - \frac{1}{\tau_r\tau_c} \tag{4.75}$$

Overall Efficiency

We know that η_o is the product of η_p and η_{th}, which can be expressed as

$$\eta_o = \eta_p \cdot \eta_{th} \tag{4.76}$$

Let us carry out the parametric performance analysis for an ideal turbofan engine by using the above equations. We have five design variables viz., compressor pressure ratio π_c, fan pressure ratio π_F, bypass ratio α, turbine inlet temperature (TIT), and Mach number, M_0 corresponding to representative altitude. Note that this engine has two more design variables at the disposal of the designer as compared to turbojet. Hence, the optimisation of this engine is quite complex and time consuming. However, the parametric studies of the performance of an ideal turbofan can be illustrated using the above equations by varying π_c and α, while keeping other variables constant. For the present analysis, we will employ constant value of TIT (1600 K), Mach number (0.85) and fan pressure ratio ($\pi_F = 2.0$). Three bypass ratios ($\alpha = 0.5$, 2.0 and 5) are considered, where $\alpha = 0.5$ represents supersonic fighter aircraft, whereas $\alpha = 2.0$ and 5.0 represent transonic aircraft and long-range subsonic aircraft, respectively.

The variation of T_s and TSFC with π_c for three values of α at $M_0 = 0.85$ (Fig. 4.15) shows that T_s varies in the lower range of π_c and remains almost constant with respect to pressure ratio in the range of 15 to 25. Interestingly, T_s decreases with increasing α. The TSFC also decreases with increasing π_c and increasing α. We know that the weight of the engine is dependent strongly on mass flow rate of air and, in turn, it substantially affects the carrying capacity of an aircraft. In comparison, higher fuel consumption affects the flight operating cost. Hence, the designer has to select the best compressor pressure ratio for which TSFC would have an optimum value. For this set of data, the optimum bypass ratio turns out to be 12.

The variation of η_p and η_{th} with π_c for all three bypass ratios (Fig. 4.16) shows that η_{th} does not vary with α. However, η_p decreases, initially, with engine bypass ratio and for higher π_c, particularly, in the range of 15 to 25 and subsequently it remains almost constant. In contrast, η_o increases with both π_c and α (not shown in Fig. 4.16).

Another parametric study is undertaken to illustrate the effects of three representative flight Mach numbers on T_s, TSFC, η_p and η_{th}. For all these cases, TIT of 1600 K, $\pi_F = 2$ and $\alpha = 5$ are kept constant. The plot of T_s vs. TSFC (Fig. 4.17) shows that T_s increases with pressure ratio for a particular flight

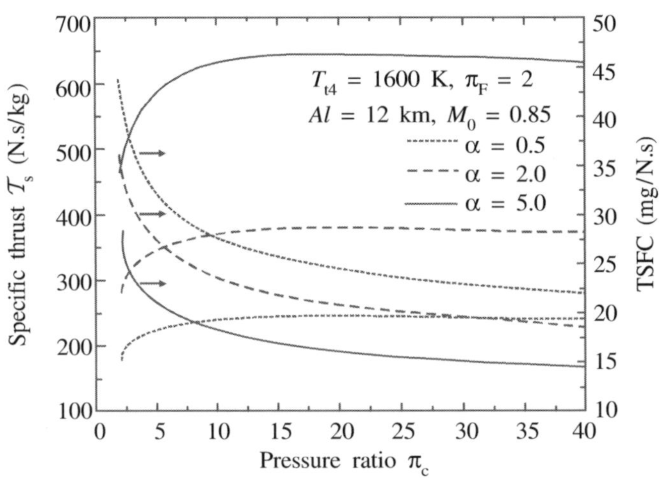

Fig. 4.15. Variation of T_s and TSFC with π_c for flight Mach number = 0.85 for turbofan engine.

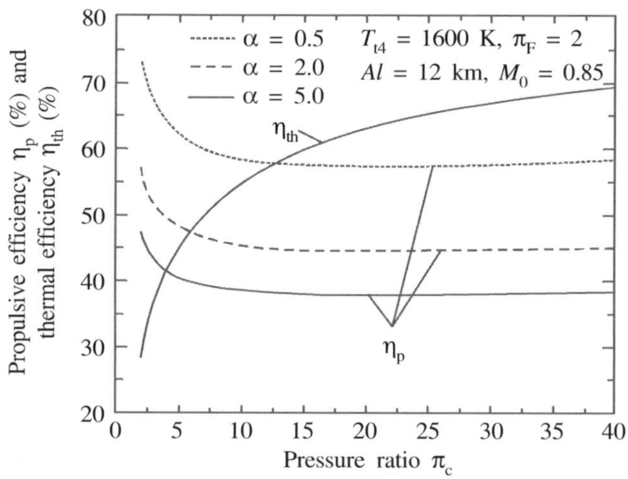

Fig. 4.16. Variation of η_p and η_{th} with π_c for an ideal turbofan engine for flight Mach number = 0.85.

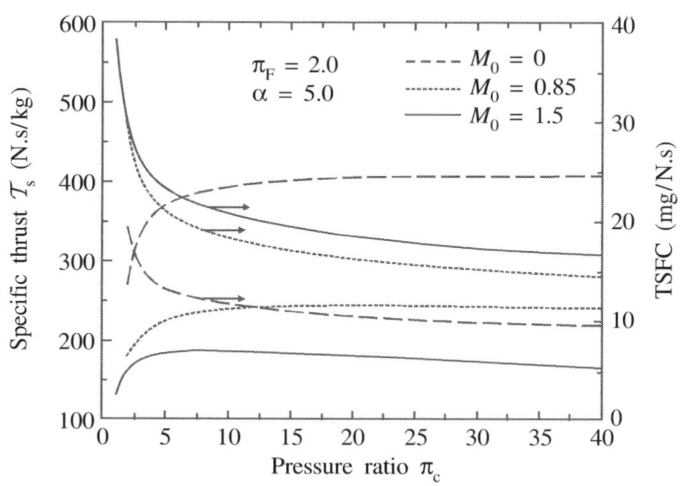

Fig. 4.17. Variation of $\mathcal{T}_s$ and TSFC with π_c for three flight Mach numbers for an ideal turbofan engine.

Mach number and attains a maximum value at a particular π_c. For example, the optimum value for π_c corresponding to maximum $\mathcal{T}_s$ is around 20 for sea level condition (M = 0.0) while for M_0 = 1.5, it becomes around 8. Note that although the same trend of $\mathcal{T}_s$ vs π_c maintains but the actual $\mathcal{T}_s$ for any π_c decreases with flight Mach number. The variations of TSFC with π_c for all three cases indicate that TSFC decreases with pressure ratio across compressor. Of course when flight Mach number increases, it exhibits similar trend with increased values of TSFC. The variation of η_p and η_{th} with π_c for all three flight Mach numbers (Fig. 4.18) shows that η_p decreases with π_c while η_{th} increases with π_c as expected for all the cases. Interestingly, η_p increases with flight Mach number. Similarly, η_{th} gets enhanced with increase in flight Mach number. Hence the overall efficiency η_0 also gets enhanced with flight Mach number M_0.

Now let us vary both π_c and π_F with constant α = 5.0. The variation of $\mathcal{T}_s$ and TSFC with π_c for three values of π_F (= 1.2, 2.0 and 4.0) at flight Mach number of 0.85 (Fig. 4.19) shows that $\mathcal{T}_s$

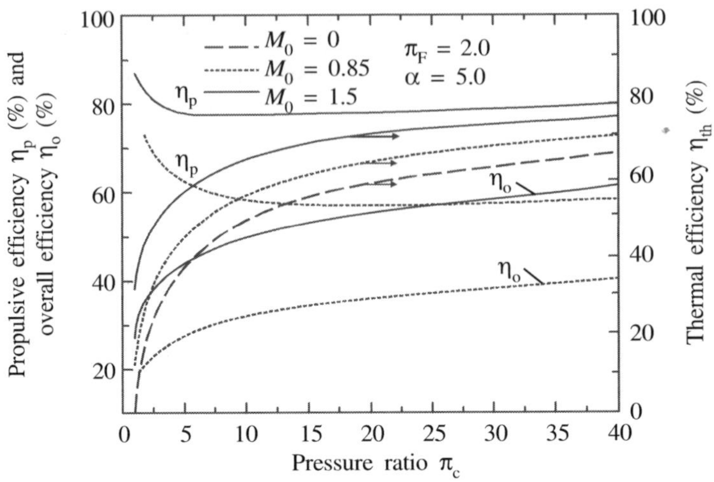

Fig. 4.18. Variation of η_p and η_{th} with π_c for an ideal turbofan engine at three flight Mach numbers for an ideal turbofan engine.

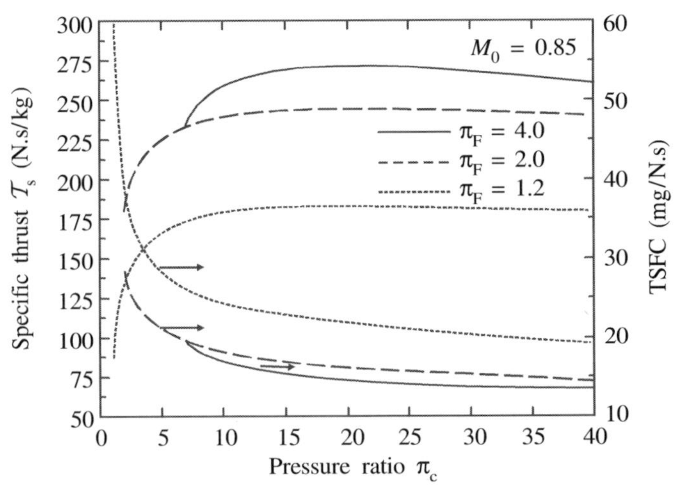

Fig. 4.19. Variation of $\mathcal{T}_s$ and TSFC with π_c for three fan pressure ratios for an ideal turbofan engine.

varies in the lower range of π_c and remains almost constant with respect to pressure ratio in range of 15 to 25. Interestingly, $\mathcal{T}_s$ decreases with increasing π_F. The TSFC also decreases with increasing π_c and increasing π_F. The designer has to select the best optimum π_F for which TSFC would be minimum. For this set of data, the optimum π_F turns out to be 4. The variation of η_p with π_c for all three π_F (Fig. 4.20) shows that η_p increases with π_F until a value of 3.5 (not shown in Fig. 4.20) and then decreases. Hence, there is a value of π_F for which η_p would attain a maximum value. The η_p varies marginally with π_c beyond the value of 15. In contrast, η_{th} increases with both π_c but remains invariant with π_F (Fig. 4.20). Hence, the overall efficiency η_o increases with both π_c and π_F (not shown in Fig. 4.20). Note that, there will be an optimum π_F corresponding to maximum value of η_o.

The qualitative results of the parametric studies carried out for a turbofan engine are summarised as follows:

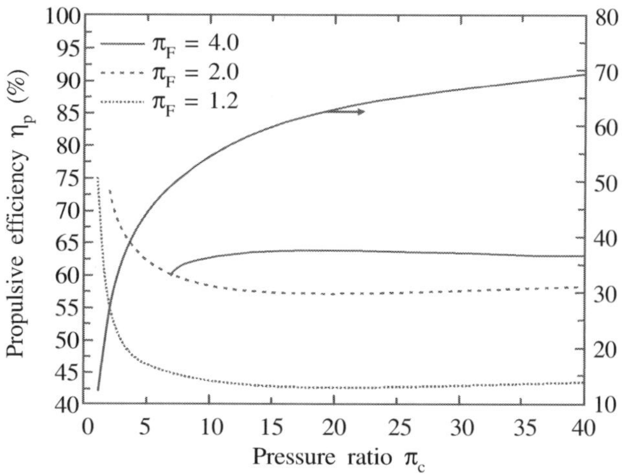

Fig. 4.20. Variation of η_p and η_{th} with π_c for three fan pressure ratios for an ideal turbofan engine.

- $\mathcal{T}_s$ increases while TSFC decreases with increase in π_c for a particular value of π_F and α.
- TSFC decreases with increasing α but at the expense of significant reduction in $\mathcal{T}_s$.
- The η_p increases with increase in α.
- For constant flight Mach number and α, the optimum pressure ratio is around 3.5.
- The optimum π_F decreases with increase in α.

4.5.1 Turbofan Configuration

We have learnt from the aforementioned parametric studies that the variation of α and π_F significantly affects the performance parameters of engines. These parameters of a turbofan engine have significant effect on the mechanical design of the engine as they dictate the overall diameter, rotational speed of the engine. In order to augment thrust level, high overall pressure ratio is used across the engine in modern days. As a result, throttling of the engine becomes difficult since flow instability, known as compressor *surge*, will be encountered across the high pressure compressor. This surging is likely to occur when some of the compressor aerofoils stall, causing a violent change in the direction of the airflow. The surging phenomena will be discussed in detail in Chapter 7. In order to avoid this problem, several configurations of turbofan engines have evolved over the past many years. For the same airflow, bypass ratio, fan pressure ratio, overall pressure ratio and HP turbine rotor inlet temperature, the choice of turbofan configuration has little impact upon the design point performance as long as overall component performance is maintained. However, off-design performance and stability get affected significantly by the engine configuration. Early development of turbofan engine from the existing turbojet engine gave rise to a new configuration known as *Aft-fan* engine, in which the fan is mounted directly onto the turbine. When the hot gas is expanded through the turbine, the fan blades being a radial extension of the turbine blades get power directly from the turbine. One of the problems with the *Aft-fan* configuration is hot gas leakage from the turbine to the fan. Another problem is the design and manufacturing of blades for the turbine fan unit. This configuration has not been used for quite some time. Around 20 years ago, *Aft-fan* configuration was exploited in the General Electric GE-36 UDF (propfan) demonstrator. However, we will restrict our discussion to five common turbofan configurations viz., single spool, two spool, boosted two spool, three spool, and geared fan.

Single Spool Turbofan

The single shaft (spool) turbofan is probably the simplest configuration, comprising a fan and high pressure compressor driven by a single turbine unit; all on the same shaft. The SNECMA M53 turbofan engine used in Mirage fighter aircraft is an example of a single spool turbofan. Although it is the simplest configuration, a variable area mixer is employed in the M53 engine to facilitate part-throttle operation.

Two Spool Turbofan

The schematic of a typical two spool turbofan engine is shown in Fig. 4.21(a). In this configuration, the fan is run by a low pressure (LP) turbine with the help of low speed shaft while the high pressure (HP) compressor is driven by a HP turbine by a separate shaft. But the problem with this kind of system is that it can cause flow instability, which may lead to surging, particularly, during off-design operating conditions. Hence, the compressor mounted on the HP rotor contributes marginally as it is operated at lower rpm. Note that it can only be used for medium level compressor ratio engines. However, it finds applications in some practical engines such as BR710, CFE738 (General Electric), PW600 (Pratt Whitney Canada) etc.

Boosted Two Spool Turbofan

If an engine needs to be operated at higher overall pressure, the simple two spool engine would not be suitable. However, the overall pressure ratios can be enhanced further by either raising the HP compressor pressure ratio or adding a low pressure (LP) compressor between the fan and the HP compressor. Note that the overall pressure ratio of the engine cycle can be enhanced even beyond 40:1 in this configuration (Fig. 4.21(b)) by carrying out compression process in two stages. Most of the large turbofans, e.g. General Electric (CF6, GE90 and GEnx), and Pratt Whitney (JT9D and PW4000) have adopted this configuration in which an IP compressor is mounted on the LP shaft and

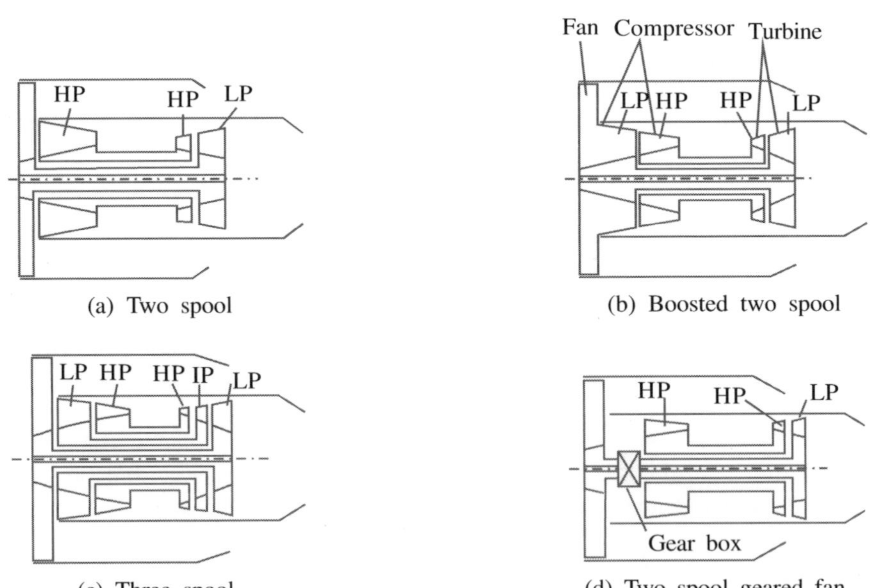

Fig. 4.21. Types of turbofan configurations (LP: Low pressure; HP: High pressure; IP: Intermediate pressure).

driven, like the fan, by the LP turbine. Note that the revolutions per minute (rpm) of the LP shaft is dictated by the tip speed and diameter of the fan.

Three Spool

The three spool turbofan configuration (Fig. 4.21(c)) is one of the appropriate concepts in which IP compressor is mounted on a separate (IP) shaft, running concentrically with the LP and HP shafts, and is driven by a separate IP turbine. As a result, the IP compressor can rotate faster than the fan and increases its mean tip speed, thereby reducing the number of IP stages required for a given IP compressor pressure ratio. The HP compressor pressure ratio becomes small resulting in a shorter and lighter compressor, leading to a more rigid engine. Hence, there is less distortion in the flow of casing under loading condition due to acceleration/deceleration during flight. Thus, there will be less blade tip rubbing and, as a result, less deterioration in the performance of each component and specific fuel consumption. However, three spool engines are quite difficult to manufacture and maintain. These configurations are being used in several engines, e.g. Rolls-Royce (RB211), Russian military Kuznetsov (NK-321), Turbo-Union (RB199) etc.

Geared Fan

For high bypass ratio engine, the mean radius ratio of the fan and LP turbine are likely to increase. As a result, the fan has to rotate at slower speed to attain optimum blade speed. So a gear box has to be employed to reduce the speed of the LP turbine. A typical configuration of the geared fan is shown in Fig. 4.21(d). As the power requirement of a fan is quite high (around 75 MW), it is quite challenging to design a lighter gear box. However, planetary gear system with a suitable gear ratio has been used between the LP turbine and the fan such that they can operate efficiently at their optimum speed. This kind of configuration has been used recently in the advanced technology fan integrator (ATFI) demonstrator engine known as *Geared Turbofan* by Pratt Whitney. Also, this kind of configuration can be adopted easily for ultra high bypass ratio (UHB) engines with variable pitch fans.

Example 4.4 An ideal turbofan engine is flying at an altitude of 10 km and flight Mach number $M_0 = 0.85$. The pressure ratio across the compressor and fan is 10 and 2.0, respectively. The exit temperature at the combustor is 1400 K. The bypass ratio of the turbofan is 5. Determine $\mathcal{T}_s$, TSFC, η_P, η_{th} and η_o. Determine $\mathcal{T}_s$ and TSFC under static condition at sea level. Take ΔH_c = 43,000 kJ/kg. Comment on your results.

Given Al = 10 km, T_0 = 223.3 K, P_0 = 26.5 kPa, $\pi_c = 10$, T_{t4} = 1400 K, $M_0 = 0.85$, $\pi_F = 2.0$ and $\alpha = 5.0$.

Solution We know that bypass ratio α is given by

$$\alpha = \frac{\dot{m}_F}{\dot{m}_c}$$

where $\dot{m}_c$ and $\dot{m}_F$ are the core and fan mass flow, respectively. Let us determine flight velocity V_0 as

$$V_0 = M_0\sqrt{\gamma R T_0} = 0.85 \times \sqrt{1.4 \times 287 \times 223.3} = 254.6 \text{ m/s}$$

From compressible flow relations we know that

$$\frac{T_{t0}}{T_0} = 1 + \frac{\gamma - 1}{2} M_0^2 \Rightarrow T_{t0} = \left[1 + \frac{(1.4 - 1)}{2} M_0^2\right] \times T_0 = 255.6 \text{ K}$$

Then

$$T_{t0} = T_{t1} = T_{t2} = 255.6 \text{ K}$$

Again from isentropic relation, we have

$$\frac{P_{t0}}{P_0} = \left(\frac{T_{t0}}{T_0}\right)^{\frac{\gamma}{\gamma-1}} = 1.6 \Rightarrow P_{t0} = 42.5 \text{ kPa}$$

For fan stream, we can have

$$T_{t2} = 255.6 \text{ K}; \quad P_{t2} = P_{t0} = 42.5 \text{ kPa}$$

The pressure ratio across the fan π_F becomes

$$\pi_F = 2.0 = \frac{P_{t13}}{P_{t2}} \Rightarrow P_{t13} = 85 \text{ kPa}$$

Now we can estimate temperature at station (13) as

$$\frac{T_{t19}}{T_{t2}} = \frac{T_{t13}}{T_{t2}} = \left(\frac{P_{t13}}{P_{t2}}\right)^{\frac{\gamma-1}{\gamma}} \Rightarrow T_{t13} = T_{t19} = 311.64 \text{ K}$$

As $P_{13} = P_9 = P_{19} = P_0$, the fan nozzle exit velocity V_{19} can be easily estimated as

$$V_{19}^2 = 2C_p\left(T_{t19} - T_{19}\right) = 2C_pT_{t19}\left[1 - \left(\frac{P_{19}}{P_{t19}}\right)^{\frac{\gamma-1}{\gamma}}\right] \Rightarrow V_{19} = 421.4 \text{ m/s}$$

For core stream

The overall pressure ratio across both compressor and fan $\pi_c = 10 \times 2 = 20$. So the total temperature and pressure at the exit of compressor is estimated as

$$T_{t3} = T_{t2}\left(\pi_c\right)^{\frac{\gamma-1}{\gamma}} = 255.6 \times (20)^{0.286} = 602.1 \text{ K}; \; P_{t3} = 20 \times 42.5 = 850 \text{ kPa} = P_{t4}$$

By energy balance across the combustor, we can estimate fuel-air ratio f as

$$f = \frac{C_p\left(T_{t4} - T_{t3}\right)}{\Delta H_C} = \frac{1005(1400 - 602.1)}{43000 \times 10^3} = 0.0186$$

By carrying out a power balance between the compressor and turbine, we can estimate the temperature at the exit of turbine T_{t5} as

$$T_{t5} = T_{t4} - (T_{t3} - T_{t2}) - \alpha(T_{t13} - T_{t2}) = 773.3 \text{ K}$$

By using the isentropic relation, we can estimate the total pressure at the exit of turbine as

$$P_{t5} = P_{t4}\left(\frac{T_{t5}}{T_{t4}}\right)^{\frac{\gamma}{\gamma-1}} = 850\left(\frac{773.3}{1400}\right)^{3.5} = 106.46 \text{ kPa}$$

In order to estimate the nozzle exit velocity, we can assume $T_{t9} = T_{t5}$ and $P_{t9} = P_{t5}$

$$V_9 = \sqrt{2C_p\left(T_{t9} - T_9\right)} = \sqrt{2C_pT_{t9}\left[1 - \left(\frac{P_9}{P_{t5}}\right)^{\frac{\gamma-1}{\gamma}}\right]} = 714 \text{ m/s}$$

The $\mathcal{T}_s$ can be estimated as

$$\mathcal{T}_s = V_9 + \alpha V_{19} - (1 + \alpha)V_0 = 714 + 5 \times 421.4 - (1 + 5)\ 254.6 = 1293.4 \text{ N.s/kg}$$

Then TSFC can be determined easily by

$$\text{TSFC} = \frac{f}{\mathcal{T}_s} = \frac{0.0186}{1293.4} = 1.44 \times 10^{-5} \text{ kg/N.s}$$

The η_p from the basic definition can be estimated as

$$\eta_p = \frac{2\left[V_9/V_0 - 1 + \alpha\left(V_{19}/V_0 - 1\right)\right]}{\left[V_9^2/V_0^2 - 1 + \alpha\left(V_{19}^2/V_0^2 - 1\right)\right]} = \frac{2\left[\frac{714}{254.6} - 1 + 5\left(\frac{421.4}{254.6} - 1\right)\right]}{\left[\left(\frac{714}{254.6}\right)^2 - 1 + 5\left\{\left(\frac{421.4}{254.6}\right)^2 - 1\right\}\right]} \times 100 = 65.32\%$$

Similarly, η_{th} from the basic definition can be estimated as

$$\eta_{th} = \frac{V_0^2\left[V_9^2/V_0^2 - 1 + \alpha\left(V_{19}^2/V_0^2 - 1\right)\right]}{2 f \Delta H_C} = \frac{254.6^2\left[\left(\frac{714}{254.6}\right)^2 - 1 + 5\left\{\left(\frac{421.4}{254.6}\right)^2 - 1\right\}\right]}{2 \times 0.0186 \times 43000 \times 10^3} \times 100 = 63.06\%$$

Then η_o for the turbofan can be determined as

$$\eta_o = \eta_{th} \cdot \eta_p = 41.19\%$$

4.6 Turboprop

Figure 4.22a shows the schematic of a turboprop engine along with the station numbers. The processes of compression, heat addition and expansion for an ideal turboprop engine are described in a *T-s* diagram (Fig. 4.22(b)). Now, let us write down an expression for the total thrust produced by an ideal turboprop, contributed by the propeller and exhaust nozzle of the core engine as

$$\mathcal{T} = \mathcal{T}_{prop} + \mathcal{T}_c \tag{4.77}$$

However, it is customary to use the work supplied to the vehicle in contrast to the thrust used in turbojet and turbofan engines. Hence, let us define a dimensionless work output coefficient C as

$$C = \frac{\text{Power production per } \dot{m}_c}{\text{Total inlet static enthalpy } h_0 \text{ per } \dot{m}_c} \tag{4.78}$$

where $\dot{m}_c$ is the mass flow rate of air through the core engine. Then C for the core stream and propeller would be expressed as

$$C_c = \frac{\mathcal{T}_c V_0}{\dot{m}_c C_p T_0}; \quad C_{prop} = \frac{\eta_{prop} \dot{W}_{prop}}{\dot{m}_c C_p T_0} \tag{4.79}$$

where η_{prop} is the efficiency of the propeller. The work output coefficient for the whole turboprop engine C_{tot} would be

$$C_{tot} = C_{prop} + C_c \tag{4.80}$$

From the definition of work output coefficient, we can rewrite the expression for the total thrust as

$$\mathcal{T} = \mathcal{T}_{prop} + \mathcal{T}_c = \frac{C_{tot} \dot{m}_0 C_p T_0}{V_0} \tag{4.81}$$

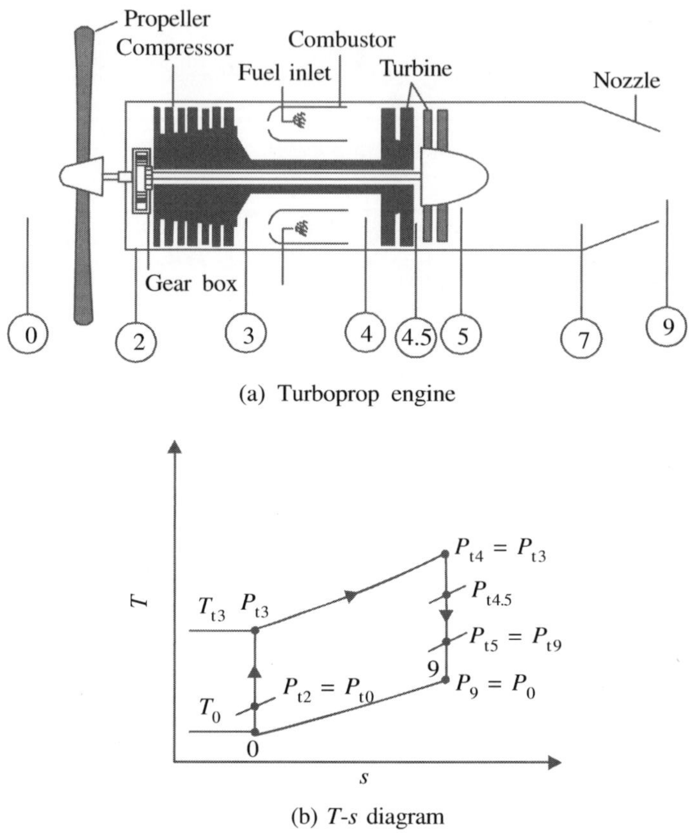

Fig. 4.22. (a) Schematic of a turboprop engine with station numbers and (b) *T-s* diagram.

Specific Thrust

Quite often, the core stream of the turboprop engine is fully expanded. Hence, the pressure term does not contribute any thrust. Then the expression for $\mathcal{T}_s$ developed by the core flow would be

$$\mathcal{T}_s = \frac{\mathcal{T}}{\dot{m}_c} = a_0\left(\frac{V_9}{a_0} - M_0\right) = a_0\left[\sqrt{\frac{2}{\gamma-1}\frac{\tau_\lambda}{\tau_c\tau_r}\left(\tau_t\tau_c\tau_r - 1\right)} - M_0\right] \tag{4.82}$$

This expression is the same as that for a turbojet engine. We can express the work output coefficient for the core engine C_c as

$$C_c = \frac{\mathcal{T}_c V_0}{\dot{m}_c C_p T_0} = \frac{V_0 a_0}{C_p T_0}\left(\frac{V_9}{a_0} - M_0\right) = (\gamma-1)M_0\left(\frac{V_9}{a_0} - M_0\right) \tag{4.83}$$

Specific Power

The specific power (SP) is another important performance parameter used for turboprop engines and it is given by

$$\text{SP} = \frac{\dot{W}_{\text{tot}}}{\dot{m}_c h_0} = C_{\text{tot}} C_p T_0 \text{ where } C_{\text{tot}} = C_{\text{prop}} + C_c \tag{4.84}$$

In order to find an expression for C_{prop}, let us strike a power balance across high pressure turbine and propeller as

$$\dot{W}_{\text{prop}} = \dot{m}_{4.5} C_p \left(T_{t4.5} - T_{t5}\right) \tag{4.85}$$

But we know by definition that C_{prop} (Eq. (4.79)) is given by

$$C_{\text{prop}} = \frac{\dot{W}_{\text{prop}}}{\dot{m}_c C_p T_0} = \eta_{\text{prop}} \tau_\lambda \tau_{tH} \left(1 - \tau_{tL}\right) \tag{4.86}$$

where τ_{tH} is the temperature ratio across the high pressure turbine $T_{t5}/T_{t4.5}$ and τ_{tL} is the temperature ratio across the low pressure turbine $T_{t4.5}/T_{t4}$. Similar to other engines, the expression for fuel-air ratio f can be obtained as

$$f = \frac{\dot{m}_f}{\dot{m}_c} = \frac{C_p T_0}{\Delta H_C} \left(\tau_b \tau_r \tau_c - \tau_r \tau_c\right) \tag{4.87}$$

Thrust Specific Fuel Consumption (TSFC)

The expression for TSFC can be written as

$$\text{TSFC} = \frac{\dot{m}_f}{\mathcal{T}} = \frac{f}{\mathcal{T}/\dot{m}_c} \tag{4.88}$$

Power Specific Fuel Consumption (PSFC)

The power specific fuel consumption is also used for turboprop engine and is given by

$$\text{PSFC} = \frac{\dot{m}_f}{\dot{W}_{\text{tot}}} = \frac{f}{C_{\text{tot}} C_p T_0} \tag{4.89}$$

Thermal Efficiency

The expression for η_{th} of the turboprop engine is same as that of turbojet engine

$$\eta_{th} = 1 - \frac{1}{\tau_r \tau_c} \tag{4.90}$$

Overall Efficiency

The η_o for a turboprop engine can be simply expressed as

$$\eta_o = \frac{\dot{W}_{\text{tot}}}{\dot{m}_f \Delta H_C} = \frac{C_{\text{tot}}}{\left(\tau_\lambda - \tau_r \tau_c\right)} \tag{4.91}$$

Propulsive Efficiency

By knowing η_{th} and η_o, we can easily determine η_p as

$$\eta_p = \frac{\eta_o}{\eta_{th}} \tag{4.92}$$

Let us now carry out a parametric analysis for an ideal turboprop engine to be employed in a smaller passenger aircraft capable of carrying 10-40 passengers, operated at flight Mach number of 0.6 at around 8 km altitude ($P_0 = 35.65$ kPa, $T_0 = 236.3$ K). The propeller efficiency is assumed to be around 80%, TIT (T_{t4}) is 1400 K, heat of combustion ΔH_C is 43,000 kJ/kg. In this study, we will vary the compressor pressure ratio π_c and turbine temperature ratio τ_t. The variation of $\mathcal{T}_s$ and TSFC with compressor pressure ratio (Fig. 4.23) shows that $\mathcal{T}_s$ seems to be quite higher as compared to that for turbojet and turbofan engines. This is a little misleading since $\mathcal{T}_s$ shown in Fig. 4.23 is obtained by dividing the total thrust by mass flow rate through the core engine ($\dot{m}_c$) only without considering the mass flow through the propeller. With increase in τ_t, $\mathcal{T}_s$ increases while TSFC decreases. Interestingly, $\mathcal{T}_s$ attains a maximum value at a particular π_c for a specific value of τ_t. In contrast, TSFC attains a minimum value at different π_c for same value of τ_t. Hence, the optimum value for π_c corresponding to minimum TSFC for $\tau_t = 0.8$ turns out to be 15. The π_c greater than 15 is desirable to achieve lower fuel consumption for the present set of operating conditions.

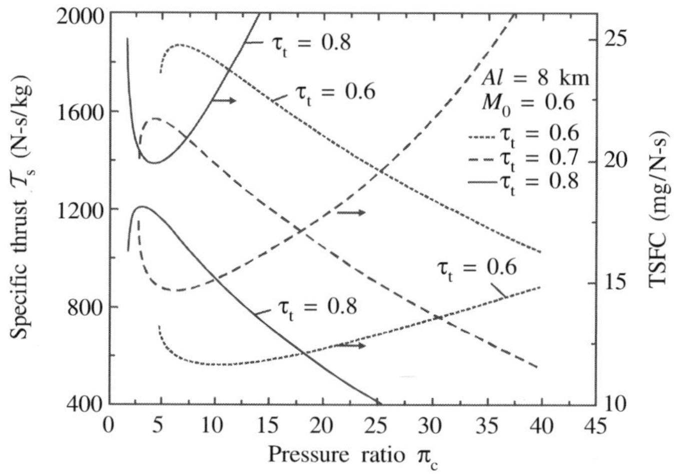

Fig. 4.23. Variation of $\mathcal{T}_s$ and TSFC with π_c for turbine temperature ratio $\tau_t = 0.6$, 0.7 and 0.8 for an ideal turboprop engine.

The variation of η_p and η_o with π_c (2 to 40) for τ_t in the range of 0.6 to 0.8 (Fig. 4.24) shows that η_p decreases from a very high value with increase in π_c in a manner similar to that of other gas turbine engines, while η_o attains a maximum value at certain pressure ratio. Hence, optimum value of τ_t corresponding to maximum η_p and η_o is around 0.6 for which $\pi_c > 15$ must be used.

Example 4.5 A turboprop engine is flying at an altitude of 7 km ($T_0 = 242$ K, $P_0 = 41.1$ kPa) with flight Mach number of 0.5. In the gas generator, 50 kg/s of air is compressed with pressure ratio of 12. The turbine inlet temperature is 1400 K. The temperature ratio across the turbine is 0.5. Determine the thrust developed by the core engine and propeller and TSFC if the efficiency of the propeller is 0.9. What will be its overall efficiency? Take $\Delta H_c = 43,000$ kJ/kg.

Given $Al = 7$ km, $T_0 = 242$ K, $P_0 = 41.1$ kPa, $T_{t4} = 1400$ K, $\pi_c = 12$, $\tau_t = 0.5$, $\eta_{prop} = 0.9$, $M_0 = 0.5$, $\dot{m}_0 =$ 50 kg/s, $V_0 = 0.5 \times \sqrt{1.4 \times 287 \times 242} = 155.9$ m/s, $C_P = 1.005$ kJ/kgK, $\Delta H_C = 43,000$ kJ/kg.

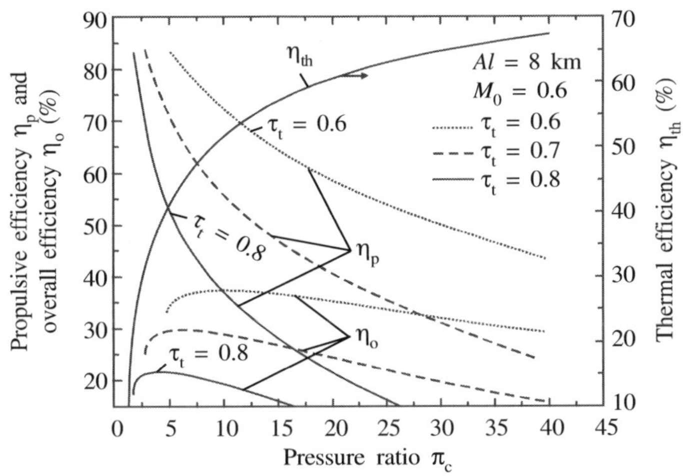

Fig. 4.24. Variation of (a) η_p with π_c and (b) η_o with π_c for an ideal turboprop engine.

Solution The processes involved during the operation of a turboprop engine are shown in Fig. 4.22 with the station numbers. We know that $\mathcal{T}_{prop}$ is expressed as

$$\mathcal{T}_{prop} = \frac{C_{prop}\dot{m}_c C_P T_0}{V_0} = \frac{\eta_{prop}\dot{W}_{prop}}{V_0} \text{ as } C_{prop} = \frac{\eta_{prop}\dot{W}_{prop}}{\dot{m}_c C_P T_0}$$

But we know that $\dot{W}_{prop}$ is given as

$$\dot{W}_{prop} = \dot{m}_c C_P (T_{t4.5} - T_{t5})$$

In order to evaluate $T_{t4.5}$ we need to carry out an energy balance across the low pressure (free) turbine as

$$C_P(T_{t3} - T_{t2}) = C_P(T_{t4} - T_{t4.5}) \Rightarrow T_{t4.5} = T_{t4} - (T_{t3} - T_{t2})$$

But the total temperature at stations (2) and (3) can be easily obtained by invoking the isentropic relationship and pressure ratio across the compressor as

$$T_{t2} = \frac{T_{t2}}{T_0}T_0 = \frac{T_{t0}}{T_0}T_0 = \left(1 + \frac{\gamma - 1}{2}M_0^2\right)T_0 \Rightarrow T_{t2} = \left(1 + 0.2 \times 0.5^2\right)242 = 254.1\text{ K}$$

So the value of T_{t3} can be estimated easily as

$$T_{t3} = \frac{T_{t3}}{T_{t2}}T_{t2} = \left(\frac{P_{t3}}{P_{t2}}\right)^{\frac{\gamma-1}{\gamma}} T_{t2} = 12^{0.286} \times 254.1 = 517.2\text{ K}$$

Then we can estimate $T_{t4.5}$ as

$$T_{t4.5} = T_{t4} - (T_{t3} - T_{t2}) = 1400 - (517.2 - 254.1) = 1137\text{ K}$$

Let us now evaluate the thrust produced by the propeller $\mathcal{T}_{prop}$ as

$$\mathcal{T}_{prop} = \frac{\eta_{prop}\dot{m}_c C_P (T_{t4.5} - T_{t5})}{V_0} = \frac{0.9 \times 50 \times 1.005(1137 - 700)}{155.9} = 126.77\text{ kN}$$

Then the thrust produced by the core engine $\mathcal{T}_c$ is estimated as

$$\mathcal{T}_c = \dot{m}_0 (V_9 - V_0)$$

But the nozzle exit velocity V_9 is estimated as

$$V_9 = \sqrt{2\times C_p (T_{t9} - T_9)} = \sqrt{2\times C_p T_{t5}\left[1-\left(\frac{P_9}{P_{t5}}\right)^{\frac{\gamma-1}{\gamma}}\right]}$$

As we know that the total pressure ratio P_9/P_{t5} can be easily estimated as

$$\frac{P_9}{P_{t5}} = \frac{P_9}{P_0}\cdot\frac{P_0}{P_{t2}}\cdot\frac{P_{t2}}{P_{t3}}\cdot\frac{P_{t3}}{P_{t4}}\cdot\frac{P_{t4}}{P_{t5}} = 1\cdot\frac{1}{\left(1+\frac{\gamma-1}{2}M_0^2\right)^{\frac{\gamma}{\gamma-1}}}\cdot\frac{1}{12}\cdot 1\cdot\frac{1}{\tau_t^{\frac{\gamma}{\gamma-1}}} = 1\cdot\frac{1}{1.186}\cdot\frac{1}{12}\cdot\frac{1}{0.0884} = 0.8$$

Substituting all the values, we can determine V_9 as

$$V_9 = \sqrt{2\times C_p T_{t5}\left[1-\left(\frac{P_9}{P_{t5}}\right)^{\frac{\gamma-1}{\gamma}}\right]} = \sqrt{2\times 1005\times 700\left(1-0.8^{0.286}\right)} = 294.94 \text{ m/s}$$

Thus, $\mathcal{T}_c$ can be evaluated as

$$\mathcal{T}_c = \dot{m}_0 (V_9 - V_0) = 50(294.94 - 155.9) = 6.95 \text{ kN}$$

Then the total thrust can be estimated as

$$\mathcal{T} = \mathcal{T}_{prop} + \mathcal{T}_c = 126.77 + 6.95 = 133.7 \text{ kN}$$

Combustor

Let us evaluate the amount of fuel consumed in the combustor by striking an energy balance across the combustor

$$\dot{m}_f \Delta H_c = \dot{m}_0 C_P (T_{t4} - T_{t3}) \Rightarrow \dot{m}_f = \frac{50\times 1.005(1400-517.2)}{43000} = 1.032 \text{ kg/s}$$

Then TSFC can be determined as

$$\text{TSFC} = \frac{\dot{m}_f}{\mathcal{T}} = \frac{1.032}{133.7} = 7.72 \text{ gm/kN}$$

The η_o can be determined as

$$\eta_o = \frac{V_0 \mathcal{T}}{\dot{m}_f \times \Delta H_c} = \frac{155.9\times 133700}{1.032\times 43000\times 10^3}\times 100 = 46.97\%$$

It can be noted that $\mathcal{T}_{prop}$ is quite high as compared to $\mathcal{T}_c$. It must also be kept in mind that TSFC is quite small as compared to that of the other jet engines.

4.7 Chapter Summary

This chapter presents the basics of parametric cycle analysis for all airbreathing engines operated under ideal conditions. All the components are assumed to be operated without incurring any losses. The performance of various types of airbreathing engines namely ramjet, turbojet, turbofan and turboprop can be easily carried out using this ideal cycle analysis. For this purpose, the International Standard is used to identify station numbers of each components. The processes in each engine configuration are described in *P-v* and *T-s* diagrams. Starting from the basic principles, the equation for major performance parameters such as $\mathcal{T}_s$, TSFC, η_p and η_{th} are derived for different types of engines. In the process, the reader is expected to appreciate the dependence of parameters related to each component on the overall performance of the whole engine. In certain cases, the methods of analytical optimisation for either maximum $\mathcal{T}_s$ or minimum TSFC are demonstrated, which are very essential for design-point analysis. The parametric studies are carried out for each engine configuration by varying operating parameters and component characteristics. The advantage of this kind of study is that it makes the reader appreciate how the performance of an engine can get influenced by various parameters. Although this parametric study is meant for an ideal engine, but the trends will definitely hold for real engines. In addition, different types of engines are compared with each other highlighting their relative advantages and disadvantages while emphasising the operational limitation of each type of engine. An attempt is made to discuss the specification of practical engines. In some case, new concepts of engine design are introduced. Thrust augmentation methods are also discussed with emphasis on afterburning. It is also shown that an afterburner in a turbojet enhances the thrust, of course, with significant increase in TSFC. Hence, it can be used for short duration, particularly, when sudden higher thrust is required. For this reason, an afterburner system is fitted with fighter aircraft. The turbofan engine is discussed extensively. A turbofan engine is shown to operate with lower TSFC as compared to turbojet engine. With bypass ratio, TSFC of a turbofan engine decreases but its engine size and weight increase for a given thrust level. In addition, η_o of the turbofan engine increases with bypass ratio. For a relatively lower flight speed, turboprop engine is preferred over other engine configurations as it has lower TSFC and higher η_o. For each engine configuration, one example is provided with an objective of highlighting some of the above observations so that the reader can appreciate the intricate processes involved during the operation of engines. In the next chapter, we will learn about the non-ideal components and their inherent losses. The methods of taking account of losses in each component will be discussed, which will be employed subsequently to carry out design-point real cycle analysis.

Nomenclature

C	Work output coefficient
C_c	Work output coefficient for care stream
C_p	Specific heat of constant pressure
C_{tot}	Work output coefficient for whole turboprop engine
F	Fuel/Air ratio
M	Mach number
$\dot{m}_a$	Mass flow rate of air
$\dot{m}_f$	Mass flow rate of fuel
$\dot{m}_F$	Mass flow rate through fan
$\dot{m}_c$	Mass flow rate through core engine
P	Pressure
R	Gas constant
T	Temperature
V	Velocity
$\dot{W}_c$	Compressor power input
$\dot{W}_t$	Turbine power output

$\dot{W}_{prop}$	Propeller power output
$\dot{W}_{tot}$	Total power output

Greek Symbols

α	Bypass ratio
$\mathcal{T}$	Thrust
$\mathcal{T}_s$	Specific thrust
π_c	Total pressure ratio
γ	Specific heat ratio
η_o	Overall efficiency
η_{prop}	Propeller efficiency
η_p	Propulsive efficiency
η_{th}	Thermal efficiency

Subscripts/Superscripts

a	Air
c	Compressor, cross stream
f	Fuel
F	Fan
o	Overall
p	Pressure, propulsive
prop	Propeller
r	Ratio
s	Specific
t	Total, turbine
th	Thermal
tot	Turboprop engine

Abbreviations

CV	Control volume
SP	Specific power
TSFC	Thrust specific fuel consumption

References and Suggested Readings

1. Mattingly, J.D., *Elements of Gas Turbine Propulsion*, McGraw-Hill Inc., 1996.
2. Oates, G.C. (Ed.), *Aerothermodynamics of Gas Turbine and Rocket Propulsion*, AIAA Education Series, 3rd Edition, 1998.
3. Hill, P.G. and Peterson, C.R., *Mechanics and Thermodynamics of Propulsion*, Addison-Wesley, 2nd Edition, 1999.
4. Saravanamuttoo, H.I.H., Rogers, G.F.C. and Cohen, H., *Gas Turbine Theory*, Pearson Education, 2001.

Review Questions

1. What is meant by on-design cycle analysis? Why is it important?
2. What is meant by off-design cycle analysis? How is it different from on-design cycle analysis?
3. What is meant by rubber engine? What kind of cycle analysis is carried out for such kind of an engine?
4. What are the applications of the ramjet engine?
5. Why cannot the ramjet engine produce static thrust?

6. Suggest methods to modify a ramjet engine such that it can produce static thrust itself.
7. Which component in a ramjet dictates the performance of the engine? Why is it so?
8. Express the thermal efficiency of a ramjet engine in terms of τ_r.
9. Derive an expression for propulsive efficiency in terms of τ_r and τ_λ for a ramjet engine.
10. Discuss how does the thrust of a ramjet engine vary with flight Mach number.
11. How does the mass flow rate per unit intake area of a ramjet engine vary with flight Mach number?
12. How is the turbojet engine different from the ramjet?
13. For flight Mach number of 2.0, if one has to choose between ramjet and turbojet engines to achieve minimum TSFC, which one should be preferred and why?
14. What are the advantages of turbojet over the ramjet engine for the same flight conditions?
15. A turbojet engine is running when the aircraft is standing on the tarmac. The $\mathcal{T}_s = \dot{m}_a V_9$. It is observed that some foreign materials such as birds etc., can be dragged into the engine. Why is it so?
16. Can you relate the exit nozzle Mach number with the flight Mach number? Compare your result for turbojet with that of ramjet engine.
17. Draw the variation of temperature, pressure, and velocity across the various components of a typical operational turbojet engine. If an afterburner is used in this engine, how does it affect these parameters?
18. Define TSFC for a turbojet engine. Why is it considered to be a good parameter to compare engine efficiencies?
19. What are the factors that affect the overall efficiency of a turbojet engine?
20. How does a pilot control the thrust in a turbojet engine?
21. How does the thrust of a particular turbojet engine vary with flight Mach number? Discuss this.
22. Explain how does the mass flow rate per unit intake area of a turbojet engine vary with flight Mach number.
23. Which component in a turbojet dictates the performance of the turbojet? Why is it so?
24. Derive an expression for the thermal efficiency of a turbojet engine in terms of τ_r, τ_c.
25. Derive an expression for propulsive efficiency for turbojet in terms of temperature ratio across various components.
26. What is bypass ratio? Why is it so important?
27. How does the bypass ratio affect the thermal efficiency and propulsive efficiency?
28. What are the advantages and disadvantages of using high bypass ratio in a turbofan engine?
29. How does the thrust of a particular turbofan engine vary with flight Mach number? Discuss it.
30. How does the thrust of high bypass turbofan engine get affected by an increase in altitude and flight speed?
31. Explain how does the mass flow rate per unit intake area of a turbofan engine vary with flight Mach number.
32. Which component in a turbofan dictates the performance of the turbofan? Why is it so?
33. Derive an expression for propulsive efficiency for turbofan in terms of temperature ratio across various components.
34. What are the factors that affect the overall efficiency of a turbofan engine?
35. Can you call turboprop engine as ultrahigh bypass ratio turbofan engine? If yes, why?
36. Define work output coefficient for propeller and core engine? How is it related to the thrust generated by the propeller and core engine?
37. What are the differences between a turboprop and a turboshaft engine?
38. What is propfan engine? How is it different from the turboprop engine?

Problems

Q1. A ramjet is to be designed to fly at an altitude of 20 km. Carry out an ideal cycle analysis and show the variation of $\mathcal{T}_s$, TSFC, η_p, η_{th} and η_o with flight Mach number M_0 by varying it from 1 to 3 at intervals of 0.5 for three combustor exit temperatures of 1400, 1800 and 2200 K, respectively. Plot the results and discuss them. Take ΔH_c = 43,000 kJ/kg, C_P = 1.005 kJ/kg.

Q2. A ramjet engine is flying at an altitude of 10 km with flight Mach number of 2.5. The total temperature at the exit of the combustor is 2000 K. If the caloric value of the special kerosene is 47,000 kJ/kg, determine $\mathcal{T}_s$, TSFC, η_p, η_{th} and η_o.

Q3. A ramjet engine flying at an altitude of 10 km with certain flight Mach number swallows 60 kg/s of air. The total temperature at the exit of the combustor is 2000 K. If the caloric value of the special kerosene is 47,000 kJ/kg, determine the flight Mach number at which TSFC will attain an optimum value. Estimate T, TSFC, η_p, and η_{th} under this condition.

Q4. In order to overcome the problem of zero static thrust in a ramjet engine, a designer is planning to add a compressor driven by an electric motor. Derive an expression for T_s in terms of τ_r, τ_c, τ_b, and M_0. If pressure ratio across the compressor $\pi_c = 5.0$, $T_{t4} = 2000$ K, and $T_0 = 217$ K, estimate T_s, η_{th} and TSFC.

Q5. Derive an expression for mass flow rate per unit intake area in terms of Mach number and mass flow parameter and area ratio, ambient temperature and pressure etc. The total temperature at the exit of the combustor of a ramjet engine is 2000 K. If the ramjet is flying at an altitude of 12 km, plot the variation of specific thrust, thrust per unit intake area and mass flow rate per unit intake area along with flight Mach number from 0 to 7 at intervals of 0.5. Discuss the results.

Q6. An ideal turbojet engine flying at an altitude of 12 km with flight Mach number of 0.8 swallows 50 kg/s of air. The pressure ratio across the compressor is 15. The total temperature at the exit of the combustor is 1700 K. If the caloric value of kerosene is 45,000 kJ/kg, determine T_s, TSFC, η_p, η_{th} and η_o.

Q7. A turbojet engine is operating at flight Mach number of 2.0 at an altitude of 20 km. The pressure ratio across the compressor is 10. The maximum allowable temperature is 1500 K. Estimate T_s, TSFC, η_p, η_{th}, and η_o considering this as an ideal engine. If a ramjet engine is operating at the same altitude and flight Mach number, compare all the performance parameters such as T_s, TSFC, η_p, η_{th} and η_o of the ramjet engine with those of the turbojet. The maximum temperature for the ramjet engine is 2200 K. Take ΔH_c = Caloric value of fuel = 43,000 kJ/kg. Comment on your results.

Q8. An ideal turbojet engine flying at an altitude of 12 km with flight Mach number of 0.85 can produce thrust of 100 kN. The pressure ratio across the compressor is 17. The total temperature at the exit of the combustor is 1700 K. If the caloric value of kerosene is 45,000 kJ/kg, determine the mass flow rate consumed by the engine, nozzle exit area. Estimate T_s, TSFC, η_p, η_{th} and η_o.

Q9. An ideal turbojet engine is to be designed for an aircraft which can fly at an altitude of 10 km with flight Mach number of 1.5. It can swallow 40 kg/s of air. The total temperature at the exit of the combustor is 2000 K. If the caloric value of kerosene is 43,000 kJ/kg, determine the compressor pressure ratio at which TSFC will attain optimum value. Estimate the T, η_p, η_{th} and η_o under this condition.

Q10. A turbojet engine is to be designed for flying with flight Mach number of 2.0 at an altitude of 15 km. Carry out an ideal cycle analysis and show the variation of T_s, TSFC, η_p, η_{th} and η_o with compressor pressure ratio π_c by varying it from 1 to 20 at intervals of 2 for three combustor exit temperatures of 1400, 1800 and 2200 K, respectively. Take ΔH_c = 45,000 kJ/kg and C_P = 1.005 kJ/kg.

Q11. A turbojet engine is flying in the stratosphere (20 km) at flight Mach number of 0.85 with propulsive efficiency of 75%. Estimate T_s.

Q12. An ideal turbojet engine flying at an altitude of 12 km with a flight Mach number of 0.85 can produce a thrust of 150 kN. The pressure ratio across the compressor is 17. The total temperature at the exit of the combustor is 1700 K. The caloric value of kerosene is 45,000 kJ/kg. If the compressor pressure ratio is increased by 20%, what will be the increase in the thrust? By how much would TSFC change? What will be change in thermal efficiency?

Q13. An ideal turbojet engine is to be designed for an aircraft which can fly at an altitude of 18 km with flight Mach number of 0.9. It can swallow 50 kg/s of air. The total temperature at the exit of the combustor is 1500 K. If the caloric value of kerosene is 43,000 kJ/kg, determine the compressor pressure ratio at which thrust will attain optimum value. Estimate the TSFC, η_p, η_{th} and η_o under this condition.

Q14. A turbojet engine fitted with an afterburner is flying at 10 km. The turbine inlet temperature is restricted to 1400 K and maximum afterburner outlet temperature is 2200 K. Carry out an ideal parametric cycle analysis by varying π_c from 1 to 30 from M_0 = 0, 0.5, 1, 2.0. Plot T, TSFC, η_p, η_{th} and η_o with π_c. Comment on your results. Indicate the optimum π_c in the T, TSFC plot. Take ΔH_c = 43,000 kJ/kg.

Q15. An ideal turbojet engine with an afterburner, flying with flight Mach number of 1.2 at an altitude of 18 km consumes 70 kg/s of air. The total pressure ratio across the compressor is 20. The total temperature at the exit of the main burner and afterburner is 1400 and 2200 K, respectively. Estimate the thrust and TSFC for a simple turbojet engine and compare their values with that of an afterburning turbojet engine. Take ΔH_c = 43,000 kJ/kg.

Q16. An ideal turbojet engine with an afterburner, flying with flight Mach number of 1.5 at an altitude of 18 km consumes 70 kg/s of air when operated without light-up of the afterburner. The total pressure ratio across the compressor is 15. The total temperature at the exit of the main burner is 1300 K. When the afterburner is lit to increase the thrust by 30%, the total temperature at the exit of afterburner becomes 2300 K. Estimate the increase in TSFC. Take ΔH_c = 43,000 kJ/kg.

Q17. An ideal turbofan engine flying at an altitude of 8 km and flight Mach number M_0 of 0.8 is to be designed with compressor pressure ratio of 17 and fan pressure ratio of 2.0. If turbine inlet temperature is 1400 K, find out $\mathcal{T}_s$, TSFC, η_p, η_{th} and η_o with bypass ratio of 3.0. Determine $\mathcal{T}_s$ and TSFC under static condition at sea level. Take ΔH_c = 43,000 kJ/kg.

Q18. A two spool turbofan engine is flying at an altitude of 10 km and flight Mach number M_0 of 0.85. The pressure ratio across the compressor and fan is 20 and 1.5, respectively. The turbine inlet temperature T_{t4} is 1500 K. The bypass ratio across the engine is 5.0. Estimate $\mathcal{T}$, TSFC, η_p, η_{th} and η_o if the total flow rate of air is 200 kg/s. If this turbofan is operated as a turbojet at M_0 = 0.85 and M_0 = 2.0, what will be $\mathcal{T}$, TSFC, η_p, η_{th} and η_o? Compare the results with turbofan and comment on it. Assume an ideal cycle analysis. Take ΔH_c = 43,000 kJ/kg.

Q19. A designer has an option of choosing one engine to produce takeoff thrust of 100 kN at sea level condition from two types of engines – turbojet and turbofan – with fan pressure of 2.0 and bypass ratio of 3.0. Both engines have same core airflow rate of 60 kg/s. Both use same aviation turbine fuel with ΔH_c = 45,000 kJ/kg and produce TIT of 1500 K. Determine the TSFC and η_{th}.

Q20. An ideal turbofan engine flying at an altitude of 8 km and flight Mach number M_0 of 0.95 is fitted with an afterburner. The pressure ratio across the compressor and fan is 18 and 2.5, respectively. The exit temperature at the combustor and afterburner is 1400 and 2200 K, respectively. The bypass ratio of the turbofan is 3.2. Determine the $\mathcal{T}_s$, TSFC, η_p, η_{th} and η_o with and without afterburner under static condition at sea level condition. Take ΔH_c = 43,000 kJ/kg. Comment on your results.

Q21. An ideal turbofan operated at an altitude of 12 km and flight Mach number M_0 of 0.75 is to be designed for producing thrust of 120 kN. The pressure ratio across the compressor and mixed fan is 15 and 2.5, respectively. The bypass ratio of the turbofan engine is 4.0. The exit temperature at the combustor is 1600 K. If the heating value of the fuel is 43,000 kJ/kg, what will be the fan pressure ratio and core flow rate? Determine TSFC, η_p, η_{th} and η_o.

Q22. A turbofan engine is flying at an altitude of 10 km. The turbine inlet temperature is restricted to 1600 K. The bypass ratio of the turbofan engine is 4.0. The fan pressure ratio across the fan is 2.5. Carry out an ideal parametric cycle analysis by varying π_c from 1 to 30 for M_0 = 0, 0.85, 1.2. Plot $\mathcal{T}_s$, TSFC, η_p, η_{th} and η_o with π_c. Comment on your results. Indicate the optimum π_c in the $\mathcal{T}_s$, TSFC plot. Take ΔH_c = 43,000 kJ/kg.

Q23. A turbofan engine is operated at sea level. The turbine inlet temperature is restricted to 1600 K. The pressure ratio across the fan is 2.5. Carry out an ideal parametric cycle analysis by varying π_c from 1 to 30 for bypass ratio α = 0.5, 1, 2, 5.0. Plot $\mathcal{T}_s$, TSFC, η_p, η_{th} and η_o with π_c. Comment on your results. Indicate the optimum π_c in plots of specific takeoff thrust and TSFC. Take ΔH_c = 43,000 kJ/kg.

Q24. A turboprop engine is being operated with flight Mach number of 0.6 at an altitude of 6 km. The propeller efficiency is 0.83. The turbine inlet temperature is 1300 K. The pressure ratio across the compressor is 12 and total temperature ratio across the turbine is 0.6. Considering an ideal cycle, estimate $\mathcal{T}_s$, TSFC, η_p, η_{th} and η_o. Take ΔH_c = Caloric value of fuel = 43,000 kJ/kg.

Q25. An ideal turboprop engine is used to propel an aircraft operated with flight Mach number of 0.3 at an altitude of 4 km. The propeller efficiency is 0.83. The turbine inlet temperature is 1300 K. The pressure ratio across the compressor is 5. The total temperature at the inlet of turbine is 1400 K. Estimate optimum thrust, TSFC, η_p, η_{th} and η_o. Take ΔH_c = Caloric value of fuel = 50,000 kJ/kg.

5. Real Cycle Analysis

In its present state and even considering the improvements possible when adopting the higher temperature proposed for the immediate future, the gas turbine could hardly be considered a feasible application to airplanes mainly because of the difficulty in complying with stringent weight requirements imposed by aeronautics.

Expert Committees on Gas Turbine Engine, National Academy of Science, January 1941

5.1 Introduction

In Chapter 4, we discussed in detail the parametric cycle analysis of an ideal engine. Although, we discovered the general performance trend of gas turbine engines, ideal analysis does not provide a realistic view about the performance of actual gas turbine engines. Hence, we must explore further to make the cycle analysis more realistic by incorporating the losses incurred in the various components of an actual gas turbine engine. Generally, the losses caused in gas turbine engines can be classified into three categories:

(i) *Aerodynamic losses*: These losses are caused due to fluid flow through the various components such as air intake, compressor, combustion chamber, turbine and nozzle. The air flow has to pass through various convergent or divergent ducts, blades and bluff bodies, which causes a decrease in stagnation pressure. The losses contributed in these components can be attributed to different effects such as boundary layer and its separation, wall friction, shock formation, mixing of fluids, and heat addition.

(ii) *Thermodynamic losses*: These losses are caused due to incomplete combustion of fuels and heat transfer from its surface to the surroundings. The thermodynamic imperfections can be incurred due to non-isentropic processes in the intake, compressor, turbine and nozzle.

(iii) *Mechanical losses*: There will be losses in mechanical energy during power transmission from the turbine to the compressor and fan. These losses may be caused due to bearing friction and windage. Although such losses are quite small in magnitude but they must be accounted while carrying out a real cycle analysis of gas turbine engines.

From the above, we must appreciate the need for real cycle analysis and make efforts to arrive at proper methods of developing analytical tools to carry out real cycle analysis. Let us now enumerate how the real cycle performance differs from that of an ideal cycle by considering the following reasons:

- Variation in gas properties throughout the engine owing to both temperature, pressure and composition.
- Non-isentropic diffusion process in the air intake and compressor.

- Non-isentropic expansion in the turbine and nozzle.
- Change in kinetic energy and potential energy between the inlet and outlet can be ignored in all components of a gas turbine engine.
- Bleed air extractions from the air intake and compressor for turbine cooling.
- Incomplete combustion of fuel.
- Gas dynamic losses in the combustor due to heat addition.
- The flow in the exhaust nozzle may not be fully expanded to the ambient pressure. Rather, the flow in the exhaust nozzle can be either over- or under-expanded.

The main objective of this chapter is to make the reader understand how to characterise the actual behaviour of each of the components of a gas turbine engine, which can be included in the ideal cycle analysis to develop an analytical tool for real engines. We will carry out detailed discussions in subsequent sections about the various mechanisms of losses in each component. Efforts will be made to take account of these losses while carrying out real cycle analysis. We will also learn about the method of reducing these unwanted losses in the various components in gas turbine engines.

5.2 Variation in Gas Properties

We have already seen that the temperature and pressure of the working fluid changes across each component of gas turbine engines. Let us recall that gas temperature, pressure, density etc., vary in almost all the components such as air intake, compressor, combustor, turbine, nozzle etc. Also, fuel is being added in the combustor which it burns to release high amount of heat per unit volume. Consequently, the chemical composition and temperature of gas are likely to change considerably. Hence, the thermodynamic properties of the working fluid change across each component of the engine. In general, the thermodynamic properties like enthalpy h, specific heat at constant pressure C_p and specific heat ratio γ (= C_p/C_v) play an important role in estimating the engine cycle analysis. So we must take into account the variation of C_p and γ while carrying out real cycle analysis. Knowing the fact that aircraft gas turbine engines operate on refined kerosene, which can be represented chemically as n-Octane (C_8H_{18}), we can estimate the combustion products for a range of fuel-air ratio considering chemical equilibrium. By estimating the chemical composition, we can evaluate the mean values of h, C_p and γ of the mixture. The variation of C_p and γ for certain fuel-air ratio vs. temperature (Fig. 5.1) shows that C_p increases while γ decreases with temperature and fuel-air ratio. In fact, such kind of calculation is quite complex and time consuming in nature. Hence, we will adopt a simpler method of taking account of variation in gas properties during real cycle analysis. In this book, we will consider constant values of C_{pa} and γ_a in all the components upstream of the combustor and different constant values of C_{pg} and γ_g in all the components downstream of the combustor. Although this simplication takes into account that variation in gas properties will cause inaccuracies in the real cycle analysis, it is well-established that these two constant properties provide sufficiently accurate results, particularly, for the preliminary design calculations.

5.3 Inlet and Diffuser Pressure Recovery

The main function of air intake is to decelerate the free stream fluid smoothly to a lower velocity either in subsonic or supersonic diffuser. In subsonic inlet, viscous shear on the wall causes the growth of boundary layers, causing losses in the stagnation pressure. In contrast, shocks are formed in a supersonic diffuser, which cause additional losses in the stagnation pressure. Let us take into account these losses in real cycle analysis by considering a term known as *Pressure Recovery Factor* (PRF)

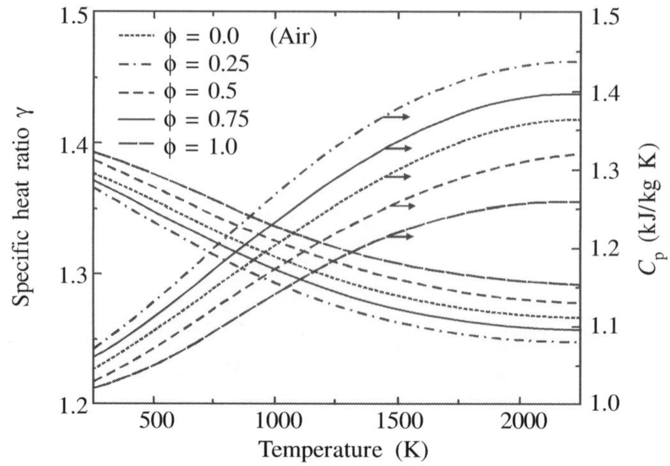

Fig. 5.1. Variation of C_p and specific heat ratio γ with temperature.

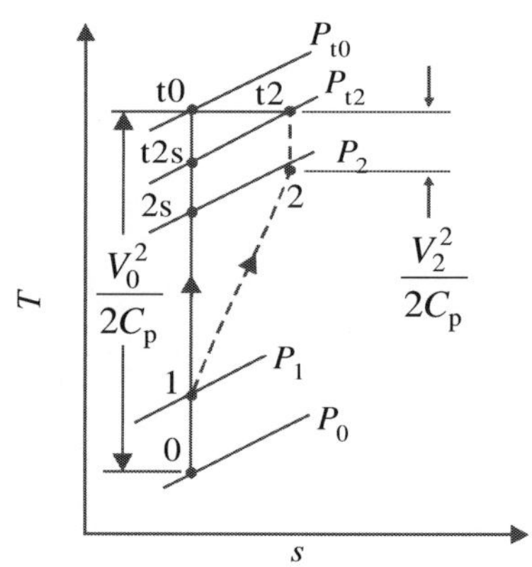

Fig. 5.2. *T-s* diagram showing the real processes involved in a subsonic air intake.

π_d, which is defined as the ratio of average stagnation pressure at the air intake exit, P_{t2} to the freestream value P_{t0} and is denoted as

$$\text{Pressure Recovery Factor} = \pi_d = \frac{P_{t2}}{P_{t0}} \tag{5.1}$$

Thus, we have learnt that the losses in subsonic air intake arise due to wall friction and in supersonic air intakes due to a combination of wall friction and shock losses. As a result, $\pi_d < 1$ always in an air intake. Of course, one can easily assume the processes in an air intake to be adiabatic as the fluid velocity will be a bit fast and the temperature gradient is quite small for heat exchange to take place between the inlet and ambient air. Therefore, the stagnation temperature ratio τ_d becomes one ($\tau_d \simeq 1$).

From Fig. 5.2 we note that a smaller ram pressure ratio develops in an actual air intake than in an ideal case. To take account of this, let us define another parameter known as isentropic efficiency η_{id} as

$$\eta_{id} = \frac{h_{t2s} - h_0}{h_{t0} - h_0} \simeq \frac{T_{t2s} - T_0}{T_{t0} - T_0} \simeq \frac{\left(\frac{P_{t2}}{P_0}\right)^{\frac{\gamma-1}{\gamma}} - 1}{\tau_r - 1} = \frac{\tau_d^{\frac{\gamma-1}{\gamma}}\left(1 + \frac{\gamma_a - 1}{2} M_0^2\right) - 1}{\left(\frac{\gamma_a - 1}{2}\right) M_0^2} \tag{5.2}$$

It can be observed from Eq. (5.2) that η_{id} is a function of flight Mach number, PRF and fluid properties (γ, T_0). Interestingly, η_{id} is related to PRF by Eq. (5.2). Figure 5.3 shows the variation of π_d and η_{id} with flight Mach number M_0 for subsonic air intake. The value of η_{id} turns out to be lower than PRF. However, both parameters decrease with M_0. The isentropic efficiency is mainly dependent on the

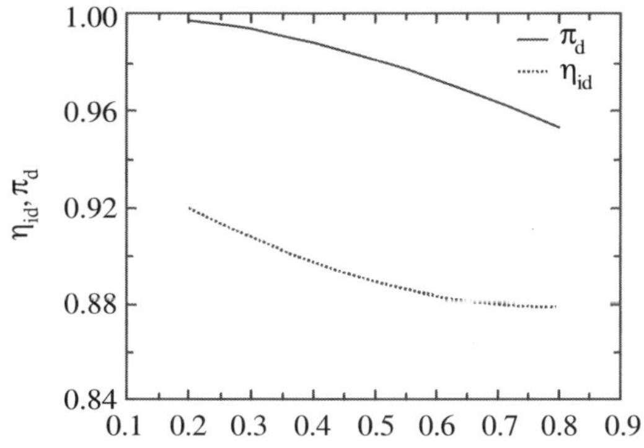

Fig. 5.3. Variation of π_d and η_{id} with flight Mach number M_0 in a real air intake.

extent of internal deceleration in the air intake. But we are interested in the overall deceleration instead of internal deceleration alone. Note that overall efficiency depends on the mass flow rate and flight Mach number. Besides this, if $M_0 > 1$, then losses are contributed by shock formation in addition to viscous effects. The overall pressure ratio of the air intake is contributed by ram pressure ratio and diffuser pressure ratio [1], which can be expressed as

$$\pi_d = \frac{P_{t2}}{P_{t0}} = \frac{P_{t2}}{P_{t1}} \cdot \frac{P_{t1}}{P_{t0}} = \pi_{d,i}\,\pi_{d,e} \tag{5.3}$$

where $\pi_{d,i}$ is a certain portion of ram total pressure recovery due to internal compression and $\pi_{d,e}$ a certain portion of total pressure recovery due to external compression. The entire portion of ram pressure cannot be recovered due to shock formation. For subsonic and supersonic flow, we can use an expression for ram recovery $\pi_{d,e}$ [1, 2] as:

$$\pi_{d,e} = \begin{cases} 1 & M_0 \le 1 \\ 1 - 0.075\left(M_0^2 - 1\right) & 1 < M_0 < 5 \\ \dfrac{800}{M_0^4 + 935} & 5 < M_0 \end{cases}$$

The isentropic efficiency is very handy for use in cycle analysis. However, it is not easy to measure total temperature used in defining isentropic efficiency in comparison to total pressure. Hence, another parameter based on pressure and known as ram efficiency η_{rd} is defined as

$$\eta_{rd} = \frac{P_{t2} - P_0}{P_{t0} - P_0} \tag{5.4}$$

Note that η_{rd} is found to be almost same as η_{id}. Although it is easier to evaluate experimentally, but it offers no specific advantages over η_{id}. We know that the location of the engine in an aircraft (in the wing, pod or fuselage) dictates the air intake efficiency. But we can take the value η_{id} to be around 0.93 for subsonic and a lower value for supersonic aircraft while solving preliminary design problems.

5.4 Compressor and Turbine Efficiencies

Again, like in the air intake, the processes in a compressor and turbine can be assumed as adiabatic to a high degree of approximation. The processes involved in a real compressor are shown on *T-s* plane in Fig. 5.4. Both ideal and actual compression processes for a given π_c are shown in Fig. 5.4. However, the losses originating primarily due to viscous shear on the blades and the walls of casings also occur in the compressor and turbine. These losses result in significant decrease in stagnation pressure as compared to air intake. Also, if shocks occur in the flow they may cause further losses in stagnation pressure. These are frequently encountered in the fan stage and in the first stage of modern transonic compressors.

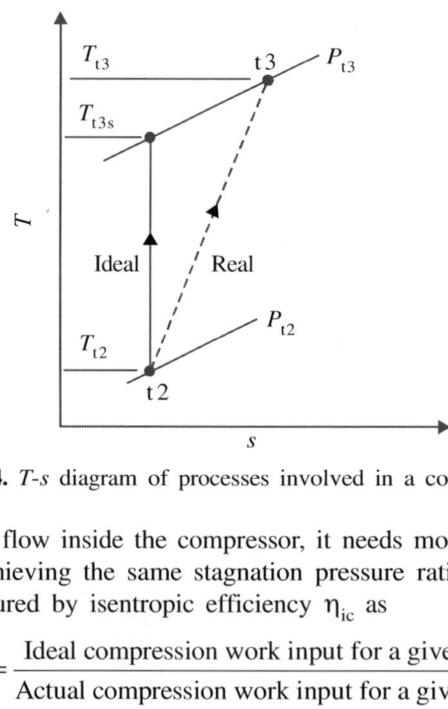

Fig. 5.4. *T-s* diagram of processes involved in a compressor.

Because of losses in the flow inside the compressor, it needs more energy input unlike an ideal isentropic compressor for achieving the same stagnation pressure ratio. The overall performance of the compressor can be measured by isentropic efficiency η_{ic} as

$$\eta_{ic} = \frac{\text{Ideal compression work input for a given } \pi_c}{\text{Actual compression work input for a given } \pi_c} \tag{5.5}$$

Actual work per unit mass $\dot{W}_c/\dot{m}$ is $h_{t3} - h_{t2}$ and ideal work per unit mass $\dot{W}_{cs}/\dot{m}$ is $h_{t3s} - h_{t2}$. So η_{ic} becomes

$$\eta_{ic} = \frac{\dot{W}_{cs}}{\dot{W}_c} = \frac{h_{t3s} - h_{t2}}{h_{t3} - h_{t2}} \simeq \frac{\frac{T_{t3s}}{T_{t2}} - 1}{\frac{T_{t3}}{T_{t2}} - 1} = \frac{\pi_c^{\frac{\gamma_a - 1}{\gamma_a}} - 1}{\tau_c - 1} \tag{5.6}$$

Similarly, we can use isentropic efficiency of a turbine η_{it} for the expansion process occurring in the turbine as shown in *T-s* diagram (Fig. 5.5), while carrying out real cycle analysis. The isentropic efficiency of a turbine is defined as

$$\eta_{it} = \frac{T_{t4} - T_{t5}}{T_{t4} - T_{t5s}} = \frac{1 - \tau_t}{1 - (\pi_t)^{\frac{\gamma_g - 1}{\gamma_g}}} \tag{5.7}$$

It must be kept in mind that losses in a turbine are, generally, lower in comparison to that of a compressor for the same pressure ratio as the flow is being accelerated in the turbine.

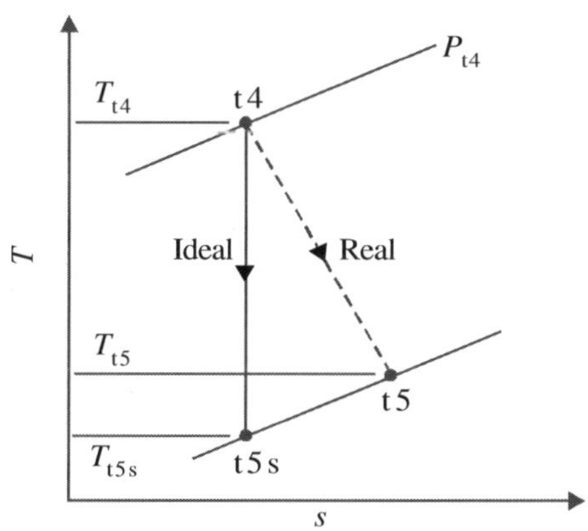

Fig. 5.5. *T-s* diagram of processes involved in a turbine.

5.5 Polytropic Efficiency

Till now we have discussed about the overall efficiency of a multistage compressor. For aircraft gas turbine engines, the pressure ratio is quite high across the compressor. Now a pertinent question arises: whether η_{ic} remains constant or can it vary considerably with change in pressure ratio across the compressor? Generally, η_{ic} cannot remain constant across the high pressure compressor. However, we can assume η_{ic} across a stage of an axial flow compressor to remain constant, as its stage pressure ratio is quite small particularly for an axial compressor. Let us approximate η_{ic} across a stage to be equal to the polytropic efficiency η_{pc} of that compressor. Hence, the polytropic efficiency of a compressor can be defined as the ratio of ideal to actual work for an infinitesimal step during the compression process given as

$$\eta_{pc} = \frac{\text{Ideal compression work input for an infinitesimal } dP_t}{\text{Actual compression work input for an infinitesimal } dP_t} \tag{5.8}$$

Let us consider an incremental pressure rise from P_t to $P_t + dP_t$ as shown in the *T-s* diagram (Fig. 5.6). Let the actual temperature rise accompanying the increase in pressure be denoted as dT_t and the ideal temperature rise as dT_{ts}. Then η_{pc} becomes

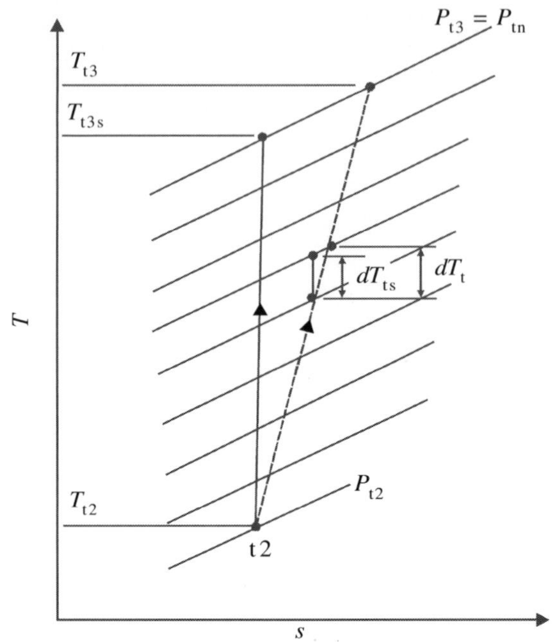

Fig. 5.6. *T-s* diagram depicting polytropic processes involved in a compressor.

$$\eta_{pc} = \frac{dT_{ts}}{dT_t} = \frac{dT_{ts}/T_t}{dT_t/T_t} \tag{5.9}$$

Note that η_{pc} is not the same as η_{ic} even for a stage although its value is quite closer particularly for an axial compressor. In order to conceive the polytropic efficiency properly, it was mentioned that it can be approximated as the isentropic efficiency across a stage. For an infinitesimal isentropic pressure change, the temperature and pressure ratio can be related as

$$\frac{T_t + dT_{ts}}{T_t} = \left(\frac{P_t + dP_t}{P_t}\right)^{\frac{\gamma_a - 1}{\gamma_a}} \tag{5.10}$$

Equation (5.10) can be rearranged as

$$\Rightarrow \frac{dT_{ts}}{T_t} = \left[\left(\frac{P_t + dP_t}{P_t}\right)^{\frac{\gamma_a - 1}{\gamma_a}} - 1\right] \tag{5.11}$$

Expanding $\left(1 + \frac{dP_t}{P_t}\right)^{\frac{\gamma_a - 1}{\gamma_a}}$ by Binomial series and neglecting higher order terms, we have

$$\left(1 + \frac{dP_t}{P_t}\right)^{\frac{\gamma - 1}{\gamma}} \simeq 1 + \frac{\gamma_a - 1}{\gamma_a}\frac{dP_t}{P_t} \tag{5.12}$$

Using Eqs. (5.9) and (5.12), we can rewrite the expression for η_{pc} as

$$\eta_{pc} = \left(\frac{\gamma_a - 1}{\gamma_a}\right)\frac{dP_t/P_t}{dT_t/T_t} \Rightarrow \frac{dT_t}{T_t} = \left(\frac{\gamma_a - 1}{\gamma_a}\right)\frac{1}{\eta_{pc}}\frac{dP_t}{P_t} \tag{5.13}$$

Now let us consider the compressor to be polytropic in each stage of a multistage compressor as described in the *T*-s diagram of Fig. 5.6. Then we can arrive at an overall compressor pressure ratio across a compressor by integrating Eq. (5.13) between stations (2) and (3), which becomes

$$\frac{T_{t3}}{T_{t2}} = \left(\frac{P_{t3}}{P_{t2}}\right)^{\frac{\gamma_a - 1}{\gamma_a \eta_{pc}}} = \pi_c^{\frac{\gamma_a - 1}{\gamma_a \eta_{pc}}} \tag{5.14}$$

Then the overall isentropic efficiency of a multistage compressor can be related to η_{pc} as

$$\eta_{ic} = \frac{\frac{T_{t3s}}{T_{t2}} - 1}{T_{t3}/T_{t2} - 1} = \frac{\pi_c^{\frac{(\gamma_a - 1)}{\gamma_a}} - 1}{\pi_c^{\frac{(\gamma_a - 1)}{\gamma_a \eta_{pc}}} - 1} \tag{5.15}$$

For a fixed value of $\eta_{pc} \simeq 0.85$ in each stage, Fig. 5.7 shows variation of overall isentropic compressor efficiency η_{ic} vs. pressure ratio π_c. The η_{ic} decreases with increase in π_c, implying that it is less than stage efficiency η_{ics} ($\simeq \eta_{pc}$). Note that the differences between η_{ic} and η_{pc} increase with the number of stages and with increase in π_c because increase in temperature caused by friction in one stage causes more compressor work in the subsequent stages. This process is also known as *preheating effect*.

Similarly, we can derive an expression relating η_{pt} and η_{it} for a turbine as

$$\eta_{it} = \frac{1 - \pi_t^{(\gamma_g - 1)\eta_{pt}/\gamma_g}}{1 - \pi_t^{(\gamma_g - 1)/\gamma_g}} \tag{5.16}$$

The variation of η_{it} against pressure ratio π_t for a fixed value of overall turbine isentropic efficiency of 0.9 (Fig. 5.7) shows that η_{it} increases with π_t. This can be explained by a similar argument, which recognises the fact that reheating of gas due to frictional loss in one stage is converted to work in subsequent stages.

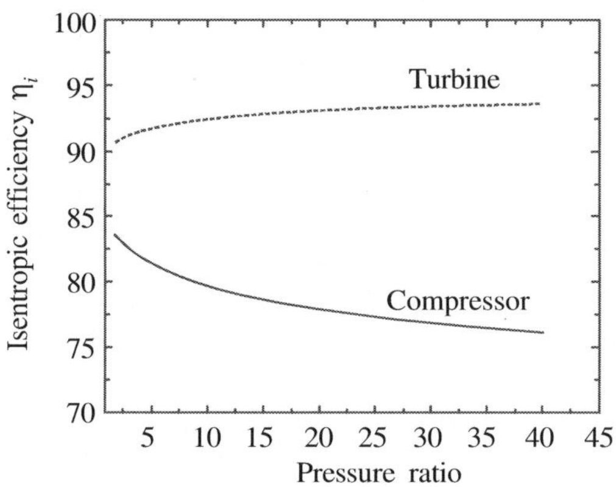

Fig. 5.7. Variation of η_i for both compressor and turbine with pressure ratio for η_{pc} = 0.85 and η_{pt} = 0.9.

5.6 Combustor Efficiency and Pressure Loss

The burning processes of fuel in a real gas turbine engine can be modelled by considering two distinct losses due to: (i) inability of combustor to convert entire chemical energy of fuel, and (ii) stagnation pressure loss across the entire combustor. The first arises due to imperfect combustion of fuel whose energy can remain as partially combusted carbon in terms of CO, unburnt hydrocarbon (HC) or soot particles. This loss is characterised as combustor (burner) efficiency η_b, which is defined as the actual change enthalpy to product of fuel flow rate and calorific value of the fuel. It is given by

$$\eta_b = \frac{(\dot{m}_a + \dot{m}_f)h_{t4} - \dot{m}_a h_{t3}}{\dot{m}_f \Delta H_C} \tag{5.17}$$

where $\dot{m}_f$ and $\dot{m}_a$ are the mass flow rate of fuel and air, respectively, and ΔH_C the heating value of fuel. Typical heating value of aviation turbine fuel (ATF) is considered to be around 43,000 kJ/kg.

The stagnation pressure loss is contributed partially due to viscous effects and gas dynamics loss, accompanied by heat addition to the flow combustor stagnation pressure loss defined as

$$\pi_b = \frac{P_{t4}}{P_{t3}} < 1.0 \tag{5.18}$$

5.7 Losses in Exhaust Nozzle

In the exhaust nozzle, the expansion of gases will be accompanied by the usual frictional losses. As a result, the gases from the nozzle will come out less hotter than in the ideal case, leading to a reduction in enthalpy drop across the nozzle with a lower exhaust jet velocity. An ideal nozzle means that the gas is expanded fully to ambient pressure P_0. But if the nozzle is not expanded completely, then the nozzle can either be over- or under-expanded, which will be discussed in subsequent chapters. Such situation will cause losses in the nozzle, leading to loss in total pressure even when the flow in the nozzle can be considered to be adiabatic. So the total pressure ratio across the nozzle π_n will be less than unity $\left(\pi_n = \frac{P_{t9}}{P_{t5}} < 1\right)$. This expansion process in the nozzle is described in a *T-s* diagram (Fig. 5.8).

To take account of the losses, we can define nozzle isentropic efficiency η_{in} similar to that of the turbine as the ratio of actual to isentropic enthalpy drops. Assuming mean C_p value for actual expansion process in a nozzle, η_{in} becomes

$$\eta_{in} = \frac{T_{t9} - T_9}{T_{t5} - T_{9s}} = \frac{1 - T_9/T_{t9}}{1 - \left(\frac{P_9}{P_{t5}}\right)^{\frac{\gamma_g - 1}{\gamma_g}}} \tag{5.19}$$

Now let us see the range of realistic parameters that can be used in real cycle analysis. Basically, we will use polytropic efficiencies for compressor and turbine η_{pc} and η_{pt}, respectively, unless otherwise stated explicitly. For well-designed engines, the data for performance parameters of various components are given in Table 5.1.

The component efficiencies and losses of a gas turbine engine are dependent on the engine installation aspects and applications of the aircraft. For example, for a long range subsonic aircraft with nacelle mounted engines and fixed area convergent nozzle, the values of π_d and π_n will be much higher than for a supersonic fighter aircraft with variable area convergent-divergent (CD) nozzle.

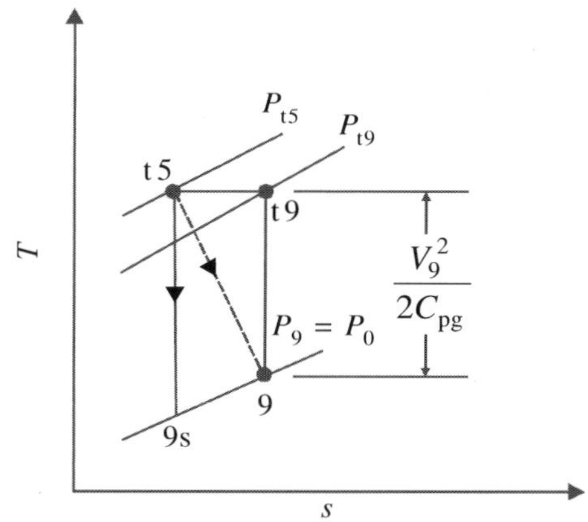

Fig. 5.8. *T-s* diagram of processes involved in the expansion in an exhaust nozzle.

Table 5.1. Parameters for various components in gas turbine engine [1-3]

Component	Parameter	Symbol	Typical range
Airintake	Total pressure recovery	π_d	0.9-0.96
	Isentropic efficiency	η_{id}	0.7-0.9
	Ram efficiency	η_{rd}	0.8-0.9
Compressor	Isentropic efficiency	η_{ic}	0.8-0.93
	Polytropic efficiency	η_{pc}	0.8-0.9
Fan	Isentropic efficiency	η_{iF}	0.8-0.93
	Polytropic efficiency	η_{pF}	0.8-0.9
	Bypass ratio	α	0-5
Combustor	Combustor efficiency	η_b	0.97-0.99
	Total pressure ratio	π_b	0.9-0.95
Turbine	Isentropic efficiency	η_{it}	0.9-0.98
	Polytropic efficiency	η_{pt}	0.85-0.95
Exhaust nozzle	Isentropic efficiency	η_{in}	0.93-0.98
	Polytropic efficiency	η_n	0.9-0.95
	Pressure ratio at exit	P_0/P_9	0.5-1.0

5.8 Turbojet

Let us consider the usual station numbering as discussed in Chapter 4 for a turbojet. The processes in a typical real turbojet engine with losses are described in detail in a *T-s* diagram (Fig. 5.9). We can use the same procedure as that of an ideal engine to derive an expression for $\mathcal{T}_s$, TSFC and different efficiencies. We know that the uninstalled thrust $\mathcal{T}$ for a turbojet engine is expressed as

$$\mathcal{T} = (\dot{m}_9 V_9 - \dot{m}_0 V_0) + A_9 (P_9 - P_0) \tag{5.20}$$

By manipulating Eq. (5.20), we can obtain an expression for specific thrust $\mathcal{T}_s$ as

$$\mathcal{T}_s = a_0 \left(\frac{\dot{m}_9}{\dot{m}_0} \frac{V_9}{a_0} - M_0 \right) + \frac{A_9 P_9}{\dot{m}_0} \left(1 - \frac{P_0}{P_9} \right) \tag{5.21}$$

where, $$\frac{A_9 P_9}{\dot{m}_0} = \left(\frac{\dot{m}_9}{\dot{m}_0} \right) + \frac{A_9 T_9}{r_0 R_0 T_0} \frac{a_0}{V_9 / a_0}$$

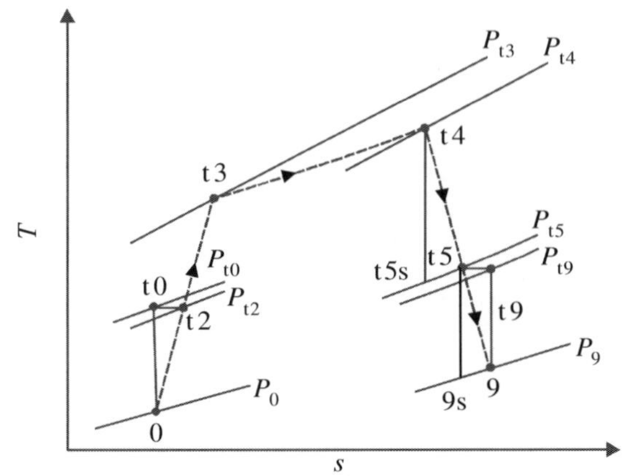

Fig. 5.9. *T-s* diagram of processes involved in a real turbojet engine.

However, we can write $\dot{m}_9/\dot{m}_0$ in terms of fuel-air ratio f by striking a mass balance across the turbojet engine as

$$\dot{m}_9 = \dot{m}_0 + \dot{m}_f \Rightarrow \frac{\dot{m}_9}{\dot{m}_0} = 1 + f \tag{5.22}$$

Similar to an ideal cycle analysis, we can express V_9/a_0 in terms of temperature ratio as

$$\frac{V_9}{a_0} = \sqrt{\frac{\gamma_g R_g T_9}{\gamma_a R_a T_0}} \cdot M_9 \tag{5.23}$$

By using the isentropic relation, we have

$$M_9 = \sqrt{\frac{2}{\gamma_g - 1}\left[\left(\frac{P_{t9}}{P_9}\right)^{\frac{\gamma_g - 1}{\gamma_g}} - 1\right]} \tag{5.24}$$

where

$$\frac{P_{t9}}{P_9} = \frac{P_{t9}}{P_{t5}} \cdot \frac{P_{t5}}{P_{t4}} \cdot \frac{P_{t4}}{P_{t3}} \cdot \frac{P_{t3}}{P_{t2}} \cdot \frac{P_{t2}}{P_{t0}} \cdot \frac{P_{t0}}{P_0} \cdot \frac{P_0}{P_9} \tag{5.25}$$

$$= \pi_n \pi_t \pi_b \pi_c \pi_d \pi_r \frac{P_0}{P_9} \tag{5.26}$$

From previous analysis we also know that

$$\frac{T_9}{T_0} = \frac{T_{t9}/T_0}{\left(P_{t9}/P_9\right)^{\frac{\gamma_g - 1}{\gamma_g}}} \tag{5.27}$$

where
$$\frac{T_{t9}}{T_0} = \frac{T_{t9}}{T_{t5}} \cdot \frac{T_{t5}}{T_{t4}} \cdot \frac{T_{t4}}{T_{t3}} \cdot \frac{T_{t3}}{T_{t2}} \cdot \frac{T_{t2}}{T_{t0}} \cdot \frac{T_{t0}}{T_0} = \tau_n \tau_t \tau_b \tau_c \tau_d \tau_r = \tau_t \tau_b \tau_c \tau_r \tag{5.28}$$

Note that in Eq. (5.28) τ_n and τ_d are equal to unity as the processes in the nozzle and diffuser are assumed to be adiabatic. Using Eqs. (5.24) to (5.28), we can easily evaluate the value of V_9/a_0.

Now let us strike an energy balance across the combustion chamber by invoking the First Law of Thermodynamics for steady flow process as

$$\dot{m}_0 C_{pa} T_{t3} + \eta_b \dot{m}_f \Delta H_C = \dot{m}_4 C_{pg} T_{t4} \tag{5.29}$$

By algebraic manipulation, we can express Eq. (5.29) as

$$f = \frac{\tau_\lambda - \tau_r \tau_c}{\dfrac{\eta_b \Delta H_C}{C_{pa} T_0} - \tau_\lambda} \tag{5.30}$$

where

$$\tau_\lambda = \frac{C_{pg} T_{t4}}{C_{pa} T_0} \tag{5.31}$$

Now we will attempt to relate the turbine temperature ratio τ_t to compressor temperature ratio τ_c and other relevant parameters. We recall that the turbine work output is meant to supply power to the compressor only in case of a turbojet engine. Hence, let us strike an energy balance between the compressor and turbine

$$\dot{W}_c = \dot{W}_t \tag{5.32}$$

$$\dot{m}_0 C_{pa} \left(T_{t3} - T_{t2}\right) = \eta_m \dot{m}_4 C_{pg} \left(T_{t4} - T_{t5}\right) \tag{5.33}$$

where η_m is the mechanical efficiency. Note that when a turbine is coupled to a compressor, there will be certain losses during power transmission. These losses can be accounted easily by the mechanical efficiency. By algebraic manipulation, we get the expression for τ_t as

$$\tau_t = 1 - \frac{1}{\eta_m (1+f)} \frac{\tau_r}{\tau_\lambda} \left(\tau_c - 1\right) \tag{5.34}$$

By knowing τ_t we can find the total pressure ratio across the turbine π_t by

$$\pi_t = \tau_t^{\frac{\gamma_g}{(\gamma_g - 1)\eta_{pt}}} \tag{5.35}$$

where η_{pt} is the polytropic efficiency of the turbine. Similarly, the total pressure ratio across the compressor π_c can be determined by

$$\pi_c = \tau_c^{\frac{\gamma_a \cdot \eta_{pc}}{(\gamma_a - 1)}} \tag{5.36}$$

Hence, T_s can be easily estimated by using the above expressions, provided we know the back pressure acting on the nozzle. For this, we need to know the pressure ratio P_9/P_0, particularly, for a convergent-divergent (CD) nozzle. If the exhaust nozzle happens to be a convergent one, then we can evaluate it directly depending on the extent of expansion. Note that the convergent nozzle is preferred for both subsonic and supersonic aircraft as it is quite efficient over a wide range of operating conditions as compared to the CD nozzle. Moreover, it is easy to manufacture and maintain.

Let us now look at how we can determine the pressure ratio P_{t5}/P_0 across the exhaust convergent nozzle, while taking into account the exhaust nozzle losses in a cycle analysis. The expansion processes in the exhaust nozzle for unchoked and choked conditions are described in a *T-s* plot (Fig. 5.10). It can be noted that if the pressure ratio across the nozzle is less than equal to critical value then the nozzle is assumed to be fully expanded. But if it is greater than the critical value, then the nozzle is choked. Under this condition, P_9 remains constant at critical pressure P_c and nozzle exhaust velocity V_9 remains equal to local speed of sound. Let us see how we can evaluate P_c for non-isentropic flow conditions in a real engine. Let us recall that critical pressure ratio across the nozzle P_{t5}/P_c in a simple turbojet/turbofan engine becomes equal to P_{t5}/P_0 when nozzle exit Mach number $M_9 = 1$. Note that critical temperature ratio T_{t5}/T_c remains same for both isentropic and non-isentropic expansion as adiabatic flow is assumed for both cases. Now we can express T_{t5}/T_c in terms of M_9 as

$$\frac{T_{t9}}{T_9} = \frac{T_{t5}}{T_c} = 1 + \frac{\gamma_g - 1}{2} M_9^2 = \frac{\gamma_g + 1}{2} \tag{5.37}$$

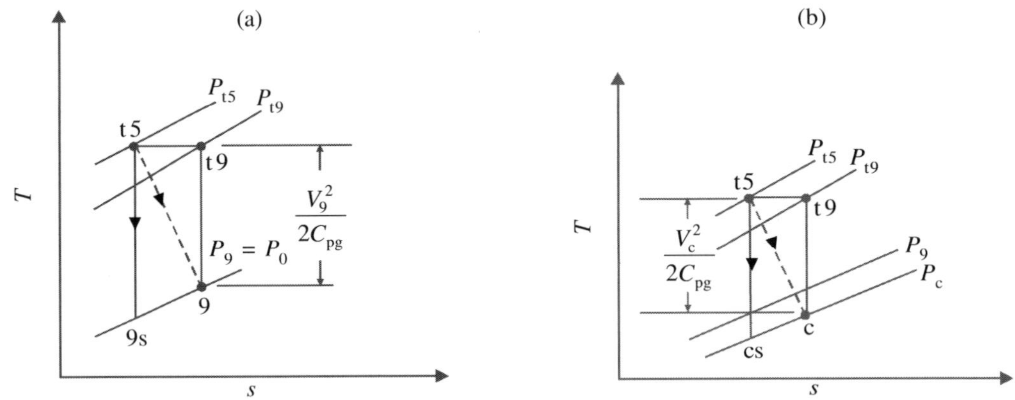

Fig. 5.10. Expansion process across a convergent nozzle at (a) unchoked and (b) choked condition ($P_{t5}/P_0 \leq P_{t5}/P_c$) on *T-s* plane.

As the expansion process is non-isentropic, the pressure ratio across nozzle P_{t5}/P_c cannot be determined from T_{t5}/T_c directly. Rather, we need to use isentropic efficiency for nozzle η_{in} to determine P_{t5}/P_c. For this, we can invoke the definition of η_{in} to derive an expression for temperature ratio T_{cs}/T_c for isentropic process in terms of non-isentropic temperature ratio T_c/T_{t5} as

$$\frac{T_{cs}}{T_{t5}} = 1 - \frac{1}{\eta_{in}}\left[1 - \frac{T_c}{T_{t5}}\right] \tag{5.38}$$

By using Eq. (5.38) and isentropic relation, we can derive an expression for P_c/P_{t5} as

$$\frac{P_c}{P_{t5}} = \left[\frac{T_{cs}}{T_{t5}}\right]^{\frac{\gamma_g}{\gamma_g - 1}} = \left[1 - \frac{1}{\eta_{in}}\left(1 - \frac{T_c}{T_{t5}}\right)\right]^{\frac{\gamma_g}{\gamma_g - 1}} \tag{5.39}$$

By substituting the expression for T_c/T_{t5} in terms of γ_g, the P_{t5}/P_c can be expressed as

$$\frac{P_{t5}}{P_c} = \frac{1}{\left[1 - \frac{1}{\eta_{in}}\left(\frac{\gamma_g - 1}{\gamma_g + 1}\right)\right]^{\frac{\gamma_g}{\gamma_g - 1}}} \tag{5.40}$$

From Eq. (5.40) we find that pressure drop across the nozzle increases due to frictional losses in the nozzle. With the help of Eq. (5.40) and other equations, we can determine the specific thrust as

$$\mathcal{T}_s = a_0 \left[(1+f)\frac{V_9}{a_0} - M_0 + (1+f)\frac{R_g\, T_9/T_0}{R_a\, V_9/V_0} \frac{\left(1 - \frac{P_0}{P_9}\right)}{\gamma_a} \right] \tag{5.41}$$

It must be noted that the equation for specific thrust is a bit complicated as compared to the ideal turbojet cycle analysis. The next step is to find out TSFC from the following

$$\text{TSFC} - \frac{f}{\mathcal{T}_s} \tag{5.42}$$

Propulsive Efficiency

Now the propulsive efficiency η_p for a real turbojet engine can be derived easily in a similar manner from its basic definition as

$$\eta_p = \frac{2V_0\mathcal{T}_s}{a_0^2\left[(1+f)(V_9/a_0)^2 - M_0^2\right]} \tag{5.43}$$

Thermal Efficiency

Similarly, we can evaluate an expression for the thermal efficiency η_{th} as

$$\eta_{th} = \frac{a_0^2\left[(1+f)(V_9/a_0)^2 - M_0^2\right]}{2f\Delta H_C} \tag{5.44}$$

The overall efficiency η_o for the turbojet engine, a product of η_p and η_{th}, is given by

$$\eta_o = \eta_p \cdot \eta_{th} \tag{5.45}$$

We will use the above expressions to carry out the parametric analysis of a real turbojet engine by varying the compressor pressure ratio π_c at a particular turbine inlet temperature (TIT) and flight Mach number M_0. Let us first consider the case of sea level flight (M_0 = 0.85) condition for TIT of 1600 K. The performance parameters considered for the analysis of real cycle for a turbojet engine are given in Table 5.2. The variation of $\mathcal{T}_s$ and TSFC against π_c in a single plot are shown in Fig. 5.11. Note that the results of an ideal cycle analysis for the same condition are also shown in the same plot (Fig. 5.11) for comparing the results of real cycles with the ideal one. Form Fig. 5.11, it is evident that the losses decrease the specific thrust, of course, while retaining the general trend of $\mathcal{T}_f$ vs. π_c. Also, with increase in π_c the differences of $\mathcal{T}_s$ between the real and ideal cycles increases considerably. Interestingly, TSFC are higher for a real engine as compared to an ideal one because there will be

Table 5.2. Loss parameters of various components for the analysis of turbojet and turbofan engines

Component	Parameter	Symbol	Typical range
Air intake	Total pressure recovery	π_d	0.9
	Isentropic efficiency	η_{id}	0.85
Compressor	Polytropic efficiency	η_{pc}	0.85
Fan	Polytropic efficiency	η_{pF}	0.85
Combustor	Combustor efficiency	η_b	0.97
	Total pressure ratio	π_b	0.9
Turbine	Polytropic efficiency	η_{pt}	0.9
Exhaust nozzle	Isentropic efficiency	η_{in}	0.95
	Pressure ratio at nozzle exit	P_0/P_9	0.9

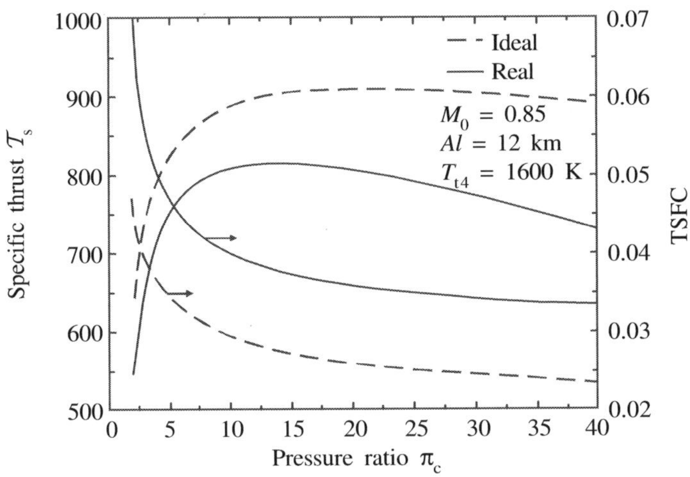

Fig. 5.11. Variation of T_s and TSFC with π_c for M_0 = 0.85 for both real and ideal turbojet engine.

an increase in specific heat losses across the main burner, leading to an inefficient combustion process and higher mass flow rate passing through the main burner. The plot of TSFC vs. π_c (Fig. 5.11) indicates that TSFC increases for a real engine. Interestingly, Fig. 5.11 shows that TSFC does not decrease with increasing π_c beyond a certain threshold value of π_c. Rather, TSFC for a real engine attains minimum value at certain π_c, particularly, for high flight Mach number M_0 and it increases further from this minimum point with increase in π_c. This is due to decrease in compressor efficiency η_{ic} at higher pressure ratio for same polytropic efficiency η_{pc}.

The variation of η_p, η_{th} and η_o with π_c for a real engine is shown in Fig. 5.12 along with that of an ideal engine. Figure 5.12 indicates that η_p for a real turbojet has higher values, which is attributed to the decrease in exit velocity of the exhaust nozzle for the same real turbojet engine. The plot of η_{th} vs. π_c shows that the real engine has lower values of η_{th}, as expected. However, η_o is found to be much lower for a real turbojet engine as compared to an ideal one, which is attributed to the decrease in thermal efficiency.

Now we will carry out an analysis for two flight Mach number cases corresponding to altitudes of 15 km. The variation of T and TSFC against π_c in a single plot (Fig. 5.13) shows that the variation of T_s decreases with π_c at high flight Mach number and differs to some extent from that of the ideal

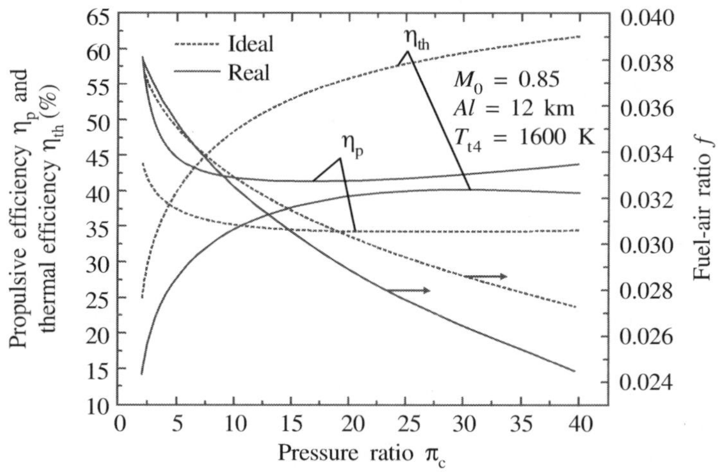

Fig. 5.12. Variation of η_p and η_{th} with π_c for both real and ideal turbojet engine.

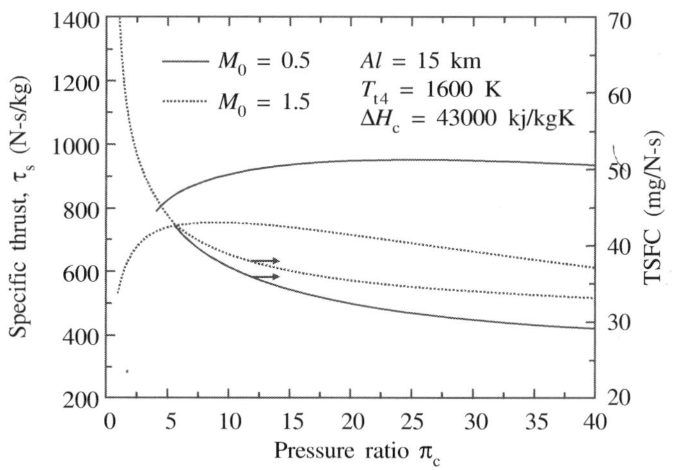

Fig. 5.13. Variation of T_s and TSFC with π_c at M_0 = 0 and 1.5 for a real turbojet engine.

cycle. Also, the thrust can be quite small at lower compressor pressure ratio due to exorbitant losses. The optimum compressor pressure ratio corresponding to maximum specific thrust has a lower value than that of the ideal turbojet engine for a particular flight Mach number.

The plot of η_o vs. π_c (Fig. 5.14) shows the general trend of η_o vs. π_c for all two flight Mach numbers. But the effects of losses for higher flight Mach number are quite significant, particularly, at higher compressor pressure ratio as compared to static condition.

Example 5.1 A turbojet is flying at flight Mach number M_0 = 0.8 at an altitude of 15 km. The pressure ratio across an axial compressor with polytropic efficiency of 85% is 12. The turbine has a polytropic efficiency of 88%. At the combustion chamber exit the temperature is 1600 K. The burner has an efficiency of 98% and a total pressure ratio of 0.95. A convergent nozzle with isentropic efficiency of 95% is used to produce thrust. The mechanical efficiency of this engine is assumed to be 98%. If the total pressure recovery factor in the air intake is 0.91, estimate

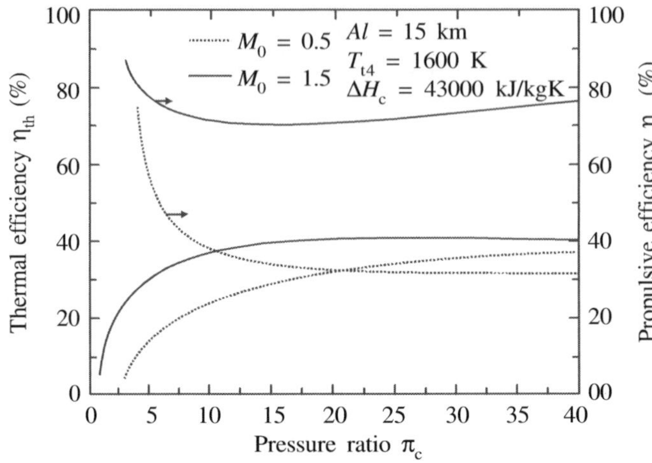

Fig. 5.14. Variation of η_p and η_{th} with π_c at two values of M_0 for a real turbojet engine.

the net thrust produced by the engine if it consumes 50 kg/s of air. Determine $\mathcal{T}_s$, TSFC, η_P and η_{th} if the heat of combustion is 43,000 kJ/kg. Compare these results with Example 4.2 (Take $\gamma_g = 1.333$, $\gamma_a = 1.4$).

Given $Al = 15$ km, $T_0 = 216.7$ K, $P_0 = 11.2$ kPa, $\pi_c = 12$, $T_{t4} = 1600$ K, $M_0 = 0.8$, $\dot{m}_0 = 50$ kg/s, $C_{Pa} = 1.005$ kJ/kgK, $C_{Pg} = 1.148$ kJ/kgK, $\Delta H_C = 43{,}000$ kJ/kg, $\eta_{pc} = 0.85$, $\eta_{pt} = 0.88$, $\eta_b = 0.98$, $\pi_b = 0.95$, $\eta_n = 0.95$, $\eta_m = 0.98$, PRF = 0.91.

Solution The processes involved during the operation of a real turbojet engine are shown in Fig. 5.9 with the station numbers. We know that specific thrust $\mathcal{T}_s$ is expressed as

$$\mathcal{T}_s = [(1+f)V_9 - V_0] + \frac{A_9}{\dot{m}_0}(P_9 - P_0)$$

We know that flight velocity is given as

$$V_0 = \sqrt{\gamma R T_0} \cdot M_0 = \sqrt{1.4 \times 287 \times 216.7} \times 0.8 = 236.06 \text{ m/s}$$

But the nozzle exit velocity is to be determined by

$$V_9 = \sqrt{2C_p(T_{t5} - T_9)}$$

In order to evaluate T_{t5} and T_9, we need to estimate the temperature and pressure at each station number.

Air Intake

The total pressure and temperature at station (0) can be evaluated as

$$T_{t0} = T_0\left(1 + \frac{\gamma - 1}{2} M_0^2\right) = 216.7\left(1 + 0.2 \times 0.8^2\right) = 244.43 \text{ K}$$

Let us now evaluate the pressure ratio P_{t0}/P_0 as

$$\frac{P_{t0}}{P_0} = \left(\frac{T_{t0}}{T_0}\right)^{\frac{\gamma_a}{(\gamma_a - 1)}} = \left(\frac{244.43}{216.7}\right)^{3.5} = 1.52 \quad \Rightarrow P_{t0} = \frac{P_{t0}}{P_0} \times P_0 = 17.07 \text{ kPa}$$

As PRF = 0.91, pressure at the air intake exit becomes

$$P_{t2} = \text{PRF} \times P_{t0} = 0.91 \times 17.07 = 15.53 \text{ kPa}$$

Compressor

The total pressure ratio across the compressor is given as

$$\frac{P_{t3}}{P_{t2}} = 12 \Rightarrow P_{t3} = \frac{P_{t3}}{P_{t2}} \times P_{t2} = 186.43 \text{ kPa}$$

Using the polytropic efficiency of the compressor, let us now evaluate the total temperature ratio T_{t3}/T_{t2} as

$$\frac{T_{t3}}{T_{t2}} = \left(\frac{P_{t3}}{P_{t2}}\right)^{\frac{\gamma_a - 1}{\gamma_a \cdot \eta_{pc}}} = 12^{\frac{1.4-1}{1.4\times 0.85}} = 2.305 \Rightarrow T_{t3} = \frac{T_{t3}}{T_{t2}} \times T_{t2} = 563.53 \text{ K}$$

Combustor

By carrying out energy balance across the combustor, we can find out the expression for fuel-air ratio f as

$$f = \frac{C_{pg}T_{t4} - C_{pa}T_{t3}}{\eta_b \Delta H_c - C_{pg}T_{t4}} = \frac{1148 \times 1600 - 1005 \times 563.63}{0.98 \times 43000 \times 10^3 - 1148 \times 1600} = 0.0315$$

Then we can determine $\dot{m}_f$ as

$$\dot{m}_f = 0.0315 \times 50 = 1.575 \text{ kg/s}$$

Turbine

As there will be pressure loss in the combustor, the pressure at the turbine inlet P_{t4} becomes

$$P_{t4} = \pi_b \times P_{t3} = 0.95 \times 186.43 = 177.11 \text{ kPa}$$

The total temperature at the turbine exit is to be evaluated. We know that the function of a turbine is to supply the requisite power to the compressor. Hence

$$\dot{W}_c = \eta_m \dot{W}_t \Rightarrow \dot{m}_a C_{pa}(T_{t3} - T_{t2}) = \eta_m(\dot{m}_a + \dot{m}_f)C_{pg}(T_{t4} - T_{t5})$$

From the above equation, we can estimate the turbine exit temperature as

$$T_{t5} = T_{t4} - \frac{C_{pa}}{C_{pg}} \frac{1}{\eta_m(1+f)}(T_{t3} - T_{t2}) = 1600 - \left(\frac{1005}{1148}\right)\frac{1}{0.98(1+0.0315)}(563.53 - 244.43) = 1323.7 \text{ K}$$

So the total pressure P_{t5} can be determined as

$$\frac{P_{t5}}{P_{t4}} = \left(\frac{T_{t5}}{T_{t4}}\right)^{\frac{\gamma_g}{(\gamma_g - 1)\eta_{pt}}} = \left(\frac{1323.7}{1600}\right)^{\frac{1.333}{(1.333-1)0.85}} = 0.422$$

$$\Rightarrow P_{t5} = \frac{P_{t5}}{P_{t4}} \times P_{t4} = 0.422 \times 177.11 = 74.76 \text{ kPa}$$

Nozzle

The critical pressure ratio across the nozzle is determined as

$$\frac{P_{t5}}{P_c} = \frac{1}{\left[1 - \frac{1}{\eta_n}\left(\frac{\gamma_g - 1}{\gamma_g + 1}\right)\right]^{\frac{\gamma_g}{\gamma_g - 1}}} = \frac{1}{\left[1 - \frac{1}{0.95}\left(\frac{1.333-1}{1.333+1}\right)\right]^{\frac{1.333}{1.333-1}}} = 1.914$$

But the actual pressure ratio across the nozzle with respect to the ambient pressure is

$$\frac{P_{t5}}{P_0} = \frac{74.76}{11.2} = 6.67$$

Since $\frac{P_{t5}}{P_0} > \frac{P_{t5}}{P_c}$, the nozzle is considered to be choked. So the static temperature at the nozzle exit T_9 becomes

$$T_9 = T_c = \left(\frac{2}{\gamma_g + 1}\right) T_{t5} = \left(\frac{2}{1.333+1}\right) \times 1323.7 = 1134.7 \text{ K}$$

Now the static pressure at the nozzle exit P_9 becomes

$$P_9 = P_c = \left(\frac{P_c}{P_{t5}}\right) P_{t5} = 38.9 \text{ kPa}$$

By using equation of state, the density of gas at the nozzle exit is estimated as

$$\rho_9 = \frac{P_9}{RT_9} = 0.119 \text{ kg/m}^3$$

Hence, we can evaluate V_9 as

$$V_9 = C_9 = \sqrt{\gamma_g R T_c} = \sqrt{1.333 \times 287 \times 1134.7} = 658.87 \text{ m/ s}$$

Let us evaluate $\frac{A_9}{\dot{m}}$ using the continuity equation as

$$\frac{A_9}{\dot{m}} = \frac{1}{\rho_9 V_9} = 0.0127 \text{ m}^2\text{s/kg}$$

$$\mathcal{T}_s = [(1+f)V_9 - V_0] + \frac{A_9}{\dot{m}_0}(P_9 - P_0) = (1.0315 \times 658.87 - 236.06) + 0.0127(38.9 - 11.2) \times 10^3 = 795.77 \text{ Ns/kg}$$

In order to estimate TSFC, we will have to evaluate the fuel-air ratio as

$$\text{TSFC} = \frac{f}{\mathcal{T}_s} = \frac{0.0315}{795.77} = 39.61 \text{ mg/N.s}$$

The propulsion efficiency can be evaluated as

$$\eta_p = \frac{2V_0(\mathcal{T}_s)}{a_0^2\left[(1+f)(V_9/a_0)^2 - M_0^2\right]}$$

$$\eta_p = \frac{2 \times 236.06(795.77)}{236.06^2\left[1.0315(658.87/295.07)^2 - 0.8^2\right]} = 95.82\%$$

The thermal efficiency can be evaluated as

$$\eta_{th} = \frac{1\left[(1+f)V_9^2 - V_0^2\right]}{2f\Delta H_C} = 14.46\%$$

5.9 Turbofan

Now we consider the real cycle analysis of a turbofan engine. The processes involved in a typical turbofan engine with losses is described in detail in a *T-s* diagram (Fig. 5.15). The station numbers shown in Fig. 5.15a are discussed in Chapter 4. We will follow a similar procedure to derive an expression for the performance parameters for a real turbofan engine as discussed for the turbojet engine. We may recall that the total thrust contributed by both primary (core) and secondary (fan) streams of a real turbofan engine is given by

$$\mathcal{T} = \dot{m}_c\left(V_9 - a_0\right) + A_9\left(P_9 - P_0\right) + \dot{m}_F\left(V_{19} - V_0\right) + A_{19}\left(P_{19} - P_0\right) \tag{5.46}$$

where the core thrust $\dot{m}_c\ (V_9 - a_0) + A_9\ (P_9 - P_0)$ is due to primary (core) stream and the fan thrust $\dot{m}_F\ (V_{19} - V_0) + A_{19}\ (P_{19} - P_0)$ is contributed by the secondary (fan) stream. So the total specific thrust $\mathcal{T}_f$ can be derived as

$$\mathcal{T}_s = \frac{\mathcal{T}}{\dot{m}_0} = \frac{a_0}{\alpha+1}\left[(1+f)\frac{V_9}{a_0} - M_0 + (1+f)\frac{R_g\left(T_9/T_0\right)}{R_a\left(V_9/a_0\right)}\frac{\left(1-\dfrac{P_0}{P_9}\right)}{\gamma_a}\right] \tag{5.47}$$

$$+\frac{a_0\alpha}{\alpha+1}\left[\frac{V_{19}}{a_0} - M_0 + \frac{T_{19}/T_0}{V_{19}/a_0}\frac{\left(1-\dfrac{P_0}{P_{19}}\right)}{\gamma_a}\right]$$

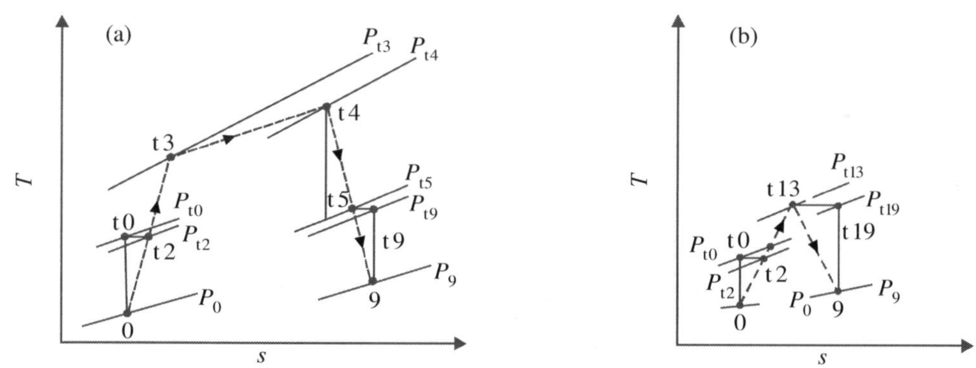

Fig. 5.15. *T-s* diagram depicting processes involved in a real turbofan engine.

By carrying out similar analysis, we can obtain an expression for both V_9/a_0 and V_{19}/a_0 as

$$\frac{V_9}{a_0} = \sqrt{\frac{2}{\gamma_g - 1}\left[\left(\pi_r, \pi_d, \pi_c, \pi_b, \pi_t, \pi_n\frac{P_0}{P_9}\right)^{\frac{\gamma_g-1}{\gamma_g}} - 1\right]} \times \sqrt{\left(\frac{\gamma_g R_g C_{pa}}{\gamma_a R_a C_{pg}}\right)\frac{\tau_\lambda \tau_t}{\left(\pi_r, \pi_d, \pi_c, \pi_b, \pi_t, \pi_n\dfrac{P_0}{P_9}\right)^{\frac{\gamma_g-1}{\gamma_s}}}} \tag{5.48}$$

$$\frac{V_{19}}{a_0} = \sqrt{\frac{2}{\gamma_a - 1}\left[\left(\pi_r, \pi_d, \pi_F, \pi_{Fn}\frac{P_0}{P_{19}}\right)^{\frac{\gamma_a - 1}{\gamma_a}} - 1\right]} \times \sqrt{\frac{\tau_r \tau_F}{\left(\pi_r \pi_d \pi_F \pi_{Fn}\frac{P_0}{P_{19}}\right)^{\frac{\gamma_a - 1}{\gamma_a}}}} \tag{5.49}$$

Of course, we will have to derive an expression for τ_t. Note that both compressor and fan are run using the power obtained by expansion of gas in the turbine. The mechanical losses in transmitting power from the turbine to the compressor and fan can be taken into account from mechanical efficiency η_m. By carrying out an energy balance between the turbine, compressor and fan, we have

$$\tau_t = 1 - \frac{1}{\eta_m(1+f)}\frac{\tau_r}{\tau_\lambda}\left[(\tau_c - 1) + \alpha(\tau_F - 1)\right] \tag{5.50}$$

where η_m is the mechanical efficiency as defined earlier. In a way similar to the real turbojet engine, we can derive an expression for fuel-air ratio f for the turbofan engine as

$$f = \frac{\tau_\lambda - \tau_r \tau_c}{\dfrac{\eta_b \Delta H_C}{(C_{pa} T_0)} - \tau_\lambda} \tag{5.51}$$

Hence, the relation for TSFC can be estimated by

$$\text{TSFC} = \frac{f}{(1+\alpha)T_s} \tag{5.52}$$

Efficiencies

The expressions for propulsive and thermal efficiencies can be derived in similar ways as

$$\eta_p = \frac{2M_0\left[(1+f)V_9/a_0 + \alpha(V_{19}/a_0) - (1+\alpha)M_0\right]}{(1+f)(V_9/a_0)^2 + \alpha(V_{19}/a_0)^2 - (1+\alpha)M_0^2} \tag{5.53}$$

$$\eta_{th} = \frac{a_0^2\left[(1+f)(V_9/a_0)^2 + \alpha(V_{19}/a_0)^2 - (1+\alpha)M_0^2\right]}{2f\Delta H_C} \tag{5.54}$$

Keep in mind we have assumed that both the nozzles are fully expanded ($P_9 = P_0$) and ($P_{19} = P_0$) while deriving the above expressions. The overall efficiency becomes

$$\eta_o = \eta_p \cdot \eta_{th} \tag{5.55}$$

Let us now carry out parametric performance analysis of a turbofan engine by using the above expressions. The turbofan is being used widely in long range passenger aircraft and for the present discussion we take its cruise flight speed M_0 as 0.85. The loss coefficients and efficiencies of various components considered in this parametric real cycle analysis are given in Table 5.1. We will also use a constant turbine inlet temperature of 1600 K and fan pressure ratio π_F of 2.0 while varying the compressor pressure ratio π_c for three bypass ratios viz., 0.5, 2.0 and 5.0.

The variation of $\mathcal{T}_s$ with π_c for all three bypass ratios viz. 0.5, 2.0 and 5.0 along with the results of ideal turbofan engines (Fig. 5.16) shows that the difference in $\mathcal{T}_s$ between real and ideal turbofan engines increases with increase in bypass ratio α. The variation of TSFC with π_c (Fig. 5.17) exhibits a similar trend as an ideal cycle but the real engine has higher TSFC than the ideal one under same condition. This difference is caused by higher consumption of fuel due to losses incurred in the various components of real engines. Similar trends are seen in the variation of η_p and η_{th} with π_c (Fig.5.18).

Now we will investigate the influence of flight Mach number M_0 and bypass ratio α on the performance of the turbofan engine for a given π_c of 12 and π_F = 2.0. The loss coefficients and efficiencies considered for this analysis are given in Table 5.2. The results of $\mathcal{T}_s$ vs. M_0 for a real turbofan engine (Fig. 5.19) show that the specific thrust decreases with flight Mach number M_0 and bypass ratio α. Moreover, the difference between the two different bypass ratio engines increases at

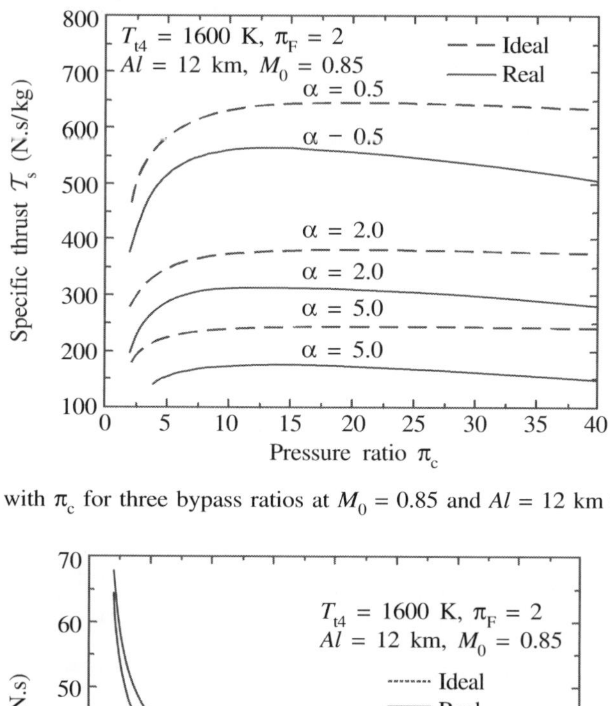

Fig. 5.16. Variation of $\mathcal{T}_s$ with π_c for three bypass ratios at M_0 = 0.85 and Al = 12 km for a real turbofan engine.

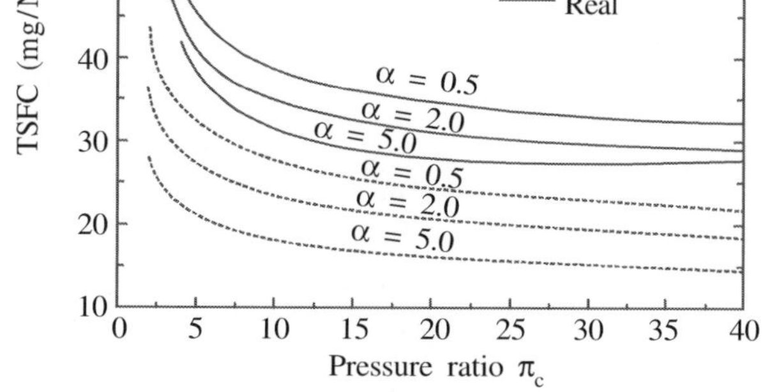

Fig. 5.17. Variation of TSFC with π_c for three bypass ratios at M_0 = 0.85 and Al = 12 km for a real turbofan engine.

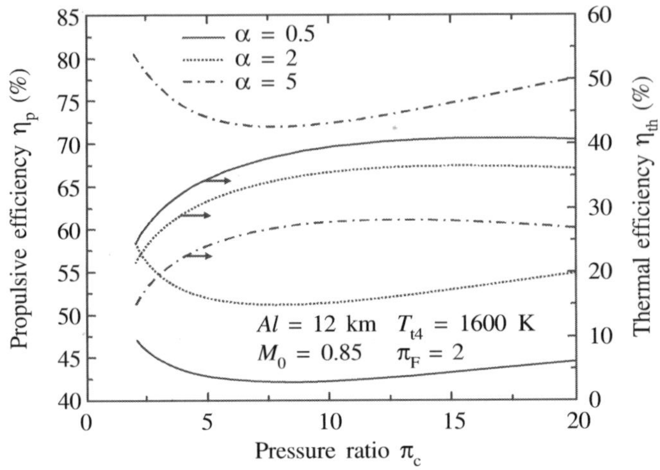

Fig. 5.18. Variation of η_p and η_{th} with π_c for three bypass ratios at $M_0 = 0.85$ for a real turbofan engine.

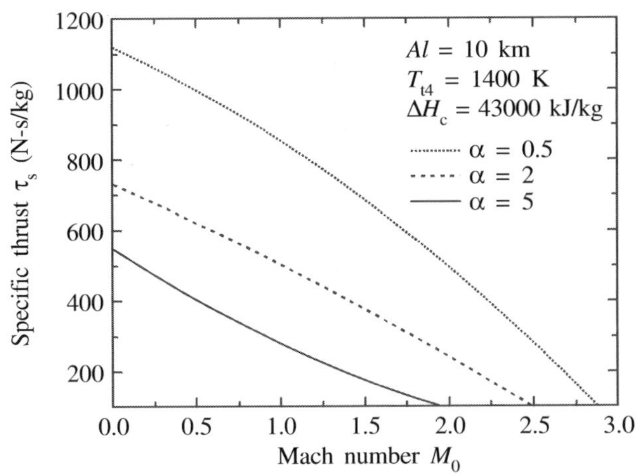

Fig. 5.19. Variation of T_s with M_0 for three bypass ratios at $Al = 10$ km for a real turbofan engine.

transonic flight Mach number due to increasing air intake total pressure loss. Interestingly, for economical operation of a turbofan engine, the limiting Mach number for a given bypass ratio in the case of a real engine is found to be lower as compared to an ideal engine. The variation of TSFC with M_0 (Fig. 5.20) indicates that TSFC increases sharply at higher Mach number. Although TSFC is lower for higher bypass ratio at low flight speed, it becomes higher at higher M_0. For real engines, this crossover of TSFC occurs for higher α (= 5.0) at lower M_0 as compared to ideal engines. The variation of η_p and η_{th} with M_0 (Fig. 5.21) shows that the real engine has higher propulsive efficiency as compared to an ideal engine due to reduction in exit velocity caused by losses in various components of the engine. In contrast, the thermal efficiency of a real engine decreases under the same flight condition.

Example 5.2 A turbofan engine is flying at an altitude of 10 km and flight Mach number M_0 of 0.85. The pressure ratio across the compressor is 10 and it has a polytropic efficiency of 85%. In contrast, the pressure ratio across the fan is 2 and it has a polytropic efficiency of 82%. The exit temperature at the combustor is 1600 K. The burner

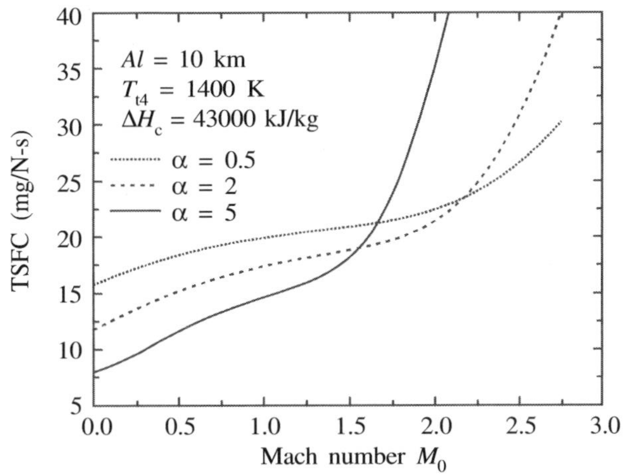

Fig. 5.20. Variation of TSFC with M_0 for three bypass ratios at Al = 10 km for a real turbofan engine.

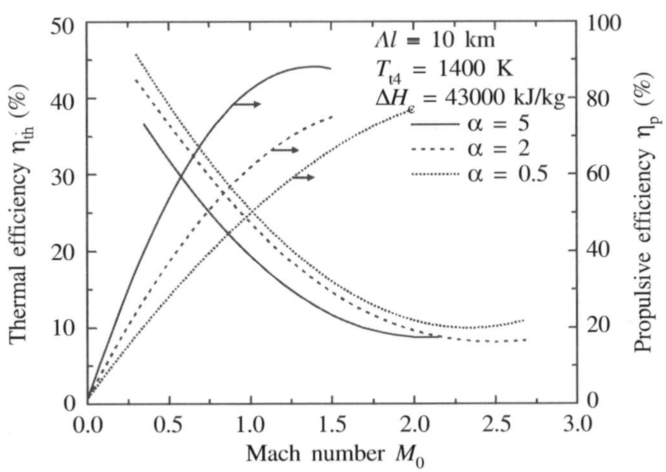

Fig. 5.21. Variation of η_p and η_{th} with M_0 for three bypass ratios at Al = 10 km for a real turbofan engine.

has an efficiency of 94% and a total pressure ratio of 0.95. The bypass ratio of the turbofan is 5. Two separate convergent nozzles with same isentropic efficiency of 95% are used to produce thrust. The mechanical efficiency of this engine is assumed to be 95%. If the total pressure recovery factor in the air intake is 0.91, determine $\mathcal{T}_s$, TSFC, η_p, η_{th} and η_o. Determine $\mathcal{T}_s$ and TSFC under static condition at sea level, if it consumes 150 kg/s of air. Compare these results with Example 4.4. (Take ΔH_c = 43,000 kJ/kg, γ_g =1.3, γ_a = 1.4).

Given Al = 10 km; T_0 = 223.3 K, P_0 = 26.5 kPa, T_{t4} = 1600 K, M_0 = 0.85, ΔH_C = 43,000 kJ/kg, η_{nF} = 0.82, α = 5.0, η_{pc} = 0.85, η_{pt} = 0.88, η_b = 0.94, π_b = 0.95, η_n = 0.95, η_m = 0.95, PRF = 0.91.

Solution $C_{p,a}$ = 1.005 kJ/kgK, $C_{p,g}$ = 1.148 kJ/kgK, γ_a = 1.4, γ_g = 1.333. The processes involved during the operation of real turbofan engine are shown in Fig. 5.15 with the station numbers. We know that $\mathcal{T}_s$ is given as

$$\mathcal{T}_s = \frac{\mathcal{T}}{\dot{m}_0} = \frac{a_0}{\alpha+1}\left[(1+f)\frac{V_9}{a_0} - M_0 + (1+f)\frac{R_g(T_9/T_0)}{R_a(V_a/a_0)}\frac{\left(1-\frac{P_0}{P_9}\right)}{\gamma_a}\right] + \frac{a_0\alpha}{\alpha+1}\left[\frac{V_{19}}{a_0} - M_0 + \frac{T_{19}/T_0}{V_{19}/a_0}\frac{\left(1-\frac{P_0}{P_{19}}\right)}{\gamma_a}\right]$$

We know that flight velocity is given as

$$V_0 = 0.85 \times \sqrt{1.4 \times 287 \times 223.3} = 254.6 \text{ m/s}$$

But we need to evaluate the nozzle exit velocity of both fan and core stream. For this we will have to evaluate the properties at each station.

Air Intake

By using the isentropic relation, we can determine the total temperature T_{t0} as

$$T_{t0} = T_{t1} = T_{t2} = \left[1 + \frac{\gamma_a - 1}{2} M_0^2\right] T_0 = 255.56 \text{ K}$$

The pressure ratio at station (0) can be determined as

$$\frac{P_{t0}}{P_0} = \left(\frac{T_{t0}}{T_0}\right)^{\frac{\gamma_a}{\gamma_a - 1}} = \left(\frac{255.56}{223.3}\right)^{\frac{1.4}{1.4-1}} = 1.6$$

As the PRF in the air intake is 0.91, the total pressure at station (2) becomes

$$P_{t2} = \frac{P_{t2}}{P_{t0}} \frac{P_{t0}}{P_0} P_0 = 1.6 \times 0.91 \times 26.5 = 38.6 \text{ kPa}$$

For Fan Stream

As π_F = 2.0, the total pressure at station (13) can be estimated as

$$P_{t13} = \frac{P_{t13}}{P_{t2}} P_{t2} = 2 \times 38.6 = 77.2 \text{ kPa}$$

By considering polytropic efficiency of the fan, the total temperature at station (8) can be estimated as

$$T_{t19} = T_{t13} = \left(\frac{P_{t13}}{P_{t2}}\right)^{\frac{\gamma_a - 1}{\gamma_a \times \eta_{pF}}} \times T_{t2} = 2^{\frac{0.4}{1.4 \times 0.82}} \times 255.56 = 324.56 \text{ K}$$

The critical pressure ratio across the fan nozzle is determined as

$$\frac{P_{t13}}{P_c} = \frac{1}{\left[1 - \frac{1}{\eta_{nF}}\left(\frac{\gamma_a - 1}{\gamma_a + 1}\right)\right]^{\frac{\gamma_a}{\gamma_a - 1}}} = \frac{1}{\left[1 - \frac{1}{0.95}\left(\frac{1.4 - 1}{1.4 + 1}\right)\right]^{\frac{1.4}{1.4-1}}} = 1.96$$

But the actual pressure ratio across the fan nozzle with respect to ambient pressure is

$$\frac{P_{t13}}{P_0} = \frac{77.2}{26.5} = 2.91$$

Since $\frac{P_{t13}}{P_0} > \frac{P_{t13}}{P_c}$, the nozzle is considered to be choked. So the static temperature at the nozzle exit T_9 becomes

$$T_{19} = T_c = \left(\frac{2}{\gamma_a + 1}\right) T_{t19} = \left(\frac{2}{1.4 + 1}\right) \times 255.56 = 213 \text{ K}$$

Then the static pressure at the nozzle exit P_9 becomes

$$P_{19} = P_c = \left(\frac{P_c}{P_{t13}}\right) P_{t13} = \frac{1}{1.92} \times 77.2 = 40.2 \text{ kPa}$$

By using equation of state, the density of gas at the nozzle exit is estimated as

$$\rho_{19} = \frac{P_{19}}{RT_{19}} = \frac{40.2\times 10^3}{287\times 213} = 0.657 \text{ kg/m}^3$$

Hence, we can evaluate V_{19} as

$$V_{19} = C_{19} = \sqrt{\gamma_g RT_c} = \sqrt{1.4\times 287\times 213} = 292.54 \text{ m/s}$$

Let us evaluate $\frac{A_{19}}{\dot{m}_F}$ using continuity equation as

$$\frac{A_{19}}{\dot{m}_F} = \frac{1}{\rho_{19}V_{19}} = \frac{1}{0.657\times 292.54} = 5.2\times 10^{-3} \text{ m}^2\text{s/kg}$$

The specific thrust with respect to total mass flow rate contributed by the fan can be evaluated as

$$T_{sF} = \frac{\alpha}{\alpha+1}\left[(V_{19}-V_0)+\frac{A_{19}}{\dot{m}}(P_{19}-P_a)\right] = \frac{5}{5+1}\left[(292.54-254.6)+5.2\times 10^{-3}(40.2-26.5)\right] = 31.67 \text{ N.s/kg}$$

For Core Stream

We will use a similar procedure to evaluate the properties in each section of the core stream as given below:

Compressor

The total pressure ratio across both the fan and the compressor can be determined as

$$\frac{P_{t3}}{P_{t2}} = 10\times 2 = 20$$

Then the total pressure at station (3) becomes

$$P_{t3} = \frac{P_{t3}}{P_{t2}}\times P_{t2} = 20\times 38.61 = 772 \text{ kPa}$$

Using the polytropic efficiency of the compressor, let us now evaluate the total temperature ratio T_{t3}/T_{t2} as

$$\frac{T_{t3}}{T_{t2}} = \left(\frac{P_{t3}}{P_{t2}}\right)^{\frac{\gamma_a-1}{\gamma_a\cdot\eta_{pc}}} = 20^{\frac{1.4-1}{1.4\times 0.85}} = 2.74 \Rightarrow T_{t3} = \frac{T_{t3}}{T_{t2}}\times T_{t2} = 700.2 \text{ K}$$

Combustor

By carrying out energy balance across the combustor, we can find out the expression for fuel-air ratio f as

$$f = \frac{C_{pg}T_{t4} - C_{pa}T_{t3}}{\eta_b\Delta H_c - C_{pg}T_{t4}} = \frac{1148\times 1600 - 1005\times 700.2}{0.94\times 43000\times 10^3 - 1148\times 1600} = 0.029$$

As the pressure ratio across the combustor is 0.95, the pressure at the turbine inlet P_{t4} becomes

$$P_{t4} = \pi_b \times P_{t3} = 0.95 \times 772 = 733.4 \text{ kPa}$$

Turbine

The total temperature at the turbine exit is to be evaluated. We know that the function of a turbine is to supply the requisite power to the compressor. Hence

$$\dot{W}_c + \dot{W}_F = \eta_m \dot{W}_t$$

From the previous equation, we can estimate exit temperature as

$$T_{t5} = T_{t4} - \frac{1}{\eta_m (1+f)} \frac{C_{pc}}{C_{pg}} [(T_{t3} - T_{t2}) + \alpha(T_{t13} - T_{t2})]$$

$$T_{t5} = 1600 - \frac{1}{0.95 \times 1.029}\left(\frac{1005}{1148}\right)[(700.2 - 255.56) + 5(324.56 - 255.56)] = 892.84 \text{ K}$$

Then the total pressure P_{t5} can be determined by using the polytropic relationship

$$\frac{P_{t5}}{P_{t4}} = \left(\frac{T_{t5}}{T_{t4}}\right)^{\frac{\gamma_g}{(\gamma_g - 1)\eta_{pt}}} = \left(\frac{892.84}{1600}\right)^{\frac{1.333}{0.333 \times 0.88}} = 0.069 \Rightarrow P_{t5} = \frac{P_{t5}}{P_{t4}} \times P_{t4} = 0.069 \times 733.4 = 50.6 \text{ kPa}$$

Nozzle

The critical pressure ratio across the nozzle is determined as

$$\frac{P_{t5}}{P_c} = \frac{1}{\left[1 - \frac{1}{\eta_n}\left(\frac{\gamma_g - 1}{\gamma_g + 1}\right)\right]^{\frac{\gamma_g}{\gamma_g - 1}}} = \frac{1}{\left[1 - \frac{1}{0.95}\left(\frac{1.333 - 1}{1.333 + 1}\right)\right]^{\frac{1.333}{1.333 - 1}}} = 1.914$$

But the actual pressure ratio across the nozzle with respect to the ambient pressure is

$$\frac{P_{t5}}{P_0} = \frac{50.6}{26.5} = 1.90$$

Since $\frac{P_{t5}}{P_0} < \frac{P_{t5}}{P_c}$, the nozzle is not choked. Hence, we can evaluate V_9 as

$$V_9 = \sqrt{2\eta_n C_{pg} T_{t5}\left[1 - \left(\frac{P_9}{P_{t5}}\right)^{\frac{\gamma_9 - 1}{\gamma_9}}\right]} = \sqrt{2 \times 0.95 \times 1148 \times 892.84 \times \left[1 - \left(\frac{26.5}{50.6}\right)^{\frac{0.333}{1.333}}\right]} = 539 \text{ m/s}$$

The specific thrust for the core steam can be estimated as

$$\mathcal{T}_{sc} = \frac{1}{1+\alpha}[(1+f)V_9 - V_0)] = \frac{1}{6}[1.029 \times 539 - 254.6] = 50 \text{ Ns/kg}$$

The total specific thrust becomes

$$\mathcal{T}_s = \mathcal{T}_{sc} + \mathcal{T}_{sc} = 50 + 31.67 = 81.67 \text{ Ns/kg}$$

In order to estimate TSFC, we will have to evaluate the fuel-air ratio as

$$\text{TSFC} = \frac{f}{\mathcal{T}_s} = \frac{0.029}{81.67} = 355.1 \text{ mg/Ns}$$

The propulsion efficiency can be evaluated as

$$\eta_P = \frac{2M_0\left[(1+f)(V_9/a_0) + \alpha\left(V_{19}/a_0\right) - (1+\alpha)M_0\right]}{(1+f)\left(\frac{V_9}{a_0}\right)^2 + \alpha\left(\frac{V_{19}}{V_0}\right)^2 - (1+\alpha)M_0^2}$$

Substituting the values in the above expression, we can have,

$$\eta_P = \frac{2 \times 0.85[1.029(1.794) + 5(0.976) - (6 \times 0.85)]}{\left[(1.029 \times 1.794^2) + (5 \times 0.976^2) - (6 \times 0.85^2)\right]} = 73.92\%$$

The thermal efficiency can be evaluated as

$$\eta_{\text{th}} = \frac{a_0^2\left[(1+f)\right](V_9/V_0)^2 + \alpha(V_{19}/V_0)^2 - (1+\alpha)M_0^2}{2f\Delta H_{\text{C}}} = 13.45\%$$

5.10 Turboprop

Let us now find out the relevant expression to carry out parametric cycle analysis for a real turboprop engine in a similar manner as we have been done for turbojet and turbofan engines. The various processes involved in a typical real turboprop engine are shown on a *T-s* plane in Fig. 5.22. We may refer to the schematic of a typical turboprop engine with proper station numbers given in Chapter 4. We will follow the usual steps to arrive at the relations for various performance parameters of a real turboprop engine. From the basic definition, the specific thrust of the core engine is given by

$$\mathcal{T}_{\text{sc}} = \frac{\mathcal{T}_{\text{c}}}{\dot{m}_{\text{c}}} = a_0\left[(1+f)\frac{V_9}{a_0} - M_0 + (1+f)\frac{R_{\text{g}}}{R_{\text{a}}}\frac{T_9/T_0}{V_9/a_0}\frac{(P_9 - P_0)}{P_9\,\gamma_{\text{a}}}\right] \qquad (5.56)$$

Note that the thrust generated by the core engine is quite low in the case of turboprop engines. The specific thrust due to the core stream will also be quite low, due to the fact that $\mathcal{T}_{\text{c}}$ is divided by total mass flow rate $\dot{m}_{\text{c}}$. From the definition of workout coefficient for the core stream C_{c} an expression for C_{c} can be derived using Eq. (5.56) as

$$C_{\text{c}} = \frac{\mathcal{T}_{\text{c}}V_0}{\dot{m}_{\text{c}}C_{\text{p}}T_0} = (\gamma_{\text{a}} - 1)M_0\left[(1+f)\left(\frac{V_9}{a_0}\right) - M_0 + (1+f)\frac{R_{\text{g}}}{R_0}\frac{(T_9/T_0)}{(V_9/a_0)}\left(\frac{P_9 - P_0}{P_9\,\gamma_{\text{a}}}\right)\right] \qquad (5.57)$$

The expression for V_9/a_0 is obtained in a similar manner as was done for turbojet engine

$$\frac{V_9}{a_0} = \sqrt{\frac{2\tau_\lambda\tau_{\text{tH}}\tau_{\text{tL}}}{\gamma_{\text{a}} - 1}\left[1 - \left(\frac{P_9}{P_{\text{t9}}}\right)^{\frac{\gamma_{\text{g}}-1}{\gamma_{\text{g}}}}\right]} \qquad (5.58)$$

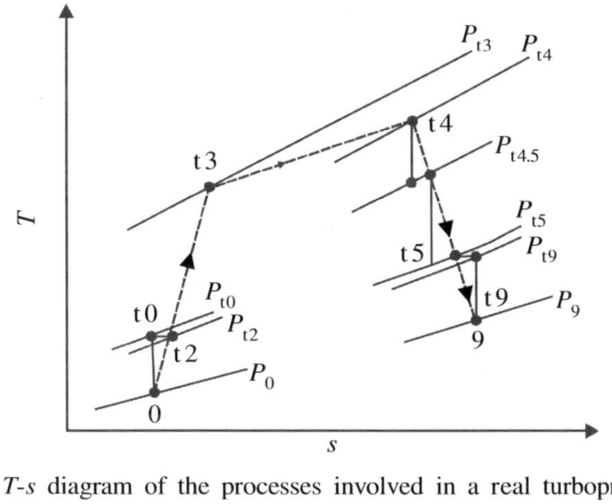

Fig. 5.22. *T-s* diagram of the processes involved in a real turboprop engine.

The above expression is valid for unchoked flow and exit pressure P_9 is equal to ambient pressure P_0.

By carrying out an energy balance for the main combustor in the core engine, we can derive

$$f = \frac{\tau_\lambda - \tau_r \tau_c}{\dfrac{\eta_b \,\Delta H_C}{C_{pa}\, T_0} - \tau_\lambda} \tag{5.59}$$

Now we must carry out an energy balance between the high pressure turbine and compressor to get

$$\tau_{tH} = 1 - \frac{1}{\eta_{mH}(1+f)} \frac{\tau_r}{\tau_\lambda}(\tau_c - 1) \tag{5.60}$$

where η_{mH} is the mechanical efficiency of the high pressure turbine shaft. But the low pressure turbine is meant to run the propeller. Hence, we must equate the power output of this low pressure turbine with the power being consumed by the propeller, which is given as

$$\dot{W}_{prop} = \eta_{gb}\ \eta_{mL} \dot{m}_{4.5} C_{pq} T_0\ \tau_{tH} \tau_{tL} \left(1 - \tau_{TL}\right) \tag{5.61}$$

where η_{mL} is the mechanical efficiency of the high pressure turbine shaft and η_{gb} the efficiency of the gear box, which transmits power from the low pressure turbine to the propeller.

By using Eq. (5.61), we can get an expression for C_{prop} as

$$C_{prop} = \frac{\eta_{prop} \dot{W}_{prop}}{\dot{m}_c C_{pa} T_0} = \eta_{prop} \eta_{gb} \eta_{mL} (1+f) \tau_{tH} \tau_\lambda \left(1 - \tau_{tL}\right) \tag{5.62}$$

Now $C_{tot} = C_{prop} + C_c$. Hence, total specific thrust and workout coefficient for the turboprop becomes

$$T_s = \frac{T}{\dot{m}_c} = \frac{C_{tot} C_{pa} T_0}{V_0} \tag{5.63}$$

The expression for specific power $\dot{W}_s$ from the turboprop engine is given by

$$\dot{W}_s = \frac{\dot{W}_{tot}}{\dot{m}_c} = C_{tot} C_{pa} T_0 \tag{5.64}$$

TSFC and PSFC

Now let us derive the expressions for TSFC and PSFC as

$$\text{TSFC} = \frac{f\, V_0}{C_{tot} C_{pa} T_0} \tag{5.65}$$

$$\text{PSFC} = \frac{\dot{m}_f}{\dot{W}_s} = \frac{f}{C_{tot} C_{pa} T_0} \tag{5.66}$$

Thermal Efficiency

The expression for the thermal efficiency η_{th} of a real turboprop engine is the same as that of a turbojet engine

$$\eta_{th} = \frac{a_0^2 \left[(1+f)(V_9/a_0)^2 - M_0^2 \right]}{2 f \Delta H_C} \tag{5.67}$$

Here we have assumed that the gas is expanded fully.

Overall Efficiency

The overall efficiency η_o of a turboprop engine becomes

$$\eta_o = \frac{\dot{W}_{tot}}{\dot{m}_f \Delta H_C} = \frac{C_{tot} C_{pc} T_0}{f \Delta H_C} \tag{5.68}$$

Propulsive Efficiency

The propulsive efficiency η_p can be found out by

$$\eta_p = \frac{\eta_0}{\eta_{th}} \tag{5.69}$$

In the usual way, we will investigate the performance of a real turboprop engine by varying several parameters. However, we will restrict our parametric analysis to flight Mach number of 0.6 and TIT = 1400 K. The loss coefficients and various efficiencies used for this analysis are given Table 5.1. In this analysis, the compressor pressure ratio π_c and turbine temperature ratio τ_t are varied over a wide range. The results of $\mathcal{T}_s$ vs. π_c are shown in Fig. 5.23 along with results of an ideal engine. Although these results show a trend similar to that of an ideal engine, the value of $\mathcal{T}_s$ of a real engine for particular π_c and τ_t is found to be lower than that of the ideal engine. Interestingly, the optimum π_c for maximum $\mathcal{T}_s$ in the real engine is bit lower as compared to the ideal engine. The results of TSFC vs. π_c are shown in Fig. 5.23. Similar trends can be observed for TSFC. However, the optimum value of τ_t corresponding to minimum TSFC is found to be 0.45, which is higher than that of the ideal engine. The variation of work output coefficient for both propeller and core stream with compressor pressure ratio π_c for three values of temperature ratio across turbine τ_t is plotted in Fig. 5.24, along with the results of an ideal cycle. It can be noted that the real cycle has lower work output coefficient than that of ideal cycle as expected due to occurence of losses in the real cycle.

Example 5.3 A turboprop engine is flying at an altitude of 7 km (T_0 = 242 K, P_0 = 41.1 kPa) with flight Mach number of 0.5. It ingests 50 kg/s of air into an axial compressor with pressure ratio of 12 and polytropic efficiency of 85%. The turbine inlet temperature is restricted to 1400 K. The burner has an efficiency of 92% and total pressure

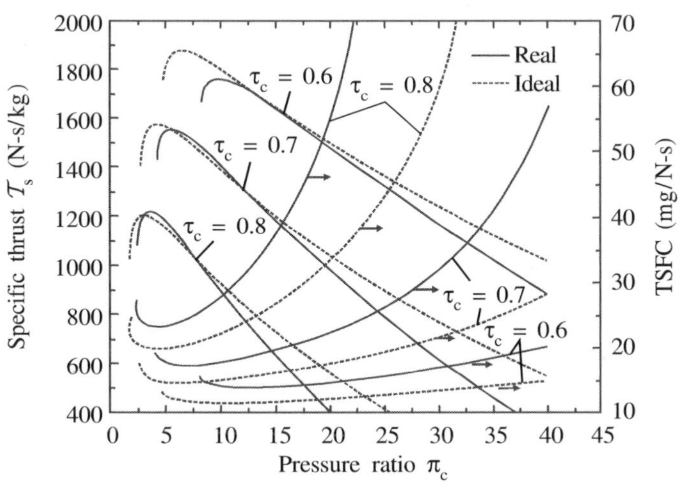

Fig. 5.23. Variation of $\mathcal{T}_s$ and TSFC with π_c for four τ_t at M_0 = 0.6 and Al = 7.5 km for a real turboprop engine.

ratio of 0.95. The temperature ratio across the turbine with a polytropic efficiency of 88% is 0.6. The overall efficiency pressure recovery factor in the air intake is 0.91. The efficiency of exhaust nozzle is 0.98. The mechanical efficiency of this engine is assumed to be 95% while gear box efficiency is 98%. If the propeller efficiency is 0.9, determine the thrust developed by both core engine and propeller, TSFC and η_0. Take ΔH_c = 43,000 kJ/kg.

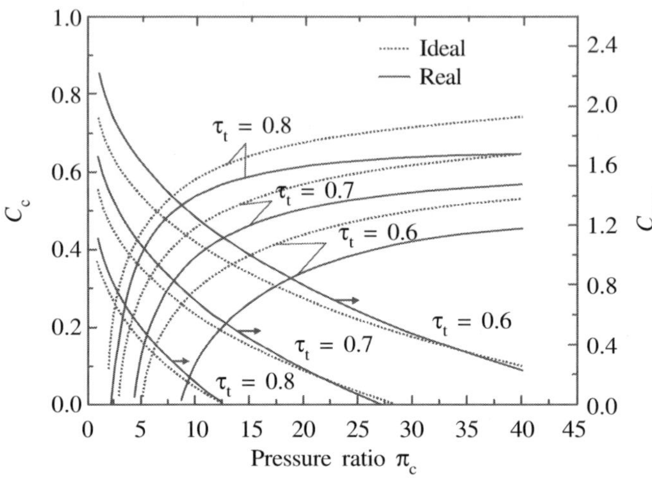

Fig. 5.24. Variation of η_p and η_{th} with π_c for four τ_t = 0.6, 0.7 and 0.8 at M_0 = 0.6 for a real turboprop engine.

Given Al = 7 km, T_0 = 242 K, P_0 = 41.1 kPa, T_{t4} = 1400 K, π = 12, τ_t = 0.6, η_{prop} = 0.9, M_0 = 0.5, $\dot{m}_0$ = 50 kg/s, η_{pc} = 0.85, η_n = 0.98, ΔH_C = 43,000 kJ/kg, $\eta_{mL} = \eta_{mH}$ = 0.95, η_{pt} = 0.88, η_b = 0.92, π_b = 0.95, η_{gb} = 0.98, PFR = $\dfrac{P_{t2}}{P_{t0}}$ = 0.95.

Assumption $C_{p,a}$ = 1.005 kJ/kgK, $C_{p,g}$ = 1.148 kJ/kgK, γ_a = 1.4, γ_g = 1.333.

Solution The processes involved during the operation of a turboprop engine are shown in Fig. 5.22 with the station numbers. We know that the specific thrust T_s is expressed as

$$T_{prop} = \frac{C_{prop}\dot{m}_c C_p T_0}{V_0} = \frac{\eta_{prop}\dot{W}_{prop}}{V_0} \text{ as } C_{prop} = \frac{\eta_{prop}\dot{W}_{prop}}{\dot{m}_c C_p T_0}$$

Let us evaluate flight velocity as

$$V_0 = a_0 M_0$$

But the speed of sound a_0 can be determined as

$$a_0 = \sqrt{\gamma R T_0} = \sqrt{1.4 \times 287 \times 242} = 311.82 \text{ m/s}$$

Then, the flight velocity is estimated as

$$V_0 = 311.82 \times 0.5 = 155.9 \text{ m/s}$$

But we know that $\dot{W}_{prop}$ is given as

$$\dot{W}_{prop} = \eta_{mL}\eta_{gb}\dot{m}_{4.5}C_{pg}\left(T_{t4.5} - T_{t5}\right)$$

In order to evaluate $T_{t4.5}$ we need to carry out an energy balance across the low pressure turbine as

$$C_{pa}(T_{t3}-T_{t2})=\eta_{mL}\eta_{gb}C_{pg}(T_{t4}-T_{t4.5}) \Rightarrow T_{t4.5}=T_{t4}-\frac{C_{pa}}{\eta_{mL}\eta_{gb}C_{pg}}(T_{t3}-T_{t2})$$

But we need to evaluate T_{t3} and T_{t2} to determine $T_{t4.5}$. Let us first evaluate T_{t2} by invoking the isentropic relationship as

$$T_{t0}=T_{t2}=\frac{T_{t2}}{T_0}T_0=\frac{T_{t0}}{T_0}T_0=\left(1+\frac{\gamma-1}{2}M_0^2\right)T_0=\left(1+0.2\times 0.5^2\right)242=254.1\text{ K}$$

Now the value of T_{t3} can be estimated easily by using the polytropic relationship as

$$T_{t3}=\frac{T_{t3}}{T_{t2}}T_{t2}=\left(\frac{P_{t3}}{P_{t2}}\right)^{\frac{\gamma_a-1}{\gamma_a\eta_{pc}}}T_{t2}=12^{0.336}\times 254.1=585.6\text{ K}$$

We can estimate $T_{t4.5}$ as

$$T_{t4.5}=T_{t4}-\frac{C_{pa}}{\eta_{mH}C_{pg}}(T_{t3}-T_{t2})=1400-\frac{1005}{1148\times 0.95}(585.6-254.1)=1094.5\text{ K}$$

The turbine exit temperature is estimated as

$$T_{t5} = 0.6 \times 1400 = 840\text{ K}$$

Since the pressure recovery factor is 0.95, the pressure at the compressor inlet is estimated to be

$$P_{t2}=\frac{P_{t2}}{P_{t0}}\cdot\frac{P_{t0}}{P_0}P_0=\frac{P_{t2}}{P_{t0}}\times\left(\frac{T_{t0}}{T_0}\right)^{\frac{\gamma_a}{\gamma_a-1}}P_0=0.95\left(\frac{254.1}{242}\right)^{\frac{1.4}{0.4}}\times 41.1=46.32\text{ kPa}$$

As the pressure ratio across the compressor is 12, the total pressure at exit compressor is 555.84 kPa. The pressure at the combustor exit becomes

$$P_{t4} = 0.95 \times 555.84 = 528.1\text{ kPa}$$

The thrust produced by the core engine is estimated as

$$\mathcal{T}_c=\dot{m}_0(V_9-V_0)$$

Assuming the nozzle to be fully expanded the nozzle exit velocity V_9 is estimated as

$$V_9=\sqrt{2\eta_n\times C_{pg}T_{t5}\left[1-\left(\frac{P_9}{P_{t5}}\right)^{\frac{\gamma_g-1}{\gamma_g}}\right]}$$

As we know that the total pressure ratio P_9/P_{t5} can be easily estimated as

$$\frac{P_9}{P_{t5}}=\frac{P_9}{P_0}\cdot\frac{P_0}{P_{t2}}\cdot\frac{P_{t2}}{P_{t3}}\cdot\frac{P_{t3}}{P_{t4}}\cdot\frac{P_{t4}}{P_{t5}}=1\times\frac{41.1}{46.32}\times\frac{1}{12}\times\frac{1}{6.95}\times\frac{1}{(0.6)^{\frac{1.333}{(1.333-1)0.88}}}=0.79\Rightarrow\frac{P_{t5}}{P_9}=1.27$$

Let us determine critical pressure ratio across the exhaust nozzle

$$\frac{P_{t5}}{P_c}=\frac{1}{\left[1-\frac{1}{\eta_{in}}\left(\frac{\gamma_g-1}{\gamma_g+1}\right)\right]^{\frac{\gamma_g}{(\gamma_g-1)}}}=\frac{1}{\left[1-\frac{1}{0.98}\left(\frac{0.333}{2.333}\right)\right]^4}=1.88$$

As $\frac{P_{t5}}{P_c}>\frac{P_{t5}}{P_9}$, the nozzle is not chocking.

Then we can determine V_9 by using the expression given above as

$$V_9 = \sqrt{2 \times 0.98 \times 1148 \times 840\left[1-(0.79)^{\frac{0.333}{1.333}}\right]} = 328.8 \text{ m/s}$$

By carrying out energy balance across combustion chamber, we will get

$$f = \frac{C_{pg}T_{t4} - C_{pa}T_{t3}}{\eta_b \Delta H_c - C_{pg}T_{t4}} = \frac{1.148 \times 1400 - 1.005 \times 585.6}{0.92 \times 43000 - 1.148 \times 1400} = 0.0268, \text{ then } \dot{m}_f = \dot{m}_0 f = 50 \times 0.0268 = 1.34 \text{ kg/s}$$

Thus, the thrust produced by the core engine can be evaluated as

$$\mathcal{T}_c = (\dot{m}_0 + \dot{m}_f)(V_9 - V_0) = (50 + 1.34)[328.8 - 155.91] = 8876 \text{ N} = 8.876 \text{ kN}$$

And the thrust produced by the propeller can be evaluated as

$$\mathcal{T}_{prop} = \frac{(\dot{m}_c + \dot{m}_f)\eta_{prop}C_{pg}\left(T_{t4.5} - T_{t5}\right)}{V_0} = \frac{51.34 \times 0.9 \times 1.148(1094.5 - 840)}{155.9} = 88.6 \text{ kN}$$

The total thrust produced by turboprop is determined as

$$\mathcal{T} = \mathcal{T}_{prop} + \mathcal{T}_c = 88.6 + 8.876 = 95.47 \text{ kN}$$

The TSFC of this turboprop engine can be determined as

$$\text{TSFC} = \frac{\dot{m}_F}{\mathcal{T}} = \frac{1.34 \times 10^3}{95.47} = 14.04 \text{ gm/kN}$$

The overall efficiency can be estimated as

$$\eta_0 = \frac{V_0 \mathcal{T}}{\dot{m}_F \Delta H_c} = \frac{155.91 \times 95.47}{1.34 \times 43000} \times 100 = 25.83\%$$

5.11 Chapter Summary

This chapter discusses how to include losses incurred in each component of the actual engine in the cycle analysis. Initially, physical processes in each component are discussed and certain terms are introduced such as efficiency and pressure loss ratio in each component to take account of their respective losses. Although thermodynamic properties vary from air intake to nozzle in a continuous manner, in this book one set of thermodynamic properties for the air intake and compressor are used while another set of properties are employed for the rest of components. The real processes in each engine configuration are described via *T*-s diagrams. Starting from the fundamental principles, the analytical equations for major performance parameters such as specific thrust, TSFC, propulsive efficiency and thermal efficiency are derived for three types of engines – turbojet, turbofan and turboprop. The parametric studies are carried out for each engine configuration by varying the operating parameters and component characteristics while taking account of the realistic losses in each component. These results for each engine are compared with ideal analysis carried out in Chapter 4. Efforts have been made to explain the difference between ideal and real cycle analysis. For each engine configuration, one example is provided to highlight some of the above observations so that the reader can appreciate the intricate processes involved during the operation of real engines. The methods of taking account of losses in each component will be useful for the preliminary design of aircraft gas turbine engine. In the subsequent chapters, we will learn about the aerothermodynamics of both axial and centrifugal turbomachines used in gas turbine engines.

Nomenclature

a_o Speed of sound in free stream (m/s)
C_P Specific heat capacity at constant pressure (kJ/kgK)
C_v Specific heat capacity at constant volume (kJ/kgK)
C_{pa} C_P for air (1.005 kJ/kgK)
C_{pg} C_P for gas (1.148 kJ/kgK)
ΔH_c Calorific value of fuel (kJ/kg)
f Fuel air ratio of main burner $(\dot{m}_f / \dot{m}_a)$
f_{AB} Fuel air ratio of after burner $(\dot{m}_{f,AB} / \dot{m}_a)$
M_0 Free stream Mach number
$\dot{m}_f$ Mass flow rate of fuel (kg/s)
$\dot{m}_a$ Mass flow rate of air (kg/s)
P_0 Free stream pressure
P_C Static pressure at exit of nozzle under choked condition
P_{t0}, T_{t0} Free stream total pressure, total temperature
R_a Molecular gas constant for air (kJ/kgK)
R_g Molecular gas constant for gas (kJ/kgK)
T_0 Free stream temperature
$\dot{W}_c$ Work done per unit time on compressor
$\dot{W}_t$ Work done per unit time by turbine
$\dot{W}_{tot}$ Total power produced by turboprop engine

Symbols

α Bypass factor
Υ_a Specific heat ratio of air = C_{pa}/C_{va} = 1.4
Υ_g Specific heat ratio of gas = C_{pg}/C_{vg} = 1.333
π_d Pressure recovery factor
π_{dw} Component of π_d due to wall friction
π_c Pressure ratio across compressor
π_b Pressure ratio across burner
π_n Total pressure ratio across nozzle
π_t Total pressure ratio across turbine
π_F Fan pressure ratio
η_{id} Isentropic efficiency of diffuser
η_{in} Isentropic efficiency of nozzle
η_{ic} Isentropic efficiency of compressor
η_{it} Isentropic efficiency of turbine
η_{Pc} Polytropic efficiency of compressor
η_{Pt} Polytropic efficiency of turbine
η_{ram} Ram efficiency
η_b Burner/combustor efficiency
η_p Propulsive efficiency
η_{th} Thermal efficiency
η_o Overall efficiency
η_m Mechanical efficiency
η_{gb} Efficiency of gear box
η_{mH} Mechanical efficiency of high pressure turbine
η_{prop} Propeller efficiency

$\mathcal{T}_s$ Specific thrust
$\mathcal{T}_{sf}$ Specific thrust for fan stream
$\mathcal{T}_{sc}$ Specific thrust for core stream
$\mathcal{T}_{tH}$ Temperature ratio across high pressure turbine
$\dot{W}_{prop}$ Power consumed by the propeller
τ_t Turbine temperature ratio
τ_{tL} Temperature ratio across free turbine
τ_λ T_{t4}/T_o

Subscripts/Superscripts

a Air
b Burner
c Compressor
d Diffuser
g Gas
o Overall
prop Propeller
p Propulsive
r Ram
s Specific
t Turbine
th Thermal
tot Total

Abbreviations

AB Afterburner
Al Altitude
PRF Pressure recovery factor
PSFC Power specific fuel consumption
TIT Turbine inlet temperature
TSFC Thrust specific fuel consumption

References and Suggested Readings

1. Mattingly, J.D., *Elements of Gas Turbine Propulsion*, McGraw-Hill, International Edition, 1996.
2. Saravanamuttoo, H.I.H., Rogers, G.F.C. and Cohen, H., *Gas Turbine Theory*, Pearson Education, 5th Edition, 2008.
3. Hill, P.G. and Peterson, C.R., *Mechanics and Thermodynamics of Propulsion*, Addison-Wesley, 2nd Edition, 1999.
4. Oates, G.C. (Ed.), *Aerothermodynamics of Gas Turbine and Rocket Propulsion*, AIAA Education Series, 3rd Edition, 1998.

Review Questions

1. What are the types of losses incurred in a gas turbine engine? Describe them briefly.
2. How is the real cycle analysis different from an ideal cycle analysis? Enumerate the differences briefly.
3. How does the variation of the properties of the fluid affect the performance analysis of a gas turbine engine?
4. What is meant by pressure recovery factor? How is it different from the isentropic efficiency of an air intake? Show that it can be related to isentropic efficiency of an air intake.

5. What is meant by the polytropic efficiency for a compressor? How can it be related to the isentropic efficiency of a compressor?
6. What is meant by combustion efficiency? What are the parameters that can influence the performance of a combustor?
7. What is meant by the polytropic efficiency for a turbine? How can it be related to the isentropic efficiency of a turbine?
8. What is the main parameter used to assess the performance of the exhaust nozzle?
9. How does the specific thrust vary with pressure ratio in case of a turbojet engine operated at sea level? Compare it with an ideal cycle analysis.
10. Does TSFC becomes higher in real cycle analysis of a turbojet engine as compared to ideal cycle analysis? Why is it so?
11. How does the takeoff thrust vary with pressure ratio in case of a turbofan engine operated at sea level? Compare it with an ideal cycle analysis.
12. Derive an expression for the optimum fan pressure ratio in a turbofan engine.
13. Why is higher bypass ratio turbofan engine preferred for long range passenger aircraft?
14. Derive an expression for the optimum turbine expansion ratio in a turboprop engine.

Problems

1. Derive an expression for specific thrust and TSFC in terms of operating parameters such as τ_b, M_0 etc., by considering the ramjet engine to be operated under real condition. State the assumptions made for a real cycle analysis. Derive an expression for TSFC in terms of M_0, a_0, T_o and also derive an expression for M_0 in terms of τ_λ, γ etc., when the specific thrust tends to zero.
2. In order to overcome the problem of zero static thrust in a ramjet engine, a designer is planning to add a compressor driven by an electric motor. Derive an expression for the specific thrust in terms of τ_r, τ_c, τ_b and M_0. Assume the total pressure recovery factor in air intake to be 0.95. The polytropic efficiency of compressor is 85%. The burner has efficiency of 97% and total pressure loss of 0.96 across the combustion chamber. If the pressure ratio across the compressor π_c = 5.0, T_{t4} = 2000 K and T_0 = 217 K, estimate the specific static thrust, thermal efficiency and TSFC.
3. A ramjet is to be designed for flying at an altitude of 20 km. Carry out a real cycle analysis and show the variation of $\mathcal{T}_s$, TSFC, η_p, η_{th} and η_o with flight Mach number M_0 by varying it from 1 to 3 at intervals of 0.5 for three combustor exit temperatures of 1400, 1800 and 2200 K, respectively. Assume the total pressure recovery factor in air intake to be 0.95. The burner has efficiency of 97% and total pressure loss of 0.96. Take ΔH_c = 43,000 kJ/kg, C_p = 1.005 kJ/kg. Use sigma plot software for making graphs, if possible.
4. A turbojet engine with a convergent nozzle is operating at flight Mach number of 2.0 at an altitude 20 km. The pressure ratio across the compressor is 10. Assume the total pressure recovery factor in air intake to be 0.95. The polytrophic efficiency of compressor and turbine is 85% and 90% respectively. The burner has efficiency of 98% and total pressure loss of 0.97. The maximum allowable temperature is 1500 K. Estimate the specific thrust, and TSFC, considering the engine as a real engine. If a ramjet engine is operating at the same altitude and flight Mach number, compare all the performance parameters such as $\mathcal{T}_s$, and TSFC, of the ramjet engine with those of the turbojet. The ramjet burner has efficiency of 96% and total pressure loss of 0.95. The maximum temperature for the ramjet engine is 2200 K. Take ΔH_c = Caloric value of fuel = 43,000 kJ/kg. Comment on your results.
5. A real turbojet engine with an afterburner, flying with Mach number of 1.2 at an altitude of 18 km consumes 70 kg/s of air. The total pressure ratio across the compressor is 20. The total temperature at the exit of the main burner and afterburner is 1400 and 2200 K, respectively. Assume the total pressure recovery factor in air intake to be 0.95. The polytrophic efficiency of compressor and turbine is 85 and 95% respectively. The efficiency of burner and afterburner is 95 and 98% respectively and total pressure loss of 0.97. Considering the nozzle efficiency of 0.95, estimate the thrust and TSFC for a simple turbojet engine and compare their values with that of an afterburning turbojet engine. Take ΔH_c = 43,000 kJ/kg.

6. A turbojet engine fitted with an afterburner and convergent nozzle is flying at 10 km. The turbine inlet temperature is restricted to 1400 K and maximum afterburner outlet temperature is 200 K. The polytropic efficiency of compressor and turbine is 85 and 90% respectively. The efficiency of burner and afterburner is 95 and 98% respectively and total pressure loss of 0.97. Carry out a real parametric analysis by varying π_c from 10 to 30 for $M_0 = 0, 0.5, 1, 2.0$. Plot specific thrust, TSFC, η_p, and η_{th} with π_c. Comment on your results. Indicate the optimum π_c in the specific thrust, TSFC plot. Take ΔH_c = 43,000 kJ/kg.
7. A two spool turbofan engine with a convergent nozzle is flying at Al = 10 km and flight Mach number M_0 of 0.85. The pressure ratio across the compressor and fan is 20 and 1.5, respectively. The turbine inlet temperature T_{t4} is 1500 K. The bypass ratio across the engine is 5.0. The polytrophic efficiency of fan, compressor and turbine are 83, 85 and 90% respectively. The burner has efficiency of 98% and total pressure loss of 0.97. Estimate the thrust, TSFC, η_p, η_{th} and η_o, if the total flow rate of air is 200 kg/s. Assume a real cycle analysis. Take ΔH_c = 43,000 kJ/kg.
8. A turbofan engine with convergent nozzle with efficiency of 0.95 for both fan and core stream is flying at Al = 12 km and flight Mach number M_0 of 0.75. The pressure ratio across the compressor and fan is 18 and 1.7, respectively. The turbine inlet temperature T_{t4} is 1600 K. The bypass ratio across the engine is 5.5. The polytrophic efficiency of fan, compressor and turbine are 90, 86 and 92% respectively. The burner has efficiency of 98% and total pressure loss of 0.95. If the mechanical efficiency is around 0.98, determine the ratio of compressor to fan power ratio, ratio of core to fan nozzle exit velocity, fan thrust and core thrust, specific impulse, TSFC, η_p, η_{th} and η_o. Assume a real cycle analysis. Take ΔH_c = 43,000 kJ/kg.
9. A turboprop engine is being operated with flight Mach number of 0.6 at an altitude of 6 km. The propeller efficiency is 0.83. The turbine inlet temperature is 1300 K. The pressure ratio across the compressor is 12 and total temperature ratio across the turbine is 0.6. The polytropic efficiency of compressor and turbine is 86 and 92% respectively. The burner has efficiency of 96% and total pressure loss of 0.98. The mechanical and gear box efficiency of this engine are 0.95 and 0.9 respectively. Considering as a real cycle, estimate specific thrust, TSFC, and η_o. Take ΔH_c = Caloric value of fuel = 43,000 kJ/kg.
10. An innovative turboprop is designed to fly at an altitude of 10 km and flight Mach number, M_0 = 0.8. The air flow rate captured by the propeller (η_{prop} = 0.98) with total pressure recovery of 0.97 is 55 kg/s. The pressure ratio across the compressor is 28 and its polytropic efficiency is 93%. The combustor has an exit temperature of 1600 K and has combustion efficiency of 97% with total pressure loss of 0.98. The polytropic efficiencies of HPT and LPT turbine are 86 and 92% respectively. The mechanical and gear box efficiency of this engine are 0.95 and 0.9 respectively. The convergent exhaust nozzle has efficiency of 0.95. Considering as a real cycle analysis, estimate total pressure and temperature at each station points, engine core and propeller thrust and total specific thrust, PSFC, TSFC and η_o if percentage of mass flow rate through core is happens to be 20%. Take ΔH_c = Caloric value of fuel = 43,000 kJ/kg.

6. Introduction to Turbomachinery

We have sought power in the same fire which serves to keep the vessel aloft. The first which presented itself to our imagination is the power reaction, which can be applied without any mechanism and without expense; it consists solely in one or more openings in the vessel on the side opposite to that in which one wishes to be conveyed.

Josesph Montgolfier (First recorded technical statement in history on jet propulsion for a flight vehicle), 1783

6.1 Introduction

In Chapter 5, we learnt about the behaviour of turbojet engines by cycle analysis without paying much attention to the internal details of turbomachinery (compressor, turbine, fan etc.). However, the cycle analysis of engines can tell us how much the overall performance of engine can be influenced by the efficiency and pressure ratio of the compressor, fan and turbine in gas turbine engines. For an aircraft engine, it is very essential to keep the performance at a high level to minimise the bulk and cross-sectional area and its weight. This requirement of high air flow rate per unit frontal area makes the design of turbomachinery more difficult. Therefore, the designer should have some insight into the complex fluid dynamic behaviour of these machines.

Now let us understand the meaning of the word 'turbomachine'. The word consists of two terms: 'turbo' and 'machine'. The word 'turbo', of Latin origin, means something that "spins or whirls around". Hence, a turbomachine can be defined as a device in which energy transfer between flowing fluid and rotating element takes place due to dynamic action of the rotating elements, which results in a change of pressure and momentum of the fluid. It implies that there will be transfer of mechanical energy into or out of the turbomachine. This definition of a turbomachine is too general. It embraces both open turbomachine (e.g. propeller, wind turbine, etc.) and enclosed turbomachine (e.g. compressor, turbine, shrouded fan, etc.). The enclosed turbomachine can be broadly divided into two types – power absorbing machine (e.g. ducted fan, blower, compressor, etc), and power producing machine (gas turbine and steam turbine). Turbomachines used in gas turbines can be further sub-divided into a radial/centrifugal machine in which fluid flow passes through the rotating element of the machine mainly in the radial direction; and an axial-flow machine in which the flow mainly occurs in the axial direction. The principal components of a turbomachine (Fig. 6.1) are:

(i) Rotating elements carrying blades.
(ii) Stationary element carrying stationary blades.
(iii) Shaft and hub.
(iv) Housing.

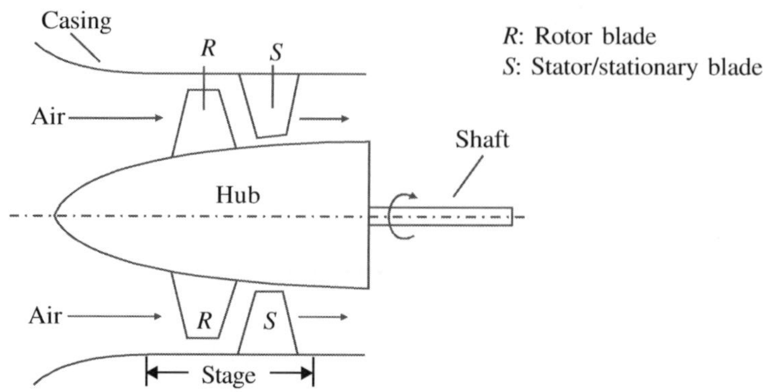

Fig. 6.1. Schematic of a typical turbomachine

Most of the enclosed turbomachines, particularly compressor and turbine, contain both rotating and stationary rows of blades as shown in Fig. 6.1. A row of rotating and stationary blades (rotor and stator) is termed as a stage (Fig. 6.1). In each turbomachine, number of stages likely to occur depends on the pressure ratio across the turbomachine. The details of both axial and centrifugal turbomachines will be discussed in subsequent chapters.

6.2 Eulerís Equation for Turbomachinery

We know that energy is either added to or extracted from the fluid by the rotating components in a turbomachine. These rotating components transfer energy to the fluid and result in change in momentum of the fluid. In this section, we will derive a relationship between the change in energy transfer and change in momentum, which is popularly known as *Euler's Equation for Turbomachinery*. We can derive this equation with the following assumptions:

- Steady and adiabatic flow.
- Uniform flow at the inlet and exit of the turbomachine.
- No viscous effects.
- Applicable to infinitesimal stream tube.
- No discontinuity of pressure in flow.

Let us consider adiabatic steady flow of a single stream tube of the fluid between inlet and outlet stations (1) and (2) respectively, of a typical turbomachine (Fig. 6.2). The fluid in a stream tube enters

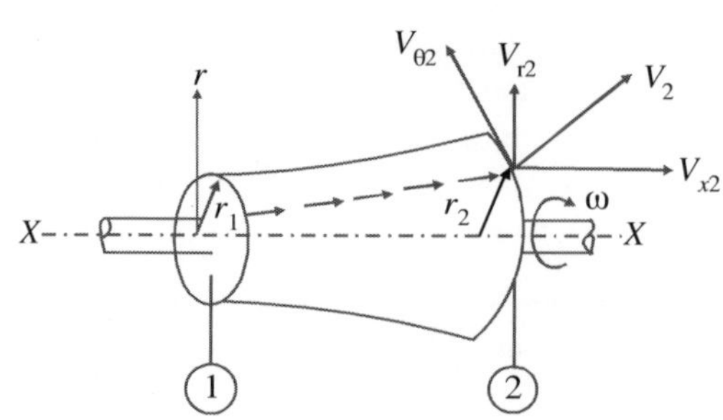

Fig. 6.2. CV for a typical turbomachine.

the control volume (CV) of the rotor of a turbomachine at radius r_1 with velocity V_1 and leaves the outlet at radius r_2 and velocity V_2. The velocity at any location can be divided into three principal components – axial V_x, radial V_r and tangential V_θ. The change in momentum of the fluid between the inlet and exit of the rotor can be used to estimate the force on the rotor, which is equal but opposite to that on the fluid. This force has three principal components – axial F_x, radial F_r and tangential F_θ. The axial and radial components do not contribute to the energy transfer between the working fluid and the rotor. Of course, one needs to estimate F_x and F_r since these are important for the design of bearings and vibrational analysis. It must be kept in mind that only tangential force F_θ can transfer energy. This energy or work can be estimated from the force and torque. We know that the tangential force F_θ on the fluid by the rotor at the inlet is $\dot{m}V_{\theta 1}$. So the tangential torque on the fluid by the rotor at the inlet is $\dot{m}V_{\theta 1}r_1$. Similarly, the tangential torque on the fluid by the rotor at its exit is $\dot{m}V_{\theta 2}r_2$.

We know that the torque applied on the fluid T_q is equal to the change in angular momentum of the fluid and that it is given by

$$T_q = \dot{m}(V_{\theta 2}r_2 - V_{\theta 1}r_1) \tag{6.1}$$

So the energy transfer $\dot{W}$(work rate) to the fluid is equal to the product of torque and angular velocity ω and it is expressed as

$$\dot{W}_c = \dot{m}\omega(V_{\theta 2}r_2 - V_{\theta 1}r_1) = \dot{m}(V_{\theta 2}U_2 - V_{\theta 1}U_1) \tag{6.2}$$

where U_1 and U_2 are the blade velocity at radius r_1 and r_2, respectively. Expression (6.2) is known as *Euler's Equation for Turbomachinery*, particularly for the compressor (fan). Similarly, *Euler's Equation for Turbine* can be derived as

$$\dot{W}_t = \dot{m}\omega(V_{\theta 1}r_1 - V_{\theta 2}r_2) = \dot{m}(V_{\theta 1}U_1 - V_{\theta 2}U_2) \tag{6.3}$$

The Euler Equation for Turbomachinery is almost universally applicable covering both compressible and incompressible flow. However, the assumption of steady flow restricts its applicability because it is quite impossible to keep the flowing fluid steady within a rotor. The flow may be considered steady near the inlet and the exit of a rotor, but flow properties at these locations are traditionally measured as average properties for steady flow performance predictions. The effect of this averaging of the fluctuating flow would not affect the performance prediction of the turbomachinery.

In the absence of large change in elevation and external torques, the application of the First Law of Thermodynamics (Eq. (2.17)) to the flow through the CV of turbomachinery will give

$$\dot{W}_c = \dot{m}(h_{t2} - h_{t1}) \tag{6.4}$$

Combining Eq. (6.4) with Euler's Equation (Eq. (6.2)), we obtain an expression for total enthalpy change across a compressor as

$$h_{t2} - h_{t1} = (U_2V_{\theta 2} - U_1V_{\theta 1}) \tag{6.5}$$

Similarly, Euler's Equation for turbine (Eq. (6.3)) can be combined with the First Law of Thermodynamics, to get

$$h_{t1} - h_{t2} = (U_1V_{\theta 1} - U_2V_{\theta 2}) \tag{6.6}$$

Equations (6.5) and (6.6) are very important relations in turbomachinery, which will be used throughout the analysis of turbomachinery.

6.3 Velocity Diagrams

We know that the velocity of a fluid element is dependent on the motion of the observer as well as on the motion of the fluid itself. Since the turbomachine generally consists of rotor and stator elements, the velocity of the fluid must be considered from two frames of references – *absolute frame* of reference, which is fixed to the frame of the turbomachine; and *relative frame* of reference, which rotates with the rotor of the turbomachine. The velocity of the fluid with respect to the absolute frame of reference is known as *absolute velocity* and denoted by V. The fluid velocity with respect to the relative frame of reference is known as *relative velocity* and denoted by V_R. The velocity of a fluid in the absolute frame of reference can be related to the relative frame of reference by

$$V = V_R + U \tag{6.7}$$

where U is the local blade velocity. Equation (6.7) states that absolute velocity is the vector sum of relative and local blade velocities. The graphical solution of Eq. (6.7) leads to the formation of a triangle (Fig. 6.3), known as *velocity triangle*. The flow angle of absolute velocity is denoted by α and the flow angle of relative velocity is denoted by β. Generally, the convention of defining these angles with respect to the axial direction will be followed throughout this book unless otherwise stated explicitly.

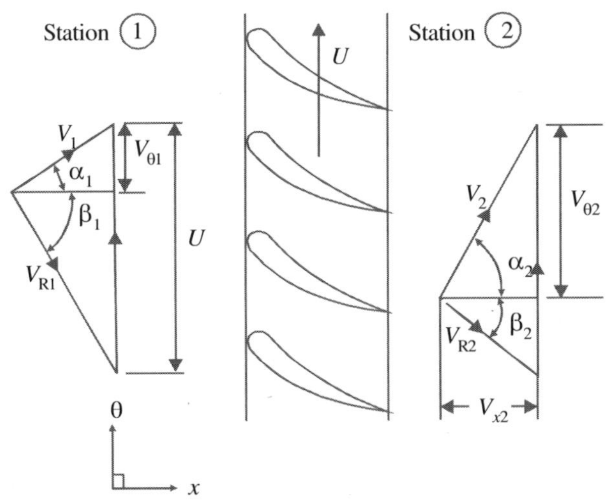

Fig. 6.3. Velocity triangle of a typical turbomachine between inlet (1) and exit (2) of rotor blade.

6.4 Physical Interpretation of Eulerís Equation

Let us now understand the physical meaning of Euler's Equation for Turbomachinery. To do so we need to express it in terms of absolute, relative and blade velocities. Therefore, we can consider a velocity triangle at the exit of a rotor in an axial compressor (Fig. 6.3). Using this velocity triangle, we can relate axial velocity in terms of absolute, tangential and blade velocities at station (2) as

$$V_{x2}^2 = V_2^2 - V_{\theta 2}^2 \text{ and } V_{x2}^2 = V_{R2}^2 - (U_2 - V_{\theta 2})^2 \tag{6.8}$$

where V_{x2} is the axial velocity. By equating the above two expressions for V_{x2} and simplifying, we get

$$U_2V_{\theta 2} = \frac{1}{2}\left(V_2^2 + U_2^2 - V_{R2}^2\right) \tag{6.9}$$

Similarly, we can obtain an expression for $U_1V_{\theta 1}$ at the inlet as

$$U_1V_{\theta 1} = \frac{1}{2}\left(V_1^2 + U_1^2 - V_{R1}^2\right) \tag{6.10}$$

Substituting the expressions for $U_2V_{\theta 2}$ and $U_1V_{\theta 1}$ in Euler's Equation for compressor (Eq. (6.2)), we get

$$\dot{W}_c = \dot{m}\frac{1}{2}\left[\underbrace{\left(V_2^2 - V_1^2\right)}_{I} + \underbrace{\left(U_2^2 - U_1^2\right)}_{II} - \underbrace{\left(V_{R2}^2 - V_{R1}^2\right)}_{III}\right] \tag{6.11}$$

It can be noted that this modified Euler's Equation contains three terms. The first term (I), $(V_2^2 - V_1^2)$ is related to the change in absolute kinetic energy of the fluid between the inlet and exit of the rotor, which is known as *impulse effect*. In the case of compressor, the absolute velocity at exit V_2 will increase as the rotor transfers energy to the fluid. But it does not ensure an increase in the fluid pressure. Hence, a diffuser or scroll has to be provided in the compressor to reduce the fluid velocity such that static pressure rise can be accomplished. The second term (II), $(U_2^2 - U_1^2)$ represents the energy transfer caused by the rotation of the fluid in the rotor, which is known as *centrifugal effect*. It can be related to the product of angular velocity and radial distance from the center of the rotor. Hence, it is prudent to allow the fluid to enter and leave at different radii, which is used in the design of centrifugal turbomachinery. The third term (III), $(V_{R2}^2 - V_{R1}^2)$ is the change in the relative kinetic energy of the fluid in the rotor. It gives rise to change in static pressure in the rotor and is often known as *reaction effect*, which will be predominant in axial compressor, fan and turbine.

6.5 Reaction Ratio

We saw in Section 6.4 that diffusion (flow deceleration) takes place both in rotor and stator blade passages. As a result, there will be an increase in static pressure through both rows of blade. But a question arises: what portion of the total and static pressure per stage will be contributed individually by the rotor and the stator? This information is very useful for designing any turbomachine, because it has an important influence on the stage efficiency. Therefore, let us define a term known as *reaction ratio* (RR), which is normally expressed as the change in static enthalpy in the rotor divided by the static enthalpy rise in the stage.

$$\text{RR} = \frac{\text{Static enthalpy rise in rotor}}{\text{Static enthalpy rise in stage}} \tag{6.12}$$

Note that RR is also known as *degree of reaction*. It is one of the important design parameters, which indicates the extent of compression or expansion shared between the rotor and stator in a stage. It means that high RR indicates that most of enthalpy change and, hence, compression or expansion occurs in the rotor. In contrast, low RR implies that the rotor effects lower change in enthalpy in the rotor. If RR = 0, there will be no change in enthalpy in the rotor. It implies that there would not be any change in the pressure in the rotor and, hence, no change in the velocity through the rotor. The work in done on the rotor just to turn the fluid in the blade passages. This kind of machine is known as an impulse type of turbomachine, which is hardly used in gas turbine engines. More details about this important parameter will be discussed in subsequent chapters.

6.6 Flow Over Aerofoil

We know that the blades of aerofoil shape are normally used in subsonic turbomachine. An aerofoil blade is a streamlined body having a thick rounded leading edge and a thin trailing edge (Fig. 6.4). The straight line joining the leading edge and the trailing edge is known as the *chord line* of the aerofoil. The length of the chord line is termed as *chord* of the aerofoil and is denoted by C. Generally, its maximum thickness location can be expressed in terms of its chord. Note that symmetric (uncambered) blades (Fig. 6.4a) are not suitable in a turbomachine as they cannot transfer energy. However, symmetric blades can be used to produce the curved (cambered) shape blades (Fig. 6.4b) by providing either circular or parabolic arc to itself. Curved bladed are preferred in case of a turbomachine due to better energy transfer. In the case of a curved (cambered) shape aerofoil, the line joining the most forward and rearward points that divide the upper and lower surface of the aerofoil into equal parts is known as the *mean camber line*. If the profile of the aerofoil is symmetric in nature (Fig. 6.4a), then the chord line coincides with the camber line. Generally, the shape of the aerofoil at the leading edge is circular with a leading edge radius of approximately 0.02 C. Note that double circular arc blade sections are being employed in transonic axial flow turbomachine due to higher performance level. Let us consider a standard camber profiled airfoil, designated by NACA 65218. In this nomenclature, the first digit when multiplied with 3/2 provides the design lift coefficient in one-tenths. The half of the next two digits indicates the location of the maximum camber along the chord from the leading edge in one-hundredth of the chord. The last two digits indicate the maximum thickness in one-hundredth of the chord. For NACA 65218 airfoil, the design lift coefficient is 0.9, the location of maximum camber is at 0.26 C and it has 18% maximum thickness.

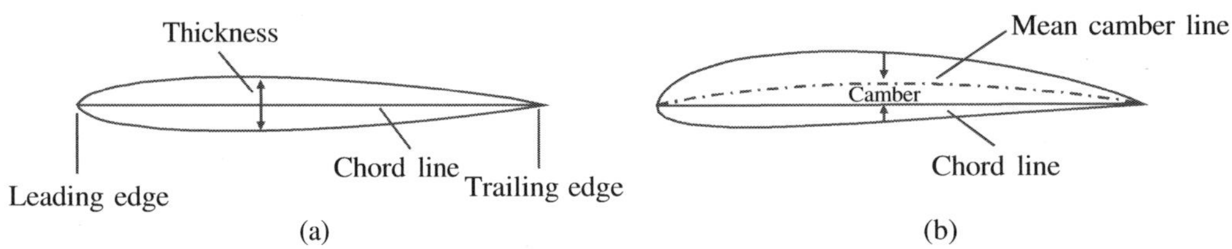

Fig. 6.4. Schematic of aerofoil shaped blades (a) uncambered blade and (b) cambered blade.

Let us consider flow over a cambered aerofoil (Fig. 6.5). We can use the frame reference in which the aerofoil is fixed such that the flow can be considered to be steady. It can be noted that the streamlines are pushed together above the upper convex surface and the fluid moves faster than the freestream velocity. Imagine that the movement of the fluid flow around the curved upper surface of the aerofoil is similar to that of a car going around a bend. As we have already learned in elementary physics, when a car turns, a force must accelerate the car toward the center of the turn. Similarly, as the fluid particle follows the cambered upper surface of the aerofoil, there must be a force acting on that little fluid particle to allow it to take a turn. In other words, pressure gradient is generated normal to these curved streamlines, leading to reduction of static pressure from free stream to the surface of the aerofoil. Similarly, the velocity gets reduced due to concave lower surface of the aerofoil, leading to an increase in static pressure from free stream to the surface of the aerofoil. These pressure changes caused due to streamline curvature created by aerofoil shape give rise to net upward force or lift. Note that energy interaction in a turbomachine takes place through the lift force. In order to enhance energy transfer in a turbomachine, an aerofoil with higher lift force is preferred. Of course, it must have lower drag. Hence, the designer has to chose an aerofoil shape blade with least drag-to-lift ratio.

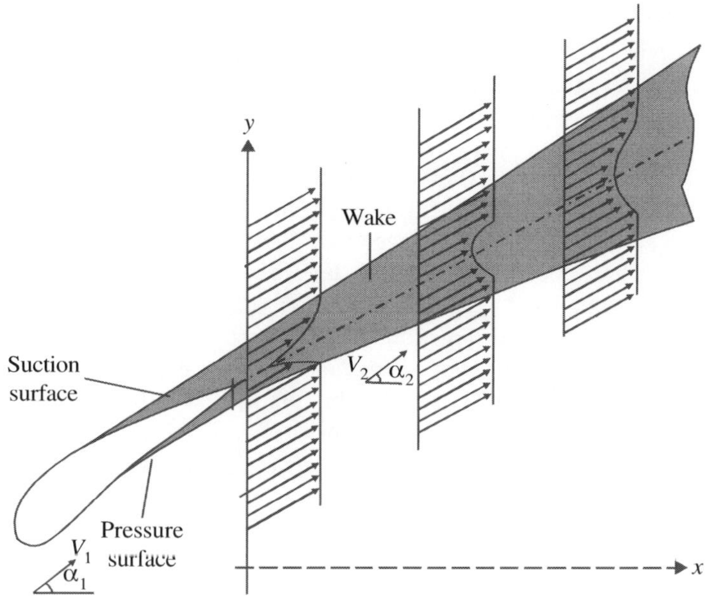

Fig. 6.5. Flow over a campered aerofoil.

6.7 Cascade Aerodynamics

We have seen in Section 6.6 that the operation of any turbomachine depends mainly on the changes in angular momentum of the fluid while passing through the rows of blades. But the fluid flow through the passages of turbomachine blade is inherently unsteady and three dimensional in nature. Hence, it is quite complicated in nature. However, further understanding of the mechanism of energy transfer can be gained by studying the flow in detail in rows of stationary blades, known as *cascade*, which resemble blade rows in the turbomachine. Basically, a cascade is an array of blades similar to the blade rows in a machine. If the blade rows are two dimensional in nature, then the cascade is known as *straight cascade*. Of course, for flow in the cascade to be truly two dimensional, the blade must be of infinite extent. However, in actual practice, the cascade is made of blades with finite blade height. Thus, care must be taken to design a straight cascade such that two dimensional flow can be maintained, at least in the central region where flow measurement are generally made. By placing rows of stationary blades in an open wind tunnel, the flow velocities and angles at the inlet and exits can be measured. Thus, forces exerted within these individual blades rows can be obtained.

Generally, a straight cascade can simulate flow conditions, particularly for axial turbomachinery of high hub/tip radius ratio. Because, the radial velocities are quite small and, hence, flow through the row of blades in an actual machine can be considered, approximately, to be two dimensional. Therefore, the information from two dimensional cascade can be directly used in designing and analysing the flow through each section of the blades of the turbomachinery. But the flow in the axial machine of low hub/tip ratio has a radial velocity component. So we may assume that the experimental data generated by the straight cascade cannot be used. The normal practice is to design the blades of low hub/tip ratio while satisfying the radial equilibrium condition, which ensures negligible radial velocity component. Hence, experimental data obtained from the straight cascade can be used to design and evaluate an axial turbomachine covering the whole range of hub/tip ratio. Moreover, the cascade results can be used to estimate the angle through which the fluid can be turned with minimum possible pressure loss and estimate the efficiency in a stage of a turbomachine.

6.7.1 Cascade Nomenclature

It is important to familiarise oneself with the terminologies of a cascade blade. In order to learn the same, let us consider a typical cascade geometry (Fig. 6.6(a)), which can be depicted by parameters such as solidity σ, stagger angle ζ, and blade shape. The solidity σ is, basically, the ratio of pitch S and chord C. The pitch S is the distance between two blades (Fig. 6.6). The stagger angle is defined as the angle between the chord line and a reference line perpendicular to the cascade front. The other angles such as blade inlet angle α_1', blade outlet angle α_2', air inlet angle α_1, and air outlet angle α_2, are defined with a reference line (axial direction in an axial compressor) (Fig. 6.6(b)). The difference between the air and blade angles at its inlet is known as *incidence angle* $i(= \alpha_1 - \alpha_1')$, which can be determined easily by the choice of air inlet angle. In contrast, the difference between the blade and air angle at its outlet is known as *deviation angle*, $\delta(= \alpha_2 - \alpha_2')$. The air *detection angle* ε $(= \alpha_1 - \alpha_2)$ is the difference between the inlet and outlet air angles, determined at the center of the cascade covering two blades, which will dictate total pressure loss in a cascade. Another important blade parameter is the *camber angle* θ, which is defined as the change in the angle of camber line between the leading and trailing edges.

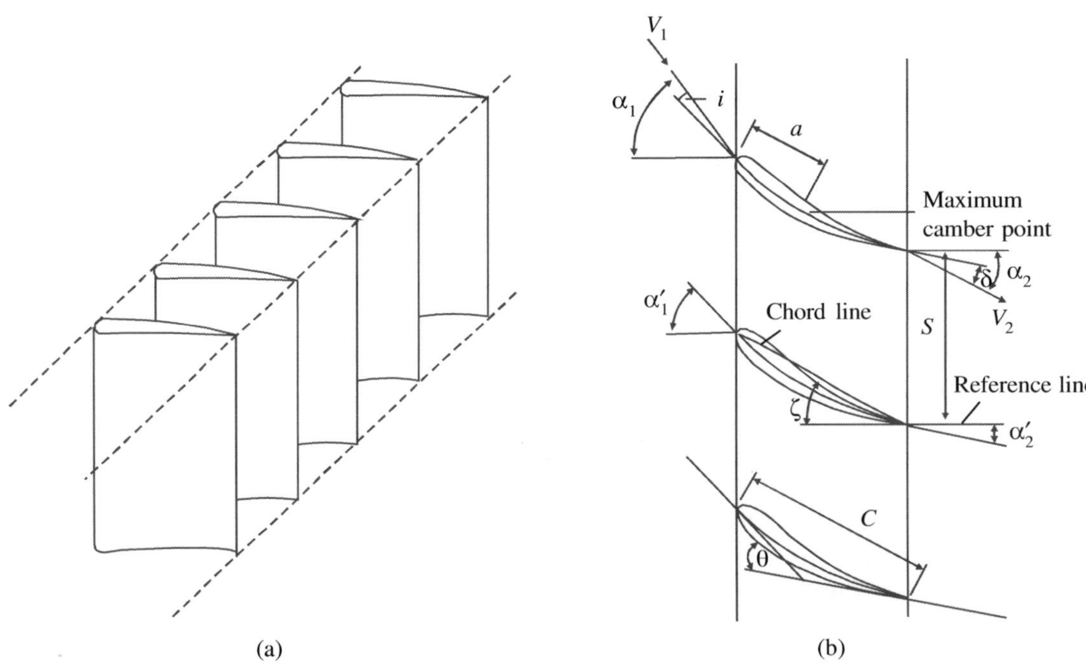

Fig. 6.6. Schematic of (a) a two-dimensional cascade geometry and (b) cascade nomenclature.

6.7.2 Cascade Flow Analysis

A portion of the isolated blade cascades for an axial compressor is shown in Fig. 6.7. The fluid approaches the cascade blades from far upstream with velocity V_1 at an angle α_1 and leaves for downstream of the cascade with velocity V_2 at an angle α_2. For analysis of the cascade flow, we can

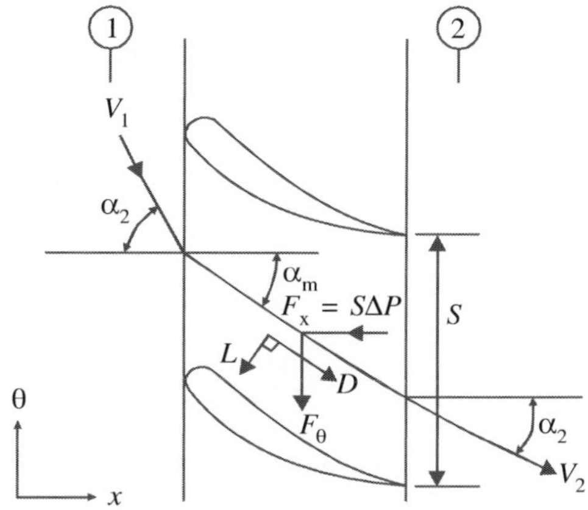

Fig. 6.7. Velocity diagram for cascade analysis.

consider a CV (Fig. 6.7), which is drawn with end boundaries for upstream and downstream of the cascade with side boundaries coinciding with the streamline at mid-plane. Although in a turbomachinery (compressor), flow is inherently unsteady, we can assume the flow to be steady in a stage for simplicity. Since the flow is low subsonic, we can assume the flow to be incompressible. The viscous effects can be neglected for simplicity. The forces F_x and F_θ exerted on the fluid by the blade of unit depth are exactly equal and opposite to the forces exerted by the fluid on the blade of unit depth. Note that we are interested in finding an expression for lift and drag coefficients. Assuming the flow to be steady and incompressible (since $M \leq 0.3$) the following expressions for continuity, momentums are derived below.

Continuity

Applying mass conservation principle to the flow on a cascade blade of unit depth of span, we get

$$\rho_1 V_1 \cos \alpha_1 S = \rho_2 V_2 \cos \alpha_2 S \Rightarrow V_1 \cos \alpha_1 = V_2 \cos \alpha_2 \tag{6.13}$$

or

$$V_{x1} = V_{x2} = V_x \tag{6.14}$$

x-momentum conservation

$$F_x - S\,(P_2 - P_1) = \rho V_x (V_{x2} - V_{x1}) = 0 \tag{6.15}$$

$$\Rightarrow F_x = S\Delta P$$

θ-momentum conservation

$$F_\theta = \rho S V_x (V_{\theta 1} - V_{\theta 2}) \tag{6.16}$$

$$F_\theta = \rho S V_x^2 (\tan \alpha_1 - \tan \alpha_2) \tag{6.17}$$

The forces F_x and F_θ are exerted on the fluid by the blades of unit depth, which are exactly equal and opposite to the forces exerted by the fluid on the cascade blade.

But $$P_2 - P_1 = \frac{1}{2}\rho\left(V_1^2 - V_2^2\right) - \Delta P_t \quad \text{where } \Delta P_t = P_{t1} - P_{t2}$$

Then the expression for F_x becomes

$$F_x = S\left(P_2 - P_2\right) = \frac{1}{2}\rho V_x^2 S\left(\tan^2\alpha_1 - \tan^2\alpha_2\right) - S\Delta P_t \tag{6.18}$$

Similarly, the forces acting in the θ direction can be expressed in terms of flow angles as

$$F_\theta = \rho V_x^2 S\left(\tan\alpha_1 - \tan\alpha_2\right) \tag{6.19}$$

Let us define the mean flow direction as

$$\tan\alpha_m = 0.5(\tan\alpha_1 + \tan\alpha_2) \tag{6.20}$$

Now we can express the drag force D along the mean flow direction as

$$D = F_\theta \sin\alpha_m - F_x \cos\alpha_m \tag{6.21}$$

Equation (6.21) can be expressed in terms of axial velocity as well as inlet and outlet velocity angles by using Eqs. (6.19) and (6.18) as

$$D = \rho V_x^2 S\left(\tan\alpha_1 - \tan\alpha_2\right)\sin\alpha_m - \frac{1}{2}\rho V_x^2 S\left(\tan^2\alpha_1 - \tan^2\alpha_2\right)\cos\alpha_m + \Delta P_t S\cos\alpha_m \tag{6.22}$$

Since we know that $\tan\alpha_m = \frac{1}{2}(\tan\alpha_1 + \tan\alpha_2)$, Eq. (6.22) becomes

$$D = \rho V_x^2 S\left(\tan\alpha_1 - \tan\alpha_2\right)\left(\sin\alpha_m - \frac{1}{2}\left(\tan\alpha_1 + \tan\alpha_2\right)\cos\alpha_m\right) + \Delta P_t S\cos\alpha_m \tag{6.23}$$

It must be noted that the first term in Eq. (6.23) turns out to be zero and so the expression for drag D would be given as

$$D = \Delta P_t\, S\cos\alpha_m \tag{6.24}$$

Hence, the drag is mainly contributed by the pressure loss in the blade. From the definition, the profile drag coefficient is given as

$$C_{DP} = \frac{D}{\frac{1}{2}\rho V_m^2 C} \Rightarrow C_{DP} = \frac{\Delta P_t\left(\frac{S}{C}\right)\cos\alpha_m}{\frac{1}{2}\rho V_m^2} \tag{6.25}$$

But as we know, $V_1\cos\alpha_1 = V_x = V_m\cos\alpha_m$. So the above equation becomes

$$C_{DP} = \frac{\Delta P_t\left(\frac{S}{C}\right)}{\frac{1}{2}\rho V_1^2}\frac{\cos^3\alpha_m}{\cos^2\alpha_1} \tag{6.26}$$

The lift force can be expressed in terms of F_θ and F_x as

$$L = F_\theta \cos\alpha_m + F_x \sin\alpha_m \tag{6.27}$$

Equation (6.27) can be expressed in terms of axial velocity as well as inlet and outlet velocity angles by using Eqs. (6.19) and (6.18) as

$$L = \rho V_x^2 S\left(\tan\alpha_1 - \tan\alpha_2\right)\cos\alpha_m + S\frac{1}{2}\rho V_x^2\left(\tan^2\alpha_1 - \tan^2\alpha_2\right)\sin\alpha_m - \Delta P_t S\sin\alpha_m \tag{6.28}$$

By simplifying Eq. (6.28), we get

$$L = \rho V_x^2 S(\tan\alpha_1 - \tan\alpha_2)/\cos\alpha_m - \Delta P_t S\sin\alpha_m \tag{6.29}$$

But we know that

$$C_L = \frac{L}{\frac{1}{2}\rho V_m^2 C} \tag{6.30}$$

So the expression for C_L would be

$$C_L = 2\left(\frac{S}{C}\right)\cos\alpha_m(\tan\alpha_1 - \tan\alpha_2) - C_{DP}\tan\alpha_m \tag{6.31}$$

Generally, the contribution of C_L from the term $C_{DP}\tan\alpha_m$ is quite small and, hence, can be neglected. Now Eq. (6.31) becomes

$$C_L = 2\left(\frac{S}{C}\right)\cos\alpha_m\left(\tan\alpha_1 - \tan\alpha_2\right) \tag{6.32}$$

While estimating the drag coefficient in a compressor stage, one needs to consider additional drags due to two factors (i) walls of the turbomachinery annulus and (ii) the secondary losses. The drag on the annular walls causes pressure loss which is dependent on the geometry of blade rows particularly pitch S and blade height h. The convenient empirical relationship used for estimation of the annulus drag is $C_{DA} = 0.02\ (S/h)$. The drag due to secondary flow losses are caused by trailing vortices and tip clearance. For an axial compressor, the general empirical relationship used for estimation of drag due to secondary flow is $C_{DS} = 0.018\ C_L^2$ [5]. Total drag will be sum of all three drags ($C_D = C_{DP} + C_{DA} + C_{DS}$) which can be represented in the similar form as that of C_{DP} (Eq. (6.26)) given as

$$C_D = \left(\frac{S}{C}\right)\left(\frac{\Delta P_t}{\frac{1}{2}\rho V_1^2}\right)\left(\frac{\cos^3\alpha_m}{\cos^2\alpha_1}\right) \tag{6.33}$$

Knowing C_D, S/C, α_m and α_1, we can determine the loss coefficient $\Delta P_t \Big/ \frac{1}{2}\rho V_1^2$.

Besides, it is important to evaluate the effectiveness of the blade as per design. To do so, we need to define the efficiency of the blade row η_b as the ratio of actual static pressure rise to the theoretical pressure rise

$$\eta_b = \frac{\text{Actual static pressure rise}}{\text{Theoretical static pressure rise}} = \frac{P_2 - P_1}{P_{2s} - P_1} = 1 - \frac{\Delta P_t}{\frac{1}{2}\rho\left(V_1^2 - V_2^2\right)} \tag{6.34}$$

We can express blade efficiency in terms of C_L and C_D. By using Eqs. (6.32) and (6.33) we get

$$\frac{C_L}{C_D} = \frac{2\left(\frac{S}{C}\right)(\tan\alpha_1 - \tan\alpha_2)\cos\alpha_m}{\left(\frac{S}{C}\right)\frac{\Delta P_t}{\frac{1}{2}\rho V_x^2}\cos^3\alpha_m} \tag{6.35}$$

$$\Delta P_t = \frac{(\tan\alpha_1 - \tan\alpha_2)}{\frac{C_L}{C_D}\cos^2\alpha_m}\frac{1}{2}\rho V_x^2 \tag{6.36}$$

Using Eq. (6.36), we can express η_b in terms of C_L and C_D as

$$\eta_b = 1 - \frac{\Delta P_t}{\frac{1}{2}\rho\left(V_1^2 - V_2^2\right)} = 1 - \frac{\Delta P_t}{\frac{1}{2}\rho V_x^2\left(\tan^2\alpha_2 - \tan^2\alpha_1\right)} = 1 - \frac{2\,C_D}{C_L\,\sin 2\,\alpha_m} \tag{6.37}$$

Assuming a constant lift-drag ratio, the optimum mean flow angle α_m can be obtained by differentiating Eq. (6.37) with α_m and equating to zero as

$$\frac{d\eta_b}{d\alpha_m} = \frac{4C_D\cos 2\alpha_m}{C_L\sin^2 2\alpha_m} = 0 \tag{6.38}$$

$$\Rightarrow \alpha_m = 45° \tag{6.39}$$

Hence, the expression of $\eta_{b,\max}$ is given by

$$\eta_{b,\max} = 1 - \frac{2C_D}{C_L} \tag{6.40}$$

We know that C_D/C_L ratio changes with variation in α_m. However, in deriving Eq. (6.40), it has been neglected. In an actual situation, η_b occurs at an angle $\alpha_m < 45°$ and the angle smaller than 45° can be considered for design purposes.

6.8 Dimensional Analysis of Turbomachinery

The performance of any turbomachine involves several parameters viz. pressure, temperature, rotational speed, fluid properties, dimension of the turbomachine etc. Hence, it will be quite cumbersome and a time consuming task to obtain the performance curves for any turbomachine, which cover full variations of the performance parameters over its working range. Even if these enormous data can be generated by some means, it will be a Herculean task to present the data often in a comprehensive format that can be deciphered by engineers. Hence, one has to resort to dimensional analysis to club the variables forming dimensionless groups that represent certain physical situations. As a result, experimental relations between independent variables can be obtained experimentally with the greatest economy of effort. Note that dimensional analysis also has two other important applications. For example, the performance of a prototype turbomachine can be predicted by conducting a test on a scaled model. It can also be used to determine suitable turbomachine based on maximum efficiency for a specified range of pressure ratio, flow rate and rotational speed. Let us now see, how we can use this analysis in reducing the dependent variables into manageable dimensionless numbers in a compressor. The pressure at the outlet of the compressor P_{t3} is dependent on several variables that can be written in functional form as

$$P_{t3} = f(P_{t2}, T_{t2}, T_{t3}, \rho_{t2}, \rho_{t3}, \dot{m}, N, D, \mu) \tag{6.41}$$

where D is the characteristics length of turbomachine which is taken as the diameter of rotor/impeller, N is the rpm of the rotor and μ is the viscosity of fluid. By examining Eq. (6.41) we can see that P, ρ and T are superfluous variables as they can be related by the equation of state ($\rho = P/RT$). Hence, by deleting density ρ and combing R with T, we can rewrite the functional relationship (Eq. (6.41)) as

$$P_{t3} = f(P_{t2}, RT_{t2}, RT_{t3}, \dot{m}, N, D, \mu) \tag{6.42}$$

As per Buckingham π-theorem, the number of independent dimensionless groups that may be used to depict a phenomena is equal to $n - k$, where n is the number of variables and k is the number of basic dimensions (mass m, length l and time t). Note that in Eq. (Eq. 6.42) there are seven variables. By using Buckingham π-theorem, the functional relationship (Eq. (6.42)) of several variables can be reduced to a function of four dimensionless variables. This reduction of three variables is due to the presence of three basic/fundamental units (m, l, t). By expressing Eq. (6.42) as a product of the terms raised to powers, we get

$$P_{t3} = C\left(P_{t2}^a, RT_{t2}^b, RT_{t3}^c, \dot{m}^d, N^e, D^f, \mu^g\right) \tag{6.43}$$

Let us express Eq. (6.43) in terms of the basic dimensions as

$$\frac{m}{lt^2} = C\left[\left(\frac{m}{lt^2}\right)^a, \left(\frac{l^2}{t^2}\right)^b, \left(\frac{l^2}{t^2}\right)^c, \left(\frac{m}{t}\right)^d, \left(\frac{1}{t}\right)^e, l^f, \left(\frac{m}{lt}\right)^g\right] \tag{6.44}$$

Equating the indices in Eq. (6.44), we get

$$m: \quad 1 = a + d + g \tag{6.45}$$

$$l: \quad -1 = -a + 2b + 2c + f - g \tag{6.46}$$

$$t: \quad -2 = -2a - 2b - 2c - d - e - g \tag{6.47}$$

Solving the above equations for a, b and f in terms of other indices such as c, d, e and g, we get

$$a = 1 - d - g \tag{6.48}$$

$$b = d/2 - c - e/2 + g/2 \tag{6.49}$$

$$f = e - 2d - g \tag{6.50}$$

Substituting the above expressions for a, b and f in Eq. (6.35), we get

$$\begin{aligned} P_{t3} &= C\left(P_{t2}^{1-d-g}, RT_{t2}^{d/2-c-e/2+g/2}, RT_{t3}^c, \dot{m}^d, N^e, D^{e-2d-g}, \mu^g\right) \\ &= CP_{t2}\left[\left(\frac{RT_{t3}}{RT_{t2}}\right)^c, \left(\frac{\dot{m}\sqrt{RT_{t2}}}{P_{t2}D^2}\right)^d, \left(\frac{ND}{\sqrt{RT_{t2}}}\right)^e \left(\frac{\mu\sqrt{RT_{t2}}}{P_{t2}D}\right)^g\right] \end{aligned} \tag{6.51}$$

By dividing $\sqrt{RT_{t2}}$ both in numerator and denominator of the last term in the bracket of Eq. (6.51), we find that this term is equal to Reynolds number, $\text{Re} = \left(\frac{\rho_{t2}V_{t2}D}{\mu}\right)$ as $P_{t2}/RT_{t2} = \rho_{t2}$ and the unit of $\sqrt{RT_{t2}}$ is equal to the units of velocity (l/t). Now the above equation can be expressed in functional form, consisting of four dimension numbers as

$$\frac{P_{t3}}{P_{t2}} = f\left[\left(\frac{T_{t3}}{T_{t2}}\right), \left(\frac{\dot{m}\sqrt{RT_{t2}}}{P_{t2}D^2}\right), \left(\frac{ND}{\sqrt{RT_{t2}}}\right), \text{Re}\right] \tag{6.52}$$

The dimensionless rotational speed $\left(\frac{ND}{\sqrt{RT_{t2}}}\right)$ can be interpreted as the rotational Mach number M_r, as the rotor velocity V is proportional to N_D while the sound speed a_{t12} is proportional to $\sqrt{RT_{t2}}$. Similarly, it can be shown that dimensionless mass flow $\left(\frac{\dot{m}\sqrt{RT_{t2}}}{P_{t2}D^2}\right)$ can be interpreted as the flow Mach number, M. Generally, the performance of a turbomachine is represented as $\frac{P_{t3}}{P_{t2}}$ vs. $\left(\frac{\dot{m}\sqrt{RT_{t2}}}{P_{t2}D^2}\right)$ for different values of non-dimensional speed for a family of machines. Generally, in a turbomachine, Re is so high that its change within the operating range of a turbomachine would not affect the performance significantly. Hence, it can be omitted from Eq. (6.52) unless there is a large change in the density of fluids. However, for a particular machine of fixed size, which uses a specified gas, the gas constant R and machine diameter D can be omitted from the dimensionless parameters. In view of these considerations, Eq. (6.52) becomes

$$\frac{P_{t3}}{P_{t2}} = f\left[\left(\frac{T_{t3}}{T_{t2}}\right), \left(\frac{\dot{m}\sqrt{T_{t2}}}{P_{t2}}\right), \left(\frac{N}{\sqrt{T_{t2}}}\right)\right] \tag{6.53}$$

We will discuss more about the above dimensionless parameters when we will learn about the performance curves of individual turbomachines in subsequent chapters.

Example 6.1 A centrifugal compressor is designed to run at 20,000 rpm while ingesting air from atmosphere at P = 101.325 kPa and T = 298 K. However, its performance characteristics is obtained at T = 45°C. What is the correct speed and mass flow rate, at which the compressor must run?

Given P_1 = 101.325 kPa, T_1 = 298 K.

Solution For this problem, the dimensionless group given in Eq. (6.53) can be used. Note that the gas constant R and diameter D have been dropped in this equation as the same machine is being considered with the same fluid for both design and operation. Considering dimensionless speed parameter, we have

$$\left(\frac{N_1}{\sqrt{T_{t1}}}\right) = \left(\frac{N_2}{\sqrt{T_{t2}}}\right)$$

Substituting the value of rpm and temperature in the above equation, we obtain

$$N_2 = N_1\left(\frac{\sqrt{T_{t2}}}{\sqrt{T_{t1}}}\right) = 20000 \times \left(\sqrt{\frac{318}{298}}\right) = 20660 \text{ rpm}$$

6.9 Chapter Summary

This chapter is devoted to the fundamentals of turbomachinery. A brief classification of turbomachinery is provided by emphasising more on the turbomachines related to gas turbine engines. The Euler's Equation of Turbomachine is derived from the basics of angular momentum, which is essential for the analysis of turbomachinery. An attempt is made to provide a physical interpretation of this important equation. Subsequently, the basic flow over an aerofoil is described briefly. As cascade (rows of blades) are employed in turbomachines, cascade nomenclature is introduced, which will be used in subsequent chapters. Aerodynamics of cascade flow analysis is carried out as lift and drag coefficients can be estimated easily by using it. It is also shown how to estimate blade efficiency by using the values of drag and lift coefficients of cascade for certain flow condition. The dimensional analysis of turbomachinery is carried out in detail by invoking Buckingham π theorem for deriving relevant non-dimensional parameters. In Chapter 7, axial compressor will be discussed in detail.

Nomenclature

C_{DP}	Profile drag coefficient
C_{DA}	Annulus drag coefficient
C_{DS}	Drag coefficient due to secondary flow
C_L	Lift coefficient
D	Drag force
F	Force
i	Incidence angle
L	Lift force
$\dot{m}$	Mass flow rate
M	Mach number
N	Speed (rpm)
P	Pressure
Q	Chamber angle
Re	Reynolds number
RR	Reaction ratio or degree of reaction
T	Temperature
T_q	Torque applied on fluid
U	Local blade velocity
V	Absolute velocity
V_R	Relative velocity
$\dot{W}_c$	Work rate for compressor
$\dot{W}_t$	Work rate for turbine

Greek Symbols

α_1	Air inlet angle
α_2	Air outlet angle
ω	Angular velocity
α_1'	Blade inlet angle
α_2'	Blade outlet angle
ε	Deflection angle
ρ	Density
δ	Deviation
η_b	Efficiency of blade rows
$\eta_{b,max}$	Maximum efficiency of blade rows at optimum mean flow angle
σ	Solidity (ratio of pitch S and chord C)

ζ Staggor angle
μ Viscosity

Subscripts

c Compressor
D Drag coefficient
L Lift
q Torque
R Relative velocity
r Rotational
t Turbine

References and Suggested Readings

1. Mattingly, J.D., *Elements of Gas Turbine Propulsion*, 1st Edn., McGraw-Hill Publishing Company Ltd., Singapore, 1996.
2. Kerrebrcok, J.L., *Aircraft Engines and Gas Turbines*, 1st Edn., MIT Press, England, 1992.
3. Gostelow, J.P., *Cascade Aerodynamics*, Pergamon Press, 1984.
4. Dixon, S.L., *Fluid Mechanics and Thermodynamics of Turbomachinery*, 5th Edition, Butterworth-Heinemann, 2005.
5. Saravanamuttoo, H.I.H., Rogers, G.F.C. and Cohen, H., *Gas Turbine Theory*, 5th Edition, Pearson Education, 2008.
6. Japikse, D. and Baines, N.C., *Introduction to Turbomachine*, Concepts ETI and Oxford, 1994.

Review Questions

1. What are types of turbomachinery that you are aware of? Discuss each turbomachine.
2. Derive the Euler's Equation for Turbomachinery by stating its assumptions.
3. How can Euler's Equation for Turbomachinery be interpreted?
4. Derive the Euler's Equation for flow over curved surface.
5. What is cascade? How can it be specified?
6. Describe flow over a cambered aerofoil and explain how lift is being generated.
7. Derive an expression for lift coefficient.
8. What is Buckingham π theorem? How can it be used to derive dimensionless number?
9. Derive dimensionless parameters for a turbine.

Problems

Q1. A car is moving at a velocity of 50 km/h. An object is dropped on it with a velocity of 20 km/h. Determine the magnitude and direction of the object with respect to the car.

Q2. A ceiling fan is running at 1,000 rpm to circulate air in a room. The blade has an angle of 75° with reference to the axis of the fan at a radius of 0.5 m. If the relative air velocity makes the same angle to the axis of the fan, determine the speed of air.

Q3. An axial compressor is designed for mass flow rate of 50 kg/s to run at 30,000 rpm under ambient conditions. During testing, if the entry pressure and temperature are 0.5 MPa and 400 K, determine mass flow rate required to simulate the design condition.

7. Axial Compressor

The real question rests now as it did before, on the possibility of providing a sufficient power with the requisite lightness. I have tried many different engines as first movers, expressly for this purpose (flight).

George Cayley, 1843

7.1 Introduction

We know that modern aircraft engines very often employ axial flow compressors in comparison to centrifugal compressors. This is due to the fact that an axial flow compressor is capable of producing higher pressure ratio and has higher efficiency than a centrifugal compressor. While carrying out parametric cycle analysis in Chapter 6, we learnt that engine performance is strongly dependent on compressor pressure ratio and efficiency. For high subsonic flight, the optimum compressor pressure ratio is quite high (~25). Such a high pressure ratio cannot be obtained in a single stage compressor. Rather, many stages have to be employed to get such a high compressor pressure ratio. In contrast, an increase in the number of stages for a centrifugal compressor results in a drastic reduction in its efficiency. As a result, axial compressor is quite suitable for multi-staging without sacrificing its efficiency. Moreover, larger flow rates of air for a given frontal area are required by modern aircraft engines. Note that an axial compressor performs efficiently for higher air flow rate in comparison to a centrifugal compressor.

Interestingly, the history of modern high performance aircraft engine is quite closely linked with the development of the axial compressor. Although the basic concepts of an axial compressor had been known for over 150 years before its advent, the Royal Aircraft Establishment (RAE) could manage to develop it only in 1941. Of course, toward 1945, the compressor efficiency was improved by using aerodynamic theory in its design and development. Later on, efforts were made to extend the allowable relative Mach number up to a high subsonic regime with improved efficiency. Recently, attempts have been made to design and develop transonic and supersonic compressors.

We have already discussed that axial compressors are mainly suitable for high pressure ratio with large mass flow rates of air. Some of the advantages of axial compressors over centrifugal compressors are enumerated briefly:

- Smaller frontal area for a given mass flow rate.
- Flow direction at discharge suitable for multi-staging.
- Efficiency does not decrease much at higher pressure ratio.
- Cascade experimental data can be used in designing an axial compressor.

The disadvantages of axial compressors as compared to centrifugal compressor are:

- Lacks simplicity and ruggedness of construction.
- Susceptible to damage by foreign object.
- Larger length, hence, heavy weight for same overall pressure ratio.
- Susceptible to loss of performance due to build-up of deposits on blade surfaces.
- Inability to operate over wide range of mass flow rates between surging and choking at a particular rotational speed.
- Flow direction at discharge not suitable for installation of intercooler and can-type combustor.

7.2 Working Principle of Axial Compressor

The sectional view of a typical axial compressor along the axial direction is shown in Fig. 7.1. This particular compressor consists of a series of rotating aerofoils, known as *rotor blades*, followed by stationary aerofoils, known as *stator blades*. One row of rotor blades together with the following row of stator blades constitutes a single stage of an axial compressor (Fig. 7.1). The rows of rotor blades are fixed to the hub and mounted on the top of shaft through which power is transmitted to the rotor. The stator blades are attached to the compressor casing. Both rotor and stator blades are generally placed as close as possible for higher efficiency. While designing and developing the rotor and hub system, care must be taken to provide least possible tip clearance at the end of rotor and stator blades to get enhanced performance. The compressor can be designed with constant casing diameter (Fig. 7.1) or constant mean blade diameter, or constant hub diameter, or with all three varying. However, the constant casing type compressor is preferred in aircraft engines. Note that the length of the blade decreases along the axial direction. This is meant to accommodate the variation of air density resulting from increased pressure such that axial velocity remains almost constant along the length of the compressor.

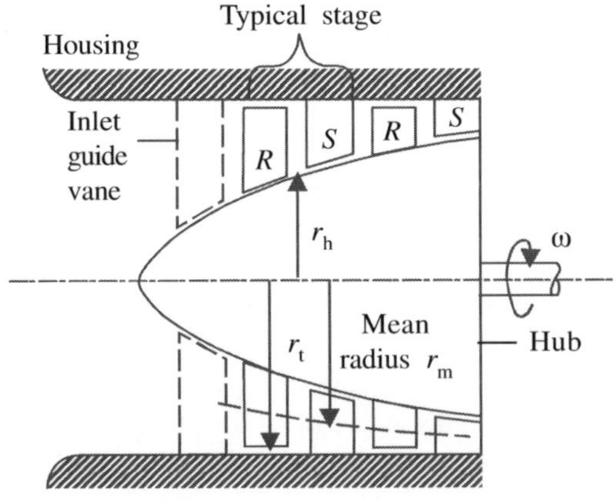

Fig. 7.1. Schematic of a typical axial compressor with constant casing diameter.

The cutaway sketch of the compressor in General Electric J85 turbojet engine is shown in Fig. 7.2. This compressor has eight stages and can produce pressure ratio of 6.5. Each row of rotor blades has been placed on a separate disc type hub (Fig. 7.2), whereas each row of stator blades is mounted on its casing. Bleed ports are provided on this compressor to take out the stagnation air in the boundary layer to avoid flow separation in the compressor. This bleed air can be utilised for the cooling of turbine blades.

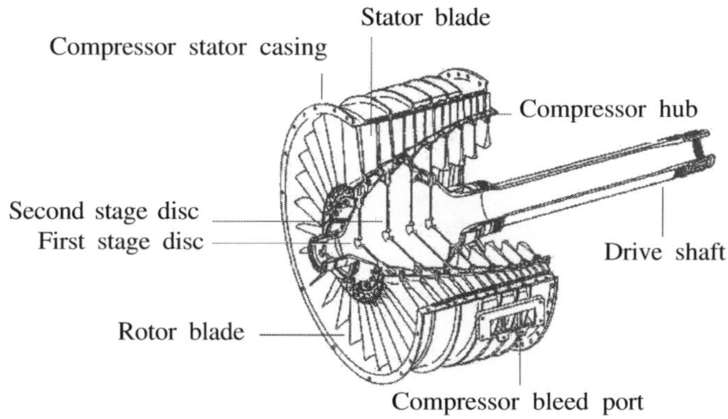

Fig. 7.2. Cutaway view of an axial compressor used in General Electric J85 engine.

Let us now look at how a compressor pressurises air to produce a certain pressure ratio. Air enters axially from left to right into the inlet guide vanes (IGVs) (Fig. 7.1), which guide the fluid entering into the row of rotating blades of the first stage. The IGVs are meant to improve the off-design performance of the compressor by varying its angle. But they are no longer used in modern aircraft engines to obtain maximum possible flow rate per unit frontal area and minimum engine weight. When air passes through the rotor, work is transferred from the moving blade to the air as per Euler's Equation of Turbomachinery, resulting in an increase in enthalpy. Also, air gets decelerated as it passes through the blades, resulting in increase in static pressure. Hence, both static enthalpy and static pressure increase due to the combined effects of work input to the air and internal conversion of kinetic energy into pressure energy. Now air enters the first row of stator blades with proper air angles such that flow separation does not occur. Here, air is decelerated and its kinetic energy is converted into internal energy with resultant rise in static pressure. However, the stator is fixed unlike the rotor and there is no transfer of work into the air. Consequently, the total enthalpy remains constant across the stator (Fig. 7.3). There will be certain losses in pressure in the stator as fluid friction and mixing lead to a decrease in total pressure. The processes occurring in the rotor and the stator for a stage in *T-s* plane (Fig. 7.4) show that increase of total pressure occurs only in the rotor.

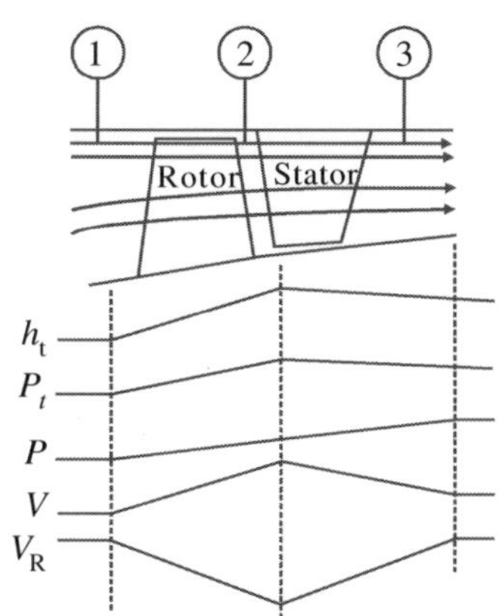

Fig. 7.3. Variation of enthalpy, pressure and velocity across an axial compressor stage.

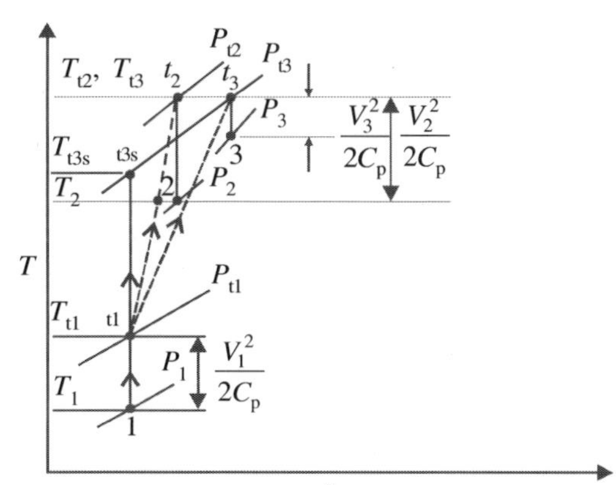

Fig. 7.4. *T-s* diagram describing the processes involved in a stage of an axial compressor.

Of course, conversion of kinetic energy into pressure energy will take place in the stator. There will also be an increase in static pressure in the rotor. After leaving the stator of the first stage, air approaches the row of rotor blades of the next stage. This process is repeated in the remaining stages of the compressor. In some compressors, one may find more number of rows of stator blades, particularly after the last stage of compressor, whose main function is to reduce the air velocity before it enters the combustor. Remember that in a compressor, the flow is always subjected to adverse pressure gradient, which makes it an arduous task for the compressor designer to achieve higher pressure ratio while maintaining higher compressor efficiency. This is so because higher diffusion rate in the row of blades may result in separation of the boundary layer, which decreases compressor efficiency. Hence, diffusion in each stage is limited, which, eventually, restricts the pressure rise per stage.

7.3 Mean Radius Analysis

The fluid flow through the compressor is inherently three dimensional in nature. Therefore, it is very difficult to analyse the fluid flow in the compressor stage. However, we can gain some insights into the details of flow phenomena by making the following assumptions:

- Steady, adiabatic, uniform flow in the axial compressor.
- Axi-symmetric 2D inviscid flow.
- Predominantly axial flow.
- Radial components of velocity can be neglected for $r_h/r_t \geq 0.8$.

In this model, flow conditions are assumed to be uniform, particularly at the inlet and exit of the rotor and stator blades. The flow can be considered as axi-symmetric only when the flow at each radial and axial station can be represented adequately by an average circumferential condition at that station. Hence, flow variations from blade to blade within a row are negligibly small. The radial component can be neglected only when the hub/tip ratio is large ($r_h/r_t \geq 0.8$). However, we have to consider the radial velocity component when the hub/tip ratio is small, which will be discussed in the subsequent section.

7.3.1 Velocity Triangle of a Single Stage

Let us analyse the flow processes in a single axial compressor stage by considering the velocity triangle at mean radius. If we unwrap the cylindrical surface at the mean radius of a typical stage, the row of rotors and consecutive stator will appear as shown in Fig. 7.5. The associated velocity triangles at the inlets of the rotor and stator are shown in Fig. 7.5. Air enters the rotor with absolute velocity V_1 making an angle α_1 with respect to the axial direction. The relative velocity V_{R_1} can then be obtained by vector subtraction of the blade velocity U as shown in the velocity triangle (Fig. 7.5). Now the flow turns through the rotor blade passage to the new relative velocity V_{R_2} with an angle β_2, which is determined by the rotor blade outlet angle. Assuming constant axial velocity V_x across the stage, the value of absolute velocity V_2 with an angle α_2 at the rotor exit (stator inlet) can be obtained by vector addition of blade velocity and relative velocity as shown in the figure of the velocity triangle. Thereafter, air enters the stator blade with absolute velocity V_2 at an angle α_2 in which it is decelerated to a velocity V_3 at an angle α_3. Generally, the axial compressor is designed such that V_3 is almost equal to V_1 and α_3 is almost equal to α_2. As a result, all the stages can be treated similar except the first and last stage. It must be recognised that there is an increase in the tangential

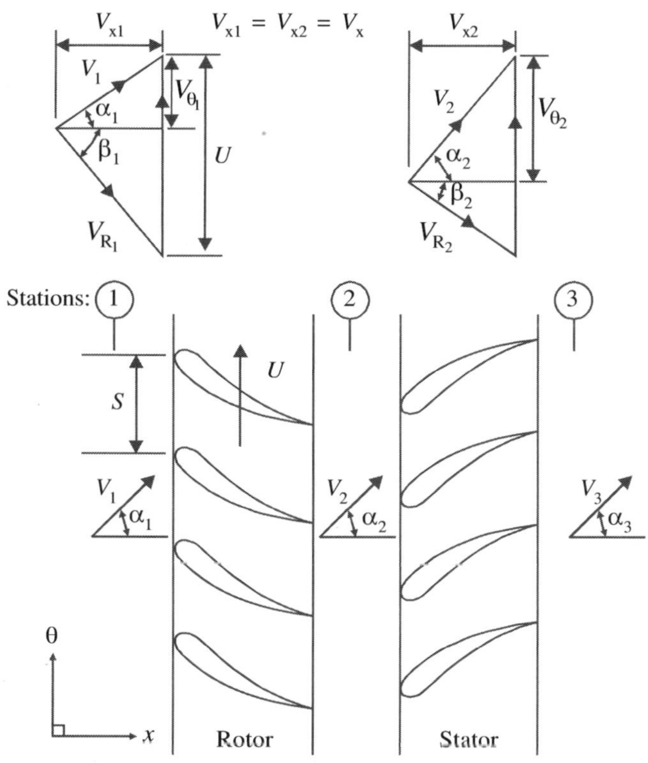

Fig. 7.5. Velocity triangles of a single stage compressor.

component of absolute velocity at the exit of the rotor, because the rotor imparts angular momentum to air. Usually, the velocity in both stator and rotor, relative to the blade passages, decreases. This results in static pressure rise in both stator and rotor blades (Fig. 7.5). Some of these concepts can be visualised well by looking at the typical variation of pressure, total pressure, absolute velocity, relative velocity and total enthalpy along the axial direction (Fig. 7.3).

If we can assume that each fluid element passing through a stage has the same velocity triangles as shown in Fig. 7.5 except for the first and last stage, then the torque and power to drive the compressor can be estimated easily. By applying Euler's Equation of Turbomachinery (Eq. (6.2)) to a control volume (CV) containing only the rotor, the power input to the stage would become

$$\dot{W}_c = \dot{m}U\left(V_{\theta_2} - V_{\theta_1}\right) = \dot{m}\left(h_{t2} - h_{t1}\right) \tag{7.1}$$

where V_{θ_1} and V_{θ_2} are the tangential components of air velocity at the inlet and exit of the rotor, respectively. To derive the above expression, flow across the stage has been assumed to be adiabatic. Equation (7.1) can be expressed easily in terms of the axial velocity considering the velocity triangle shown in Fig. 7.5 as

$$\dot{W}_c = \dot{m}UV_x\left(\tan\alpha_2 - \tan\alpha_1\right) \tag{7.2}$$

Note that the same axial velocity ($V_{x_1} = V_{x_2} = V_x$) is assumed between stations (1) and (2). From Eq. (7.2), we observe that the power input to the air is expressed in terms of stator blade air angles.

However, it will be useful to express Eq. (7.2) in terms of rotor blade air angles. For this, let us first derive a relation between blade velocity and axial velocity. From the velocity triangle (Fig. 7.5), we have

$$\frac{U}{V_x} = \tan\alpha_1 + \tan\beta_1 = \tan\alpha_2 + \tan\beta_2 \tag{7.3}$$

With the help of Eq. (7.3), we can obtain an expression for power input by invoking Eq. (7.2) in terms of rotor blade air angles as

$$\dot{W}_c = \dot{m}UV_x\left(\tan\beta_1 - \tan\beta_2\right) \tag{7.4}$$

We know that the energy transferred to air would raise the pressure while overcoming the frictional losses in the compressor. Hence, there would be rise in stagnation temperature of air per stage. Combining Eqs. (7.4) and (7.2), we can express the rise in total temperature per stage in terms of velocities and air angles as

$$\left(T_{t2} - T_{t1}\right) = \frac{UV_x}{C_p}\left(\tan\beta_1 - \tan\beta_2\right) \tag{7.5}$$

where C_p is assumed to be constant. Let us recall that adiabatic process is assumed in deriving the above equation. Keep in mind that there is no work input in the stator. Thus, total change in total temperature per stage would be

$$\Delta T_{ts} = \left(T_{t3} - T_{t1}\right) = \frac{UV_x}{C_p}\left(\tan\beta_1 - \tan\beta_2\right) \tag{7.6}$$

where ΔT_{ts} is the total temperature rise per stage. However, in an actual compressor, there will be some losses in each stage of the compressor (Fig. 7.4). These losses can be accounted by considering the polytropic efficiency η_{pc}, which is defined as

$$\eta_{pc} = \frac{T_{t3s} - T_{t1}}{T_{t3} - T_{t1}} \tag{7.7}$$

Now, the pressure rise across the stage would be

$$\frac{P_{t3}}{P_{t1}} = \left[1 + \frac{\left(T_{t3s} - T_{t1}\right)}{T_{t1}}\right]^{\frac{\gamma}{\gamma-1}} = \left[1 + \frac{\eta_{pc}UV_x\left(\tan\beta_1 - \tan\beta_2\right)}{C_pT_{t1}}\right]^{\frac{\gamma}{\gamma-1}} \tag{7.8}$$

7.4 Factors Affecting Stage Pressure Ratio

It is evident from Eq. (7.8) that pressure rise per stage is dependent mainly on four parameters viz. blade speed, axial velocity, air deflection angle ($\beta_1 - \beta_2$), and inlet temperature T_{t1}. Hence, the designer may be tempted to use high values of the first three parameters to achieve high pressure rise per stage, as the same is very desirable to reduce the number of stages for a given overall pressure ratio. However, high values of these three parameters cannot be used in practice due to certain limitations, which are discussed briefly.

7.4.1 Blade Tip Velocity

We know that the compressor rotor blade is subjected to centrifugal stress, which is dependent on the rotational speed, the blade length and the blade material. It can be found out from a simplified analysis that the maximum centrifugal stress, occurring at the root of a blade is proportional to square of rotational speed. Generally, the first stage rotor blades are the longest and, hence, are highly stressed as compared to the blades at later stages. Thus, maximum blade tip velocity is restricted to 350 m/s for compressors while tip velocity of 450 m/s is generally used for fans of turbofan engines.

7.4.2 Axial Velocity

We can achieve higher pressure rise by using air with high axial velocity. High axial velocity also ensures high flow rate per unit frontal area, which is very desirable for turbojet and turbofan engines. However, the use of higher axial velocity is restricted due to aerodynamic problems, particularly at the first stage of the compressor in engines without any IGVs. In this case, air will approach only in an axial direction into the rotor blade. So the relative velocity of air will be

$$V_{R_1} = \sqrt{V_{x_1}^2 + U_1^2} \tag{7.9}$$

Let us recall that we have assumed constant axial velocity from the root to the tip. Hence, maximum blade velocity is likely to occur at the blade tip. Let us now calculate the relative rotor Mach number for a given blade velocity by estimating the local speed of sound ($a = \sqrt{\gamma R T_1}$, where T_1 is the static temperature at the rotor inlet). The static temperature can be found from inlet total temperature T_{t1} by

$$T_1 = T_{t1} - \frac{V_{R_1}^2}{2C_p} \tag{7.10}$$

Now the expression for relative rotor Mach number would be

$$M_{R_1} = \frac{V_{R_1}}{\sqrt{\gamma R T_1}} = \frac{V_{R_1}}{\sqrt{\gamma R \left(T_{t1} - \dfrac{V_{R_1}^2}{2C_p} \right)}} = \frac{V_{x_1}^2 + U^2}{\sqrt{\gamma R \left(T_{t1} - \dfrac{V_{x_1}^2 + U^2}{2C_p} \right)}} \tag{7.11}$$

Let us see how M_{R_1} varies with blade velocity for a given value of axial velocity and inlet total temperature T_{t1}. The variation of M_{R_1} with blade velocity is shown in Fig. 7.6 for three values of axial velocity. Axial velocity up to 100 m/s and 200 m/s is used in industrial gas turbine engines and high performance aircraft engines, respectively. It can be noted that the low axial velocity flow is less susceptible to the problem of shock formation even at higher blade velocity. In earlier days, IGVs were used to reduce the problem of shock formation at the initial stage rotor blade tips. However, in recent times, transonic compressors are being designed and developed, which can handle even relative Mach number of 1.2 at the rotor blade tip without incurring excessive losses. Of course, relative Mach number as high as 1.6 is being employed for modern high bypass ratio fans. The problem of high tip Mach number can also be reduced to some extent by using two spool engine as the first stage (LP compressor) of the compressor and running it at lower speed as compared to a high pressure compressor. It must be kept in mind that this problem is not very critical in later stages as the density of the fluid element increases, leading to increase in the sound speed.

7.4.3 Air Deflection in the Rotor

We know from Eq. (7.8) that high stage pressure rise can be achieved by using higher deflection

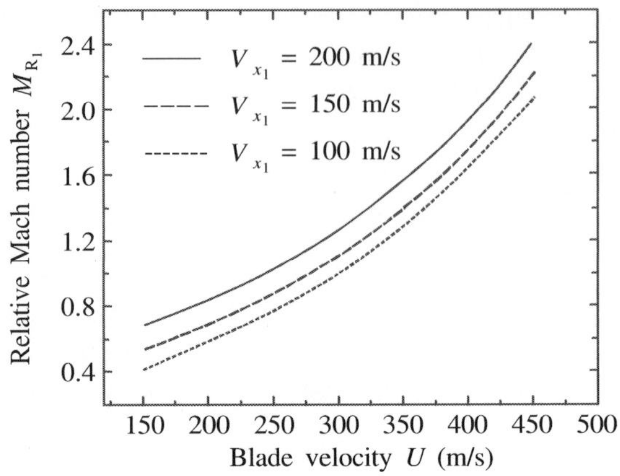

Fig. 7.6. Variation of relative Mach number at rotor entry with blade velocity.

angles. But a question arises: what will happen if high values of deflection angles are used? Let us understand, physically, by looking at the velocity triangles at the inlet and outlet of the rotor, which are drawn on a common base (Fig. 7.7). Note that we have assumed same blade velocity U at the inlet and outlet of the rotor. If we increase the deflection angle by reducing β_2 only for a fixed value of β_1 (Fig. 7.7), there will be a reduction in the relative velocity at the outlet V_{R_2}. Hence, there will be a higher rate of diffusion in the rotor as air is decelerated at a faster rate due to increase in deflection angle. Under extreme conditions, the flow is likely to get separated, incurring excessive losses, which must be avoided. Therefore, lower deflection angle must be used, which will ensure optimum rate of diffusion of air. Now another question arises: what is the method of assessing the allowable diffusion in the rotor blade? For this purpose, the *de Haller number*, defined as V_{R_2}/V_{R_1} is used as the criterion for preliminary design calculation. The de Haller number must be less than 0.72 for a normal design. This is not based on any basic principle but on a thumb rule. It does not seem to be very accurate. Hence, another factor, known as *diffusion factor D* is used for the final design. It is defined as

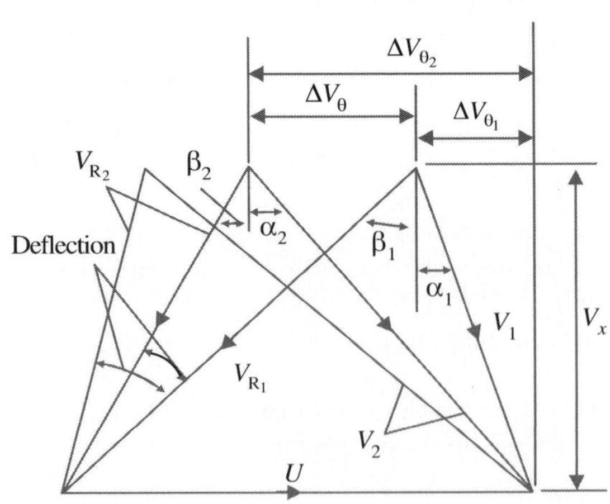

Fig. 7.7. Velocity triangle showing the effect of increasing fluid deflection.

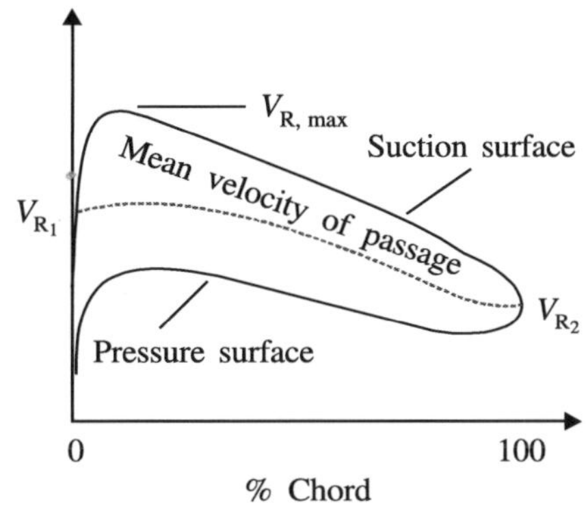

Fig. 7.8. Velocity distribution over the upper and lower surfaces of the cascade blade.

$$D \approx \frac{V_{R,\max} - V_{R_1}}{V_{R_1}} \approx \frac{V_{R_1} + \dfrac{\Delta V_\theta}{2}\dfrac{S}{C} - V_{R_2}}{V_{R_1}} \approx 1 - \frac{V_{R_2}}{V_{R_1}} + \frac{\Delta V_\theta}{2V_{R_1}} \cdot \frac{S}{C} \tag{7.12}$$

where S is the pitch distance between two blades in the cascade and C the chord of the blade. The variation of relative velocities $V_{R,\max}$, V_{R_1}, V_{R_2} in rotor cascade are shown in Fig. 7.8. It has been found out from experimental data that for subsonic flow, the variation of diffusion factor up to a value of 0.6 would not affect frictional losses in the rotor hub and stator region. However, the diffusion factor for the rotor tip region must be less than 0.4, as the losses increase rapidly beyond this value.

Example 7.1 Air at total pressure of 0.3 MPa and temperature of 350 K enters the fourth stage of an axial flow compressor at an angle of 15° to the axial direction and gets deflected through an angle 30° toward the axial direction. Then it enters the stator and exits at an angle of 15°. The axial velocity of air is 150 m/s, which remains constant through this stage. If the blade velocity at mid-radius is 200 m/s, determine pressure ratio across this stage. Assume its polytropic efficiency to be 85%.

Given T_{t1} = 350 K, P_{t1} = 0.3 MPa, $\alpha_1 = \alpha_3 = 15°$, $\varepsilon = \beta_1 - \beta_2 = 30°$, U = 200 m/s, V_x = 150 m/s.

Solution We will solve this problem by using mean line analysis as covered in this chapter. Refer Fig. 7.5 for velocity triangle to solve this problem. We need to determine pressure ratio across this stage using Eq. (7.8) as

$$\frac{P_{t3}}{P_{t1}} = \left[1 + \frac{\eta_{pc} U V_x (\tan\beta_1 - \tan\beta_2)}{C_p T_{t1}}\right]^{\frac{\gamma}{\gamma-1}}$$

Hence, we need to estimate β_1 and β_2. From the velocity triangle shown in Fig. 7.5, we have

$$\tan\beta_1 = \frac{U - V_x \tan\alpha_1}{V_x} = \frac{200 - 150 \times \tan 15°}{150} = 1.065 \;\Rightarrow\; \beta_1 = 46.81°$$

Now we can evaluate $\beta_2 = \beta_1 - \varepsilon = 46.81 - 30 = 16.81°$. Substituting the values in the equation for pressure ratio given above, we can evaluate pressure ratio across the stage as

$$\frac{P_{t3}}{P_{t1}} = \left[1 + \frac{0.85 \times 200 \times 150 (\tan 46.81° - \tan 16.81°)}{1005 \times 350}\right]^{3.5} = 1.207$$

It can be noted that pressure ratio across a stage of axial compressor is quite low as compared to that of a centrifugal compressor, which turns around to be 4.

7.5 Reaction Ratio

We have seen in Section 7.4 that diffusion (flow deceleration) takes place both in rotor and stator blade passages. As a result, there will be an increase in static pressure through both rows of the blade. Let us recall that reaction ratio (RR) is normally expressed as the change in static enthalpy in the rotor divided by static enthalpy rise in the stage and is given as

$$\mathrm{RR} = \frac{\text{Static enthalpy rise in rotor}}{\text{Static enthalpy rise in stage}} \tag{7.13}$$

For an axial compressor stage the change in density is quite small. Hence, RR can be approximated as the ratio of static pressure rise divided by the change in static pressure in the stage. However, we will use RR in terms of enthalpy. Assuming the variation of C_p over temperature change per stage to be negligible small, let us express RR in terms of temperature as

$$\mathrm{RR} = \frac{\Delta T_{\mathrm{Ro}}}{\Delta T_{\mathrm{Ro}} + \Delta T_{\mathrm{St}}} = \frac{T_2 - T_1}{T_3 - T_1} \tag{7.14}$$

where ΔT_{Ro} and ΔT_{St} are the static temperature rise across the rotor and stator, respectively. Now let us express Eq. (7.14) in terms of air angles (Fig. 7.9). Here, we assume that V_x is constant throughout the stage and that air leaves the stage with the same absolute velocity with which it enters ($V_3 = V_1$). As a result, the total temperature rise in a stage is same as the static temperature rise per stage.

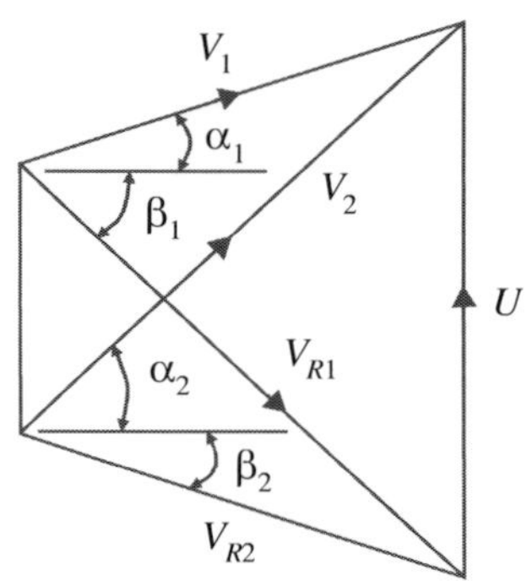

Fig. 7.9. Velocity triangles for 50% reaction ratio.

From the definition of total temperature we obtain an expression for total temperature change in a stage as

$$\Delta T_{\mathrm{stage}} = T_{t3} - T_{t1} = (T_3 - T_1) + \frac{1}{2C_p}\left(V_3^2 - V_1^2\right) = T_3 - T_1 \tag{7.15}$$

since in a stage V_3 is assumed to be equal to V_1.

By using Euler's Equation for Compressor and the velocity triangle shown in Fig. 7.9, Eq. (7.15) becomes

$$\Delta T_{\mathrm{stage}} = T_{t3} - T_{t1} = \frac{U}{C_p}\left(V_{\theta_2} - V_{\theta_1}\right) = \frac{UV_x}{C_p}(\tan\alpha_2 - \tan\alpha_1) = (T_3 - T_1) = \Delta T_{\mathrm{St}} \tag{7.16}$$

Equation (7.16) expresses the static temperature rise across a stator in a stage. Now let us see how the total temperature rise in the rotor can be expressed in terms of static temperature and absolute velocity from the basic definition

$$T_{t2} - T_{t1} = (T_2 - T_1) + \frac{1}{2C_p}\left(V_2^2 - V_1^2\right) \tag{7.17}$$

So the static temperature rise in the rotor becomes

$$\Delta T_{Ro} = T_2 - T_1 = \frac{UV_x}{C_p}(\tan\alpha_2 - \tan\alpha_1) - \frac{1}{2C_p}\left(V_2^2 - V_1^2\right) \tag{7.18}$$

From the velocity triangle shown in Fig. 7.10, $V_1 = V_x \sec\alpha_1$ and $V_2 = V_x \sec\alpha_2$ and Eq. (7.18) becomes

$$\Delta T_{Ro} = T_2 - T_1 = \frac{UV_x}{C_p}(\tan\alpha_2 - \tan\alpha_1) - \frac{V_x^2}{2C_p}\left(\tan^2\alpha_2 - \tan^2\alpha_1\right) \tag{7.19}$$

By combining Eqs. (7.14), (7.16) and (7.19), we can derive an expression for RR as

$$\text{RR} = 1 - \frac{V_x}{2U}(\tan\alpha_2 + \tan\alpha_1) \tag{7.20}$$

$$= \frac{V_x}{2U}\left[\frac{2U}{V_x} - (\tan\alpha_2 + \tan\alpha_1)\right]$$

From the velocity triangle, we know that

$$\frac{2U}{V_x} = \tan\alpha_1 + \tan\alpha_2 + \tan\beta_1 + \tan\beta_2 \tag{7.21}$$

The, the expression for RR would be

$$\text{RR} = \frac{V_x}{2U}(\tan\beta_1 + \tan\beta_2) \tag{7.22}$$

7.5.1 Choice of Reaction Ratio

In practical axial compressors, particularly for aircraft gas turbine engines, 50% reaction blades are widely used as they provide less twisted blades with higher blade efficiency. Let us now see the conditions under which we can achieve 50% RR blade. We can invoke Eq. (7.22) and get

$$\frac{U}{V_x} = (\tan\beta_1 + \tan\beta_2) \tag{7.23}$$

From the velocity triangle, we know that

$$\frac{U}{V_x} = (\tan\alpha_1 + \tan\beta_1) = (\tan\alpha_2 + \tan\beta_2) \tag{7.24}$$

From Eqs. (7.23) and (7.24) and the velocity triangle, we find that $\tan\alpha_1 = \tan\beta_2$ and $\tan\alpha_2 = \tan\beta_1$ for RR of 0.5. Because of the above relations, the velocity triangles will be symmetrical for RR of 0.5 (Fig. 7.9). Hence, this type of blade is known as *symmetrical blade.* This choice of 50% RR reduces the tendency of boundary layer separation from the blade surfaces. Thus, large losses in total pressure can be avoided. As a result, best efficiency can be attained with this design and, therefore, it is popularly used in aircraft engines. But if RR is quite different from 50%, then one blade row would be subjected to higher pressure gradient, resulting in higher pressure losses as it is excessively loaded.

Example 7.2 Determine degree of reaction at mid-radius for third stage of an axial compressor of Example 7.1.

Solution We solve this problem by using meanline analysis as covered in this chapter. Refer Fig. 7.5 for velocity triangle to solve this problem. Since angles such as β_1 and β_2 are already determined in Example 7.1, the degree of reaction (RR) can be determined by using Eq. (7.22) directly as

$$\text{RR} = \frac{V_x}{2U}\left(\tan\beta_1 + \tan\beta_2\right) = \frac{150}{2\times 200}\left(\tan 48.18° + \tan 16.81°\right) = 0.53$$

It can be noted that the velocity triangle will be almost symmetric in nature as RR is closer to 0.5.

7.6 Three Dimensional Analysis

The assumptions of two dimensional flow can be considered to be reasonable if the blade height is relatively small in comparison to the mean diameter of the annulus. This is possible when the hub-tip ratio (r_h/r_t) is greater than about 0.8, which would be true for the later stages of a typical compressor. However, the front stages have lower values of hub-tip ratio. Generally, the first stage of an aeroengine compressor uses a hub-tip ratio of around 0.4 to accommodate high mass flow rate. When this situation of low hub-tip ratio in the first stage and high hub-tip ratio in the last stage occurs, the streamlines will not lie on a surface of revolution parallel to the axis of the rotor, as previously assumed. Under these conditions, the flow must have a radial component of velocity, although it is generally small as compared to the axial and tangential components.

There is another reason why the radial component of velocity can prevail in long blade passages. As the tangential component of velocity increases with the radius for constant inlet velocity, the pressure must increase with the radius to provide the force associated with centrifugal acceleration of the fluid. The tangential velocity can cause a centrifugal acceleration even as high as 1,000 times of gravity, particularly for longer blades. Under these circumstances, strong radial component of flow velocity is likely to be set up in the rotor blade passages. This may lead to maldistribution of flow with radius, which may result in an increase in relative Mach number of air, particularly at blade tip even beyond 1.0. Keep in mind that this situation must be avoided at any cost. The radial flow can be reduced if a suitable static pressure variation along the blade length is provided and the condition of radial equilibrium is maintained.

In order to achieve radial equilibrium, it is necessary to use appropriate compressor blades that, to some extent, are twisted in shape. Of course, due to this special blade design, the degree of reaction may vary along the blade height. Now let us consider the condition under which radial equilibrium can be maintained in the rotor blade.

If air has no radial component of velocity, then for a fluid element the difference in static pressure force must be equal to the centrifugal force. Let us consider a fluid element of unit depth subtending an angle at the x-axis (Fig. 7.10a). Note that this fluid element is rotating about x-axis with tangential velocity V_θ at certain radius r and is subjected to centrifugal force. For the fluid element to be in radial equilibrium, the pressure must be balanced by the centrifugal forces. It can be noted that the pressure force F_p, acting on the fluid element (Fig. 7.10b) can be expressed as

$$F_p = (P + dP)(r + dr)\,d\theta - Prd\theta - 2\left(P + \frac{dP}{2}\right)dr \times \sin\left(\frac{d\theta}{2}\right) \tag{7.25}$$

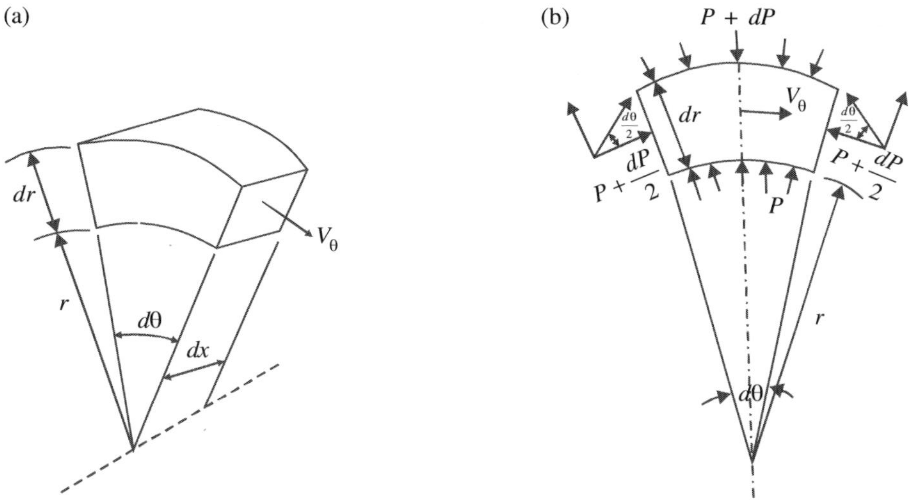

Fig. 7.10. Schematic diagram of a fluid element in a centrifugal force field (a) 3D and (b) 2D.

Simplifying and neglecting higher order terms, we get

$$F_{\mathrm{p}} = r\, dP\, d\theta \tag{7.26}$$

So the mass of this fluid element $\rho r d\theta dr$, is subjected to centrifugal force F_{c}, which is given by

$$F_{\mathrm{c}} = \frac{dm V_\theta^2}{r} = \rho\left(r d\theta\, dr\right)\frac{V_\theta^2}{r} = \rho V_\theta^2\, dr\, d\theta \tag{7.27}$$

But we know that $F_{\mathrm{P}} = F_{\mathrm{c}}$. Hence, we can get a relation for radial equilibrium as

$$\frac{1}{\rho}\frac{dP}{dr} = \frac{V_\theta^2}{r} \tag{7.28}$$

If this condition can be satisfied, then the smaller radial component of velocity V_{r} is ensured, and can be neglected easily. Let us club this radial equilibrium condition with the energy equation. For this we can invoke the basic definition of total enthalpy per unit mass as

$$h_{\mathrm{t}} = h + \frac{V^2}{2} = h + \frac{1}{2}\left(V_x^2 + V_\theta^2\right) \tag{7.29}$$

Differentiating the above expression with respect to radius, we get

$$\frac{dh_{\mathrm{t}}}{dr} = \frac{dh}{dr} + V_x\frac{dV_x}{dr} + V_\theta\frac{dV_\theta}{dr} \tag{7.30}$$

From the thermodynamic relation, we know that

$$dh = Tds + \frac{dP}{\rho} \tag{7.31}$$

Differentiating Eq. (7.31) with respect to radius r, we get

$$\frac{dh}{dr} = T\frac{ds}{dr} + ds\frac{dT}{dr} + \frac{1}{\rho}\frac{dP}{dr} - \frac{1}{\rho^2}\frac{d\rho}{dr}dP \tag{7.32}$$

Dropping the higher order terms, we get

$$\frac{dh}{dr} = T\frac{ds}{dr} + \frac{1}{\rho}\frac{dP}{dr} \tag{7.33}$$

By using Eq. (7.33) in the expression of total enthalpy (Eq. (7.30)), we get

$$\frac{dh_t}{dr} = T\frac{ds}{dr} + \frac{1}{\rho}\frac{dP}{dr} + V_x\frac{dV_x}{dr} + V_\theta\frac{dV_\theta}{dr} \tag{7.34}$$

By substituting $\frac{1}{\rho}\frac{dP}{dr}$ in Eq. (7.34) by $\frac{V_\theta^2}{r}$ from the radial equilibrium equation (Eq. (7.28)), we have

$$\frac{dh_t}{dr} = T\frac{ds}{dr} + \frac{V_\theta^2}{r} + V_x\frac{dV_x}{dr} + V_\theta\frac{dV_\theta}{dr} \tag{7.35}$$

The radial equilibrium energy equation can be used for two kinds of problems viz. the design problem and the direct problem. In case of a design problem, the tangential velocity distribution along the radius is specified and axial velocity variation can be obtained. But in case of a direct problem, swirl angle distribution is specified. Hence, the axial and tangential velocities can be found out easily. Note that the term $T\frac{ds}{dr}$ is basically the radial variation of loss across the annulus and can be neglected for subsonic flow in the compressor. In contrast, the contribution of the entropy term will be significant, when the flow in the rotor blade attains supersonic speed, resulting in the formation of shocks. In such a situation, losses due to entropy term in the rotor blade cannot be neglected. However, the entropy term in Eq. (7.35) can be neglected for subsonic flow in the rotor blade. Hence, Eq. (7.35) becomes

$$\frac{dh_t}{dr} = V_x\frac{dV_x}{dr} + V_\theta\frac{dV_\theta}{dr} + \frac{V_\theta^2}{r} \tag{7.36}$$

This expression is often referred to as *radial equilibrium energy equation*. In order to satisfy this equation, it is necessary to specify two of the three variables, h_t, V_x and V_θ as a function of the radius. Then only we can find the variation of the remaining variables with radius from Eq. (7.36). Generally, total enthalpy can be easily considered to be uniform along the blade height, except near the wall regions. Hence, the terms dh_t/dr can be neglected in Eq. (7.36), which becomes

$$V_x\frac{dV_x}{dr} + V_\theta\frac{dV_\theta}{dr} + \frac{V_\theta^2}{r} = 0 \tag{7.37}$$

Note that Eq. (7.37) is valid for the flow between the rows of blades in an isentropic compressor in which each rotor blade row delivers equal work at all radii. It must be kept in mind that the total pressure losses across a row remains uniform with radius in this condition. Note that this radial equilibrium analysis can also be used for the design of turbine blades.

7.6.1 Free Vortex Flow

Let us consider a case for which V_x and blade height remain constant. Hence, $dV_x/dr = 0$. Therefore, Eq. (7.37) becomes

$$V_\theta\frac{dV_\theta}{dr} + \frac{V_\theta^2}{r} = 0 \tag{7.38}$$

Then

$$\frac{dV_\theta}{V_\theta} = -\frac{dr}{r} \tag{7.39}$$

By integrating Eq. (7.38), we get

$$V_\theta\, r = \text{Constant} \tag{7.40}$$

This expression indicates that tangential velocity varies inversely with radius. The flow which satisfies this condition is known as *free vortex flow*. Note that such kind of flow occurs under natural phenomena like whirlpools and cyclones. The inlets of centrifugal compressor impellers and the exists of radial turbine rotors are usually designed according to free vortex principles. The advantage of using a free vortex blade is that the work distribution along with the radius remains constant. But RR will vary along the blade height. Let us see how RR varies with blade height. We have

$$\text{RR} = 1 - \frac{V_x}{2U}\left(\tan\alpha_2 + \tan\alpha_1\right) \tag{7.41}$$

We can rewrite Eq. (7.41) in terms of tangential velocities as

$$\text{RR} = 1 - \frac{V_{\theta_2} + V_{\theta_1}}{2U} \tag{7.42}$$

Recall that $U = \omega r$. Hence, Eq. (7.42) becomes

$$\text{RR} = 1 - \frac{V_{\theta_2} r + V_{\theta_1} r}{2\omega r^2} = 1 - \frac{\text{constant}}{r^2} \tag{7.43}$$

For a free vortex design $V_\theta r$ remains constant. It can be noted from the above expression that RR increases from the root to tip of the blade even for constant angular velocity ω. Of course, we can choose 50% RR at the mean radius for higher efficiency. But RR becomes very low at the root section. As a result, static enthalpy (pressure) rise in the rotor is quite low as compared to the stator. Hence, this condition may demand high diffusion of flow in the stator, leading to greater losses in the stage. In contrast, the blade velocity U at the root is low and the fluid has to be deflected more for a given work input. In other words, low RR is not desirable in the root. This problem becomes more severe when the blade becomes longer. However, at the tip of the blade, the degree of reaction becomes larger, which is also undesirable, because the fluid has to deflect less for given work input as the blade velocity U is large at the blade tip. These anomalies cause heavy pressure losses in the blades. The blade angle also varies from the hub to tip. As a result, the rotor blade is highly twisted from the hub (root) to the tip and, hence, is very expensive to manufacture. To appreciate this point further, let us consider the following example:

Example 7.3 A single stage axial flow compressor with blade height of 10 cm is running at 10,000 rpm. Air enters the first stage of the compressor with pressure of 0.1 MPa and temperature of 288 K. The hub diameter of the compressor is 30 cm. The total temperature rise across the stage is restricted to 22 K. The axial velocity is 140 m/s, which remains constant from the root to tip. This compressor is to be designed by the free vortex method. Assuming a symmetric blade at its mean blade height, determine the stage air angles and reaction ratio at the hub and the tip.

Given T_{t1} = 288 K, P_{t1} = 0.1 MPa, N = 10000 rpm, V_x = 140 m/s, ΔT_s = 22 K, RR_m = 0.5, h = 10 cm, r_h = 15 cm.

Solution For air angles at the hub, mean and tip and RR at the hub and tip we know that for a free vortex design, the following relation has to be used

$$V_\theta r = C \text{ and } \mathrm{RR} = 1 - \frac{C}{r^2}$$

where C is the constant. Refer Fig. 7.5 for the velocity triangle for solving this problem. It can be noted from this velocity triangle that blade velocity U can be expressed in terms of air angles as

$$U = V_x (\tan \alpha_1 + \tan \beta_1) = V_x (\tan \beta_2 + \tan \alpha_2)$$

The mean radius r_m can be determined as

$$r_m = r_h + \frac{h}{2} = 0.15 + \frac{0.1}{2} = 0.2 \text{ m}$$

Similarly we can determine r_t as

$$r_t = r_h + h = 0.15 + 0.1 = 0.25 \text{ m}$$

But the blade velocity U_m can be determined as

$$U_m = 2\pi r_m N/60 = 2\pi \times 0.2 \times 10000/60 = 209.44 \text{ m/s}$$

Similarly the blade velocity at the hub and tip can be estimated as

$$U_h = 2\pi r_h N/60 = 2\pi \times 0.15 \times 10000/60 = 157.08 \text{ m/s and } U_t = 2\pi r_t N/60 = 261.8 \text{ m/s}$$

We know that RR and the work done W can be related to air angles as

$$\mathrm{RR} = 1 - \frac{V_x}{U}(\tan \alpha_2 + \tan \alpha_1) = \frac{V_x}{U}(\tan \beta_1 + \tan \beta_2)$$

$$\frac{\dot{W}_c}{\dot{m}} = C_p \Delta T = V_x U (\tan \alpha_2 - \tan \alpha_1) = V_x U (\tan \beta_1 - \tan \beta_2)$$

At mean blade height Note that RR = 0.5 at mean blade height, which means that $\alpha_1 = \beta_2$ and $\alpha_2 = \beta_1$. By using the above equations, we can determine air angles at mean blade height as

$$\frac{209.44}{140} = (\tan \alpha_1 + \tan \alpha_2)$$

$$\frac{C_p \Delta T}{V_x U} = \frac{1005 \times 22}{140 \times 209.44} = (\tan \alpha_2 - \tan \alpha_1)$$

By solving the above two equations, we can determine air angle as

$$\alpha_1 = 20.5° = \beta_2 \text{ and } \alpha_2 = 48.42° = \beta_1$$

We need to determine tangential velocity V_θ at stations (1) and (2) as

$$V_{\theta_1 m} = V_x \tan \alpha_1 = 52.3 \text{ m/s and } V_{\theta_2 m} = V_x \tan \alpha_2 = 157.8 \text{ m/s}$$

At hub By using the free vortex condition, we can determine tangential velocity at hub as

$$V_{\theta_1 h} = \frac{V_{\theta_1 m} r_m}{r_h} = \frac{52.34 \times 0.2}{0.15} = 69.8 \text{ m/s}$$

Similarly tangential velocity is estimated at station (2)

$$V_{\theta_2 h} = \frac{V_{\theta_2 m} r_m}{r_h} = \frac{157.8 \times 0.2}{0.15} = 210.4 \text{ m/s}$$

The air angles can be evaluated as

$$\alpha_{1h} = \tan^{-1}\frac{V_{\theta_1 h}}{V_x} = 26.5° \text{ and } \alpha_{2h} = \tan^{-1}\frac{V_{\theta_2 h}}{V_x} = 56.36°$$

We can estimate the reaction ratio RR_h as

$$RR_h = 1 - \frac{V_x}{2U}(\tan\alpha_{1h} + \tan\alpha_{2h}) = 1 - \frac{140}{2\times 157.08}(\tan 26.5^\circ + \tan 56.36^\circ) = 0.11$$

By using the expression for RR and W we can determine relative air angle β_{h1} as

$$\frac{2RR_h U_h}{V_x} = 0.247 = (\tan\beta_{1h} + \tan\beta_{2h}) \text{ and } \frac{C_p\Delta T}{V_x U} = 1.005 = (\tan\beta_{1h} - \tan\beta_{2h})$$

Solving the above two equations, we can determine relative air angles as

$$\beta_{1h} = 32.05^\circ \text{ and } \beta_{2h} = 20.76^\circ$$

At tip blade height By following similar procedure, air angles and RR at blade tip can be determined easily as

$$\alpha_{1t} = 16.7^\circ \text{ and } \alpha_{2t} = 56.7^\circ;\ \beta_{1t} = 54.88^\circ \text{ and } \beta_{2t} = 39.42^\circ;\ RR_t = 0.68$$

It can be noted from these data that the blade is twisted from the hub to tip, which also causes unequal diffusion distribution between the rotor and stator. This condition can be improved for constant work input by satisfying constant $r\Delta V_\theta$ instead of rV_θ condition along the blade height. If such a design approach is considered, the problem of twisted blade and unequal diffusion distribution between the rotor and stator can be reduced drastically. But this may not satisfy the radial equilibrium condition. However, we will restrict our discussion to the situation where radial equilibrium can be maintained. We will now discuss how to maintain radial equilibrium condition by considering generalised tangential velocity distribution.

7.6.2 Generalised Tangential Velocity Distribution

In order to maintain constant pressure ratio, it is desirable to maintain a constant specific work input along the rotor blade height. By choosing variation of V_θ along the radius, we can determine the variation of axial velocity by using Eq. (7.37). Hence, we can choose an arbitrary tangential velocity distribution while maintaining same specific work input along blade height. Euler's Equation of Turbomachinery states that $V_{\theta_2} - V_{\theta_1}$ has to be a constant to satisfy constant specific work input condition in a stage of axial turbomachinery. Let us consider the generalised tangential velocity distribution before and after the rotor blade, respectively, as

$$V_{\theta_1} = ar^n - \frac{b}{r} \quad \text{(before rotor)} \tag{7.44}$$

$$V_{\theta_2} = ar^n + \frac{b}{r} \quad \text{(after rotor)} \tag{7.45}$$

where n is a constant specified by the designer, a and b can be determined from the value of V_{θ_1} and V_{θ_2} at mean radius. The distribution of work for all values of index n is constant with radius so that both h_{t1} and h_{t2} are uniform with blade height. Let us see this by invoking Eqs. (7.44) and (7.45) as

$$V_{\theta_2} - V_{\theta_1} = \frac{2b}{r} \tag{7.46}$$

Hence

$$\dot{W}_{12} = (h_{t2} - h_{t1}) = U(V_{\theta_2} - V_{\theta_1}) = \frac{2b\,\omega r}{r} = 2b\omega \tag{7.47}$$

where ω is the angular velocity. Note that specific work input is independent of radius. By selecting different values of n, several tangential velocity profiles, that can satisfy the radial equilibrium condition, can be easily obtained. However, we restrict our discussion to three special cases namely $n = 1$, $n = 0$ and $n = -1$.

For unity index (n = 1) case, the tangential velocity distribution before and after the rotor is

$$V_{\theta_1} = ar - \frac{b}{r};\; V_{\theta_2} = ar + \frac{b}{r} \tag{7.48}$$

This is often referred to as the *first power blade design* concept. In this case, for constant axial velocity, RR remains constant, which is shown below by invoking Eq. (7.21) for the RR relationship

$$\text{RR} = 1 - \frac{V_x}{2U}(\tan\alpha_1 + \tan\alpha_2) = 1 - \frac{V_{\theta_1} + V_{\theta_2}}{2U} = 1 - \frac{ar}{U} = 1 - \frac{a}{\omega} = \text{constant} \tag{7.49}$$

Hence, this can also be called as *constant reaction blade design* concept. But it does not satisfy the radial equilibrium condition. As a result, actual air angles will not agree with the design value, which leads to reduction in efficiency of the turbomachinery. However, it also leads to having a more twisted blade as compared to the free vortex design. In order to satisfy the radial equilibrium, the axial velocity can be varied, which results in variation in RR along the blade height.

Let us consider another case corresponding to zero index (n = 0). Here, the tangential velocity distribution before and after rotor is

$$V_{\theta_1} = a - \frac{b}{r};\; V_{\theta_1} = a + \frac{b}{r} \tag{7.50}$$

It can be shown that RR expression for this case is

$$RR = 1 - \frac{a}{U} = 1 - \frac{a}{\omega r} = 1 - \frac{\text{constant}}{r} \tag{7.51}$$

This reaction rate varies from the root to tip of the rotor blade. This kind of concept is known as *exponential blade design*. But in the case where exponent n = – 1, RR is given by

$$RR = 1 - \frac{a}{\omega r^2} = 1 - \frac{\text{constant}}{r^2} \tag{7.52}$$

It is evident from Eq. (7.52) that RR varies from the root to tip even though the most desirable RR of 50% can be chosen at the mean radius. It may lead to very low RR at the root and very high RR at the tip of the blade. We know that blade velocity is lower at the root as compared to the tip. Air has to deflect more to have greater diffusion rate for a given work input, which may eventually result in lower efficiency. Hence, it is undesirable to have lower RR at the root. The problem will be more acute for higher hub-tip ratio. This case corresponds to the *free vortex blade design*, as described earlier. For each case, the radial equilibrium equation can be solved to find distribution of axial velocity along the radius. Several other variations in the design of blades are possible, for which the reader may refer other advanced books [3, 7].

7.7 Aerodynamic Losses in an Axial Compressor

The main cause of aerodynamic losses in an axial compressor is attributed to the growth of boundary layer and its separation over the blade surfaces. Other losses in the compressor are due to vortex flows and formation of shock waves. Besides these, mixing losses also occur due to non-uniform velocity profile at the exit of the rows of blades. Generally, the losses in a compressor can be broadly divided into two categories – primary losses and secondary losses. Primary losses are caused due to boundary layer growth over the blade surface and shock formation. Secondary losses are caused by distortion of fluid flow during turning process in blades, which are three-dimensional in nature. Between the two types of losses, secondary losses are much more difficult to analyse.

7.7.1 Primary Losses

These can be further classified into two kinds of losses – profile and annulus.

(i) Profile Losses

The profile losses associated with the boundary layer growth depend upon the geometries of the blade and its operating conditions. A lot of useful information has been obtained from *cascade* testing of blades. We know that the suction surface of a blade is more prone to boundary layer separation. Let us recall that the separation point in the boundary layer depends on factors like degree of turbulence, Reynolds number and angle of incidence, apart from the boundary layer growth. Also, an expression for profile drag can be obtained from the cascade analysis, covered in Chapter 6, as

$$C_{DP} = \left(\frac{S}{C}\right)\left(\frac{\Delta P_{t12}}{\frac{1}{2}\rho V_1^2}\right)\left(\frac{\cos^3 \beta_m}{\cos^2 \beta_1}\right) \tag{7.53}$$

where ΔP_{t12} is the average total pressure loss across the blade rows, β_m is the mean rotor blade air angle, S is the pitch and C is the chord. We have already carried out the analysis for cascade in Chapter 6.

(ii) Annulus Loss

Both rotor and stator blades of an axial compressor are housed in a casing. There will be growth of boundary layers both on the wall of the shroud (casing) and the hub (Fig. 7.11). In stationary blade rows, there will be loss of fluid pressure due to growth of the boundary layer. However, loss is associated with the rotation in the rotor blades in which the boundary layer of the blade passages is subjected to centrifugal forces.

This loss due to annulus drag is dependent on the blade height h and pitch of the blade rows S. The loss increases when blades that are shorter, relative to chord length, are used. Hence, the experimental empirical expression for the annulus drag is [3, 5] as mentioned in Chapter 6.

$$C_{DA} = 0.02\left(\frac{S}{h}\right) \tag{7.54}$$

7.7.2 Secondary Losses

We know that when a blade is rotating, along with it air is also subjected to rotational motion. At the same time, this rotating air is turned through the blade rows (Fig. 7.12). As a result, its axis of

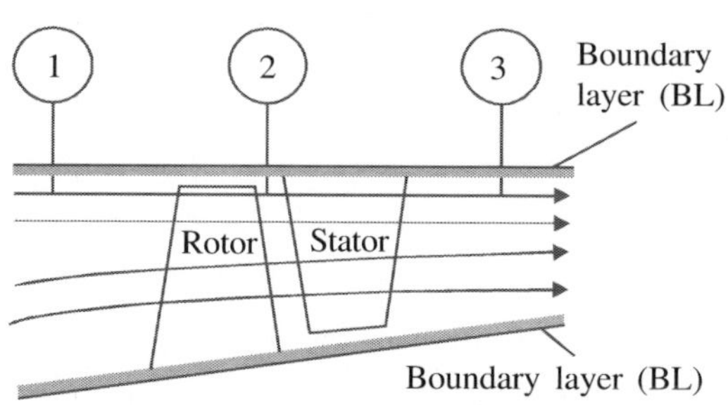

Fig. 7.11. Primary losses comprising of profile and annular losses in an axial compressor.

rotation is deflected in a manner analogous to the motion of a gyroscope in a direction perpendicular to the axis of turning. Due to turning of air through the blade passage, a pressure gradient is set up across the blade passage normal to the streamlines of a given magnitude that can be expressed by Euler's n-equation as

$$\frac{1}{\rho}\frac{\partial P}{\partial n} = \frac{V_\theta^2}{r} \tag{7.55}$$

where r is the local radius of curvature in the normal direction. As we know, there will be boundary layer growth both in the hub and shroud (casing) surfaces. The velocity V, thus, is larger in the middle passage. Hence, the local radius of curvature must be large enough to equalise the pressure gradient. As a result, twin circulation are created, often known as *passage vortex* (Fig. 7.12). Apart from this, the end wall (hub) boundary layers itself roll up into the vortex at the leading edges of the blade, which, subsequently, spreads around both sides of a blade. This kind of vortex is often called *horseshoe vortex*. Interestingly, this vortex amplifies the passage vortex on the pressure surface of the blade while on the suction surface it rotates in the opposite direction and, thus, suppresses the passage vortex. These secondary flows are important source of losses, because they not only reduce kinetic energy of air, but also transport low momentum fluid from the pressure to the suction side of the blade. Because of this phenomenon, not only the boundary layer on the suction surface is more liable to separate but it also causes losses due to intense mixing of fluid.

The magnitude of secondary losses depends on the portion of the total blade passage height, affected by the secondary flow. Hence, the blade passages with high hub/tip ratio (small height) are subjected to more losses (Fig. 7.13) as the whole passage is affected by the secondary flow. However, the blade with low hub/tip ratio (larger height) experiences lower secondary losses. This aspect can be understood clearly by looking at the experimental data of total pressure loss along the blade height (Fig. 7.13). It can be noted from Fig. 7.13 that this loss attains maximum value near both the hub and the tip. The losses shown in the central region of this plot are due to profile drag.

Generally, the magnitude of secondary losses in a high performance compressor, particularly in later stages, is comparable to profile losses. Therefore, it is important for the designer to take account of these losses while designing an axial compressor. As mentioned earlier, secondary losses are very much influenced by the tip clearance between the blade and the casing. Hence, tip clearance of a typical axial compressor is restricted to 1-2% of the blade height. As the secondary flow is inherently

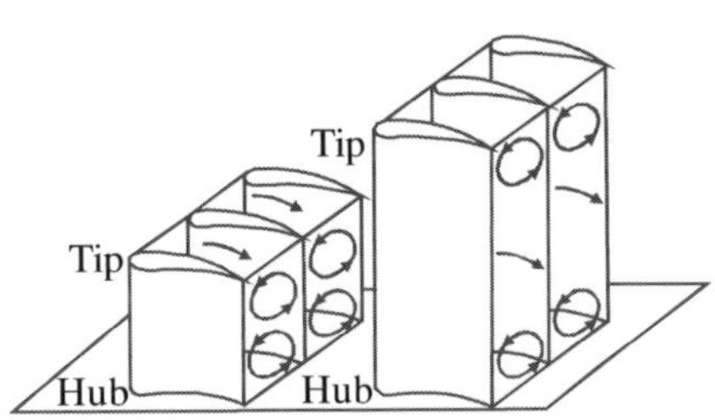

Fig. 7.12. Secondary flow losses in the short and long blades in an axial compressor.

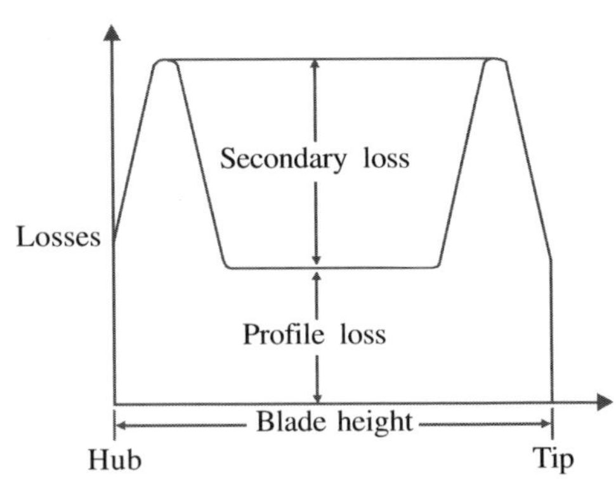

Fig. 7.13. Secondary flow losses in an axial compressor.

three dimensional, it would not be possible to predict the drag due to secondary flow. Rather, the designer has to rely on the empirical relation obtained from experimental data. The most common empirical relation for drag due to secondary flow [3, 5] is

$$C_{DS} = 0.018\, C_L^2 \tag{7.56}$$

where C_L is the lift coefficient.

7.8 Compressor Blade Design

In an earlier section we learnt how to determine the air angles for a given stage work input. It is important to design the blade (shape and size) such that the rows of blades can perform the task of raising the pressure for a given work input. The first and foremost function of a blade is to turn the air through the required angles, such as $(\beta_1 - \beta_2)$ for the rotor blade and $(\alpha_2 - \alpha_3)$ for the stator blade rows. In addition, air must be decelerated, causing minimum stagnation pressure losses in blade rows. To achieve these objectives, air angles must be same as blade angles, particularly at the inlet of the blades. In contrast, air angles may not be same as blade angles during the whole range of operation as the compressor has to operate under situations other than the design condition. Of course, the chosen blade angles are different from air angles estimated at design rpm and pressure ratio. Once the blade angles are chosen, they are fixed for a compressor. But when the compressor is operated under off-design condition, there will be change of both fluid velocities and blade velocity, which can cause changes in air angles. Hence, it is a common practice to use different blade angles than the design value of air angles. Now the question arises: how do we choose blade angles? The blade can be specified in terms of various angles, chord, pitch, etc., which has been discussed in Chapter 6. Readers are urged to revisit this portion before proceeding further. Note that the correct design of the geometry of blade shape is very important and essential for efficient operation of an axial compressor (turbomachine).

7.8.1 Estimation of Stage Efficiency

We had defined, earlier, the stage of efficiency in terms of temperature rise. It is easy to estimate stage efficiency if we can relate it to blade efficiency. Note that blade efficiency can be estimated easily from cascade data, as discussed in Chapter 6. Let us consider that the compressor stage has RR = 50%. Since symmetric blades are used in this stage, therefore, it is expected to have same efficiency η_b for both the rotor and stator rows of the blade. By definition, the blade efficiency will be

$$\eta_b = \frac{P_2 - P_1}{P_{2s} - P_1} \tag{7.57}$$

where P_{2s} is the ideal pressure outlet without any loss. In contrast, the stage efficiency is

$$\eta_{stage} = \frac{T_{3s} - T_1}{T_3 - T_1} = \frac{\Delta T_{is}}{\Delta T_{stage}} \tag{7.58}$$

Let us assume that equal losses occur in each row of stator and rotor blades, since we are using symmetric blade (RR = 50%). Hence, we consider pressure rise to be half of static temperature rise, i.e. $\Delta T/2$ in a single row of blades. Now the stage efficiency across a single row of blades becomes

$$\eta_{stage} = \frac{T_{2s} - T_1}{T_2 - T_1} = \frac{\Delta T_{is}}{\Delta T_{stage}} \tag{7.59}$$

Let us relate the actual static pressure rise in terms of static temperature as

$$\frac{P_2}{P_1} = \left(\frac{T_{2s}}{T_1}\right)^{\frac{\gamma}{\gamma-1}} = \left[1 + \frac{T_{2s} - T_1}{T_1}\right]^{\frac{\gamma}{\gamma-1}} \Rightarrow \frac{P_2}{P_1} = \left[1 + \eta_{\text{stage}} \frac{\Delta T_{\text{stage}}}{2T_1}\right]^{\frac{\gamma}{\gamma-1}} \tag{7.60}$$

As ideal static pressure ratio $P_{2s}/P_1 = \left[1 + \frac{\Delta T_{\text{stage}}}{2T_1}\right]^{\frac{\gamma}{\gamma-1}}$, the blade efficiency η_b can be expressed as

$$\eta_b = \frac{P_2 - P_1}{P_{2s} - P_1} = \frac{\frac{P_2}{P_1} - 1}{\frac{P_{2s}}{P_1} - 1} = \frac{\left[\left(1 + \eta_{\text{stage}} \frac{\Delta T_{\text{st}}}{2T_1}\right)^{\frac{\gamma}{\gamma-1}} - 1\right]}{\left[\left(1 + \frac{\Delta T_s}{2T_1}\right)^{\frac{\gamma}{\gamma-1}} - 1\right]} \simeq \eta_{\text{stage}} \tag{7.61}$$

By expanding the terms in Eq. (7.61) by Binomial series and neglecting second order term, we get

$$\eta_b = \eta_{\text{stage}} \left[1 - \frac{1}{\gamma - 1} \times \frac{\Delta T_{\text{st}}}{4T_1} (1 - \eta_{\text{st}})\right] \approx \eta_{\text{stage}} \tag{7.62}$$

which is approximately equal to η_{stage}. If the degree of reaction is quite different from 50%, the stage efficiency can be expressed as

$$\eta_s = RR\ \eta_{b,Ro} + (1 - RR)\ \eta_{b,st} \tag{7.63}$$

7.9 Stage Off design Performance

So far, we have discussed the characteristics of a single stage compressor at the design condition. But the compressor is also required to operate at conditions quite different from the design conditions, since the engine has to operate under various conditions such as starting, idling, acceleration, deceleration etc. Therefore, let us consider the off-design performance of a single stage compressor. We know that the enthalpy rise across a stage is given by

$$\Delta h_{ts} = U\ \Delta V_\theta = U\ [U - V_x\ (\tan\alpha_1 - \tan\beta_2)] \tag{7.64}$$

Dividing Eq. (7.64) by U^2, we get

$$\frac{\Delta h_{ts}}{U^2} = 1 - \frac{V_x}{U}(\tan\alpha_1 - \tan\beta_2)$$

$$\psi = 1 - \Phi\ (\tan\alpha_1 - \tan\beta_2) \tag{7.65}$$

where $\psi = \frac{\Delta h_{ts}}{U^2}$ is known as *blade loading coefficient* and $\Phi = \frac{V_x}{U}$ as *flow coefficient*.

Normally, ψ in the range of 0.3 to 0.45 and Φ in the range of 0.45 to 0.55 are employed in the design of modern axial compressor in aircraft gas turbine engines.

We know that when air flow rate is different from the design value, V_x will change for the same annulus area. Also, the blade velocity U will definitely change with change in rpm. Thus, the angle β_1 with which air approaches the rotor will change when either V_x or U changes from its design value.

Let us understand what happens when there is a small change in these variables from their respective design conditions. For this, we can assume that a small change in β_1 would not affect β_2. From cascade test, it has also been found out that α_1 (= α_3) from the preceding stator, do not change appreciably for a small change in incidence angle. Hence, we can consider ($\tan \alpha_1 + \tan \beta_2$) in Eq. (7.65) to be a constant k at design condition. Now Eq. (7.65) can be rewritten as

$$\psi = 1 - k\Phi \tag{7.66}$$

The variation of blade loading coefficient ψ with flow coefficient Φ (dashed line, Fig. 7.14) shows that ψ increases as Φ decreases and reaches a value of 1 at $\Phi = 0$, which is not feasible in practice. Keep in mind that this relationship provides stage characteristics under an ideal condition.

However, the actual stage characteristics would be quite different from the ideal one (Fig. 7.14). It can be seen from Fig. 7.14 that the actual characteristic is quite closer to the ideal one at design point in comparison to off-design conditions. These differences are expected, because of fact that the values of α_1 and β_2 will not remain constant in practice under off-design conditions. Also, the boundary layer is likely to separate from the blade surface for large departure from the designed V_x/U value. Hence, there will be considerable departure from the ideal linear relationship as shown in Fig. 7.14. Similarly, the stage efficiency has a maximum value at design condition and drops on both sides (Fig. 7.14). Besides this, choking is likely to occur at higher flow coefficient, resulting in a sharp decrease in efficiency. Keep in mind that beyond this flow coefficient, in practice, the compressor cannot be operated.

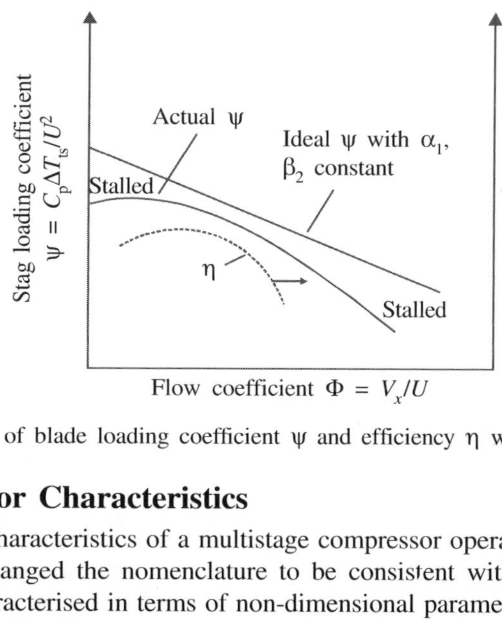

Fig. 7.14. Variation of blade loading coefficient ψ and efficiency η with flow coefficient Φ.

7.10 Axial Compressor Characteristics

Now we will discuss the characteristics of a multistage compressor operating between stations (2) and (3). Note that we have changed the nomenclature to be consistent with cycle analysis. Also, recall that turbomachinery is characterised in terms of non-dimensional parameters. From Chapter 6, we also know that four non-dimensional parameters viz. total pressure ratio across the compressor P_{t3}/P_{t2}, isentropic efficiency η_{ic}, non-dimensional flow rate $\left(\dot{m}\sqrt{T_{t2}}\right)/P_{t2}$, and non-dimensional speed $N/\sqrt{T_{t2}}$ are used to characterise the performance of an axial flow compressor.

A typical characteristics curve of P_{t3}/P_{t2} vs. $\left(\dot{m}\sqrt{T_{t2}}\right)/P_{t2}$ for certain values of $N/\sqrt{T_{t2}}$ for an axial compressor obtained from the testing data (Fig. 7.15(a)) shows that the variation of P_{t3}/P_{t2} at a fixed value of $N/\sqrt{T_{t2}}$ is shown in Fig. 7.15(a) for a narrow range of $\left(\dot{m}\sqrt{T_{t2}}\right)/P_{t2}$. Above this curve, stable operation is limited by the phenomenon of surging, which is discussed in Section 7.10.1. The last line joining this terminating point for all constant speed lines shown in Fig. 7.15 is known as the *surge line*. Generally, this limit is approached by reduction in mass flow at a particular throttle value for a constant rotational speed. Unfortunately, the surge line is reached before the curves reach maximum value under normal operating condition. As the design operating point is supposed to be closer at peak value, the surge line is quite close to this design point as shown in Fig. 7.15(a).

Interestingly, it can be observed in Fig. 7.15(a) that the constant speed lines for high mass flow rate have very steep slopes and becomes almost vertical, particularly at higher rotational speed. Under this condition the compressor is said to be choked, which means that the mass flow rate leaving the compressor does not depend on the corresponding exit pressure. The choking in the compressor does not always mean that flow velocity at compressor exit has reached a sonic value. However, it has been revealed recently by Cumpsty [9] that the front stages of the compressor continue to function normally under choked conditions. The flow may get separated due to negative flow incidence angle at the rear stages and become so severe that the last few stages act like throttle valves instead of compressor stages. As a result, choking may occur in the axial compressor. Consequently, the range of stable operation of an axial compressor is quite limited (Fig. 7.15(a)). As a result, greater care must be taken for matching individual components of the gas turbine for ensuring stable operation of the engine, particularly when the compressor is operated far away from design conditions. Figure 7.15(b) also shows that the peak isentropic efficiency of the compressor η_{ic} remains almost constant for the whole range of constant speed operation.

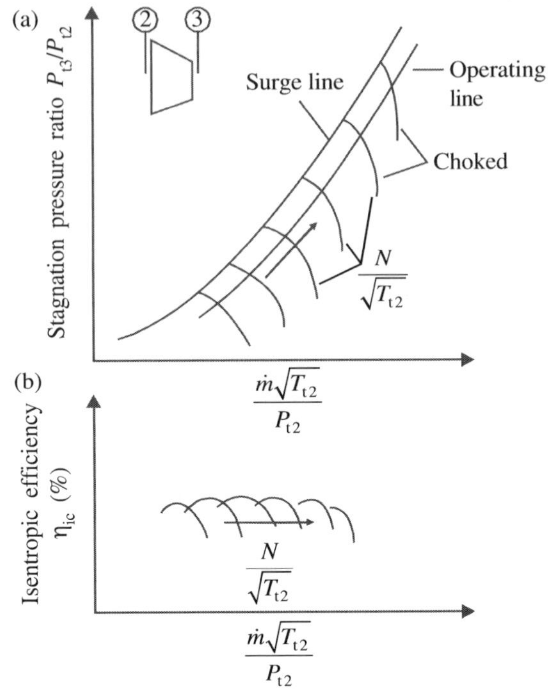

Fig. 7.15. Characteristics of a typical axial compressor for: (a) $\frac{P_{t3}}{P_{t2}}$ vs $\frac{\dot{m}\sqrt{T_{t2}}}{P_{t2}}$ and (b) η_{ic} vs $\frac{\dot{m}\sqrt{T_{t2}}}{P_{t2}}$.

7.10.1 Compressor Surge and Choking

We have seen that the surge line indicates the locus of unstable operation of a compressor, which is mainly associated with large fluctuations in flow rate and delivery pressure. As a result, the stages will interact with the system in a violent manner, giving rise to a phenomena known as *surge* or *surging*. Hence, surge can be conceived as a phenomenon in which the compressor interacts with other components of the whole system in a violent unstable manner with sudden drop in delivery pressure. This phenomenon of surging is accompanied with cyclic violent aerodynamic pulsation, which is transmitted through the whole compressor. Hence, the compressor must be operated below the surge lines; otherwise this surge may destroy not only the compressor but also other components of the engine. The mechanism of surging in an axial compressor is highly complex, which is even now far from being fully understood. Of course, the flow instability of a single stage compressor can be understood without too much difficulty. With increase in the number of stages, a variety of interactions may be possible. Although surging phenomenon is quite complex involving several intricate physical processes, but we will attempt to understand it by considering a single stage whose constant speed pressure rise characteristics, under no surging condition, are shown in Fig. 7.16.

Let us understand how the surging phenomena occurs in a compressor. Let us consider a typical performance characteristics curve of a compressor operating at a particular speed N. We can also assume that at any instant of time the compressor is operating smoothly corresponding to a point A (P_A, $\dot{m}_A$). The flow rate through the compressor can be decreased or increased depending on the swallowing capacity of the turbine and the nozzle in a gas turbine engine. For the present discussion, imagine that the downstream component of the gas turbine engine is replaced by a closing valve, which can control the mass flow rate through the compressor. If the flow rate through the compressor is decreased from $\dot{m}_A$ to $\dot{m}_B$ (Fig. 7.16), the static pressure upstream of the throttle valve will increase. However, this decrease in mass flow rate can cause an increase in pressure at the compressor exit. Hence, the compressor will run smoothly without any problem. With further decrease in mass flow rate to $\dot{m}_C$ and, subsequently, to $\dot{m}_D$ due to closing of the throttling valve, would not cause any problem in the operation of the compressor. However, if the flow rate happens to decrease beyond point D and, subsequently, point E, then pressure delivered by the compressor turns out to be lower than that pressure prevailing at the upstream of the valve. This situation may cause reversal of flow in the delivery line leading to complete breakdown of the normal steady flow from the compressor to the delivery pipe. This phenomena of breaking down of the flow can occur for a short period of time, because the pressure in the delivery line is likely to fall off after a short duration of time and the compressor will regain its normal stable operating condition. In the present case, its operating point will shift suddenly from point E to the stable operating point B. But the valve position (the condition downstream of the delivery line) may still be at the mass flow rate $\dot{m}_D$ corresponding to point D on the characteristic curve. Hence, the compressor operating condition will have to pass through points B, C, D, E and B. As a result, breakdown of flow will occur leading to a shift in the operating point E and, subsequently, to B. The surging cycle, EBCSDE is repeated again and again.

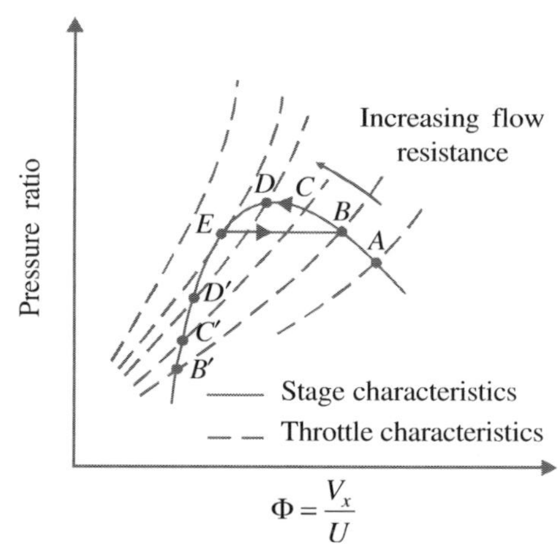

Fig. 7.16. Description of surging phenomenon in an axial compressor.

The frequency and magnitude of this surging cycle depends on the swallowing capacity of the turbine, delivery pipe size, and flow rate below $\dot{m}_S$. This phenomena, known as surging can lead to vibration of the entire gas turbine engine and, eventually, damage the entire engine. Hence, the operation of the compressor must be restricted up to the surge line that is passing through the point $\dot{m}_D$ in Fig. 7.16). It must be noted that, in practice, the compressor is operated much below the surge line.

In contrast, if the mass flow rate increases beyond a certain limit, there will be a decrease in static pressure. As a result, there will be a decrease in the speed of sound. Hence, due to high increase in mass flow rate, the velocity of air flow will increase for the same cross-sectional area. Beyond a certain high mass flow rate, the flow velocity may attain sonic speed leading to choking of the flow. Thus, the operating condition of the compressor must be restricted between the choking condition and the surging condition.

7.10.2 Rotating Stall

Another unsteady flow phenomenon that may occur in an axial compressor is known as *rotating stall.* This term often refers to the stall patterns in the blade passages, which move progressively from one passage to the other, sequentially. The instability of flow associated with rotating stall may lead to surging but it can exist of its own during stable operation of the compressor. Hence, it is important to understand why it occurs at all. Let us consider a few blade rows (Fig. 7.17) in which air flow direction is illustrated in the number of blade passages. When the incidence angle of air into the blade passage B becomes excessive due to either non-uniform flow or partial blockage in the blade passage, then the flow is likely to separate on the suction surface (Fig. 7.17). Now the blade is said to be under *stall.* This can cause a blockage in the blade passage causing air to be deflected in such a way that the blade passage A receives air at reduced incidence angle and passage C receives air at an increased incidence angle. As a result, the blade passage C will get stalled while flow on the blade surface B will be recovered to the normal condition. This process of sequential stalling will pass from one blade passage to another in the direction opposite to the direction of rotation (Fig. 7.17). This phenomena is known as *rotating stall.* The cyclic, prolonged loading and unloading of rotor blades causes aerodynamic induced vibration, which may lead to fatigue failure or even immediate catastrophic failure of gas turbine engines. The frequency and amplitude of vibrations depend on the extent of loading and unloading of the blades. If the frequency of the rotating stall coincides with its natural frequency, it may lead to resonance. During compressor testing, the rotating stall is generally

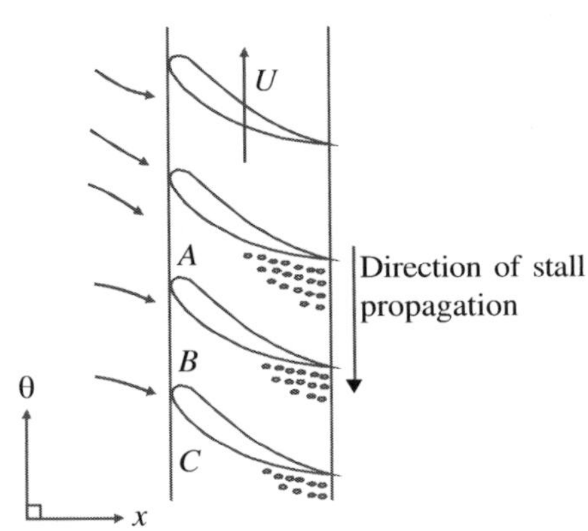

Fig. 7.17. Description of rotating stall phenomenon in an axial compressor.

recognised by the high frequency sound known as *screech*. Due to occurrence of rotating stall in the compressor, there can be considerable reduction in its delivery pressure and efficiency. Hence, it is essential to avoid operation of compressor during occurrence of rotating stall.

7.11 Off design Performance

Generally, the compressor is designed for a particular mass flow rate and rotational speed, as shown in Fig. 7.18. But, in practice, the actual compressor may not operate at the design condition all the time. So we must understand what happens within a multistage compressor when the mass flow rate and operating speed deviate from the design condition. Normally, the design point is chosen corresponding to maximum efficiency, which happens to be closer to the surge line (Fig. 7.18). We know that, at the design condition, all stages will operate at the correct value of V_x/U and, hence, at the correct incidence angle. Now if the mass flow is slightly reduced then the operating point will shift to point A (Fig. 7.18), resulting in an increase of pressure and density in the last stage. Thus, the axial velocity will decrease as $V_x = \dot{m}/\rho A$. This will lead to an increase of incidence angle, as shown in Fig. 7.18 and, thereby, cause a stall in the last stage. However, a small drop in mass flow rate will result in a sharp drop in pressure ratio to point B. Hence, there will be a significant drop in density such that V_x will increase, resulting in a large decrease in the incidence angle in the last few stages. Therefore, stalling is likely to occur in these stages, when the speed of the compressor decreases to such an extent that the compressor will now operate at point D (Fig. 7.18). However, the mass flow rate falls more sharply than the speed. Therefore, axial velocity falls than the blade speed, resulting in an increased incidence angle (Fig. 7.18). A little further reduction in mass flow along the constant speed line will cause further increase in incidence angle, leading to stalling in the first stage itself. Hence, surging at low flow rate may be caused due to stall in the first stage itself. Of course, if the main flow rate increases, then, there will be less chance of stalling in first stage. But there will be decrease in density in the last few stages. Consequently, V_x increases, causing the incidence angle to decrease till its choking point.

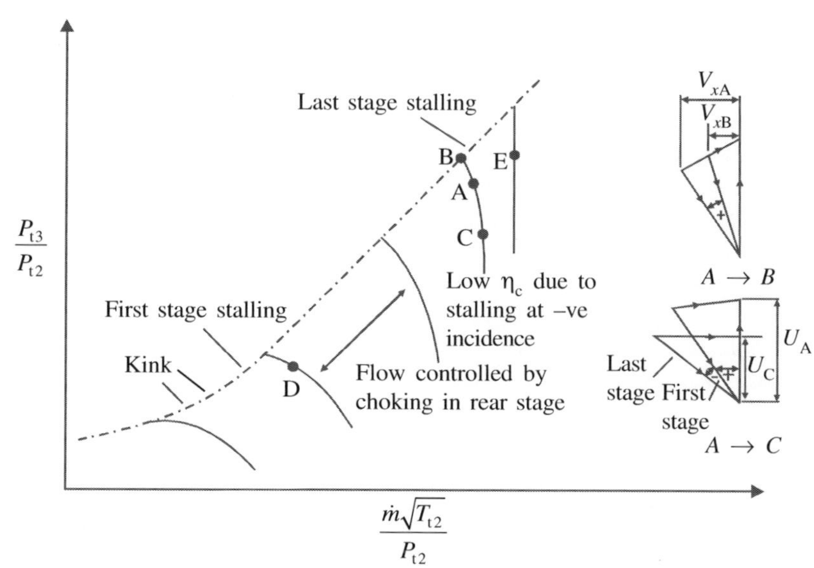

Fig. 7.18. Phenomenon involved in off-design conditions.

In summary, surging is likely to occur at low speeds due to stalling in the first stages, while it occurs at high speeds due to stalling in the last stages. Interestingly, it has been experienced that an axial compressor can operate with several early stages being stalled, which is thought to create a 'kink' in the surge line (Fig. 7.18). When the compressor is operated beyond the design speed (point E; Fig. 7.18), eventually, the characteristics become almost vertical because with the increase in speed, more air passes through the compressor at higher density and pressure ratio. However, the mass flow is likely to increase at a faster rate than density, causing the flow to be choked at the first stage. All the limiting conditions discussed above can lead to unstable operation of the compressor, causing a decrease in efficiency and, thus, must be avoided during its actual operation.

7.12 Chapter Summary

The main objective of this chapter is to familiarise the reader with the essentials of aerothermodynamics that govern the performance of an axial compressor. The basic principle of axial compressor operation is introduced in the beginning while identifying its various hardware parts. When air passes through different stages of the compressor, which is comprised of stator and rotor, its kinetic energy is enhanced by means of rotating blades and then it is converted into potential energy by diffusion process in the rotor blades, resulting in an increase in total pressure. Velocity triangles are employed to discuss the processes involved in a single stage. Two dimensional analysis of a compressor stage at mean line is carried out to derive pressure ratio relationship in terms of air angles, blade velocity and inlet air total temperature. The blade velocity seems to affect the stage pressure ratio significantly. The deflection angle has to be used judiciously to enhance the pressure ratio per stage while avoiding excessive pressure losses due to flow separations over the blade surfaces. The concept of reaction ratio is discussed emphasising its importance on the performance of the stage. In order to take account of three dimensional flow in an axial compressor, the radial equilibrium equation is derived from the basic principle. This concept is being employed to devise various design concepts viz. free vortex, first power, constant reaction etc., which are also introduced. A few examples are considered to demonstrate their uses. Both primary and secondary losses in the axial compressor are discussed in detail. Secondary flow losses are caused due to turning of air flow through curved blade passages under adverse pressure gradient. The compressor blade design is also discussed briefly, mentioning the methods of estimating stage efficiency. The off-design performance of a stage in the compressor is analysed in term of stage loading coefficient and flow coefficient. The overall performance curve of a typical multistage axial compressor is presented in terms of non-dimensional pressure ratio, rotational speed and isentropic efficiency. The phenomena of surge in an axial compressor is discussed adequately to provide its simplistic but conceptual understanding to the reader. The operating lines, as mentioned in the performance plot, must be below the surge line. It must be kept in mind that the axial compressor is more prone to surge as compared to the centrifugal compressor. Another interesting phenomena known as rotating stall is discussed. An account of off-design performance of an axial compressor is provided to highlight its certain operational aspects. In Chapter 8, the centrifugal compressor is covered in detail.

Nomenclature

a	Speed of sound
C_{DA}	Annulus drag coefficient
C_{DP}	Profile drag coefficient
C_{DS}	Secondary drag coefficient
C_P	Specific heat

D	Diffusion factor
F_c	Centrifugal force
F_P	Pressure force
h_t	Total entropy (stagnation enthalpy)
M_R	Relative Mach number
P	Static pressure
P_t	Stagnation pressure
r_n	Hub radius
r_m	Mean radius
r_t	Blade tip radius
R	Rotor
RR	Reaction ratio
RT	Product of gas constant and temperature
S	Entropy
S	Stator
T	Temperature (static temp)
U	Blade velocity
V_{R_2}/V_{R_1}	de Haller number
V_θ	Tangential velocity
V_x	Axial velocity
$\dot{W}_c$	Power input to compressor

Greek Symbols

ΔT_{ts}	Total change in temperature per stage
η_b	Blade efficiency
η_{pc}	Polytropic efficiency
η_{Stage}	Stage efficiency
γ	Specific heat ratio
ψ	Blade loading coefficient $\psi = \Delta h_{ts}/u^2$
Φ	Flow coefficient $\Phi = V_x/U$

Subscripts

AD	Annulus drag
b	Blade
c	Centrifugal
DP	Profile drag
DS	Secondary drag
n	Hub
pc	Polytropic
p	Pressure
r	Relative number
Ro	Rotor
St	Stator
t	Stagnation
ts	Temperature per stage
θ	Tangential

References and Suggested Readings

1. Mattingly, J.D., *Elements of Gas Turbine Propulsion*, 1st Edn., McGraw-Hill Publishing Company Ltd., Singapore, 1996.
2. Kerrebrcok, J.L., *Aircraft Engines and Gas Turbines*, 1st Edn., MIT Press, England, 1992.
3. Cohen, H. Rogers, G.F.C. and Saravanamuttoo, H.I.H., *Gas Turbine Theory*, 4th Edn., Addison-Wesley Longman Ltd., England, 1998.
4. Hill, P.G. and Peterson, C.R., *Mechanics of Thermodynamics of Propulsion*, 2nd Edn., Addison-Wesley Publishing Company, Inc., Reading, MA, 1992.
5. Horlock, J.H., *Axial Flow Compressor*, Robert E. Kreiger Publishing Co., Melbourne, Florida, 1973.
6. Lieblein, S. and Johansen, I.A., "Resume of Transonic Compressor Research at NACA Lewis laboratory", *Transaction of American Society of Mechanical Engineers, Journal of Engineering for Power*, 83, pp. 219-234, 1961.
7. Howell, A.R., "Fluid Dynamics of Axial Compressor and Design of Axial Compressor", *Proc. Institute of Mech. Engineers*, 12, p. 153, 1945.
8. Gostelow, J.P., *Cascade Aerodynamics*, Pergamon Press, 1984.
9. Cumpsty, N.A., *Compressor Aerodynamics*, Longman Group, UK, 1988.

Review Questions

1. Draw the schematic of a typical axial compressor and describe its operating principle. How is it different from a centrifugal compressor?
2. What are the advantages and disadvantages of an axial compressor over centrifugal compressor?
3. What is the maximum pressure ratio across a single stage for an axial compressor?
4. What are the applications of axial compressor? Compare it with that of the centrifugal compressor.
5. Show schematically the variation of total enthalpy, pressure and temperature across a single stage of an axial compressor. How are these comparable with their respective static quantities?
6. Is the pressure rise per stage higher in the rear or front stages of an axial compressor. Explain why is it so?
7. Can you suggest methods of reducing the flow area toward the rear of an axial compressor? Why is it needed at all?
8. How does axial velocity vary along the blade height for the first and last stages in an axial compressor?
9. What are the factors that affect the stage pressure ratio?
10. What is diffusion factor? How is it different from the de Haller number? Which is preferred for the final design of a compressor blade? Why is it so?
11. What is reaction ratio (RR)? Why is it so important? Derive an expression for RR and show that 50% RR produces a symmetric blade.
12. Why is symmetric blade design preferred in an aircraft gas turbine engine?
13. Classify the compressor blade stages on the basis of reaction ratio? Explain it with the help of velocity diagram.
14. When a designer is asked to keep the axial velocity constant through a stage, he states that the Mach number will decrease across the stage. He further argues that if more stages follow, then it will result in poor design. Is it really so? Explain it by providing suitable arguments.
15. What is radial equilibrium? Derive an expression for the radial equilibrium energy equation.
16. What is free vortex design concept? How is it different from the exponential design concept?
17. What is solidity? Discuss how do these two quantities influence the blade design of an axial compressor.
18. Why is polytropic efficiency used in the stage design of an axial compressor?
19. Define lift and drag coefficients for an axial compressor blade. Indicate how do lift and drag coefficients vary with angle of incidence?
20. Define blade efficiency and show that it can be expressed in terms of drag and lift. Find an expression for the blade efficiency for a symmetrical blade when the flow coefficient is 0.5.
21. What is rotating stall? Under what conditions is an axial compressor likely to get stalled? What are the ways of reducing the stall in an axial compressor?

22. What is surging? Explain this phenomena in detail. Under what conditions is an axial compressor likely to get surged?
23. What is choking in a stage? Explain it in detail. Under what conditions is an axial compressor likely to get choked?

Problems

1. A single stage compressor, with reaction ratio of 0.5, which consumes 50 kg/s of air at 18,000 rpm is to be designed. Air at pressure of 1 MPa and 500 K enters the rotor blade with an absolute angle of 18° and leaves the stator blade with the same angle at mean radius of 15 cm. If the inlet rotor blade height is 4 cm and the stage efficiency is 0.88, determine (i) stator turning angle, (ii) rotor turning angle, (iii) static pressure at the rotor and stator exits, (iv) total pressure ratio, abd, and (v) power input to the stage.
2. An axial flow compressor is designed for pressure ratio of 3 with 50% reaction ratio and an isentropic efficiency of 85%. The inlet and outlet air angles from the rotor blades are 15° and 60°, respectively. The mean blade velocity is 400 m/s. Assuming axial velocity to remain constant and total temperature change to be 30°C per stage find out the number of stages. Determine the relative velocity Mach number at mean blade height of the first stage assuming that total inlet temperature is 300 K. Estimate the polytropic efficiency. Draw the velocity triangles.
3. A single stage axial compressor with hub-tip ratio of 0.7, operated at 15,000 rpm is designed on the free vortex principle. The stagnation temperature rise is 25 K. If inlet velocity is 120 m/s and stage efficiency is 0.9, estimate (i) rotor blade air angles at the tip if relative Mach number is restricted to 0.95, (ii) mass flow rate, (iii) rotor air angles at root section, and (iv) stage pressure ratio and power required. Assume ambient condition to be at 1 bar and 288 K.
4. Air enters the repetitive stage of an axial flow compressor with total pressure 200 kPa and 400 K. At the mean blade height, the blade velocity U is 200 m/s. Air enters at the mean blade height at an angle of 15° with the axial direction. But in the stator blades, air gets turned and leaves the blade at 65° with respect to its axis. Air velocity in the axial direction is 150 m/s, which remains constant through the stage. If the diffusion factor is restricted to 0.4 and polytropic efficiency is 0.88, estimate (i) air angles, (ii) pressure ratio across the stage, (iii) pitch-chord ratio, and (iv) reaction ratio.
5. The flow rate per unit frontal area can be increased by (i) reducing the hub/tip ratio from 0.6 to 0.5 at the same axial velocity and duct diameter, or (ii) by increasing the axial velocity from 100 to 150 m/s. Which method is better and explain why is it so?
6. A 12-stage axial compressor produces an overall pressure ratio of 8:1 with polytropic efficiency of 88%. Air enters the first stage of the compressor with pressure 0.1 MPa and temperature 288 K. It can be assumed that the work is equally consumed by each stage. Blade and axial velocities are 200 and 150 m/s, respectively. Assuming 50% reaction ratio, determine the blade angles.
7. A repetitive stage axial compressor takes air from the atmosphere with an axial velocity of 200 m/s. The air is turned through 25° as it passes through a rotor with tip and hub diameter of 60 and 40 cm, respectively. If the compressor is rotating at 16,000 rpm, determine (i) mass flow rate, (ii) pressure ratio, and (iii) reaction ratio. Assume polytropic efficiency to be 85%.
8. A single stage axial compressor with blade height of 20 cm is running at 10,000 rpm. The hub diameter of the compressor is 30 cm. The total temperature rise across the stage is restricted to 22 K. The axial velocity of air is 140 m/s, which remains constant from the root to tip. Assuming reaction ratio of 0.5, at mean radius, determine (i) stage air angles at root, mean and tip, and (ii) reaction ratio at root and tip for free vortex design condition.
9. A stage of an axial compressor is designed on the principle of free vortex design for a total temperature rise of 25 K. The blade velocities at the root, mean and tip of the rotor blade are 150, 200 and 250 m/s, respectively. The axial velocity of 120 m/s remains constant from the root to tip. The reaction ratio at the mean radius is considered to be 0.5. Determine the air angles at the root, mean and tip of the rotor. Estimate the reaction ratio at its root and tip.

10. A stage of an axial flow compressor is designed with symmetric blades. The axial velocity of 120 m/s remains constant from the root to tip. Air enters the first stage of the compressor at angle of 35° with pressure of 0.1 MPa and temperature of 288 K. The blade velocity at mean radius is 120 m/s. A rotor blade with solidity of 1.0, lift coefficient of 0.75 and *L/D* ratio of 20 is used in this compressor. Determine the stage pressure ratio across the compressor.
11. An axial flow compressor has an overall pressure ratio of 4.0 and mass flow of 4 kg/s. Air enters the first stage of the compressor with pressure of 0.1 MPa and temperature of 288 K. If the polytropic efficiency is 88% and per stage must be less than equal to 28 K, estimate the number of stages at pressure ratio of the first and last stages. Assume equal temperature rise in all the stages. If the absolute velocity approaching the last rotor is 175 m/s at an angle 23° from the axial direction, the velocity diagram is symmetrical and mean diameter of the last stage rotor is 20 cm. Calculate rpm of the rotor and height of the last stage rotor blade at inlet stage.
12. A stage of an axial compressor operating at 12,000 rpm consumes 2,000 kW power. It has a mean radius of 20 cm with blade height of 2.5 cm. A mass flow rate of 50 kg/s enters this stage at pressure of 0.4 MPa and temperature of 450 K. Air enters the rotor blade with an absolute angle of 10° with the axis and leaves the stator with the same absolute exit angle. The air turning angle across the rotor is 28°. The compressor efficiency is 0.89. Air enters the rotor blade with axial velocity of 300 m/s, which is increased by 3% each both in the rotor and stator. Estimate (i) total pressure ratio, (ii) pressure coefficient for rotor blades, (iii) reaction ratio, and (iv) stator turning angle. If the lift and drag coefficients of stator blades are 0.67 and 0.05, respectively, and the lift and drag coefficients of rotor blades are 0.64 and 0.08, respectively, determine the stage efficiency for this compressor.
13. Aerofoil shape blades with lift to drag ratio of 20 and lift coefficient of 0.6 with solidity ratio of 0.92 are used for the rotor blades in a stage of an axial compressor. Air enters the rotor blade with an absolute velocity of 200 m/s at mean radius at pressure of 0.3 MPa and temperature of 350 K. If the axial velocity of air is 120 m/s, determine the pressure ratio across this stage.
14. The first stage of an axial compressor operating at 10,000 rpm is designed on the free vortex concept. The total temperature rise across a stage is 25 K. Air at pressure of 0.2 MPa and temperature of 350 K enters this stage with an absolute velocity of 150 m/s at mean radius. The blade with hub-tip ratio of 0.5 is used to design the compressor. If the polytropic efficiency of compressor is 0.85, determine (i) air angles at rotor tip when its relative Mach number is 0.8, (ii) total pressure ratio, (iii) mass flow rate, (iv) rotor air angles at its root, and (v) stage efficiency.
15. Design an axial compressor for pressure ratio of 5 with reaction ratio of 0.5 at all radii. Air enters this stage at pressure of 0.1 MPa and temperature of 300 K. It has a mean radius of 20 cm with blade height of 2.5 cm. The axial velocity of 250 m/s remains constant from root to tip. Determine mass flow rate of the air.

8. Centrifugal Compressor

We had been unable to find anything of value in any of the works to which we had access. So we worked out a theory of our own on the subject and soon discovered as we usually do that all propellers built heretofore are all wrong and then built a pair. ....based on our theory which are all right.

Oliver Wright and Wilbur Wright, March 1903

8.1 Introduction

In small high speed turbochargers of reciprocating engines, initially, centrifugal compressor were employed. During Second World War, a lot of developmental work on the centrifugal compressor was carried out in Britain to develop simple turbojet engines, whereas the Germans were concentrating their research on the development of axial compressors. But you will be surprised to know that the first two turbojet engines were independently developed, one by Whittle (a British Engineer) and another by von Ohain (a German Engineer) using centrifugal compressor. Note that the technology for centrifugal compressors were sufficiently developed in the beginning of the turbojet era. However, with the growing power requirements of engines, more research funds were diverted for the development of axial compressor for turbojet engines. Although most recent turbojet engines employ axial compressors, but centrifugal compressors are still used in some turboprops, turboshafts, superchargers and auxiliary power units. The centrifugal compressor are also used for high pressure turbofan engines. When high pressure is required, the centrifugal compressor is used in conjunction with axial compressor. Generally a pressure ratio of around 4:1 per stage can be easily obtained in centrifugal compressor by using aluminum alloys. With the use of exotic materials like titanium alloys and proper aerodynamic design, the pressure ratio can be easily enhanced to 8:1 per single stage. But the centrifugal compressors are mainly suitable for handling small volume flow rates. The advantages of centrifugal compressor over an axial compressor are:

- Simplicity and ruggedness of construction.
- Better resistance to damage due to foreign objects.
- Shorter length for same overall pressure ratio.
- Less susceptibility to loss of performance by build-up deposits on blade surfaces.
- Ability to operate over wide range of mass flow rate between surging and choking at a particular rotational speed.
- Flow direction at discharge more suitable for installation of intercooler and can type combustor.
- Wide range of performance by use of adjustable prewheel and diffuser blades.

The disadvantages of centrifugal compressor over the axial compressor are

- Larger frontal area for a given mass flow rate.
- Flow direction at discharge unsuitable for multistaging and unsuitable for annular and can annular gas turbine combustor.
- Lower efficiency for pressure ratio.

8.1.1 Elements of Centrifugal Compressor

The elements of a typical single stage centrifugal compressor are shown in Fig. 8.1. The main component of centrifugal compressor is the rotating element known as *impeller* (radial vane type) shown in Fig. 8.1 which is mounted on top of the hub (impeller disk) through which power is transmitted to the impeller. The impeller can be a single or double entry type depending on the pressure ratio and mass flow requirement. When the impeller is rotated at high speed, air is drawn into the impeller eye whose function is to turn the flow relative to the impeller as it enters it. Generally the impeller eye is designed in such a way as to provide shock-free entrance for the incoming air. The rotating impeller imparts a very high velocity to the incoming air. The air then moves radially through the impeller blade passages by virtue of the centrifugal force, giving rise to an increase in static pressure. Note that this rise of static pressure is quite different from that of the pressure rise in axial compressor rotor and stator. Therefore, the pressure rise per stage is less limited by the problem of boundary layer growth and its separation under the adverse pressure gradient as generally encountered in case of axial compressor. This is the reason why higher pressure ratio in a single stage can be easily achieved by the centrifugal compressor in comparison to axial compressor. However, the pressure rise in centrifugal compressor is limited by the centrifugal stresses developed in the impeller. The air then leaves the impeller tip at a high velocity and enters into the radial diffuser where it is decelerated, resulting an increase in static pressure. Generally, the radial diffusers consists of vaneless space as shown in Fig. 8.1. The air then passes through the vaned diffuser where major portion of kinetic energy is converted into static pressure. Typical variation of pressure and velocity along the radial direction is shown in Fig. 8.2. Generally, the velocity of air in the diffuser is reduced to the initial air velocity with which it enters into the impeller eye. There will be loss of stagnation pressure in the diffuser due to frictional effect. Note that it is a common practice to design the centrifugal compressor in such way that equal pressure rise takes place in the impeller and diffuser corresponding to reaction ratio, RR of 50%. After leaving the vaned diffuser, the air is collected in a volute and delivered to the combustion chamber in case of gas turbine application.

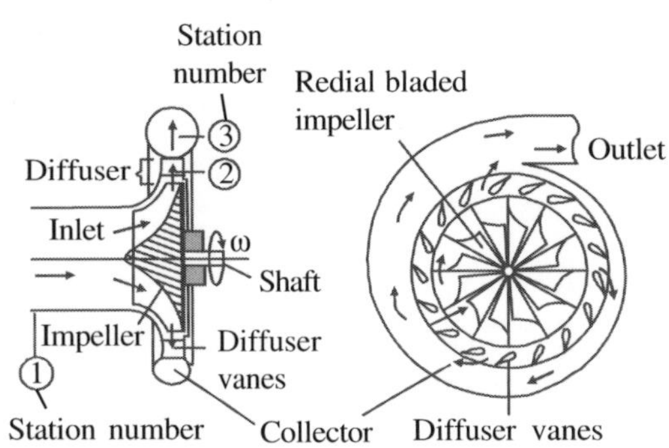

Fig. 8.1. Schematic of a typical centrifugal compressor.

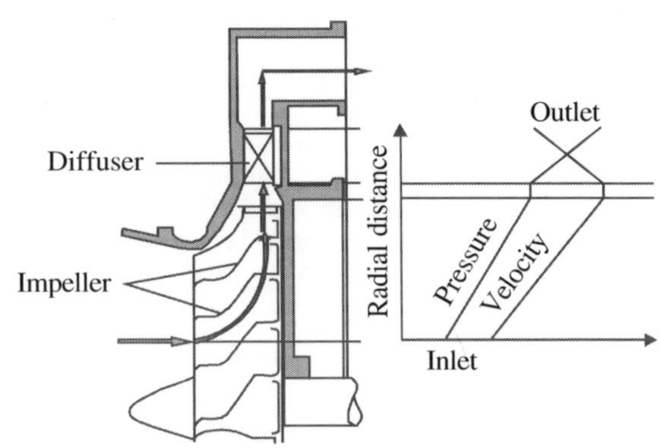

Fig. 8.2. Variation of pressure and velocity across the radial distance of a typical centrifugal compressor.

The air passing through the annulus area between hub (r_{h1}) and impeller tip (r_{t1}) enters the eye (inducer) section of the impeller at station (1) (see Fig. 8.1) with a uniform axial velocity of V_1 (assumed). The main function of the inducer (eye) is to turn the flow smoothly into the impeller passages. The velocity diagrams for an axial flow entering into impeller inlet is shown in Fig. 8.3. The impeller speed, U_1 increases from hub to tip. As a result, the relative flow angle, α with respect to the axial direction also increases from hub to tip giving rise to a twist in the inducer blade at the entrance. This twist in vane at entrance is provided to avoid flow separation in the leading edge itself. Generally, an inducer is not used in the modern centrifugal compressors/fans which produce a lot of noise due to the flow separation and violent mixing at the entrance of the impeller vane. After passing through the impeller, the air leaves the impeller tip (2) (see Fig. 8.1) with a radial velocity of V_{r2} and a tangential component $V_{\theta 2}$ as shown in Fig. 8.3. Thus the air ultimately leaves the impeller tip with an absolute velocity of V_2 which is vector summation of V_{r2} and $V_{\theta 2}$. Generally the compressor is designed in such a way that the radial velocity V_{r2} is comparatively small and approximately equal to inlet velocity V_{x1}. Note that under an ideal condition, the tangential velocity $V_{\theta 2}$ will be equal to the impeller tip speed U_2 as shown in Fig. 8.3.

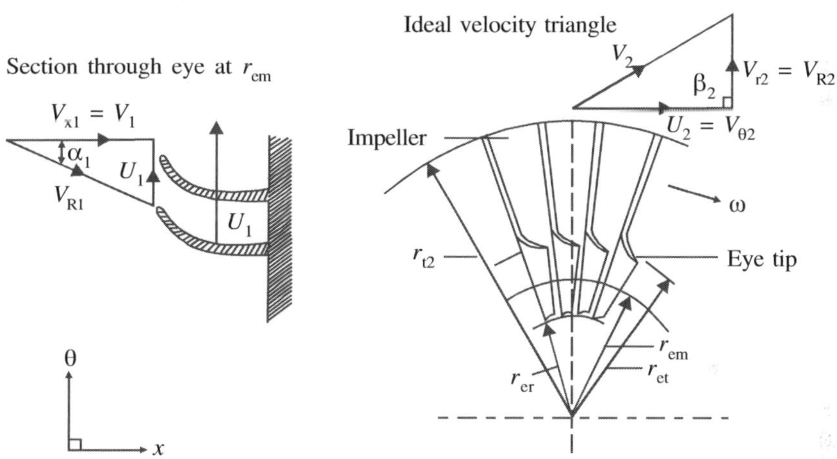

Fig. 8.3. The velocity triangle of a typical centrifugal compressor.

8.2 Theoretical Analysis

The fluid flow through the compressor is inherently three dimensional in nature. Therefore it is very difficult to analyze fluid flow in a compressor stage. However, one can gain some insight into the details of flow phenomena by adopting a simple two dimensional flow model. In this model, flow conditions are assumed to be uniform over certain cross sections of the compressor, usually before and after the impeller and diffuser. The predominant flow is considered to be in radial and tangential directions. The flow in the axial direction can be neglected in the analysis of the centrifugal compressor.

8.2.1 Ideal Energy Transfer

Now we will discuss about the energy transfer that takes place between impeller and the fluid while operating under an ideal condition. As a result there will be an increase in pressure of the incoming air due to this energy transfer. We shall derive an expression for the pressure rise across a compressor.

We have already assumed that the air is entering into the impeller eye along the axial direction only. That means, air velocity has no tangential component at its inlet. As a result, the angular momentum of the air at the inlet is zero. Of course, the impeller vanes at the inlet must be curved making a suitable angle with the tangential direction as shown in Fig. 8.3 such that air can enter into the impeller vanes smoothly without causing any losses due to flow separation and violent mixing.

It is assumed that under ideal conditions the air leaving the impeller tip will have a tangential velocity $V_{\theta 2}$, equal to the impeller tip speed U_2 as shown in Fig. 8.3. Besides this condition, it can be assumed that no losses will be incurred due to the friction and there would not be any energy loss/gain either from or to the working fluid throughout the compressor under these ideal conditions.

Let us see how energy transfer takes place with the help of the impeller, by considering the expression for the work done per unit mass flow rate by the impeller on the fluid.

$$\frac{\dot{W}_c}{\dot{m}} = h_{t2} - h_{t1} = U_2 V_{\theta 2} - U_1 V_{\theta 1} \tag{8.1}$$

Generally, air enters into the centrifugal compressor axially without any tangential velocity component as shown in Fig. 8.3. Since $U_1 V_{\theta 1}$ is zero, the specific work done on the fluid from Eq. 8.1 will be

$$\frac{\dot{W}_c}{\dot{m}} = h_{t2} - h_{t1} = U_2 V_{\theta 2} \tag{8.2}$$

Since no energy is added in the diffuser, the total enthalpy rise across the whole compressor, $(h_{t3} - h_{t1})$ must be equal to total enthalpy rise $(h_{t2} - h_{t1})$ across the impeller alone. Thus, the above equation will be

$$h_{t3} - h_{t1} = C_p (T_{t3} - T_{t1}) = U_2 V_{\theta 2} \tag{8.3}$$

where, C_p is the mean specific heat over the temperature rise. Under an ideal situation, since $V_{\theta 2}$ is equal to U_2, as can be seen in velocity diagram shown in Fig. 8.3, then the above equation becomes

$$(T_{t3} - T_{t1}) = (U_2)^2 / C_p \tag{8.4}$$

Assuming the compression to be isentropic across the compressor, we can derive an expression for the pressure ratio by using the above relation which is given below.

$$\frac{P_{t3}}{P_{t1}} = \left[1 + \frac{U_2^2}{C_p T_{t1}}\right]^{\frac{\gamma}{\gamma - 1}} = \left[1 + \frac{(\omega r_2)^2}{C_p T_{t1}}\right]^{\frac{\gamma}{\gamma - 1}} \tag{8.5}$$

From this equation, we can note that for a given working fluid, the pressure ratio $\frac{P_{t3}}{P_{t1}}$ is influenced by impeller tip speed, U^2 and inlet temperature T_{t1}. For a given work input, the pressure ratio, $\frac{P_{t3}}{P_{t1}}$ will increase with decrease in inlet temperature T_{t1}. The inlet temperature can be decreased by employing intercooler between stages of compressor. The variation of pressure ratio, $\frac{P_{t3}}{P_{t1}}$ with impeller tip speed, U_2 for a typical case is shown in Fig. 8.4. It can be observed that the pressure ratio increases as expected with increase in impeller speed. Thus, the designer may be tempted to use higher impeller speed. But the maximum impeller speed is limited by the centrifugal stress which is proportional to the square of the impeller tip speed, U_2. For single side impeller, made of aluminum alloy, U_2 is limited to about 460 m/s and to 650 m/s for a titanium alloy impeller. For double sided impeller, lower impeller speed must be used because of additional disc loading. Another way to increase the pressure ratio, is to use the impeller with higher diameter. Again it is limited by the centrifugal stress and shock formation at high impeller speed. Note that larger size impeller may result in an adverse increase in frontal area leading to higher aerodynamic drag.

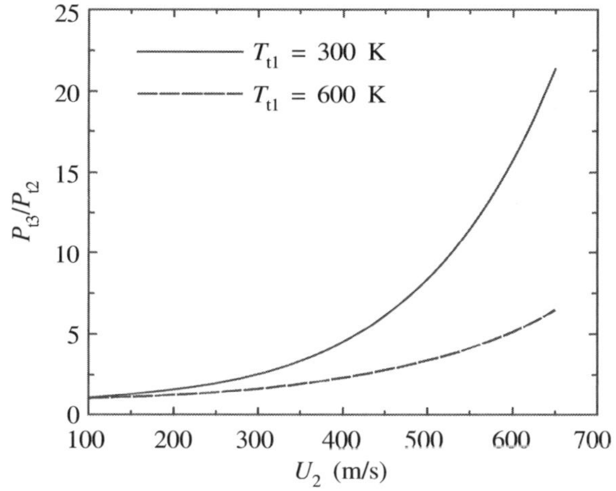

Fig. 8.4. The variation of pressure ratio with impeller tip speed for an ideal centrifugal compressor.

8.3 Effect of Impeller Blade Shape

So far, we have only discussed about the radial swept blade impeller which is generally preferred because of its simplicity in construction. But, there are other two types of impeller namely (i) forward swept and (ii) backward swept blade which are also used in the centrifugal compressors. All three types of impeller are shown in Fig. 8.5 with respective velocity triangles in the radial plane at the exit of the impeller. Let us assume that all three impeller have same blade velocity at the tip of the impeller and the same radial velocity component. The angle of the relative velocity with respect to tangential direction is denoted by α. By first law of thermodynamics, the energy transfer across the impeller would be

$$\dot{W}_c = \dot{m}C_P\left(T_{t2} - T_{t1}\right) \tag{8.6}$$

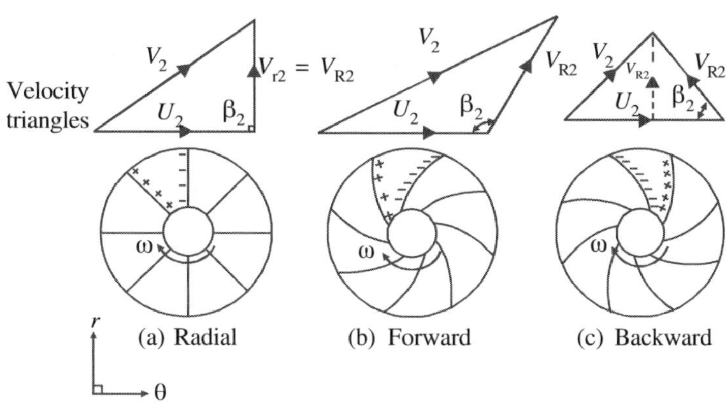

Fig. 8.5. Three kinds of impellers (a) radial, (b) forward and (c) backward.

By using Euler equation for turbomachinery, the above equation becomes

$$\dot{W}_c = V_{\theta 2}U_2 - V_{\theta 1}U_1 \tag{8.7}$$

For an axial flow entry, we can have

$$T_{t2} - T_{t1} = \frac{V_{\theta 2}\, U_2}{C_P} \tag{8.8}$$

But, from the velocity triangle as shown in Fig. 8.5, we can have

$$V_{\theta 2} = U_2 - V_{r2} \cot \beta_2 \tag{8.9}$$

By using the above two equations (Eq. 8.8 and Eq. 8.9), we can get;

$$\frac{T_{t2} - T_{t1}}{T_{t1}} = \frac{V_{\theta 2}U_2}{C_P T_{t1}} = \frac{(\gamma - 1)\, U_2\,(U_2 - V_{r2}\cot\beta_2)}{\gamma R T_{t1}} \tag{8.10}$$

By simplifying the above equation, we can have

$$\frac{T_{t2} - T_{t1}}{T_{t1}(\gamma - 1)\left(\dfrac{U_2}{a_{t1}}\right)^2} = 1 - \frac{V_{r2}}{U_2}\cot\beta_2 \tag{8.11}$$

From the continuity equation, we can have

$$\dot{m}_2 = \rho_2\, \pi\, D_2\, bV_{r2} \;\Rightarrow\; V_{r2} = \frac{\dot{m}}{\rho_2 \pi D_2 b} \tag{8.12}$$

where, b is depth of the impeller channel, ρ_2 is density at the exit of impeller, D_2 is diameter of the impeller at exit.

For the centrifugal compressor with no swirl flow at the entrance of impeller, the ratio of $\left(\dfrac{V_{x2}}{U_2}\right)$ is commonly referred as flow coefficient, which is directly proportional to mass flow rate and inversely proportional to rotational speed (rpm) of the impeller. Note that the relative temperature rise is linear function of this flow coefficient.

For the same mass flow rate, and tip speed, U_2 the rotor with forward swept blades will do more work on the fluid. Then, one may be tempted to conclude that it produces higher pressure rise in comparison to other two impellers. From the velocity diagram, we know that this leads to higher value of V_2 and high kinetic energy is seldom desired as its reduction to static pressure by diffusion is difficult to carry out efficiently in a reasonable size of casing. Thus, the overall compressor efficiency will be low. Besides this a positive slope as in the curve in Fig. 8.6 is generally associated with

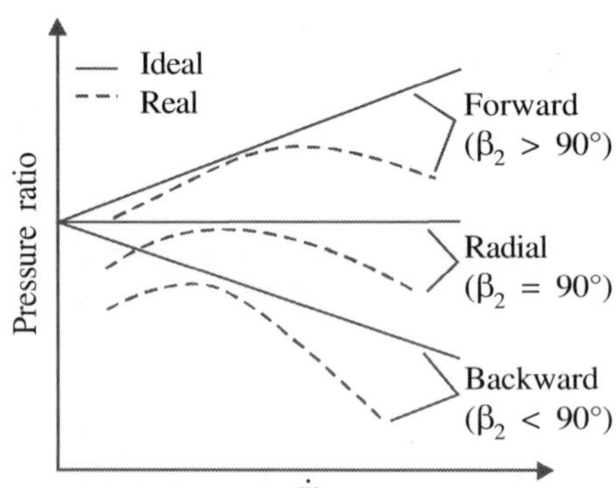

Fig. 8.6. Variation of pressure ratio with mass flow rate for forward, radial and backward swept vanes.

dynamic instability, which will be discussed in subsequent sections. Due to these associated dynamic instabilities, the forward swept vanes are never used in the centrifugal compressor for aircraft engines.

Now let us compare the radial blade and backward swept vaned impellers by considering the velocity diagram shown in Fig. 8.5. We can note that the radial blade impeller is capable of greater pressure rise at a given impeller tip speed. But it also produces higher velocities with which the fluid will be entering into the diffuser. Thus, the diffusion process will be inefficient. However the straight radial vaned impellers were used profusely for many years, because of its simpler construction. But with recent development in improved material strength, one may prefer backward swept vane particularly for higher stage pressure. Modern centrifugal compressors with backward swept blades as much as 30° are used. However, the centrifugal bending stresses are developed in backward swept blades. It has been shown that the effect of backswept vane can actually be obtained in an impeller in which the outlet blade geometry consists of radial line elements and thus, entails no additional bending stress. By using this method, one can get higher efficiency and high pressure ratio which is more desirable. The use of backswept impeller also provides the compressor a wider operating range of air flow at a given rotational speed which is important for matching of the compressor with its driving turbine.

8.4 Actual Energy Transfer

We have already discussed the case of an ideal energy transfer to the fluid by an impeller in which air enters the impeller eye in the axial direction. As a result, initial angular momentum of air will be zero. Of course, one can assume the air flow to follow the impeller vane contours in an ideal case and leave the impeller tip with a tangential velocity, $V_{\theta 2}$ as shown in Fig. 8.7. But in actual situation, the tangential velocity component of air is not equal to impeller tip speed, U_2. Let us see why is it so. Air trapped between impeller vanes resists its motion along with the impeller vanes due to inertia. This results in a higher static pressure on the leading face of vane than on the trailing edge face as shown in Fig. 8.5. Thus, air tends to flow around the edges of vanes in the clearance space between the casing and the impeller. In this process, the air fails to acquire the actual tangential velocity of the impeller. Due to pressure gradients within the impeller passages and the tangential momentum of the fluid, there will be difference in blade and air angle at the exit of the impeller. The air angle and blade angle with respect to tangential direction at the exit of the impeller are β_2 and β_2' respectively as shown in velocity triangle in Fig. 8.7. This effect can be termed as a *slip*, which is basically a measure of how well air flow adheres to the blade profile. The slip is defined as the difference between ideal and real absolute tangential velocities. In other words, the slip factor σ is defined as the ratio of tangential velocity to impeller tip speed.

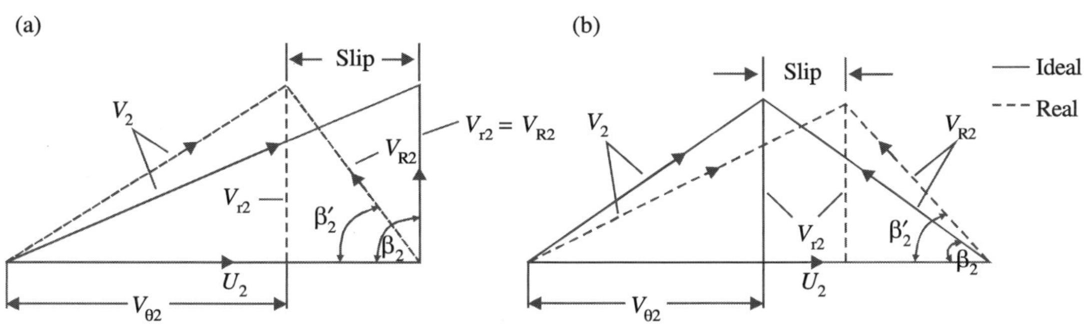

Fig. 8.7. Schematic of ideal and real velocity triangles (a) radial swept and (b) forward swept impeller.

$$\sigma = \frac{V_{\theta 2}}{U_2} \tag{8.13}$$

Note that σ is always less than or equal to unity. Of course, if these two angles are identical, then there would not be any slip between air and impeller. In any centrifugal compressor, zero slip is most desirable, as the air can acquire the actual tangential velocity at the impeller tip which ensures efficient energy transfer into air, leading to higher pressure rise per unit work input. Higher slip factor can be obtained in real compressor by increasing the number of vanes in the impeller as these will ensure the adherence of air flow with the vane profile. However, more number of vanes incur more frictional losses leading to reduction in energy transfer to the fluid. Hence, a compromise needs to be made to choose optimum number of vanes and amount of slip by the use of appropriate slip factor σ. It is to be noted that σ is very important that enumerates the accurate knowledge of the compressor's ability to transfer energy between the impeller and the fluid. Several attempts have been made in the past to arrive at a suitable expression for the slip factor, σ. For radial vane impeller, a very widely used correlation [1, 3, 6] between the slip factor, σ and the number of vanes, n is given below

$$\sigma = 1 - \frac{2}{n} \tag{8.14}$$

This expression for σ seems to agree well with the experimental data of Stanitz [6]. It can be noted from this expression that the slip factor, σ approaches 1 with increase in number of vanes. This is more desirable for 100% transfer of energy into fluid, but the frictional losses in an impeller also increases exorbitantly. Therefore, a balance must be maintained between the number of blades and frictional losses. Usually a slip factor of 0.9 is used with 19 to 20 vanes.

Besides this, losses are incurred while air is moving in between the vane and casing due to friction. There will be braking effects such as disc friction or windage. As a result, the applied torque and therefore the actual work input will be higher than the theoretical work input. In order to take account of these losses, generally a power input factor, ψ is introduced, so that actual temperature rise will be

$$\left(T_{t3} - T_{t1}\right) = \sigma\psi\,\frac{U_2^2}{C_p} \tag{8.15}$$

The power input factor, ψ represents an increase in the work input to the fluid. This extra energy is absorbed in overcoming the frictional losses and is converted into thermal energy resulting in increase of outlet temperature of the compressor. Thus, some amount of fuel can be saved in raising the final temperature that is to be feed into the turbine. Although it looks to have beneficial effects, in realising higher cycle efficiency, but more work is used in driving the compressor. Therefore, it is desirable to keep the power input factor, ψ as low as possible. Typical value of ψ lies in the range of 1.03 to 1.04.

8.4.1 Pressure Rise

Now we need to derive an expression for actual stagnation pressure ratio across the compressor by taking account of losses both in impeller and diffuser. The processes involved in the energy transfer in the impeller and diffuser of a centrifugal compressor are shown in Fig. 8.8 on a T-s diagram. An expression for actual pressure ratio can be derived by invoking isentropic relation given as:

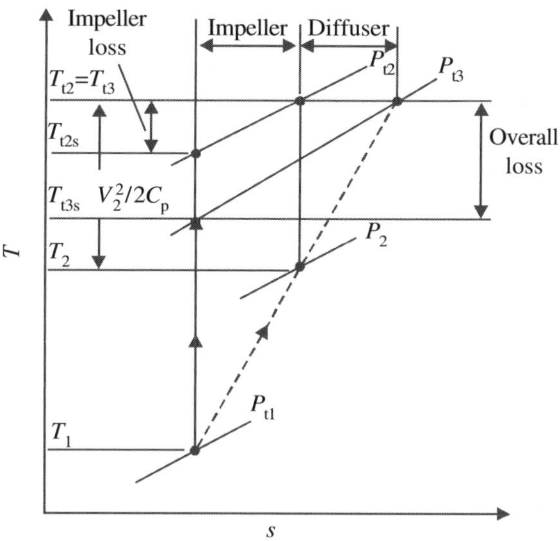

Fig. 8.8. *T-s* diagram across a compressor describing the division of loss between impeller and diffuser.

$$\frac{P_{t3}}{P_{t1}} = \left(\frac{T_{t3s}}{T_{t1}}\right)^{\frac{\gamma}{\gamma-1}} = \left(1 + \frac{T_{t3s} - T_{t1}}{T_{t1}}\right)^{\frac{\gamma}{\gamma-1}} \tag{8.16}$$

But the actual process is non isentropic as shown in Fig. 8.8. Thus, we can handle this by invoking the definition of overall isentropic efficiency that is

$$\eta_{ic} = \frac{h_{t3s} - h_{t1}}{h_{t3} - h_{t1}} \tag{8.17}$$

But the actual work input to the compressor can be evaluated by taking losses into account by invoking the parameters such as overall efficiency, η_{ic}, slip factor, σ and power input factor, ψ as given by

$$h_{t3} - h_{t1} = C_p\left(T_{t3} - T_{t1}\right) = \eta_{ic}\psi\sigma U_2^2 \tag{8.18}$$

Then the expression for actual pressure ratio across the centrifugal compressor can be derived by invoking the above equation as given below

$$\frac{P_{t3}}{P_{t1}} = \left(1 + \frac{\eta_{ic}\psi\sigma U_2^2}{C_p T_{t1}}\right)^{\frac{\gamma}{\gamma-1}} \tag{8.19}$$

It can be observed from the above equation that the actual pressure ratio will be lower than that of an ideal one for same work input due to slip and efficiency. It must also be noted that the slip factor takes into account the effectiveness with which momentum can be transferred from the impeller vanes to the air. In contrast, the overall efficiency incorporates losses occurred both in the impeller and diffuser of the compressor. It must be kept in mind that the above expression for pressure ratio is only valid for radial vaned impeller with no prewhirl. However, a more general expression for pressure ratio can be easily obtained for all three kinds of impeller by taking account of prewhirl given as:

$$\frac{P_{t3}}{P_{t1}} = \left[1 + \frac{\eta_{ic}\psi\left(\sigma U_2 V'_{\theta 2} - U_1 V_{\theta 1}\right)}{C_p}\right]^{\frac{\gamma}{\gamma-1}} \tag{8.20}$$

The above derivation for pressure ratio is left as an exercise for readers. By knowing velocity at both inlet and outlet of the impeller and considering suitable slip factor, power input factor and overall efficiency, we can find out actual total pressure ratio across a centrifugal compressor. It can be observed that the above expression for total pressure ratio across a centrifugal compressor is similar to that of axial compressor in the previous chapter. However, velocity triangles and the application of equations are quite different for both kinds of compressors. For example, the blade velocity at inlet of rotor, U_1 is approximately equal to U_2 for axial compressor. In contrast, for centrifugal compressor, U_2 is quite larger than U_1. That is the reason why the centrifugal compressor can produce higher pressure ratio per stage as compared to axial compressor.

Example 8.1 A single sided impeller of 0.5 m diameter in a centrifugal compressor is designed to rotate at speed of 15000 rpm. Air enters into the eye of the impeller axially at 100 m/s with total pressure of 0.1 MPa and total temperature of 298 K. If the slip factor is 0.89 and power input factor is 1.04, determine pressure ratio assuming the isentropic efficiency to be 0.85. (Take $\gamma = 1.4$)

Given $T_{t1} = 298$ K, $P_{t1} = 0.1$ MPa, $D_2 = 0.5$ m, $\sigma = 0.89$, $\psi = 1.04$, $\eta_{ic} = 0.85$, $N = 15000$ rpm.

Solution By using Eq. 7.20, we can determine pressure ratio as

$$\frac{P_{t3}}{P_{t1}} = \left(1 + \frac{\eta_{ic}\psi\sigma U_2^2}{C_p T_{t1}}\right)^{\frac{\gamma}{\gamma-1}}$$

But we need to find out impeller velocity at its exit as

$$U_2 = \frac{\pi D_2 N}{60} = \frac{\pi \times 0.5 \times 15000}{60} = 392.7 \text{ m/s}$$

Substituting the value of impeller velocity and other known parameters in Eq. 7.20, we have

$$\frac{P_{t3}}{P_{t1}} = \left(1 + \frac{0.85 \times 1.04 \times 392.7^2}{1005 \times 298}\right)^{\frac{1.4}{1.4-1}} = 3.76$$

Example 8.2 A centrifugal compressor is designed to operate at 15000 rpm without any prewhirl. Air at 10 kg/s enters into the impeller of diameter of 0.6 m with total pressure 100 kPa and 298 K. The root and tip diameter of the impeller eye are 0.1 and 0.3 m, respectively. Assuming no losses in the intake duct to the compressor, determine blade angle at root and tip of the eye, and Mach number at the tip of the eye.

Given $T_{t1} = 298$ K, $P_{t1} = 0.1$ MPa, $D_2 = 0.6$ m, $D_h = 0.1$ m, $D_t = 0.3$ m, $\dot{m} = 10$ kg/s, $N = 15000$ rpm.

Solution To Find: α_{r1}, α_{t1}, M_{r1} and M_{t1}.

In order to evaluate blade inlet angle at root and tip, we need to determine tangential and axial velocities. Let us first estimate the tangential velocities at root and tip of the eye. This is given as

$$U_{r1} = \frac{\pi \times 0.1 \times 15000}{60} = 78.53 \text{ m/s}$$

$$U_{t1} = \frac{\pi \times 0.3 \times 15000}{60} = 235.6 \text{ m/s}$$

Annulus area of impeller eye is determined as

$$A_1 = \frac{\pi}{4}\left(0.3^2 - 0.1^2\right) = 0.063 \text{ m}^2$$

We know that mass flow rate through the inlet of eye can be expressed as

$$\dot{m} = \rho_1 A_1 V_{x1}$$

Although we know mass flow rate and annular area, but density, ρ and axial velocity V_{x1} are unknown. Hence, we will resort to an iterative method for the determination of axial velocity. As a first approximation, let us estimate density based on total condition as

$$\rho_1 = \frac{P_{t1}}{R\,T_{t1}} = \frac{1 \times 10^5}{287 \times 298} = 1.17 \text{ kg/m}^3$$

The axial velocity can be evaluated as

$$V_{x1} = \frac{\dot{m}}{\rho_1 A_1} = \frac{10}{1.17 \times 0.063} = 135.7 \text{ m/s}$$

As there is no prewhirl in the entrance of the compressor, V_{x1} is equal to V_1. Then we can easily estimate static temperature as follows

$$T_1 = T_{t1} - \frac{V_1^2}{2C_p} = 298 - \frac{135.7^2}{2 \times 1005} = 288.84 \text{ K}$$

By using isentropic relation, static pressure can be determined as

$$P_1 = \frac{P_{t1}}{\left(T_{t1}/T_1\right)^{\frac{\gamma}{\gamma-1}}} = 90 \text{ kPa}$$

Then, we can evaluate density at inlet as

$$\rho_1 = \frac{P_1}{R\,T_1} = \frac{0.9 \times 10^5}{287 \times 288.4} = 1.09 \text{ kg/m}^3$$

Then, we can check axial velocity as

$$V_{x1,\,\text{new}} = \frac{\dot{m}}{\rho_1 A_1} = \frac{10}{1.09 \times 0.063} = 145.6 \text{ m/s}$$

Note that first guessed axial velocity does not match with the new axial velocity. Hence, we need to try next value for axial velocity. Let us have V_{x1} = 149 m/s as final trial. Then, we can evaluate

$$T_1 = T_{t1} - \frac{V_1^2}{2C_p} = 298 - \frac{149^2}{2 \times 1005} = 286.95 \text{ K}$$

By using isentropic relation, static pressure can be determined as

$$P_1 = \frac{P_{t1}}{(T_{t1}/T_1)^{\frac{\gamma}{\gamma-1}}} = 87.6 \text{ kPa}$$

Then, we can evaluate density at inlet as

$$\rho_1 = \frac{P_1}{R\,T_1} = \frac{0.876 \times 10^5}{287 \times 286.95} = 1.064 \text{ kg/m}^3$$

Axial velocity can be checked again as

$$V_{x1,\text{new}} = \frac{\dot{m}}{\rho_1 A_1} = \frac{10}{1.064 \times 0.063} = 149.17 \text{ m/s}$$

It can be noted that the value of axial velocity matches well with the final guess value. Hence, we will use this value to evaluate blade angles at root and tip of the eye as

$$\alpha_{r1} = \tan^{-1}\frac{149.17}{78.53} = 62.22°$$

$$\alpha_{t1} = \tan^{-1}\frac{149.17}{235.6} = 32.2°$$

In order to estimate Mach number at tip of the eye, we need to estimate relative velocities V_{R1r} and V_{R1t} as

$$V_{R1r} = \sqrt{149.17^2 + 78.53^2} = 168.59 \text{ m/s}$$

$$V_{R1t} = \sqrt{149.17^2 + 235.6^2} = 278.85 \text{ m/s}$$

Let us evaluate Mach number at root and tip of the eye, as

$$M_{1r} = \frac{168.59}{\sqrt{1.4 \times 287 \times 286.95}} = 0.5$$

$$M_{1t} = \frac{278.76}{\sqrt{1.4 \times 287 \times 286.95}} = 0.82$$

Note that Mach number is subsonic which is quite reasonable to accept for this design. However, if relative velocity exceeds sonic conditions, then prewhirl of inlet guide vanes can be used to reduce it. This will be discussed in subsequent sections.

8.5 Diffuser

Generally the absolute velocity with which air leaves the impeller is quite high particularly for entering into combustion chamber in the gas turbine engine. For stabilisation of flame inside the combustor, air velocity entering into the combustor must be around 70 m/s. Therefore, it is very essential to reduce the velocity to about 70 m/s so that it would not pose any serious problem for the combustor chamber design. Thus, one can design a diffuser to convert the kinetic energy into pressure energy with a minimum possible stagnation pressure loss in order to reduce the outlet absolute velocity of air leaving the compressor. Keep in mind that the main purpose of a diffuser is to enhance the static pressure of the air as it is being obtained in a stator of an axial compressor. Generally, the diffusion of air is accomplished in the centrifugal compressor either by (i) vaneless diffuser (Fig. 8.9(a)) or (ii) by providing a vaneless space followed by vaned diffuser as shown in Fig. 8.9(b).

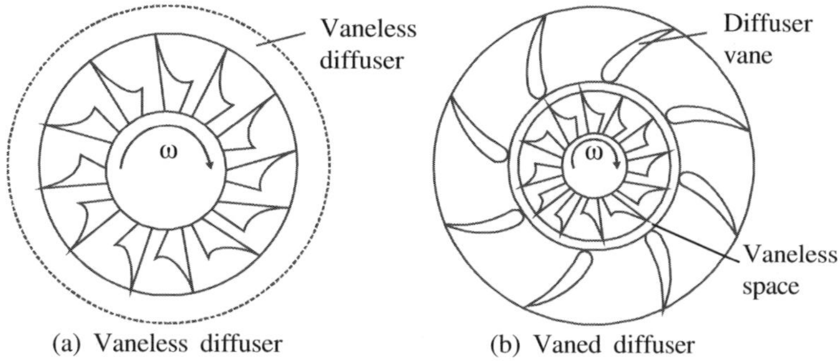

Fig. 8.9. A schematic of (a) vaneless diffuser and (b) vaned diffuser.

8.5.1 Vaneless Diffuser

The air in the vaneless diffuser is decelerated in the vaneless space around the exit of impeller before leaving the volute casing of a stage in centrifugal compressor (Fig. 8.9(a)). It can be noted that there is no addition of energy to the air during its travel in the vaneless space. This is the simplest concept where an increase in radius reduces the tangential velocity based on the principle of conservation of angular momentum ($V_\theta r$ = const.), if we neglect the frictional effects. Using subscripted variable at impeller tip outlet and unsubscripted variable at any radius between impeller tip and vaneless diffuser, V_θ will be equal to $(V_{\theta 2}\ r_2)/r$. Usually, the tangential velocity, V_θ is much larger than the radial velocity. Then *the absolute velocity V can be approximately equal to* V_θ. Although this method is quite simple and economical particularly for large flow rate compressor, but it leads to larger size diffuser. Hence, it is not suitable for aerospace applications. Besides this, the vaneless diffuser can be used for a small pressure ratio. However, it is quite suitable for industrial application where large size compressors are required. The advantage of this kind of diffuser is that even supersonic flow can be decelerated without formation of any shock. Additionally, there would not be any loss due to stalling. As a result, the performance does not deteriorate under off-design conditions. In other words, the efficiency of the compressor does not vary much when it is operated away from the design mass flow condition as shown in Fig. 8.10.

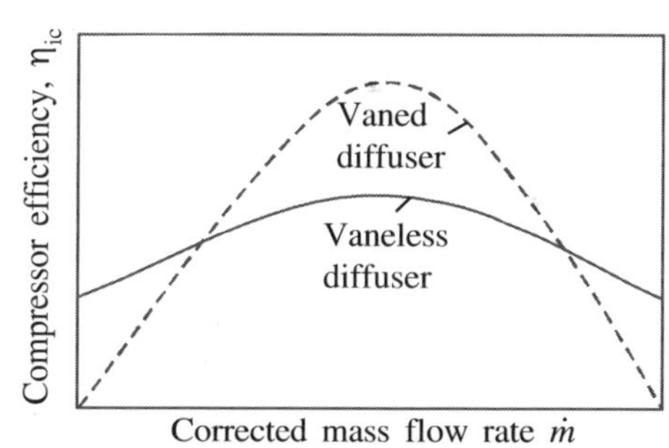

Fig. 8.10. The variation of efficiency with mass flow rate for vaneless diffuser and vaned diffuser.

8.5.2 Vaned Diffuser

We know that for a higher pressure ratio across the centrifugal compressor, the vaneless diffuser is not the right choice. Rather a vaned diffuser must be used to achieve higher rate of diffusion within a short radial distance. In fact, a vaneless space is always provided between the vane diffuser and impeller as shown in Fig. 8.9(b) to avoid steep velocity gradient at diffuser entry. As a result, the circumferential pressure gradient at the impeller tip is minimised which smoothens the outlet velocity variation between the impeller tip and fixed vanes. This vaneless space also reduces the Mach number at leading edges of the diffuser vanes. The width of this vaneless space is varied from 0.05 D_2 to 0.2 D_2; where D_2 is the exit diameter of the impeller. It must be kept in mind that if the vaneless space is too small, considerable amount of vibration and noise can occur due to interaction of flow between the vanes of the diffuser and the impeller blades. In order to carry out deceleration of high velocity air flow from the vaneless space within a short distance, the air is divided into a number of streams by using a series of stationary vanes known as vaned diffuser. This also helps in deswirling the air flow leaving from the vaneless space. Generally, these passages formed by vanes have constant depth with divergence width. In earlier days, curved aerofoil shaped vanes were used while in modern times, wedge-shaped vanes are preferred due to their higher performance particularly under off-design conditions. A schematic of a typical aerofoil vaned diffuser is shown in Fig. 8.9(b). It is usual practice to have these vanes initially in the radial plane as shown in Fig. 8.9(b) and wrapped into the axial direction at the outer radius of the diffuser (not clearly shown in Fig. 8.9(b)) which helps in controlling the three dimensional flow. It must be kept in mind that the angle of these vanes particularly at the leading edge is such that the air can enter into these vanes smoothly at least under design conditions. Besides this, to avoid flow separation in diffuser passages, their divergence angle is generally restricted to around 7°. However, higher divergence angle even up to 14° can be used without flow separation particularly for higher radius ratio. If the smaller divergence angle is used, then longer diffuser vanes are to be employed which may result in higher skin friction loss. In contrast, if larger divergence angles are to be used, then more number of vanes passages can be used to avoid flow separation which can also incur higher skin frictional losses. Of course, the angle of divergence becomes smaller when more number of vanes are employed in the diffuser. Then, the vaned diffuser becomes more efficient till losses caused by friction and blockage due to use of more number of vanes outweighs the benefits of the gradual deceleration of air flow. Note that higher number of vanes lead to reduction in radius ratio as the pitch-chord ratio is almost kept constant during the design process. When the flow from the vaneless space enters into large number of diffuser passages, uneven velocity distribution may occur in each diffuser passages. As a result, some of the vanes may experience flow separation leading to stalling of the vanes and poor performance. In order to avoid these problems, it is prudent to use less number of vanes than that of the impeller. As a thumb rule for design, generally *the number of diffuser vanes is restricted to one third of the impeller vanes*. If such design is adopted, then the nature of the flow in most of the diffuser passages would not vary significantly. Hence the efficiency of the centrifugal compressor depends on the number of diffuser vanes and its performance during its operation. The size and weight of the compressor are also dictated by the number and size of the diffuser vanes.

We must keep in mind that the diffuser passages are designed to be operated for a particular flow rate for which it can provide optimum performance. Hence, it is expected that the diffuser can perform well when it is operated at design conditions. However, vaned diffuser may perform poorly due to mismatching of flow when it is operated under off-design conditions (see Fig. 8.10). This problem can be alleviated significantly by using movable diffuser vanes whose direction can be adjusted according to changed incidence angle of the flow under off-design conditions. Of course, such arrangement has not been adopted for aircraft application due to its complex mechanism. However, it can be easily used for stationary applications.

The performance of the diffuser can be characterised by the effectiveness with which it can convert kinetic energy into static pressure rise. Hence, diffuser efficiency can be defined as

$$\eta_d = \frac{P_2 - P_3}{\rho\left(V_3^2 - V_2^2\right)/2} \tag{8.21}$$

where ρ is the density of air, the subscript 2 and 3 denote the inlet and outlet of the diffuser. The efficiency of the diffuser can be increased if the following aspects are considered during the design of the diffuser

- The vane angle at the inlet of diffuser must be closer to the air inlet angle. Of course small deviation must be provided to achieve higher performance even under off-design condition.
- The diffuser passage area must be large enough to handle air flow rate and must have optimum divergence angle.
- Any sudden change of flow must be avoided. The air from the diffuser exit must be turned through 90° smoothly for which vanes may be used in the elbow of an inlet to the combustion chamber.

8.5.3 Volute Casing

After the air leaves the diffuser passages, it is collected in the volute (scroll) which guides the flow to the delivery pipe of the combustor. A schematic of volute casing along with the impeller, vaneless diffuser and vaned diffuser is shown in Fig. 8.11. A vaneless space is provided even at the trailing edge of the diffuser vanes in order to decrease the non-uniformities and turbulence level of the flow. It can be noted that generally the volute circle is located little away from the diffuser vanes (0.1-0.2 D_3). A gradual increase in cross-sectional area of the volute is provided to decelerate the flow further as shown in Fig. 8.11. The cross-sectional area, A_θ of the volute at any angle θ can be estimated easily by assuming free vortex flow (constant angular momentum) and neglecting frictional effect as:

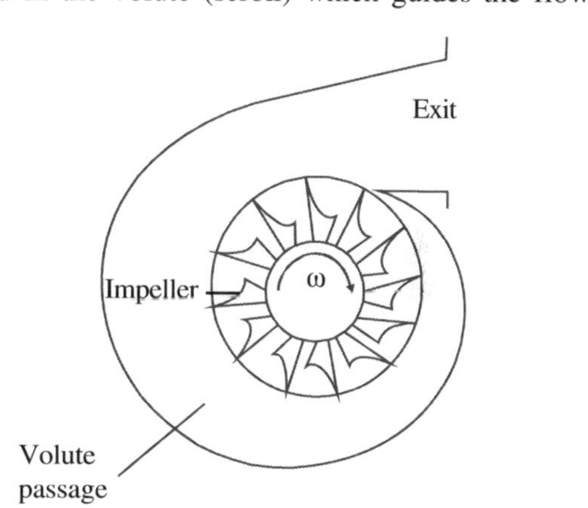

Fig. 8.11. Schematic of volute casing along with impeller, diffusers of a centrifugal compressor.

$$A_\theta = \frac{\dot{Q}R\theta}{2\pi V_\theta r} \tag{8.22}$$

where $\dot{Q}$ is the volume flow rate, $V_\theta r$ is the constant angular momentum, R is the radius of the centroid of the cross-sectional area of the volute. It has been reported in literature that the shape of the cross-section has not much effect on the frictional losses in the volute. Hence, several cross-sectional volutes are being evolved for centrifugal compressors. The schematic diagram of three types of volute shapes are shown in Fig. 8.12. The rectangular shape volute is simple and convenient to fabricate particularly from sheet metal and being profusely used in centrifugal blowers. However, the circular shape volute is being preferred for centrifugal compressor. It must be kept in mind that different shapes of volutes can have almost similar efficiency. Note that the performance of the volute depends on the quality of the flow it receives from the diffuser. The non-uniform pressure distribution around the diffuser can also be influenced by the flow in the volute which may give rise to radial thrust and bearing pressure. This undesirable effect can be avoided during the design of the volute.

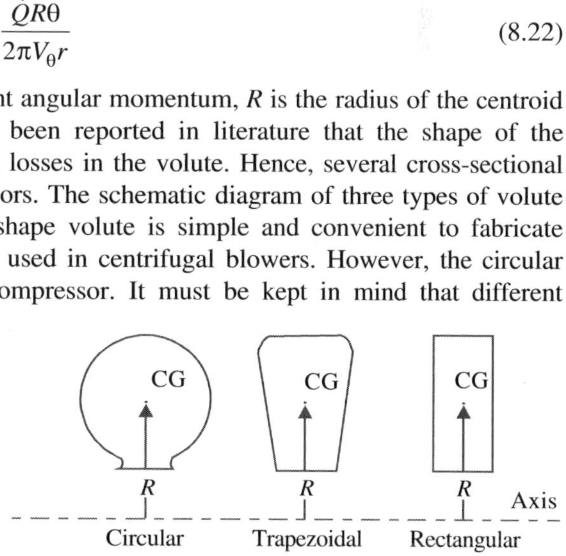

Fig. 8.12. Schematic of three types of volutes.

Example 8.3 Consider the same data given in Example 8.2. The radial width of vaneless space between impeller and diffuser in 6 cm. It has 15 diffuser vanes (2d) with depth of 2 cm. The approximate mean radius of the diffuser throat is 0.4 m. Assume that 50% losses occur upto the diffuser throat. Considering isentropic efficiency to be 0.85 and $\sigma = 0.98$ determine (a) inlet angle of the diffuser and (b) throat width of the diffuser.

Given $T_{t1} = 298$ K, $P_{t1} = 0.1$ MPa, $D_2 = 0.6$ m, $D_h = 0.1$ m, $D_t = 0.3$ m, $\dot{m} = 10$ kg/s, $N = 15000$ rpm, $r_{21} = 0.3$ m, $\Delta R = 0.06$ m, $R_{th} = 0.4$, $n_V = 15$, $H_d = 0.02$ m, $\eta_{ic} = 0.85$, $\sigma = 0.9$.

Solution To Find: α_2, and W_{th}

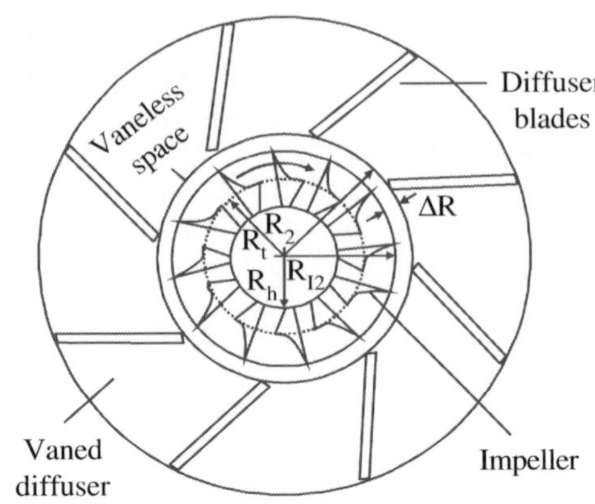

Fig. E8.1. Vaned diffuser.

(a) Let us first estimate the tangential velocity at exit of impeller as

$$U_2 = V_{\theta 21} = \frac{\pi \times 0.6 \times 15000}{60} = 471.2 \text{ m/s}$$

We know that free vortex flow prevails in the vaneless space between impeller and diffuser inlet. We can easily determine the tangential velocity at inlet of diffuser by using free vortex condition. But for this we will have to determine mean radius of diffuser inlet, R_2 as given below;

$$R_2 = R_{12} + \Delta R = 0.3 + 0.06 = 0.36 \text{ m}$$

By using free vortex theory, the tangential velocity at inlet of diffuser, $V_{\theta 2}$ is evaluated as

$$V_{\theta 2} = V_{\theta 12}\left(\frac{R_{12}}{R_2}\right) = 471.2\left(\frac{0.3}{0.36}\right) = 392.7 \text{ m/s}$$

We need to find out radial velocity at the inlet of diffuser. It can not be calculated directly as density is not known. Hence iterative method is to be used as discussed in Example 8.2. Let us consider radial velocity, $V_{r2} = 70$ m/s as the final value. Then the temperature rise due to absolute velocity, V_2 becomes

$$\frac{V_2}{2C_p} = \frac{392.7^2 + 70^2}{2 \times 1005} = 79.1 \text{ K}$$

As half of the losses occur in impeller, $\eta_{icI} = 1 - 0.5\,(1 - \eta_{ic}) = 1 - 0.5\,(1 - 0.85) = 0.925$. Assuming no additional losses between impeller and inlet of diffuser vane, we can evaluate pressure ratio, $\frac{P_{t2}}{P_{t1}}$, as

$$\frac{P_{t2}}{P_{t1}} = \left(1 + \frac{0.925 \times 0.9 \times 471.2^2}{1005 \times 298}\right)^{3.5} = 5.38$$

By similar procedure illustrated in Example 8.2, we can determine density by estimating temperature, pressure as

$$T_{t2} = T_{t1}\left(\frac{P_{t2}}{P_{t1}}\right)^{\frac{\gamma-1}{\gamma}} = 298(5.38)^{0.286} = 482.19 \text{ K}$$

$$T_2 = T_{t2} - \frac{V_2}{2C_p} = 482.19 - 79.1 = 403.09 \text{ K}$$

$$P_2 = P_{t1}\,\frac{P_2}{P_{t2}} \cdot \frac{P_{t2}}{P_{t1}} = 0.1\left(1.62\,\frac{403.09}{482.19}\right)^{3.5} = 0.30 \text{ MPa}$$

$$\rho_2 = \frac{P_2}{RT_2} = \frac{0.30 \times 10^6}{287 \times 403.09} = 2.59 \text{ kg/m}^3$$

In order to check, let us evaluate the radial velocity, V_{r2}, by using mass conservation as

$$V_{r2} = \frac{\dot{m}}{\rho_2 2\pi \times R_2 \times Hd} = \frac{10}{2.59 \times 2\pi \times 0.36 \times 0.02} = 85.34 \text{ m/s}$$

It can be noted that the above estimated radial velocity is very closer to assumed value. Hence, the assumed value can be considered to be the correct value for the present calculation. Then we can evaluate angle at inlet of diffuser using this value of radial velocity as

$$\alpha_2 = \tan^{-1}\left(\frac{85.34}{392.7}\right) = 12.26°$$

(b) We can estimate throat width, W_{th} of the diffuser channel by following similar procedure as illustrated above. The tangential velocity at radius of 0.4 becomes

$$V_{\theta 2} = 471.2\left(\frac{0.3}{0.4}\right) = 353.4 \text{ m/s}$$

We need to find out radial velocity at the throat of diffuser by iterative method. Let us consider radial velocity, V_{r2} = 59 m/s as the final value. Then the temperature rise due to absolute velocity, V_2 becomes

$$\frac{V_2^2}{2C_p} = \frac{V_{\theta 2}^2 + V_{r2}^2}{2C_p} = \frac{353.4^2 + 59^2}{2 \times 1005} = 63.87 \text{ K}$$

By similar procedure illustrated above, we can determine density by estimating temperature, pressure as

$$T_2 = T_{t2} - \frac{V_2^2}{2C_p} = 482.19 - 63.87 - 418.32 \text{ K}$$

$$P_2 = P_{t1}\frac{P_2}{P_{t1}} = 0.1\left(1.62 \times \frac{418.32}{482.19}\right)^{3.5} = 0.329 \text{ MPa}$$

$$\rho_2 = \frac{P_2}{RT_2} = \frac{0.329 \times 10^6}{287 \times 418.32} = 2.74 \text{ kg/m}^3$$

Neglecting the thickness of the diffuser vanes, throat area can be determined as

$$A_{th} = 2\pi \times 0.4 \times 0.02 = 0.05026 \text{ m}^2$$

In order to check, let us evaluate the radial velocity, V_{r2}, by using mass conservation as

$$V_{r2} = \frac{\dot{m}}{\rho_2 A_{th}} = \frac{10}{2.74 \times 0.05026} = 72.62 \text{ m/s}$$

Let us assume V_{r2} = 73 m/s. Then we can determine $\frac{V_2}{2C_p}$ as

$$\frac{V_2}{2C_p} = \frac{V_{\theta 2}^2 + V_{r2}^2}{2C_p} = \frac{353.4^2 + 73^2}{2 \times 1005} = 64.78$$

By similar procedure mentioned above, we can determine density by estimating temperature, pressure as

$$T_2 = T_{t2} - \frac{V_2^2}{2C_p} = 482.19 - 64.78 = 417.4 \text{ K}$$

$$P_2 = P_{t1}\frac{P_2}{P_{t1}} = 0.1\left(1.62\times\frac{417.4}{482.19}\right)^{3.5} = 0.327 \text{ MPa}$$

$$\rho_2 = \frac{P_2}{RT_2} = \frac{0.327\times10^6}{287\times417.4} = 2.73$$

We can evaluate V_{r2} by using mass conservation as

$$V_{r2} = \frac{\dot{m}}{\rho_2 A_{th}} = \frac{10}{2.73\times0.05026} = 72.88 \simeq 73 \text{ m/s}$$

It can be noted that the above estimated radial velocity at the throat of diffuser is very closer to assumed value. Then the angle of flow at the throat is determined as

$$\alpha_2 = \tan^{-1}\frac{73}{353.4} = 11.67°$$

The total area of throat perpendicular to the absolute velocity is estimated by mass conservation as

$$A_2 = \frac{A_{r2}V_{r2}}{V_2} = A_{r2}\times\sin\alpha_2 = 0.05026\times\sin(11.67°) = 0.01 \text{ m}^2$$

The width of throat in each passage is estimated as

$$W_{th} = \frac{A_2}{n_V H_d} = \frac{0.01}{15\times0.02} = 0.033 \text{ m} = 3.3 \text{ cm}$$

8.6 Aerodynamic Losses in Centrifugal Compressor

The nature of the aerodynamic losses in the centrifugal compressor is quite different from those of the axial compressor as the nature of flow differs significantly in impeller of the centrifugal compressor. Of course these losses can be attributed to the growth of the boundary layer and its separation over vane surfaces, fluid friction, circulatory motion and shock wave formation. Besides these, the mixing losses also occur at the exit of rows of blades due to non-uniform velocity profile. Generally, the losses in the compressor can be broadly divided into two categories namely (i) Primary losses, and (ii) Secondary losses. As we have already discussed earlier that Primary losses are caused due to fluid friction, boundary layer growths over blade surface and shock formations. On the other hand, secondary losses are caused by distortion of fluid flow during the turning process in blades, which is three-dimensional in nature. Of these two types, the secondary losses are much more difficult to analyze. The primary losses can again be classified into (i) Friction losses, and (ii) Shock losses.

8.6.1 Friction Losses

The fluid friction causes pressure losses both in impeller passages and the stationary blade passages of the diffuser. We know that the flow is subjected to deceleration while passing through both the impeller and diffuser passages under adverse pressure gradient. As a result, flow is more prone to get separated from the blade surfaces incurring losses leading to wasteful expenditure of energy. This may cause formation of vortices in the diffuser passages leading to additional losses. It must be kept in mind that boundary growth on the surface of the impeller is being attenuated due to centrifugal force field prevailing in the passage of the impeller. Hence, higher pressure ratio can be obtained by efficient diffusion of flow in the impeller passage as compared to axial compressor. That is the reason why high reaction ratio blades such as backward swept are employed in the impeller of centrifugal compressor to achieve higher efficiency.

We know that the losses due to fluid friction are dependent on the friction factor, passage length and velocity. Hence, the compressor with a longer impeller, diffuser, and volute passages can incur higher losses leading to the decline of its performance.

8.6.2 Shock Losses

We know that if the fluid flow velocity relative to the blade surface exceeds the speed of sound, shock is likely to be formed incurring exorbitant pressure loss. In the presence of the shock, the flow is more prone to get separated from the surface causing additional loss. Such kind of flow situation leading to the shock formation is likely to occur at the inlet of either impeller or diffuser particularly when it is operated under off-design conditions.

At the entry of the impeller vane, air has to take a sharp turn while passing through the curved passages of the impeller. As a result, the flow is accelerated while passing through the convex surface of the impeller vane which may lead to the formation of a shock as shown in Fig. 8.13b. The formation of shock must be avoided during the operation of the compressor. Let us explore how to avoid the formation of shock by looking at velocity triangle at the eye of the impeller carefully as shown in Fig. 8.13a. We can assume that axial velocity remains constant from the root to the tip of the eye. Then, the relative air velocity can attain a maximum value at its tip as it has maximum blade velocity. We must evaluate the inlet Mach number at the tip of the eye which is expressed as

$$M_{1t} = \frac{V_{R1t}}{\sqrt{\gamma R T_1}} \tag{8.23}$$

where T_1 is the static temperature of inlet of the impeller, V_{R1} is the relative air velocity at the tip of the eye. If this Mach number reaches its critical value of unity, shock is likely to be formed. Generally, the Mach number at the tip of the eye is restricted to 0.8 as a general rule during design process. However, flow may attain a higher Mach number when operated under off-design conditions particularly at high altitude. In order to avoid the formation of shock even under adverse conditions, inlet guide vanes are attached to the casing of the compressor. Mostly, arrangements are made to vary the angle of these vanes to accommodate a wide range of flow conditions. When air passes through these inlet guide vanes, its direction gets changed as shown in Fig. 8.13a. This helps in imparting prewhirl to the air and makes the air to be drawn smoothly into the convex surface of the impeller vanes. As a result, the relative velocity to the impeller vanes gets decreased as shown by faint lines in the velocity triangle (Fig. 8.13a) for the same axial velocity (around same mass flow rate). Hence shock formation can be avoided easily. Besides this, it can be observed in Fig. 8.13a that the incidence angle, α_1 is increased with the use of prewhirl blades leading the use of less curved vanes

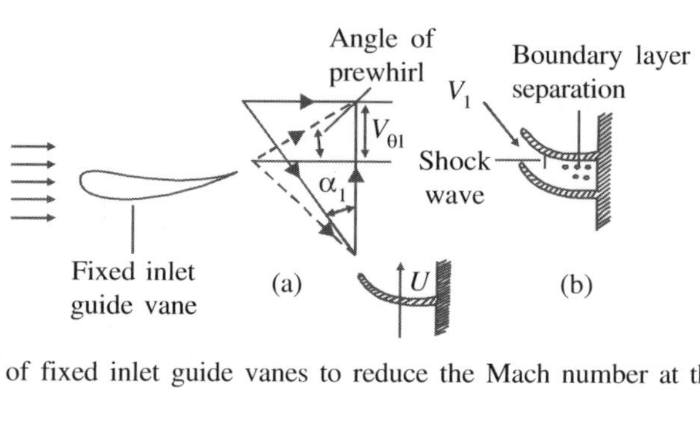

Fig. 8.13. Schematic of fixed inlet guide vanes to reduce the Mach number at the entry of impeller.

at the eye of the impeller. Of course, this use of prewhirl blades reduces work capacity of the compressor due to parasitic initial angular momentum as shown in the equation below:

$$\dot{W}_c = V_{02}U_2 - V_{01}U_1 \tag{8.24}$$

As mentioned earlier, prewhirl angle must be varied from the maximum value at the tip to zero value at the root of the eye to minimise reduction in capacity of the compressor for which the twisted inlet guide vanes have to be employed.

The formation of shock is also likely to occur at the entry of the diffuser vanes as air leaves from the impeller at a very high velocity. This assumes importance as the flow has a tendency to get separated from the surface of the diffuser even at low speed and the flow is subjected to adverse pressure gradient. This situation will be more rampant when a smaller compressor is used to deliver higher mass flow rate as required in aerospace applications, particularly when it is operated under off-design conditions. This problem can be alleviated by controlling the incidence angle of air flow in a proper manner.

8.6.3 Secondary Losses

As in the axial compressor, the secondary losses are incurred in the centrifugal compressor due to vortex formation when the air is also subjected to rotational motion along with the rotation of impeller. The eddies formed in the impeller passages as shown in Fig. 8.14 cause secondary losses. Besides this, when air is turned through the curved vane passages of the diffuser, a pressure gradient is set up across the blade passage normal to the streamlines as discussed earlier in the axial compressor. This process can cause secondary flow leading to the formation of eddies which also incur secondary losses.

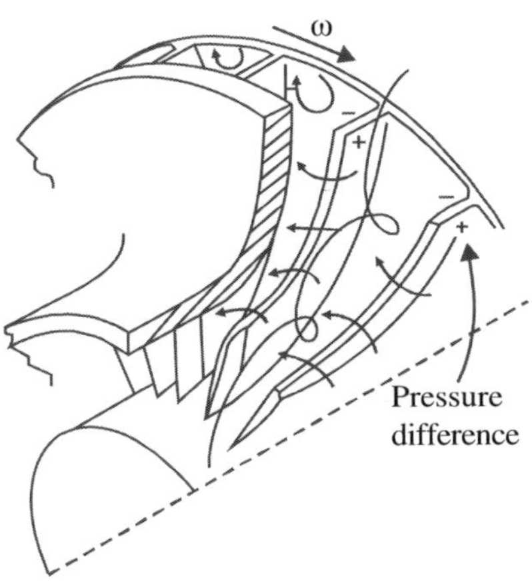

Fig. 8.14. Schematic showing the formation of eddies in the impeller.

The magnitude of secondary losses depends on the portion of the total blade passage length, affected by secondary flow. Generally, the magnitude of the secondary losses is high in high performance compressor. Therefore, it is important for the designer to take account of these losses while designing the centrifugal compressor. As discussed earlier, the secondary flow is inherently three dimensional in nature and it would not be possible to predict the drag easily due to secondary flow. Rather, the designer has to rely on the empirical relation obtained from experimental data.

8.7 Centrifugal Compressor Characteristics

The non-dimensional parameters used for describing the performance of axial compressor can also be employed to describe the performance characteristics of the centrifugal compressor. At any given speed, the pressure ratio across the compressor can be related to the non-dimensional mass flow rate as

$$\pi_c = \frac{P_{t3}}{P_{t2}} = f\left(\frac{\dot{m}\sqrt{T_{t2}}}{P_{t2}}\right) \tag{8.25}$$

Let us consider ideal performance characteristics of a radial vaned centrifugal compressor as shown in Fig. 8.15 which indicates that pressure ratio does not change with variation in mass flow rate. However, under real situation, the compressor will have both frictional and incidence losses as discussed earlier. As a result, the constant pressure ratio line becomes a curved one (Fig. 8.15). It can be observed that the pressure ratio increases from low mass flow rate till it attains a maximum value at a certain mass flow rate. Then it drops down with further increase in mass flow rate. In other words, the pressure ratio has a maximum value at a particular mass flow rate. It can also be noted that this variation will decide the actual range of the compressor operation. As discussed earlier in the axial compressor, the range of stable operation is restricted by surging and choking conditions corresponding to certain range of mass flow rate.

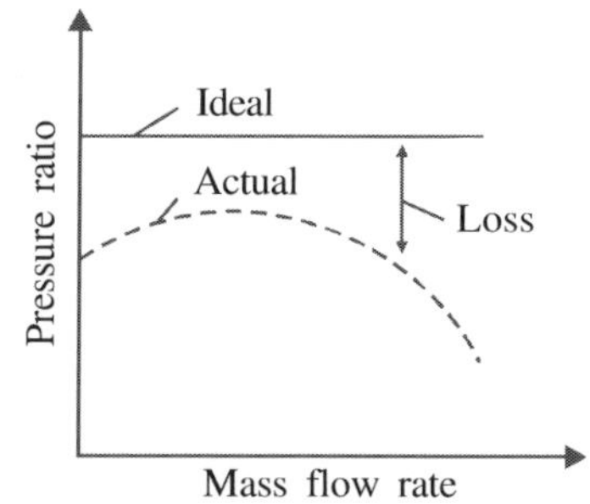

Fig. 8.15. The variation of pressure ratio with mass flow rate of a radial vaned centrifugal compressor at a particular rotational speed.

The performance curves of a typical centrifugal compressor in terms of pressure ratio with non-dimensional mass flow rate at several non-dimensional rotational speed is shown in Fig. 8.16 which seems to be identical to that of an axial flow compressor. It must also be noted that these curves at different rotational speed can be obtained by following similar testing procedures. These curves are restricted by surging and choking phenomena. Of course, surging can occur only when pressure ratio across the compressor becomes quite large at lower mass flow rate. Under such a condition, the flow momentum and centrifugal forces are not enough to maintain the desired pressure gradient which results in massive flow reversal in the compressor outlet pipe or in the diffuser itself leading to the formation of surging in the compressor. The surging point in each curve has been joined as a surge line as indicated in Fig. 8.16(a). Generally, corresponding to maximum efficiency, the operating line is located at a lower level than the surge line as shown in Fig. 8.16(a). Besides this, the variation of the pressure line with mass flow rate at a fixed rotational speed becomes flatter due to choking of flow at either diffuser passage or deliver pipeline. In contrast to axial compressor, the centrifugal constant speed lines are much flatter. In case of centrifugal compressor, speed lines determine the pressure ratio, whereas for axial compressor at high speed, rotational speed determines mass flow rate. Hence, the compressor has to be operated between the surge point and choking point for a fixed rotational speed. The effective operating pressure ratio for a fixed value of rotational speed has a wider range as compared to an axial compressor.

The variation of isentropic efficiency η_{ic} for several non-dimensional rotational speed is plotted in Fig. 8.16(b). It can be observed that efficiency varies with mass flow rate in a similar way as the pressure ratio. The efficiency attains a maximum value at certain mass flow rate. It is interesting to note that the maximum efficiency remains approximately same for all rotational speed cases. Corresponding to maximum efficiency for all rotational speeds, operating points are being located in the pressure ratio vs mass flow rate curve as shown in Fig. 8.16(a). It must also be noted that peak efficiency at each rotational speed is quite closer to the surging line.

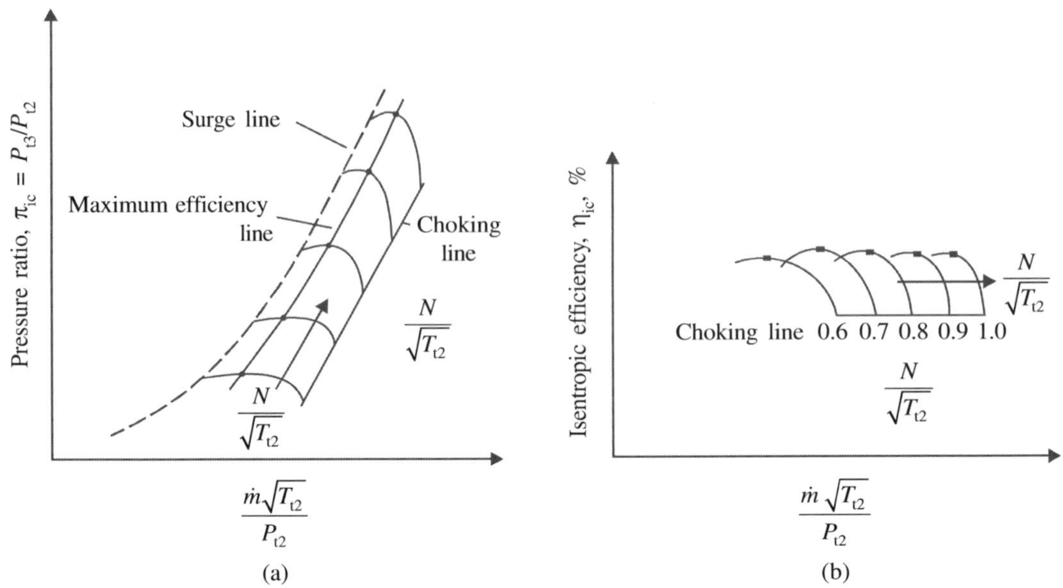

Fig. 8.16. Performance of a typical centrifugal (a) Pressure ratio vs non-dimensional mass flow rate (b) Isentropic efficiency vs non-dimensional compressor.

8.8 Chapter Summary

The basic principle of centrifugal compressor operation is introduced in the beginning of this chapter while identifying its various hardware components. When air is entered axially into the impeller of the centrifugal compressor, it gets compressed while moving against the centrifugal force field. Subsequently it is passed through the diffuser section in which kinetic energy is converted into potential energy by diffusion process, resulting in an increase in its static pressure. The velocity triangles are employed to discuss the processes involved in both impeller and diffuser in a single stage. Two dimensional analysis of an ideal centrifugal compressor stage is carried out to derive pressure ratio relationship in terms of air outlet angles, impeller velocity and inlet air total temperature. The blade velocity seems to affect stage pressure ratio significantly. It is shown that the pressure ratio across a single stage centrifugal compressor is quite high as compared to that of axial compressor due to large difference between inlet and exit diameter of the impeller. Three types of impeller vane shapes namely (i) forward swept (ii) radial (iii) backward swept are discussed briefly highlighting their respective merits and demerits. The concept of slip factor is introduced which is dependent on the number of vanes in an impeller. Generally radial impeller is preferred for aircraft engines due to larger capacity compressor with higher centrifugal stress, higher efficiency and lower manufacturing cost. However, in recent times, backward swept impeller is advocated with improved performance level over radial impeller. An expression for pressure ratio across an actual stage of the centrifugal compressor is derived by introducing the isentropic efficiency, slip factor and power input factor to ideal pressure ratio relationship. Both primary and secondary losses are mentioned in the axial compressor while discussing centrifugal compressor. Apart from frictional losses, flow losses due to shock formation at the eye of the impeller is likely to be formed in case of centrifugal compressor which can be avoided by using prewhirl at inlet of the impeller. However, there will be loss of pressure ratio for same workinput. An example is discussed to reinforce the idea of using pre whirling in reducing shock losses. The secondary flow losses is caused due to complex flow structure in the centrifugal compressor. The overall performance curve of a typical centrifugal compressor is presented highlighting the importance of surge and operating lines. It can be noted that the centrifugal compressor is less prone to surge as compared to the axial compressor. In the next chapter, the axial turbine is discussed in detail.

Nomenclature

A	Area
B	Depth of the impeller channel
C_P	Mean specific heat over temperature rise
D_r	Root diameter of impeller eye
d_t	Tip diameter of impeller eye
H_d	Diffuser vanes depth
M	Mach number
$\dot{m}$	Mass flow rate
n_V	Number of vanes
P	Static pressure type equation here.
P_t	Stagnation pressure
$\dot{Q}$	Volume flow rate
R	Specific gas constant
R_u	Universal gas constant
T	Static temperature
T_t	Stagnation temperature
U_2	Blade velocity at exit of impeller
V_r	Radial velocity
V_R	Relative velocity
$\dot{W}_c$	Power input to compressor
$\dot{W}_{th}$	Width of throat

Greek Symbols

σ	Slip factor
ψ	Power input factor
η	Efficiency
η_{ic}	Isentropic efficiency
η_d	Diffuser efficiency

Subscripts

o	Overall
d	Diffuser
d	Depth
r	Root
t	Tip
t	Stagnation
th	Throat
u	Universal

References and Suggested Readings

1. Mattingly, J.D., *Elements of Gas Turbine Propulsion*, 1st Edn, McGraw-Hill Publishing Company Ltd, Singapore, 1996.
2. Kerrebrcok, J.L., *Aircraft Engines and Gas Turbines*, 1st Edn, MIT Press, England, 1992.
3. Cohen, H., Rogers, G.F.C. and Saravanamuttoo, H.I.H., *Gas Turbine Theory*, 4th Edn, Addison Wesley Longman Ltd, England, 1998.
4. Hill, P.G. and Peterson, C.R., *Mechanics of Thermodynamics of Propulsion*, 2nd Edn, Addison Wesley Publishing Company, Inc. Reading, MA, 1992.
5. Cumpsty, N.A., *Compressor Aerodynamics*, Longman Group, Essex, UK, 1988.
6. Stanitz, J.D., *Some Theoretical Aerodynamic Investigation of Impellers in Radial and Mixed Flow Centrifugal Compressors*, Transactions of the American Society of Mechanical Engineers, 74, pp. 473-497, 1952.

Review Questions

1. Draw schematic of a typical centrifugal compressor and describe its operating principle. How is it different from the axial compressor?
2. What are the advantages and disadvantages of a centrifugal compressor?
3. What is the maximum pressure ratio across a single stage centrifugal compressor?
4. What are the applications of centrifugal compressor? Compare it with that of an axial compressor. Why is it so?
5. Show the variation of total enthalpy, pressure and temperature across a centrifugal compressor. How these are comparable with their respective static quantities?
6. What is slip factor? Why is it included in the analysis of the centrifugal compressor?
7. Show with the help of a diagram, how does the impeller shape affect the pressure ratio across the compressor.
8. What are the factors that affect the stage pressure ratio?
9. Which type of impeller vane can be preferred for high efficient centrifugal compressor? Why is it so?
10. Why is a diffuser required for a centrifugal compressor?
11. What is vaned and vaneless diffuser?
12. What is volute and why is it required at all?
13. What is prewheel? Explain it with a velocity triangle. How does it affect the pressure ratio?
14. What happens when a centrifugal compressor is operated at lower mass flow rate than its design value?
15. What is rotating stall? Under what conditions, a centrifugal compressor is likely to get stalled? What are the ways of reducing the rotating stall in a centrifugal compressor?

Problems

1. A centrifugal compressor is to be designed for pressure ratio of 4 which consumes 50 kg/s of air at 10,000 rpm. The air at pressure of 0.1 MPa and 300 K enters into the impeller with outlet diameter of 80 cm. If the slip factor is 0.95 determine the efficiency of the compressor.
2. A centrifugal compressor is designed which is operated at 15000 rpm without any prewhirl. Air at 20 kg/s enters into the impeller of diameter of 0.6 m with total pressure 200 kPa and 350 K. If the slip factor is 0.92 and the polytropic efficiency is 0.88, determine (i) pressure ratio, and (ii) power input.
3. Air at 20 kg/s enters axially into the impeller of a centrifugal compressor with total pressure 300 kPa and 400 K and leaves at the exit impeller diameter of 0.4 m with a relative velocity of 120 m/s at an angle of 75°. The slip factor is 0.92 and the polytropic efficiency is 0.88. The compressor is operated at 12000 rpm without any prewhirl. Estimate (i) the torque (ii) the power to drive this compressor, and (iii) exit temperature at its exit.

4. A centrifugal compressor is designed for pressure ratio of 4 which consumes 20 kg/s of air at 15,000 rpm. The air at pressure of 0.1 MPa and 300 K enters with an axial velocity of 120 m/s into the impeller with an inlet eye diameter of 25 cm. If the polytropic efficiency is 0.88 and pressure coefficient, $\Delta P_t/(\rho U^2)$ is 0.75, determine (i) static pressure ratio (ii) static pressure and temperature at inlet and the exit of compressor, and (ii) work input into the compressor.
5. A centrifugal compressor without any prewhirl is operated at 15,000 rpm which sucks 15 kg/s of air at pressure of 0.1 MPa and 300 K. If the inner and outer diameter of inlet eye of the impeller are 0.15 and 0.3 m respectively, determine (i) air inlet angle at the root and tip of the inlet eye, and (ii) relative Mach number at eye tip.
6. In a single sided centrifugal compressor, 30 kg/s of air at pressure of 0.05 MPa and 223 K enters with a velocity of 200 m/s into the impeller with an inner and outer eye diameter of 15 and 30 cm respectively. The impeller is rotating at 10,000 rpm. The outer exit diameter of the impeller is 55 cm. The slip factor is 0.92 and the polytropic efficiency is 0.88. Determine the stagnation pressure at the exit of compressor, if the same air prewhirl angle of 22° is maintained at all radii, of the inlet eye. Assume no losses in inlet and fixed vanes of the impeller.
7. Air at rate of 2.5 kg/s enters axially into an impeller of a centrifugal compressor with total pressure 100 kPa and 300 K and leaves at the exit impeller diameter of 0.5 m. The mechanical and isentropic efficiencies are 0.98 and 0.88 respectively. The compressor is operated at 10,000 rpm without any prewhirl. The impeller with 21 radial vanes is designed with axial width of 15 mm at its exit. The stagnation pressure ratio between the diffuser outlet and impeller exit is 1.5. Estimate power required to drive this compressor and absolute Mach number at impeller exit.
8. In order to reduce the air velocity at the eye of the impeller, fixed guide vanes based on free vortex principle are designed. Air at the rate of 25 kg/s enters axially into a fixed guide vanes with total pressure 100 kPa and 300 K and then enters into the eye with a velocity of 100 m/s at an angle of 25° w.r.t the axial direction. The root and tip diameter of the impeller eye are 0.1 and 0.3 m respectively. Air leaves with a radial velocity of 45 m/s and static pressure of 250 kPa from the exit of the impeller of diameter of 0.6 m, rotating at 12,000 rpm. Determine (i) the inlet relative Mach number assuming no losses in the guide vanes, and (ii) the efficiency of the compressor.
9. Air with total pressure 100 kPa and 300 K enters axially into a 21 radial vaned impeller of a centrifugal compressor and leaves at the exit impeller diameter of 0.5 m. The impeller tip speed is 450 m/s and the mechanical efficiency is 0.98. The compressor is designed for a reaction ratio of 0.5. The absolute air velocity at the impeller entry is 120 m/s. If the velocity at the diffuser exit is 120 m/s and the overall efficiency is 0.84, determine total and static pressure at diffuser exit. Assuming the conservation of angular momentum and diffuser efficiency of 0.86, determine the (i) total, static pressure, radial component of velocity at impeller exit and (ii) inlet and outlet radius ratio for the diffuser (iii) rotational speed if axial width at the impeller exit is 10 mm.
10. Design a centrifugal compressor for mass flow rate of 25 kg/s and pressure ratio of 5 with reaction ratio of 0.5. Air enters into this stage at pressure of 0.1 MPa and temperature of 300 K. It has a mean radius of 20 cm at the impeller eye and 50 cm at the exit of the impeller. The compressor is operated at 10,000 rpm with overall efficiency of 0.85.

9. Axial Turbine

This conquest of the air will prove ultimately, to be man's greatest and most glorious triumph. What railways have done for nations, airways will do for the world.

Claude Grahame White, British Aviator, 1914

We know that the compressor, which raises the pressure of air before combustion, is one of the important component of the gas turbine engine. But the compressor must be supplied with the requisite power to carry out its function of compressing air. This work input is supplied by the turbine. Of course, the turbine generates power using hot pressurised gases from the combustor. Hence, both the compressor and the turbine are very important components of a gas turbine engine.

In contrary to the compression process in the compressor, gas is expanded in the turbine. It must be recalled that adverse pressure gradient in an axial compressor is likely to result in flow separation and thus restricting its operation within a narrow zone. Hence, the pressure ratio per stage particularly in axial compressor is restricted to a smaller value. Therefore, ten to twenty stages are being used to achieve high pressure ratio in an axial compressor of modern gas turbine engines. In contrast, pressure falls in the flow direction in a axial turbine. Hence, it is quite possible to use higher pressure ratio across a turbine stage as compared to a compressor stage, because, the accelerating flow is less likely to separate. Normally, a single stage turbine can be employed to drive even six or seven compressor stages on the same shaft. Furthermore, very few number of turbine stages are used in turbojet engines because, the hot pressurised gas can be expanded further in an exhaust nozzle to produce the requisite thrust. Because of the above reasons, it has been observed that a well-designed axial turbine has a higher efficiency as compared to an axial compressor under normal circumstances. Besides this the design process for a turbine is much easier than that of a compressor. Moreover, it is quite easier to understand the operation of a turbine as compared to a compressor. Interestingly, the turbine design is not limited by aerodynamic problems unlike a compressor. Rather, high thermal stress of turbine blades pose challenging problems to the designer.

Based on the nature of flow, the turbines can be broadly divided into three types namely (i) axial, (ii) radial (centrifugal), and (iii) mixed flow turbo machines. Recall that in an axial machine, the predominant flow is in the axial direction whereas the fluid motion is in predominantly radial direction in case of centrifugal type. And in case of mixed type, both axial and radial flow motion occurs. Of course, the choice of turbine types depends mostly on its application. Since in modern aircraft engines, axial turbines are used profusely, we will be devoting this chapter only to the axial flow turbine. In aircraft application, a small turbine of low weight with smaller frontal area is preferred. In fact the requirement of low weight and smaller frontal area can be achieved by raising the allowable stresses but with a shorter working life. In order to make it more compact, generally higher axial velocity flow is used, which can also provide higher useful work output in the form of jet thrust.

9.1 Description of Operation

A single stage of axial turbine comprised of a row of stator blades/nozzles followed by a row of rotor blades similar to that of a compressor as shown in Fig. 9.1. A multistage turbine consists of a sequence of such stages as can be seen in Fig. 9.2. The stator blades are usually shrouded particularly when higher pressure drop occurs per stage. The main function of the shroud is to reduce blade vibration and control tip leakage of air passed over blade tips and across blade rows. Besides these, the shrouds provide resistance to flow distortion under high loads. A schematic of shroud arrangement in a typical turbine blade is illustrated in Fig. 9.2. In this case, the shrouds are formed by the tips of the blades touching each other. However, the use of shroud demands the turbine to be run at lower operating temperatures or at a lower rpm due to the presence of extra mass at the tip. Note that the use of shrouds is more common in turbines than in compressors, due to higher stage pressure ratios (hence greater leakage) under stress conditions, which is generally more critical because of high operating temperature. The combination of stator blades and the shroud is often referred to as *turbine nozzle*. A nozzle is followed by a rotor as shown in Fig. 9.1. A stator followed by a nozzle is commonly known as *stage* in a turbine. It must be observed that the nozzle and rotor blade is of increasing length as shown in this figure, to accommodate the rapidly expanding gases, while maintaining uniform axial velocity through the stage.

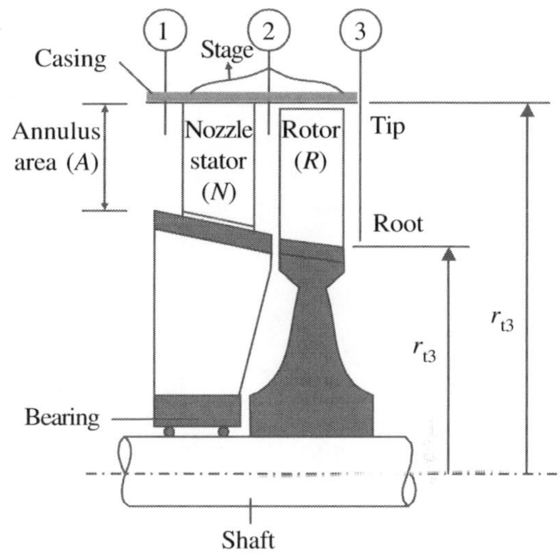

Fig. 9.1. Schematic of a single stage axial flow turbine.

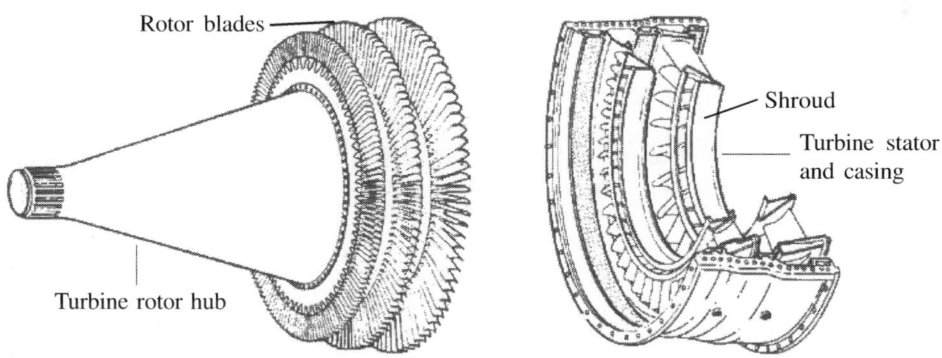

Fig. 9.2. Cutaway view of a multistage axial turbine [9].

Gas at high temperature and relatively high pressure from combustion chamber is expanded through the nozzles with an increase in kinetic energy (KE) and a decrease in static enthalpy. As the gas flows through the rows of rotating blades, the turbine wheel converts a part of this kinetic energy, KE into shaft work. In addition, further expansion may occur in rotating blades themselves with an increase in kinetic energy of fluid relative to the blades, coupled with decrease in enthalpy and a part of this KE being converted into shaft work. The changes in pressure, temperature, enthalpy across a single stage turbine is shown in Fig. 9.3. This is a simplistic picture of how the turbine functions.

Let us now consider the velocity triangles at mean radius of a single stage axial turbine shown in Fig. 9.4. The gas enters the row of nozzle blades with a static pressure of P_1, T_1, V_1 (at an angle α_1 with respect to the axis) and is expanded to P_2, T_2 and leaves with an increased velocity V_2 at

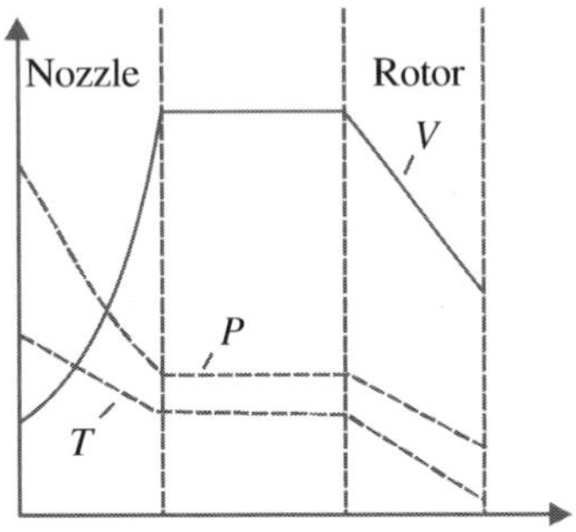

Fig. 9.3. Variation of pressure, temperature, enthalpy across a stage.

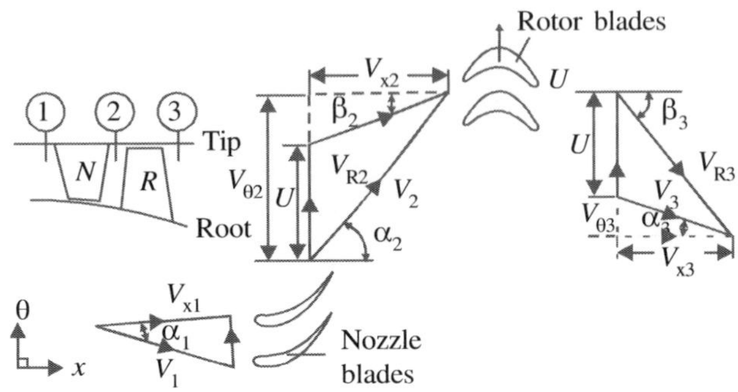

Fig. 9.4. Velocity triangle in a single stage axial turbine.

an angle α_2 with respect to the axis. The rotor blade angle is such that fluid enters the rotor with V_{R2} with an angle of β_2 with respect to the axis. Then, as discussed earlier, the relative velocity would be:

$$V_{R2} = V_2 - U \tag{9.1}$$

The gas after being deflected and further expanded in the rotor, leaves the rotor at P_3, T_3, V_{R3} at an angle β_3 with respect to the axis. One can find out V_3 from vector addition of U and V_{R3}, which makes an angle α_3 with respect to the axis. This angle α_3 is called the *swirl* angle.

For a single stage turbine, we can assume the inlet gas velocity, V_1 to be axial ($\alpha_1 = 0$) under normal conditions. But for a multistage turbine, the gas velocity V_1 and its angle α_1 are taken to be equal to V_3 and α_3 respectively, such that the same blade profiles can be employed in successive stages. Note that this is being done for design convenience. Keep in mind that at present we have considered only the mean radius analysis. However, in reality, the shapes of velocity triangle varies from root to tip of the rotor blade because of increasing blade speed, U from root to tip. It must be kept in mind that the mean radius analysis is quite suitable for higher hub-tip ratio blades (short blades).

It is interesting to note that the velocity triangle as shown in Fig. 9.4 of axial turbine is different from that of the axial compressor, in terms of change in tangential velocity, ΔV_θ. In turbine, the change in tangential velocity, $\Delta V_{\theta 2}$ (= $V_{\theta 1} + V_{\theta 3}$) produces useful torque in the direction opposite to the blade speed, U. This causes the change in the tangential momentum of the fluid, which produces a torque on the rotor in the direction of the motion. Thus, the rotor extracts the work from the gas.

Now let us invoke the familiar Euler's equation to the turbine to get an expression for work output per unit mass that is given by:

$$\frac{\dot{W}_t}{\dot{m}} = \left(U_2 V_{\theta 2} - U_3 V_{\theta 3}\right) \tag{9.2}$$

Since, we are using mean radius analysis, U_2 is equal to $U_3(= U)$. Now, considering the sign convention in the velocity triangle, the turbine work output would be

$$\frac{\dot{W}_t}{\dot{m}} = U\left(V_{\theta 2} + V_{\theta 3}\right) \tag{9.3}$$

Note that $V_{\theta 3}$ is in negative direction. Let us now use velocity triangle given in Fig. 9.4 to express V_θ in the above equation in terms of axial velocity by the following relations.

$$V_{\theta 2} = V_{x2} \tan \alpha_2;\ V_{\theta 3} = V_{x3} \tan \alpha_3 \tag{9.4}$$

However from the velocity triangle shown in Fig. 9.4, we can have

$$\frac{U}{V_x} = \tan \alpha_2 - \tan \beta_2 = \tan \beta_3 - \tan \alpha_3 \tag{9.5}$$

Here, we have assumed that V_{x2} is equal to $V_{x3}(= V_x)$. Then, using the above equation, the expression for turbine work output would be

$$\frac{\dot{W}_t}{\dot{m}} = UV_x\left(\tan \alpha_2 + \tan \alpha_3\right) \tag{9.6}$$

By considering a steady flow energy equation, we can have

$$\frac{\dot{W}_t}{\dot{m}} = C_{pg}\Delta T_{ts} = UV_x\left(\tan \alpha_2 + \tan \alpha_3\right) \tag{9.7}$$

where, ΔT_{ts} is the total temperature drop across a stage. It is generally suitable to express the turbine work output in terms of gas blade angle, β. Thus, the above equation can be rewritten in terms of β using Eq. 9.5 as:

$$\frac{\dot{W}_t}{\dot{m}} = C_{Pg}\Delta T_{ts} = UV_x\left(\tan \beta_2 + \tan \beta_3\right) \tag{9.8}$$

Note that the above work output can be obtained under ideal conditions. However, in actual turbine there will be losses which can be accounted by using polytropic efficiency. The process involved in a typical turbine stage is described in a *T-s* diagram of Fig. 9.5.

Using this *T-s* diagram, the polytropic efficiency, η_{pt} for a single stage turbine becomes:

$$\eta_{pt} = \frac{T_{t1} - T_{t3}}{T_{t1} - T_{t3s}} \tag{9.9}$$

Then, we can find an expression for the stage pressure ratio, P_{t1}/P_{t3} as

$$\frac{P_{t1}}{P_{t3}} = \frac{1}{\left[1 - \dfrac{UV_x\left(\tan \beta_2 + \tan \beta_3\right)}{\eta_{pt}T_{t1}C_{Pg}}\right]^{\frac{\gamma_g}{\gamma_g - 1}}} \tag{9.10}$$

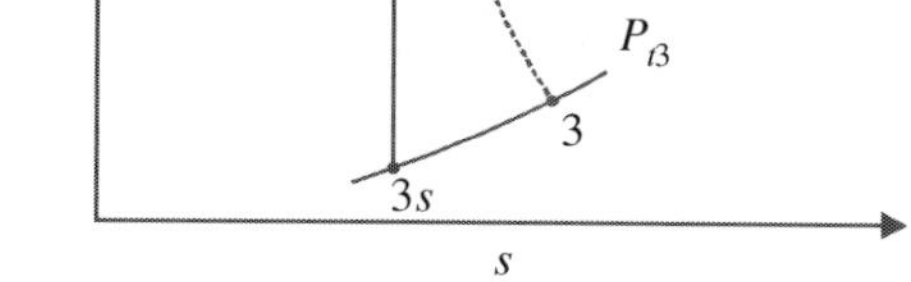

Fig. 9.5. Processes in a turbine stage described on *T-s* diagram.

It can be noted that by increasing blade velocity, U and axial velocity, V_x, the pressure ratio across a stage can be enhanced. By increasing the blade angles, we can also enhance expansion pressure ratio across the turbine stage. Of course, higher inlet temperature is required to have higher-pressure ratio and higher work output. But, the inlet temperature and blade speed cannot be increased beyond certain limit due to thermal stress problem. But, larger turning angles (β_2, β_3) can be used in turbine blades in comparison to that of compressor due to accelerating flow conditions. Since, rows of the turbine blade performance is not very sensitive to the flow conditions on its blade surface, we can use higher-pressure drop across the stage. Note that losses incurred in turbine blade are quite low as compared to compressor since the flow is accelerating in the turbine.

Let us now define three non-dimensional parameters, which are used very often for preliminary design of turbine. These three parameters are:

(i) Stage Loading Coefficient (ψ)
(ii) Flow Coefficient (Φ)
(iii) Reaction Ratio, *RR*.

Recall from the previous chapter, the definition of blade loading coefficient ψ as:

$$\psi = \frac{C_P \Delta T_{ts}}{U^2} = \frac{V_x}{U}\left(\tan\beta_2 + \tan\beta_3\right) = \Phi\,(\tan\beta_2 + \tan\beta_3) \tag{9.11}$$

This stage loading coefficient indicates the amount of stage specific work done per blade speed squared. Another important non-dimensional parameter, called flow coefficient is defined as the ratio of axial to blade velocity $\left(\Phi = \frac{V_x}{U}\right)$. The blade loading coefficient, ψ can be related to the flow coefficient, Φ for a turbine stage. The variation of ψ with Φ for several types of turbines is shown

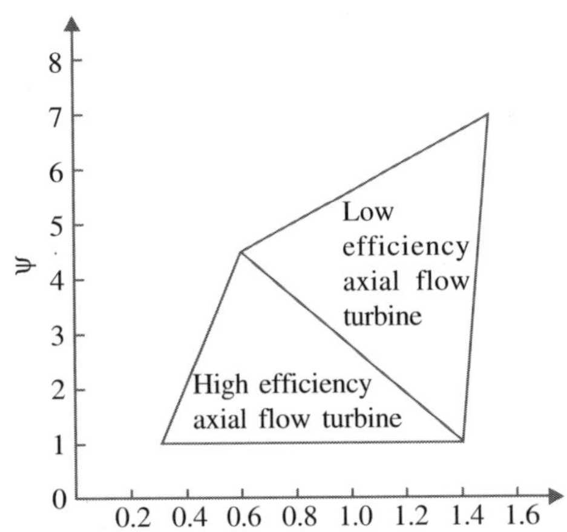

Fig. 9.6. The variation of stage loading coefficient ψ with flow coefficient Φ for different types of turbines [8].

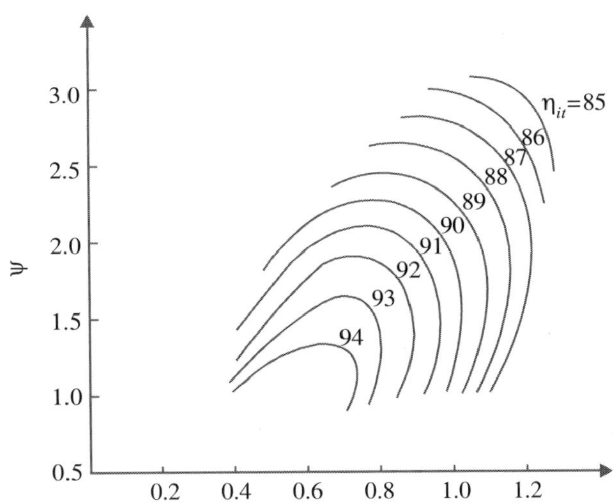

Fig. 9.7. The variation of stage coefficient η_{it} with ψ and Φ [8].

in Fig. 9.6. It can be noted that turbine with high efficiency uses lower values of stage loading and flow coefficients. Keep in mind that the turbine stage efficiency is affected by both blade loading coefficient, ψ and flow coefficient, Φ, as can be seen in Fig. 9.7.

The typical values of stage loading coefficients for axial turbine lie in the range of 1.5 to 2.5 and flow coefficient, Φ is in the range of 0.5 to 1.2 for modern high pressure turbines. The variation of ψ is plotted in Fig. 9.7 for a turbine stage indicating constant contours of turbine efficiency, η_{it}. This clearly indicates that turbine with high stage loading coefficient has lower efficiency. Of course, the actual figures on a plot such as this may be improved in future but neral trend may remain unchanged.

Let us understand what is the implication of high values of ψ and Φ for a constant blade speed. Let us see comparative velocity diagram shown in Fig. 9.8 to arrive at a qualitative interpretation of higher values of ψ and Φ. It can be observed that high values of ψ and Φ mean high gas velocities and thus, higher frictional losses. The number of stages will be lower for the same power output. As a result, minimum weight and frontal area can be achieved using higher values of ψ and Φ, which is very desirable for aircraft gas turbine engine. But the turbine efficiency will be lowered and overall TSFC will definitely increase. However, the most efficient stage design can be accomplished by using optimum ψ and Φ which of course is difficult to obtain in practice without getting detailed performance of the aircraft as a whole.

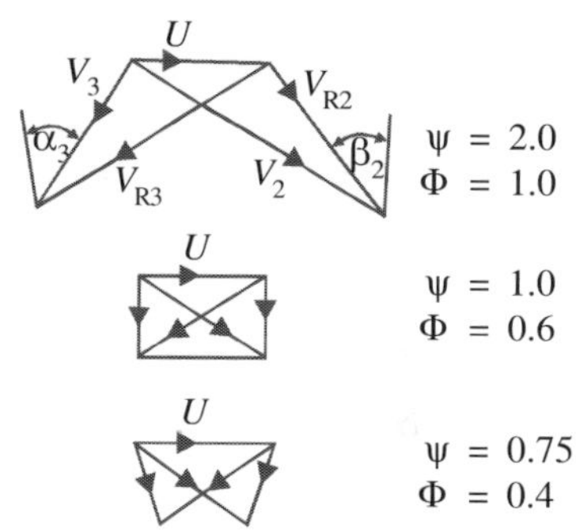

Fig. 9.8. Comparative velocity triangles for different ψ and Φ values.

Now, let us consider another important parameter called reaction ratio. Recall that the reaction ratio, *RR* for a turbine stage is defined as

$$RR = \frac{h_2 - h_3}{h_1 - h_3} \approx \frac{T_2 - T_3}{T_1 - T_3} \tag{9.12}$$

where C_p can be assumed to be constant. For a typical turbine stage, we have already assumed earlier that the axial velocity remains to be constant across the stage i.e. ($V_{x2} = V_{x3} = V_x$) and we can also assume that the velocity V_3 is same as V_1 since the stages are repetitive. Then, we can have

$$C_P\,(T_{t1} - T_{t3}) = C_P\,(T_1 - T_3) = UV_x\,(\tan\beta_2 + \tan\beta_3) \tag{9.13}$$

Recalling that no work is obtained, relative to rotor blade fluid, we can have

$$C_P\,(T_2 - T_3) = \frac{1}{2}\left(V_{3R}^2 - V_{2R}^2\right) = \frac{1}{2}\,V_x^2\left(\tan^2\beta_2 - \tan^2\beta_3\right) \tag{9.14}$$

Then using Eq. 9.13 and Eq. 9.14 in the expression of *RR* (Eq. 9.12), we can show that

$$RR = \frac{T_2 - T_3}{T_1 - T_3} = \frac{V_x}{2U}\left(\tan\beta_3 - \tan\beta_2\right) = \frac{\Phi}{2}\left(\tan\beta_3 - \tan\beta_2\right) \tag{9.15}$$

The reaction ratio, *RR* can be expressed in terms of flow coefficient Φ and stage loading coefficient, ψ as was done in case of the axial compressor. Now we can express gas angles in terms of ψ, *RR* and Φ as

$$\tan\beta_2 = \frac{1}{2\Phi}\left(\psi - 2RR\right) \tag{9.16}$$

$$\tan \beta_3 = \frac{1}{2\Phi}(\psi + 2RR)$$

Using the relation (Eq. (9.5)), we can get

$$\tan \alpha_3 = \tan \beta_3 - \frac{1}{\Phi}$$

$$\tan \alpha_2 = \tan \beta_2 + \frac{1}{\Phi} \qquad (9.17)$$

We can use the above expressions for preliminary design of axial turbine by choosing certain values of Φ and ψ and *RR*. Even with this simplification of design procedure, the designer can have several choices at hand. For example, the designer can choose small number of stages with higher ψ and Φ or large number of stages with lower ψ and Φ. Of course, the reaction ratio must be chosen properly depending on its nature of application. Generally the extreme reaction ratio corresponding to impulse stages ($RR = 0$) is not preferred in gas turbine engines, because of the fact that the pressure ratio across the turbine is quite low as compared to steam turbine for which impulse turbine is very much required. Since heavy leakage losses are incurred at higher pressure in case of reaction turbine, generally, the 50% reaction ratio turbine is preferred for aircraft to have higher turbine efficiency.

Based on the reaction ratio criteria, three basic turbine blade designs are generally used in practice. These are:

(i) Impulse (Zero Reaction)
(ii) 50% Reaction
(iii) Variable Reaction

It must be kept in mind that the designers need not to be constrained by these three types of turbine blades. They should have an open mind in choosing the type of blades. In fact, the reaction ratio may vary continuously along the blades in a three-dimensional turbine design.

Example 9.1 A single stage turbine with reaction ratio of 0.5 is designed to run at 12000 rpm by taking 25 kg/s of hot gases with axial velocity of 150 m/s which enters into its stator at inlet total temperature and pressure of 1400 K and 0.8 MPa, respectively and leaves with an angle of 60° at mean radius of 0.15 m. If the stator blade height is 0.1 m and polytropic efficiency, η_{pt} is 0.88, determine (i) rotor turning angle, (ii) static pressure at stator and rotor exit, and (iii) power output.

Given $RR = 0.5$; $T_{t1} = 1400$ K, $P_{t1} = 0.8$ MPa, $N = 12000$ rpm, $\dot{m} = 25$ kg/s, $\alpha_2 = 60°$, $r_m = 0.15$ m, $h_2 = 0.1$ m, $V_x = 150$ m/s, $\eta_{pt} = 0.88$.

To find $\beta_2 + \beta_3$, P_2, P_3 and $\dot{W}_t$.

Solution As reaction ratio, *RR* is equal to 0.5, then

$$\alpha_2 = \beta_3 \text{ and } \alpha_3 = \beta_2$$

However, we can determine angle, tan β_2 by using the relation for the reaction ratio, *RR* as given below

$$\tan \beta_2 = \tan \beta_3 - \frac{2RR \times U}{V_x}$$

For this, we need to evaluate blade velocity at its mean radius as

$$U = 2\pi r_m N = 2 \times \pi \times 0.15 \times \frac{12000}{60} = 188.5 \text{ m/s}$$

Then, by substituting the respective values in the expression of tan β_2, we can evaluate it as

$$\tan\beta_2 = \tan 60^\circ - \frac{2\times 0.5\times 188.5}{150} = 0.4753 \quad \beta_2 = 25.42^\circ$$

Then, the rotor turning angle becomes

$$\beta_2 + \beta_3 = 85.42^\circ$$

By using Eq. (9.8), we can determine T_{t3} as

$$T_{t3} = T_{t1} - \frac{UV_x}{C_{Pg}}(\tan\beta_2 + \tan\beta_3^3) = 1400 - \frac{188.5\times 150}{1148}\times(\tan 25.42 + \tan 60) = 1345.63\ \text{K}$$

Now we need to evaluate static pressure at the exit of stator and rotor for which we need to evaluate absolute velocity at station (2) and (3) as given below

$$V_2 = \frac{V_x}{\cos\alpha_2} = \frac{150}{\cos 60^\circ} = 300\ \text{m/s} \quad V_3 = \frac{V_x}{\cos\alpha_3} = \frac{150}{\cos 25.42^\circ} = 165.8\ \text{m/s}$$

Considering the flow to be isentropic, we can evaluate static temperature at station (2) and (3) as given below

$$T_2 = T_{t2} - \frac{V_2^2}{2C_{p,g}} = 1400 - \frac{300^2}{2\times 1148} = 1360.8\ \text{K}$$

$$T_3 = T_{t3} - \frac{V_3^2}{2C_{p,g}} = 1345.63 - \frac{165.8^2}{2\times 1148} = 1333.65\ \text{K}$$

By using Equation of state, we can determine static pressure at station (2) as

$$P_2 = \rho_2 RT_2 = \frac{\dot{m}\ RT_2}{2\pi r_m h_2 V_x} = \frac{25\times 287\times 1360.8}{1\times 0.15\times 0.1\times 150} = 0.691\ \text{MPa}$$

Then, let us first evaluate total pressure at station (3) as

$$P_{t3} = P_{t1}\left[1 - \frac{1}{\eta_{pt}}\left(\frac{T_{t1} - T_{t3}}{T_{t1}}\right)\right]^{\frac{\gamma_g}{\gamma_g - 1}} = 0.8\left[1 - \frac{1}{0.88}\left(\frac{1400 - 1345.63}{1400}\right)\right]^{\frac{1.33}{1.33-1}} = 0.67\ \text{MPa}$$

By using isentropic relation we can evaluate static pressure at station (3)

$$P_3 = \frac{P_{t3}}{\left(\frac{T_{t3}}{T_3}\right)^{\frac{\gamma_g}{\gamma_g - 1}}} = \frac{0.67}{\left(\frac{1345.63}{1333.65}\right)^{\frac{1.33}{1.33-1}}} = 0.646\ \text{MPa}$$

The power output of the turbine is determined as

$$\dot{W} = \dot{m}\ C_{pg}\left(T_{t1} - T_{t3}\right) = 25\times 1.148\times\left(1400 - 1345.63\right) = 1.56\ \text{MW}$$

9.1.1 Impulse Blade Design

In this case, entire enthalpy (pressure) drop occurs in the nozzle. Then for constant axial velocity case, tan β_2 becomes equal to tan β_3 as shown in velocity triangle (Fig. 9.9(a)) because the static enthalpy at inlet and outlet of rotor blade remains constant.

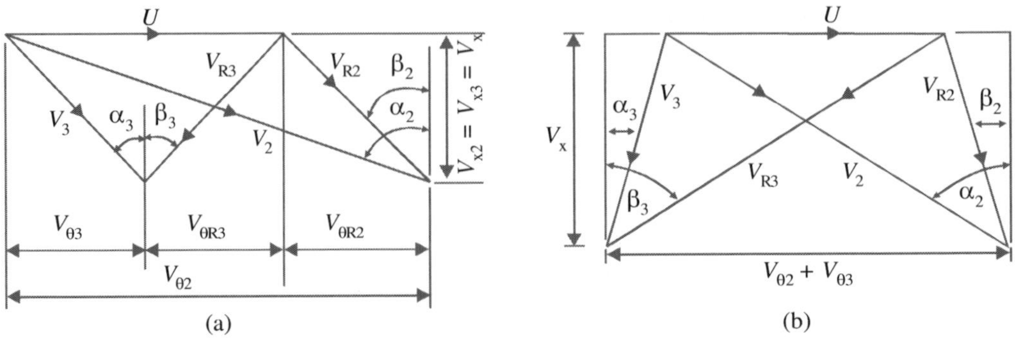

Fig. 9.9. Velocity triangles for (a) impulse, and (b) 50% reaction turbine stator.

The variations of pressure, temperature and velocity in an impulse turbine are shown in Fig. 9.10. It can be noted that all pressure changes occurs only in the nozzle.

Interestingly, it can be seen in this figure that the area between the inlet and exit of rotor blades is equal. Then the relative velocity, V_R must be equal at inlet and exit of the rotor blade. However, the absolute velocity decreases both in the nozzle and rotor. The stage loading for the impulse turbine stage with constant axial velocity would be

$$\psi = \frac{C_P \Delta T_{ts}}{U^2} = 2\Phi \tan \beta_2 = 2\Phi \tan \alpha_2 - 1 \tag{9.18}$$

For a higher blade loading, we should have larger α_2 at a constant flow coefficient. Unfortunately it leads to large V_2 and hence large V_{2R}. As a result the losses will be exorbitant. Therefore, the absolute velocity angle, α_2 is restricted in practice to 70°.

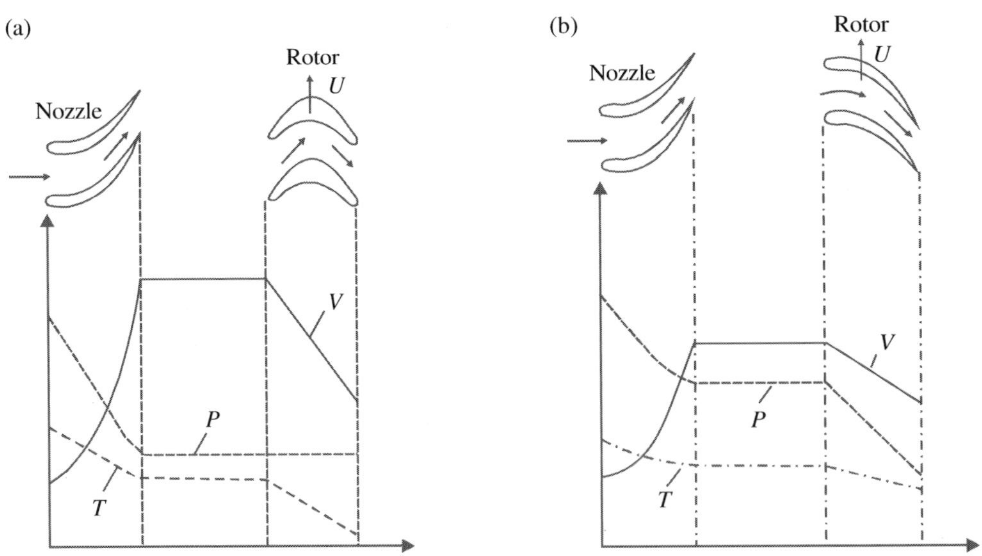

Fig. 9.10. Variation of pressure, temperature and velocity in (a) impulse (b) reaction turbine.

9.1.2 Fifty Percent Reaction Ratio

We know that in 50% reaction ratio turbine, half of the static enthalpy change occurs in the nozzle and other half in the rotor blades. The corresponding velocity triangles as shown in Fig. 9.9(b) are symmetrical. It can be shown from Eqs. (9.5) and (9.15) and velocity triangle (Fig. 9.11) that

$$\alpha_2 = \beta_3 \text{ and } \alpha_3 = \beta_2$$

Then, the stage loading coefficient, ψ of this kind of turbine with axial velocity would be:

$$\psi = 2\Phi \tan \alpha_2 - 1 = 2\Phi \tan \beta_3 - 1 \tag{9.19}$$

Recall, we desire to have higher α_2 for getting higher workoutput, but again its value is limited to 70° due to occurrence of exorbitant losses.

9.1.3 Zero Swirl Exit Velocity

Let us consider another type of turbine blade in which zero swirl exit velocity can be assumed. Then we can have

$$\alpha_3 = 0$$

$$\tan \beta_3 = \frac{U}{V_{3R}} \tag{9.20}$$

It can be shown that using Eqs. (9.16) and (9.17), we can have expression for ψ in terms of RR as

$$\psi = 2(1 - RR) \tag{9.21}$$

This expression indicates that high stage loading blades will provide low reaction ratio. Of course, it is desirable in an aircraft engine to have higher stage loading blades and lower weight. But one has to pay penalty in terms of lower stage efficiency leading to higher TSFC.

9.2 Flow Losses in Turbine Blades

The aerodynamic losses occurring in the turbine arise from the following sources is similar to that of the axial compressor. However a brief discussion is provided below pertaining to flow losses in the axial turbine for completeness.

9.2.1 Profile Loss

This loss is caused due to the growth of boundary layer on the blade profile. This also includes separation of boundary layer caused due to higher incidence angle or high inlet Mach number. We know that the boundary layer growth and its separation basically depends on the actual geometry of blade and the nature of flow.

The suction surface of the blade is more vulnerable to the separation of the boundary layer. Besides this, the separation of boundary layer depends on the angle of incidence and the degree of turbulence. The flow may separate due to positive or negative incidence angles of the flow. We have already discussed extensively about this kind of loss in the chapter of axial compressor.

9.2.2 Annulus Loss

We know that the boundary layer grows in both inner wall (hub) and outer wall (casing) of the annulus. Recall from previous chapter, that the energy loss occurring due to growth of boundary layer on these walls of annulus is known as *annulus loss*. Generally the contribution of flow losses in annular region of turbine is insignificant as the flow is accelerated in the annular region of the turbine.

9.2.3 Secondary Flow Loss

The losses occurred due to secondary flows always arise when the wall boundary layer is turned through an angle by an adjacent curved surface. This gives rise to circulatory flows generating secondary vortices in the stream wise direction. These vortices not only contribute to energy loss but also cause early separation of boundary layer due to transport of low energy fluid from pressure side to suction side of the blade. Recall that a very elaborate discussion on this topic has already been provided in the axial compressor chapter. You are urged to review these portion to reinforce these ideas in your mind.

9.2.4 Tip Clearance Loss

Particularly for the rotor blade, the gas deviates from its intended path near the blade tip and interacts with the boundary layer on the turbine casing. Thus, the requisite contribution of work output is not realised. This loss is known as *tip clearance loss* which depends on the clearance depth between the blade and the casing. Generally shrouds are used in turbine blades to minimise this loss.

Let us see how to take account of the four losses during the design of the axial turbine. Of course, the profile loss coefficient, Y_P can be measured directly in cascade test. Generally the annulus loss Y_A and secondary losses coefficient, Y_S can be estimated by empirical relationship. The tip clearance loss Y_T can also be accounted using empirical relationship, which is considered only for rotor. Then, the total loss coefficient would be

$$Y = Y_P + Y_A + Y_S + Y_T \tag{9.22}$$

Now the question arises how could we incorporate these four losses while designing stages of a turbine. Generally two main parameters namely (i) based on temperature drops and (ii) pressure drops are used to characterise the overall losses in both nozzle and stator of turbine stage. These parameters can be defined for both nozzle and stator using detailed *T-s* process diagram (Fig. 9.11) in a typical reaction turbine stage. Note the losses shown in this figure are intentionally exaggerated to depict the actual process in a better manner.

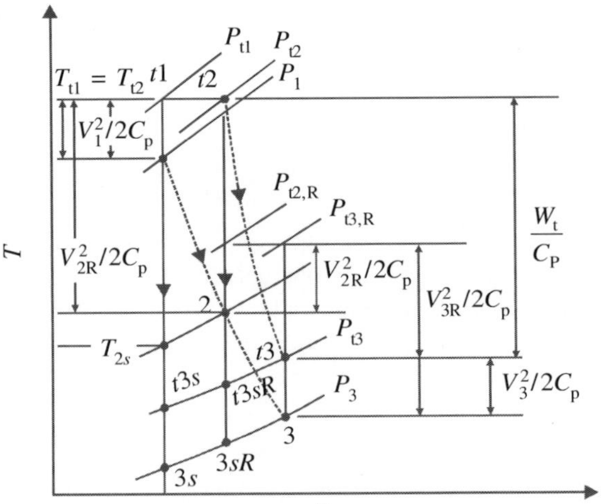

Fig. 9.11. *T-s* diagram depicting the losses in an axial turbine.

When the hot gas passes through the nozzle blades of a turbine stage, ideally it will be expanded from temperature T_{t2} to T_{t3}, but due to friction, the actual temperature at nozzle exit would be T_2 which is higher that T_{2s}. Then the loss coefficient based on temperature, λ_N can be defined as

$$\lambda_N = \frac{T_2 - T_{2s}}{\frac{V_2^2}{2C_P}} \tag{9.23}$$

where λ_N indicates the proportion of leaving energy being degraded by friction. This coefficient is often quite handy to be used during design phase. However, the losses are expressed in terms of pressure in cascade tests. Then, let us define the loss coefficient based on the pressure for nozzle blades, Y_N as

$$Y_N = \frac{P_{t1} - P_{t2}}{P_{t2} - P_2} \tag{9.24}$$

It can be shown easily that λ_N and Y_N are not very different in terms of actual numbers. Now we can define the loss coefficients for the rotor, which are

$$Y_R = \frac{P_{t2R} - P_{t3R}}{P_{t3R} - P_3} \qquad \lambda_R = \frac{T_{3R} - T_{3sR}}{\frac{V_{3R}^2}{2C_P}} \tag{9.25}$$

It must be kept in mind that the ideal temperature that could be obtained by ideal expansion in the rotor is T_{3SR} is shown in Fig. 9.11. For rotor, we are using the properties with respect to relative frame of reference. Since no work is done by the gas relative to blades, T_{t3R} is equal to T_{t2R}. Now we must use these loss coefficient data to predict the stage efficiency. Recall the definition of stage efficiency, η_{ts} given by

$$\eta_{ts} = \frac{T_{t1} - T_{t3}}{T_{t1} - T_{t3s}} = \frac{1}{1 + \frac{T_{t3} - T_{t3s}}{T_{t1} - T_{t3}}} \tag{9.26}$$

From the *T-s* diagram shown in Fig. 9.11, we know that

$$T_{t3} - T_{t3s} \approx T_3 - T_{3s} = (T_3 - T_{3sR}) + (T_{3sR} - T_{3s}) \tag{9.27}$$

From Fig. 9.11 for the isentropic process between pressure P_2 and P_3, we can show that

$$\frac{T_{2s}}{T_{3s}} = \frac{T_2}{T_{3sR}} = \left(\frac{P_2}{P_3}\right)^{\frac{\gamma_g - 1}{\gamma_g}} \tag{9.28}$$

Recasting the above equation, we get

$$\frac{T_{3sR} - T_{3s}}{T_{3sR}} = \frac{T_2 - T_{2s}}{T_2}$$

$$\Rightarrow \qquad T_{3sR} - T_{3s} = \frac{T_{3sR}}{T_2}\left(T_2 - T_{2s}\right) \approx \frac{T_3}{T_2}\left(T_2 - T_{2s}\right) \tag{9.29}$$

Then, stage efficiency for a turbine would be

$$\eta_{ts} \approx \frac{1}{1+\left[\left(T_3 - T_{3sR}\right)+\frac{T_3}{T_2}\left(T_2 - T_{2s}\right)\right]\Big/\left(T_{t1} - T_{t3}\right)} \tag{9.30}$$

Using the expression for λ_N and λ_R, we will get

$$\eta_{ts} \approx \frac{1}{1+\left[\lambda_R \cdot \frac{V_{3R}^2}{2C_P} + \frac{T_3}{T_2}\cdot\lambda_N \cdot \frac{V_2^2}{2C_P}\right]\Big/\left(T_{t1} - T_{t3}\right)} \tag{9.31}$$

From the velocity triangle of a stage, we know that V_{3R} is equal to V_x sec β_3 and V_2 is equal to V_x sec α_2. Then the work output from the stage of a turbine can be expressed as

$$C_P\left(T_{t1} - T_{t3}\right) = UV_x\left(\tan\beta_2 + \tan\beta_3\right) = UV_x\left(\tan\beta_3 + \tan\alpha_2 - \frac{U}{V_x}\right) \tag{9.32}$$

Substituting the above expressions in Eq. 9.31, we can rewrite η_{ts} as:

$$\eta_{ts} \approx \frac{1}{1+\frac{V_x}{2U}\left(\frac{\left[\lambda_R \cdot \sec^2\beta_3 + \frac{T_3}{T_2}\cdot\lambda_N \cdot \sec^2\alpha_2\right]}{\tan\beta_3 + \tan\alpha_2 - \frac{U}{V_x}}\right)} \tag{9.33}$$

Since, $\lambda \approx Y$, the loss coefficients, Y_R and Y_N obtained from the cascade testing can be used directly in the above expression, to estimate stage efficiency during design calculation.

9.3 Vortex Design Concept

In the previous section, we have assumed that two dimensional flow will prevail in the axial turbine. We have already discussed in the axial flow compressor chapter that flow in turbomachinery would be three dimensional in nature particularly for larger blade height (smaller hub-tip ratio). In order to neglect radial component velocity, the fluid elements must be in radial equilibrium which was discussed in axial flow compressor chapter. This is possible only when static pressure is increased from root to tip of the turbine blade. Thus, one can resort to the free vortex design for which following conditions are to be satisfied:

(i) Constant specific work along the blade height ($dh_t/dr = 0$)
(ii) Constant axial velocity along the blade height ($dV_x/dr = 0$)
(iii) The tangential velocity is inversely proportional to radius ($V_\theta r$ = constant)

Note that radial equilibrium will be satisfied only when the above conditions are met. If the design of turbine stage complies with the above conditions, it is known as *free vortex design concept*. Let us consider a single stage of an axial turbine. As no work is carried out in the stator blade (nozzle), total enthalpy, h_t remains constant at the inlet of rotor blade. If the nozzle is designed in such a way that axial velocity (V_x) and $V_\theta r$ remains constant along the blade height, then the above three conditions for radial equilibrium are satisfied. Generally rotor blades are designed for constant axial velocity. If rotor blades are designed for constant $V_\theta r$, it can be shown that specific work output from a turbine stage remains constant along blade height as evident in the following equation

$$\dot{W}_{st} = (U_2 V_{\theta 2} + U_3 V_{\theta 3}) = \omega(V_{\theta 2} r + V_{\theta 3} r) = \text{constant} \tag{9.34}$$

where $\dot{W}_{st}$ is the work output per stage per unit mass of gas. As work output remains constant along blade height, total enthalpy remains constant. Note that $\dot{W}_{st}$ is generally evaluated at mean blade height. One can try to evaluate total work output by multiplying it by the mass flow rate. But the mass of the gas will vary along the blade height as pressure is increased from the root and to tip. In order to estimate total mass flow rate across the annular area, one may need to use numerical integration method. However for simplicity we will consider the mass flow rate per unit area at mean blade height in this book. Let us see now how does the nozzle angle, α_2 for free vortex blade design will vary along the blade height as given below

$$V_{\theta 2} r = V_{x2} \tan \alpha_2 r = \text{constant} \tag{9.35}$$

As V_{x2} remains constant along blade height, α_2 at any radius can be related to α_{2m} at mean blade height as given below

$$\tan \alpha_2 = \tan \alpha_{2m} \left(\frac{r_m}{r}\right)_2 \tag{9.36}$$

In the similar way, the gas angle at rotor outlet can be determined as

$$\tan \alpha_3 = \tan \alpha_{3m} \left(\frac{r_m}{r}\right)_3 \tag{9.37}$$

By using the velocity triangle, the gas inlet angle to the rotor blade, β_2 can be obtained as

$$\tan \beta_2 = \tan \alpha_2 - \frac{U}{V_{x2}} = \tan \alpha_{2m} \left(\frac{r_m}{r}\right)_2 - \frac{U_m}{V_{x2}} \left(\frac{r_m}{r}\right)_2 \tag{9.38}$$

In the similarly way, we can derive an expression for gas outlet angle from the rotor blade, β_3

$$\tan \beta_3 = \tan \alpha_3 - \frac{U}{V_{x3}} = \tan \alpha_{3m} \left(\frac{r_m}{r}\right)_3 - \frac{U_m}{V_{x3}} \left(\frac{r_m}{r}\right)_3 \tag{9.39}$$

From the above expressions, it can be appreciated that rotor blade angles vary considerably from root to tip, resulting in highly twisted turbine blades. Note that it is quite cumbersome to manufacture highly twisted turbine blades. Besides this, twisted turbine blades are subjected to uneven stresses. As a result they may not withstand high thermal and centrifugal stresses particularly under higher power loading conditions. Hence, there is a need to explore other design methods to design better turbine blades.

9.3.1 Constant Nozzle Angle Design

We need not restrict to the free vortex design conditions. Rather we can explore other forms of vortex flow which can satisfy radial equilibrium conditions for the fluid element. It is desirable to have a constant nozzle blade angle in order to avoid the manufacturing problem of twisted blade that is generally obtained using free vortex design procedure. Then, let us look at how we can achieve constant nozzle angle while satisfying radial equilibrium. For this purpose let us invoke vortex energy equation as given below

$$V_x \frac{dV_x}{dr} + V_\theta \frac{dV_\theta}{dr} + \frac{V_\theta^2}{r} = \frac{dh_t}{dr} \tag{9.40}$$

Note that for vortex flow, total enthalpy remains constant along the blade height ($dh_t/dr = 0$). For this present design concept, as the nozzle angle remains constant, we can get a relations between axial velocity and tangential velocity at inlet in terms of angle by using velocity triangle as given below

$$\frac{V_{x2}}{V_{\theta 2}} = \cot \alpha_2 = \text{constant}; \quad \Rightarrow \quad V_{x2} = V_{\theta 2} \cot \alpha_2 \tag{9.41}$$

By differentiating the above equation with respect to r, we can have

$$\frac{dV_{x2}}{dr} = \cot \alpha_2 \frac{dV_{\theta 2}}{dr} \tag{9.42}$$

For constant total enthalpy condition, using the above expression in Eq. 9.40, we can have

$$V_{\theta 2} \cot^2 \alpha_2 \frac{dV_{\theta 2}}{dr} + V_{\theta 2} \frac{dV_{\theta 2}}{dr} + \frac{V_\theta^2}{r} = 0 \tag{9.43}$$

$$\frac{dV_{\theta 2}}{V_{\theta 2}} = -\sin^2 \alpha_2 \frac{dr}{r} \tag{9.44}$$

By integrating the above equation, we can have

$$V_{\theta 2} r^{\sin^2 \alpha_2} = \text{constant} \tag{9.45}$$

For constant α_2, the variation of V_{x2} along with blade height can be determined easily as given below

$$V_{x2} r^{\sin^2 \alpha_2} = \text{constant} \tag{9.46}$$

Note that higher nozzle angles greater than 60° is used for the turbine design. For constant nozzle angle design, radial equilibrium condition is satisfied easily. If the rotor blades are also twisted to provide constant angular momentum, work output remains constant along the blade height and hence, also satisfies the radial equilibrium condition. Keep in mind that apart from free vortex and constant nozzle angle design method, there are several other design concepts, one can conceive and use in design of turbine satisfying radial equilibrium condition.

Example 9.2 A typical stage in an axial turbine is designed on the free vortex concept for 25 kg/s of hot gases which enters into its stator at inlet total temperature of 1200 K and leaves with an angle of 58° with nozzle loss coefficient of 0.12. The total and static pressure at the exit of nozzle is 0.5 and 0.3 MPa respectively. The mean blade diameter to blade height is 6. Determine (i) gas static temperature and (ii) discharge angle of stator blade at root and tip radii.

Given T_{t1} = 1200 K, P_{t2} = 0.5 MPa, P_2 = 0.3 MPa, $\dot{m}$ = 25 kg/s, α_2 = 58°, D_m/h = 6.

To find T_1 and β_2 at root, mean and tip radii.

Solution We can solve this problem considering the velocity triangle at mean radius shown in Fig. 9.4. By using isentropic relation, we can evaluate temperature at station (2), T_{2sN} given as:

$$T_{2s} = T_{t2}\left(\frac{P_2}{P_{t2}}\right)^{\frac{\gamma_g - 1}{\gamma_g}} = 1200 \times \left(\frac{0.3}{0.5}\right)^{\frac{1.33-1}{1.33}} = 1056.2 \text{ K}$$

Note that total temperature at station (2) is equal to total temperature at station (1) as the process is considered to be adiabatic. We can evaluate the actual static temperature at nozzle exit, T_2 as

$$T_2 = T_{2s} + \lambda_N \frac{V_2^2}{2C_{pg}}$$

But we know that $\frac{V_2^2}{2C_{pg}}$ is expressed in terms of temperature as

$$\frac{V_2^2}{2C_{pg}} = T_{t2} - T_2$$

Solving the above two equations, we can evaluate T_2 as

$$T_2 = \frac{\lambda_N T_{t2} + T_{2s}}{1+\lambda_N} = \frac{0.12 \times 1200 + 1056.2}{1 + 0.12} = 1071.6 \text{ K}$$

Then the velocity at station (2), V_2 can be determined as

$$V_2 = \sqrt{2C_{pg}(T_{t2} - T_2)} = \sqrt{2 \times 1148\,(1200 - 1071.6)} = 543 \text{ m/s}$$

We can determine the mean diameter of rotor blade, D_m by carrying out mass balance at station (2) as given below

$$\dot{m} = \rho_2 \pi D_m h V_x = \rho_2 \pi \frac{D_m^2}{6} V_x \Rightarrow D_m = \sqrt{\frac{6\dot{m}}{\pi \rho_2 V_2 \cos\alpha_2}}$$

But the density at station (2) can be determined by using equation of state as given below

$$\rho_2 = \frac{P_2}{RT_2} = \frac{0.3 \times 10^6}{287 \times 1071.6} = 0.975 \text{ kg/s}$$

Substituting the value of ρ_2 in the above equation for mass flow rate, the value of D_m is found to be 0.41 m and blade height becomes 0.069 m. Then the radius at root and tip becomes

$$r_r = \frac{D_m}{2} - \frac{h}{2} = 0.171\, m \qquad r_t = \frac{D_m}{2} + \frac{h}{2} = 0.24\, m$$

For free vortex design, we know that $V_\theta r$ remains constant at any plane.

At root: By using free vortex design concept, the tangential velocity at root can be estimated as

$$V_{\theta,r} = V_{\theta,m}\frac{r_m}{r_r} = V_2 \sin\alpha_2 \frac{r_m}{r_r} = 543 \sin 58° \times \frac{0.205}{0.171} = 552.05 \text{ m/s}$$

Then the blade angle at root can be determined as

$$\alpha_{2,r} = \tan^{-1}\left(\frac{552.05}{543\cos 58°}\right) = 62.47°$$

We can evaluate the static temperature at root as

$$T_{2,r} = T_{t2} - \frac{V_{2,r}^2}{2C_{pg}} = 1200 - \frac{552.05^2}{\sin^2 62.74° \times 2 \times 1148} = 1032.02 \text{ K}$$

At tip: In the similar way we can determine at the blade's tip below.

$$V_{\theta,t} = V_{\theta,m}\frac{r_m}{r_t} = 543\sin 58° \times \frac{0.205}{0.24} = 393.3 \text{ m/s}$$

Then the blade angle at tip can be determined as

$$\alpha_{2,t} = \tan^{-1}\left(\frac{393.3}{543\cos 58°}\right) = 53.8°$$

We can evaluate the static temperature at root as

$$T_{2,t} = T_{t2} - \frac{V_{2,t}^2}{2C_{pg}} = 1200 - \frac{393.3^2}{\sin^2 53.8° \times 2 \times 1148} = 1096.5 \text{ K}$$

9.4 Turbine Cooling

Gas turbine engine designers always strive for increasing thrust (power output) for a given engine while reducing its weight or decreasing TSFC. One can achieve these improvements by operating at maximum possible turbine inlet temperature. But the maximum turbine inlet temperature is limited by current available materials. Improvements in materials with time have been a bit slow to allow high turbine inlet temperature. Approximately, 10°C per year could be achieved only by present material improvement trend. But turbine cooling could allow us to use much higher TIT (turbine inlet temperature). When turbine cooling was introduced, TIT could be jumped by approximately double. Historically turbine cooling was introduced by Germans in 1935 during World War II. Keep in mind that 1% increase in turbine inlet temperature can produce around 3 or 4% increase in engine output.

While considering turbine cooling the first question is to decide which fluid is the natural choice for coolant. Of course, air as a coolant will immediately come to our mind, since it is available abundantly in nature. But its specific heat value is quite low, thus, heat carrying capacity will also be eventually low. Other viable coolant fluid is water. Although, its physical properties are very suitable for use as a coolant but it has certain limitations and thus, cannot be considered as a coolant particularly for an aircraft engine. We can conclude that two main natural fluids, air and water cannot be employed as a coolant for gas turbine blades. Now we will see what are the methods of cooling that can be employed for gas turbine applications. Broadly four methods are being utilised for cooling turbine blades which are:

(i) Convective cooling
(ii) Impingement cooling
(iii) Film cooling
(iv) Transpiration cooling

9.4.1 Convective Cooling

Among all four methods, the convective cooling is the simplest one. In fact, it was the first method to be adopted for cooling of the turbine blades. In this cooling method, the coolant particularly air is allowed to flow through the passages as shown in Fig. 9.12 from the hub to tip of a turbine blade. As a result, heat is extracted from the turbine blade and thus reducing its temperature. The effectiveness of convective cooling is basically restricted by the size of internal passages in the blade and the availability of cooling air. Since air has a low specific heat, it requires large internal surface area

$$q = h_c A \Delta T = hA(T_s - T_c) \qquad (9.47)$$

where, h is the convective heat transfer coefficient whose value varies in the range of 25-200 N/m^{2}K for air, and 500-20,000 N/m^{2}K for liquid, T_s is the surface temperature of the blade and T_c is the coolant temperature. While designing the internal passageway, from best heat transfer stand point, one has to take care of strength and manufacturing considerations of blades. The schematic of a typical turbine blades with passages for convective cooling is shown in Fig. 9.12. It can be noted from this figure that no cooling air passage can be provided in trailing edge of the blade since it is very thin. Besides this, the leading edge of the turbine blade which is subjected to highest temperature can only be cooled effectively by the following impingement cooling method.

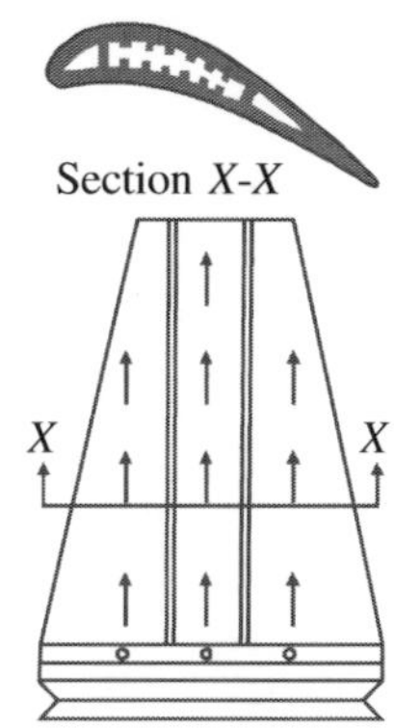

Fig. 9.12. Schematic of a convective cooling turbine blades.

9.4.2 Impingement Cooling

The schematic of impingement cooling is shown in Fig. 9.13. In the impingement cooling system, the basic process of cooling is same as convective cooling. But the main difference is that the air instead of flowing radially through the blade is allowed to turn normal to the radial direction so that it can impinge onto the metal surface of blade usually around the stagnation point of blade as shown in Fig. 9.13. It is a very effective method of reducing thermal stress in local areas and is easily adapted to stator (nozzle) blades. It can also be used for rotor blades if sufficient space is available.

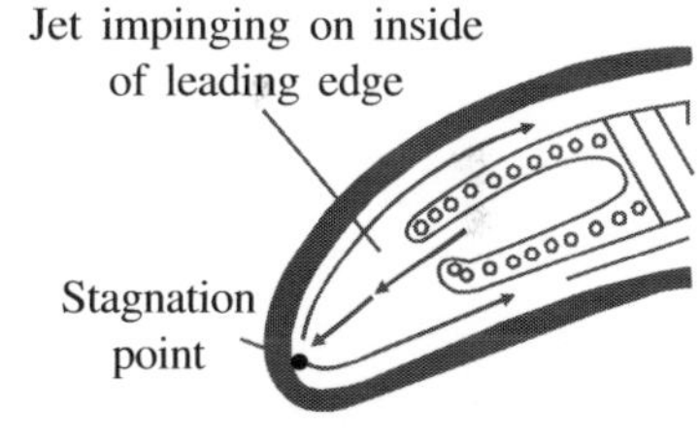

Fig. 9.13. Schematic of an impingement cooling turbine blades.

9.4.3 Film Cooling

The film cooling method is described in Fig. 9.14. Note that this method of cooling involves the injection of a secondary fluid (air) into the boundary layer of the primary fluid. This is an effective way to protect the blade surface from hot gases by directing the cooling air into boundary layer to provide a protective cool film along the surface.

The injection of coolant air into the boundary layer although causes loss in turbine work but lowers

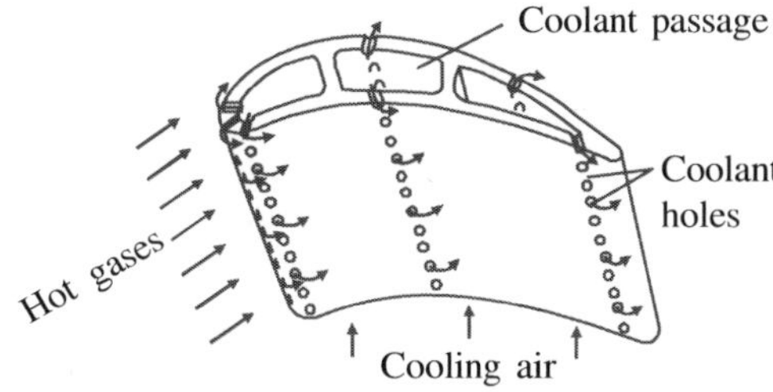

Fig. 9.14. Schematic of a film cooling turbine blades.

the turbine surface temperature considerably leading to the use of higher turbine inlet temperatures. If too much air is injected into boundary layer or its velocity is too high, then the cooling air may penetrate into the boundary layer defeating the main purpose of using film cooling. But it may not always be possible to get high pressure air supply especially for the cooling of the first nozzle blades around the stagnation point by this method.

9.4.4 Transpiration Cooling

It is most efficient air-cooling technique available. This is also known as transpiration/porous blade cooling. This method is described in Fig. 9.15 in which the cooling air is forced through porous materials of turbine blade into boundary layer to form relatively cool, insulating film. For effective cooling by this method, the pores should be small. This can lead to pore blockage due to soot particles or foreign material. It is also more economical as only 1.5-2% of the total air mass flow rate can reduce, the blade temperature in the range of 200-300°C.

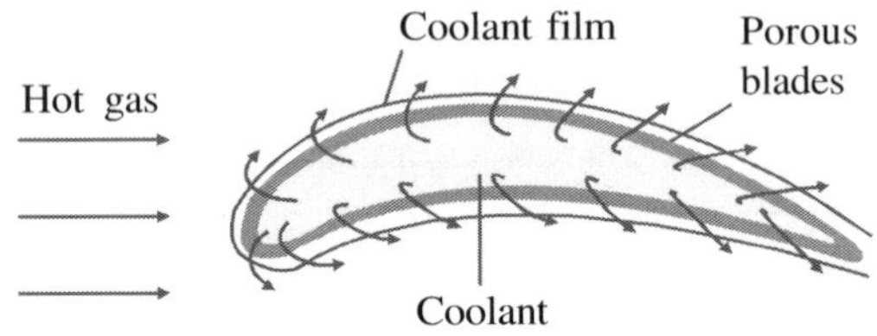

Fig. 9.15. Schematic of a transpiration cooling turbine blades.

We have discussed about individual methods of turbine blade cooling. But in actual practice, all three methods namely, convective, impingement and film cooling, have been used in designing and developing turbine blades to enhance the effectiveness of cooling of the turbine blades. The schematic of a typical cooled turbine blade is shown in Fig. 9.16. It can be noted from this figure that impingement cooling is used in leading edge of the blade while main portion blade is cooled by convective cooling. A film is being formed over the blade surface such that the heat from hot gas would not penetrate into the blade surface. However, the trailing edge is cooled by convection method.

We know that higher turbine inlet temperature can be employed which provides higher stage loading coefficient ψ and a higher flow coefficient Φ. Hence, there will be a considerable increase

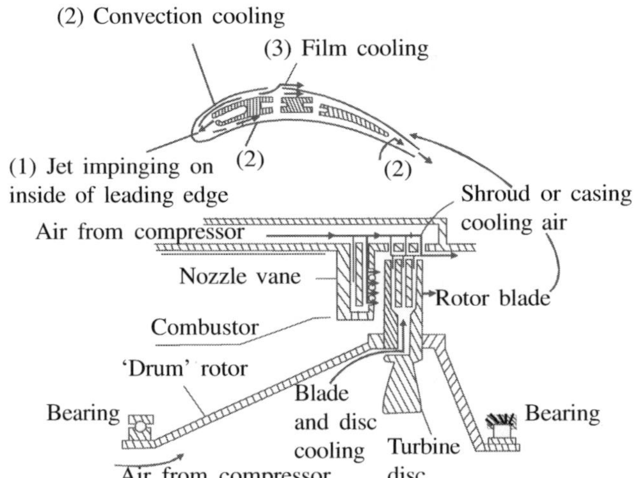

Fig. 9.16. Schematic of an air-cooled turbine blades [2].

in thrust. But the cooling of the turbine blade has several detrimental effects on the overall performance of the engine. These disadvantages are:

(i) Loss of turbine work output because of reduction of mass flow through the turbine.
(ii) Non-adiabatic expansion.
(iii) Pressure loss accompanied with reduction on enthalpy due to mixing of coolant with main gas streams.
(iv) Additional pumping work is required to supply air for cooling turbine blades.
(v) Increase in cost of production of the engines.
(vi) Hampers the reliability of turbine blades.
(vii) Decrease in efficiency of the turbine.

9.5 Overall Performance of a Multistage Turbine

The methods described in the previous section make it possible to estimate design performance of a single stage turbine. For a multistage turbine, assuming equal contribution of pressure expansion and stage efficiency from each stage, we can find out its performance easily. By similar procedure, the performance of an axial turbine can be obtained over a wide range of operating conditions different from design conditions. Of course, the actual performance of a particular turbine can be obtained by performing the testing of the turbine over a wide range of operating conditions.

Generally, the performance of an axial turbine is characterised by expressing the pressure ratio, P_{t5}/P_{t4} and η_{it} in terms of non-dimensional mass flow rate, $\dfrac{\dot{m}\sqrt{T_{t4}}}{P_{t4}}$ and non-dimensional rotational speed, $\dfrac{N}{\sqrt{T_{t4}}}$. Note that we have been using the cycle notation suffixes 4 and 5 to denote the turbine inlet and outlet conditions, respectively. The performance plot of a typical turbine is shown in Fig. 9.17. It can be seen that the efficiency, η_{it} does not vary much over the range of mass flow and

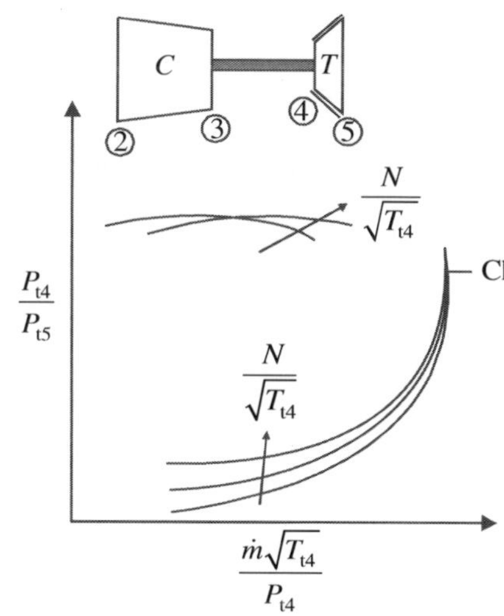

Fig. 9.17. Overall performance characteristics of a typical axial turbine.

rotor speeds. This is so because the flow is accelerated in the turbine, which allows the turbine blades to operate over a wide range of flow rate without incurring heavy losses.

It is interesting to note from this figure that the pressure ratio, P_{t5}/P_{t4} vs. mass flow rate curves for all constant speed seen to be merging into a single line at a maximum mass flow rate, which is termed as choking condition. This may be caused due to the occurrence of high velocities at the nozzle throats or at the rotor outlet, attaining a sonic condition. Normally, choking in nozzle throats occurs very often. When the turbine is not even operating at choked condition, the difference between constant speed lines are not quite large. Another interesting observation that can be made is that the pressure ratio across a single stage turbine is much greater than that for a compressor with satisfactory efficiency, which is due to accelerating flow condition in the turbine.

9.6 Chapter Summary

The aerothermodyamic analysis of axial flow turbine is covered in this chapter. As we know that the turbine is meant to convert thermal energy of hot gas into mechanical energy to drive the compressor, fan and propeller. In this axial turbine, high temperature and pressure air from the combustor is allowed to pass through a series of stators and rotors and gets expanded to produce work output. The hardware of axial turbine is discussed briefly to get students acquainted about it. The requisite velocity triangles are employed to analyze processes involved during expansion through turbine. Based on two dimensional flow assumption, an expression for pressure ratio across turbine is derived from the Euler's turbomachinery equation in terms of blade velocity, axial velocity, inlet temperature and air angles. Although this fundamental equation is quite similar to the axial compressor, its interpretation for axial turbine is quite different. The pressure ratio across an axial turbine stage is quite higher as compared to that of an axial compressor as axial turbine is operated under favourable pressure gradient. Hence, few number of stages of axial turbine is required for same pressure ratio as compared to axial compressor. As the pressure ratio across a stage is quite higher, it is likely that flow will be chocked particularly at last stage. Three nondimensional parameters such as blade loading coefficient, flow coefficient and reaction ratio are employed to discuss about the design aspects of the turbine blades briefly. The free vorte design concept is introduced briefly highlighting the importance of the constant nozzle angle design principle. The flow losses in the turbine blades are discussed adequately. An attempt is made to derive stage efficiency in terms of rotor and stator loss coefficients. The overall performance of multistage turbine is characterised in terms of non-dimensional pressure ratio, rotational speed, and isentropic efficiency. Turbine cooling is essential for high performance gas turbine engine as requisite high temperature materials for turbine blade is not available currently, although new high temperature materials are getting improved continuously over past decades. The basic principles of turbine cooling methods such as convective, film and transpiration are discussed briefly with their relative merits and demerits.

Nomenclature

A	Annular area
C_{Pg}	Specific heat of hot gas
D_m	Mean blade diameter
h	Blade height
h_c	Convective heat transfer coefficient
$\dot{m}$	Mass flow rate
N	Nozzle stator
R	Rotor

S	Entropy
T_c	Coolant temperature
T_s	Surface temperature of blade
U	Blade velocity
V_R	Relative velocity
V_x	Axial velocity
W_t	Work output of turbine
Y_A	Annulus loss coefficient
Y_P	Profile loss coefficient
Y_R	Loss coefficient for the rotor, based on pressure
Y_S	Secondary loss coefficient
Y_T	Tip clearance loss coefficient

Greek Symbols

α	Swirl angle
β	Gas blade angle
ΔT_{ts}	Total temperature drop across a stage
Φ	Flow coefficient
γ_g	Specific heat ratio for hot gas
λ	Loss coefficient
λ_N	Loss coefficient based on temperature for nozzle
λ_R	Loss coefficient for the rotor, based on temperature
η	Efficiency
η_{ts}	Turbine stage efficiency
η_{it}	Isentropic turbine efficiency
η_{pt}	Polytropic efficiency for a single stage turbine
ψ	Stage load coefficient

Subscripts

A	Annulus loss
m	Mean
N	Nozzle
P	Profile loss
R	Rotor
S	Secondary loss
st	Stage
T	Tip clearance loss
t	Turbine
ts	Stage
x	Axial

Abbreviations

KE	Kinetic energy
RR	Reaction ratio

References and Suggested Readings

1. Mattingly, J.D., *Elements of Gas Turbine Propulsion*, 1st Edn., McGraw-Hill Publishing Company Ltd., Singapore, 1996.
2. Kerrebrcok, J.L., *Aircraft Engines and Gas Turbines*, 1st Edn., MIT Press, England, 1992.

3. Cohen, H., Rogers, G.F.C. and Saravanamuttoo, H.I.H., *Gas Turbine Theory*, 4th Edn., Addison Wesley Longman Ltd., England, 1998.
4. Hill, P.G. and Peterson, C.R., *Mechanics of Thermodynamics of Propulsion*, 2nd Edn., Addison-wesley Publishing Company, Inc. Reading, MA, 1992.
5. Dixon, S.L., *Fluid Mechanics, Thermodynamics of Turbodynamics*, Butterworth-Heinemann, 3rd Edn., 1995.
6. Cumpsty, N., *Jet Propulsion*, Cambridge University Press, 1997.
7. *The Jet Engine, Rolls Royec*, 4th Edn., 1986.
8. Horlock, J.H., *Axial Flow Turbines*, Robert E. Kreiger Publishing Co., Melbourne, FL, 1973.
9. Treager, I.E., *Aircraft Gas Turbine Engine Technology*, 3rd Edition, Tata McGraw-Hill, 1996.

Review Questions

1. Draw schematic of a typical axial turbine and describe its operating principle. How is it different from an axial compressor?
2. Draw the velocity triangle for a single stage turbine.
3. Derive an expression for workoutput for a single stage turbine.
4. Why is the shroud being used very often in a turbine than in a compressor?
5. What is the function of nozzle guide vanes? How does the nozzle influence the engine performance?
6. What is the maximum pressure ratio across a single stage turbine?
7. What are the applications of the axial turbine?
8. Show schematically the variation of total enthalpy, pressure and temperature across a single stage of an axial turbine. How are these comparable with their respective static quantities?
9. What are the factors that affect stage pressure ratio?
10. What is reaction ratio, *RR*? Why is it important? Derive an expression for it and show that 50% *RR* produces a symmetric blade.
11. Why is symmetric blade design preferred in an aircraft gas turbine engine?
12. What is impulse and reaction blading? Explain it by drawing the respective velocity triangles.
13. Can you classify the turbine blade stages on the basis of reaction ratio? If it is so, explain it with the help of velocity triangles.
14. Can the free vortex design concept be used for an axial turbine? How is it different from exponential design concept?
15. What is the constant nozzle design concept? Why is it preferred in turbine design? Under what condition, this design concept can also satisfy the radial equilibrium in the blade.
16. Define stage efficiency and show that it can be expressed in terms of rotor and nozzle blade loss coefficient. Find an expression for the blade efficiency for a symmetrical blade when the flow coefficient is 0.5.
17. What is choking in a stage? Explain this phenomena in details. Under what conditions, an axial turbine is likely to get choked?
18. What are the methods being used for turbine cooling? Which is the more efficient one?
19. Draw a typical modern turbine blade in which most of the cooling methods are used to cool it. Explain advantages and disadvantages of each methods over other methods?

Problems

1. A single stage turbine with reaction ratio of 0.5 is to be designed which consumes 50 kg/s of gas at 10,000 rpm. The gas at total pressure of 8 MPa and 1500 K enters into the nozzle blade and leaves it with the same V_x = 125 m/s, at an angle of 60° at the mean radius of 15 cm. If the inlet rotor blade height is 4 cm and the efficiency of the stage is 0.88, determine (i) stator turning angle (ii) rotor turning angle (iii) static pressure at the rotor and stator exit (iv) blade height at rotor and stator, and (v) power output.

2. A mass of 20 kg/s gas with an axial velocity of 250 m/s enters into the repetitive stage of an axial flow turbine with total pressure 900 kPa and 1700 K. A mean line design analysis of an axial turbine stage with equal inlet and outlet velocity is carried out. The stage exit swirl number is 10°. The blade velocity, U at mean radius is 360 m/s. Determine the rotor blade gas angles, reaction ratio, temperature drop coefficient and power output. Assuming $N = 0.05$, determine the nozzle throat area ignoring the effect of friction on critical condition. Take $T_{t1} = 1000$ K, $P_{t1} = 0.4$ MPa.
3. The gas at the total pressure of 0.6 MPa and 1600 K enters into a repetitive stage of turbine of 50% reaction ratio and leaves its exit at an angle of 20° with an axial direction with static pressure of 200 kPa. If the total to static efficiency of the stage is 90%. Determine (i) the blade speed at mean radius, and (ii) gas velocity and its direction at the exit of stage.
4. A single stage high pressure axial turbine is operated at 15000 rpm to expand 50 kg/s of hot gas at total pressure of 1 MPa and 1500 K. The mean radius of the blade is 200 mm with inlet blade height of 40 mm. The absolute inlet angle of the gas to the rotor is 50° which is also same for stator exit flow angle. The turning of the gas flow in the rotor blade is turned out be 40°. Assuming the axial velocity to remain constant and stage efficiency as 0.85%, estimate the following: (i) Blade heights at rotor and stator exits (ii) static pressure at rotor and stator exits, (iii) stator flow turning angle, and (iv) reaction ratio.
5. A single stage high pressure axial turbine with reaction of 0.4 and $V_x = 180$ m/s is operated at 12000 rpm to expand 70 kg/s of hot gas at total pressure of 0.8 MPa and 1700 K. The mean radius of the blade is 150 mm with inlet blade height of 30 mm. The absolute inlet angle of the gas to the rotor is 55° which is also same for stator exit flow angle. Assuming the axial velocity to remain constant and stage efficiency as 0.85%, estimate the following: (i) Blade heights at rotor and stator exits (ii) Static pressure at rotor and stator exits, (iii) Stator flow turning angle (iv) Total pressure ratio, and (v) Power output.
6. A single stage axial turbine with reaction of 0.3 is operated at 10000 rpm which expands 70 kg/s of hot gas by pressure ratio of 0.7. The gas enters into the rotor blade with total pressure of 0.8 MPa and 1700 K at absolute inlet angle of 56° which is turned by relative angle of 70°. The mean radius of the blade is 180 mm with inlet blade height of 40 mm. The axial velocity remains constant at 100 m/s. Determine the stage efficiency? Explain how to improve it further?
7. In a single stage low pressure axial turbine, 100 kg/s of hot gas at total pressure of 0.8 MPa and 1700 K. Gas enters into the nozzle blade with an absolute inlet angle to the nozzle is 52° which is also same for stator exit flow angle. The turning of the gas flow in the rotor blade is turned out be 70°. The mean radius of the blade is 250 mm with inlet blade height of 60 mm. The axial velocity and speed of sound at inlet are 220 m/s and 550 m/s. If the tangential blade velocity is 270 m/s determine the following: (i) rotor and stator flow turning angle (ii) velocity triangle, (iii) efficiency of the stage, and (iv) reaction ratio.
8. A single stage axial turbine is designed on the free vortex concept, which is to expand 25 kg/s of hot gas at total pressure of 0.5 MPa and 1200 K. The static pressure at mean radius is 0.3 MPa. The blade diameter and height ratio at mean radius is 12. If the nozzle outlet angle is 32° and nozzle loss coefficient is 0.12, then estimate (i) gas temperature, and (ii) discharge angle at rotor blade root and tip radii.
9. In a single stage axial turbine, 25 kg/s of hot gas at total pressure of 0.5 MPa and 1200 K enters into the stator blade and leaves with an absolute inlet angle of 62°. The absolute angle of gas from the rotor exit is 12°. The axial velocity and blade speed at inlet are 200 m/s and 350 m/s. The turbine is designed on the principle of free vortex concept. If the nozzle loss coefficient is 0.07, then determine following: (i) rotor flow turning angle (ii) velocity triangle, (iii) temperature drop coefficient, (iv) nozzle throat area, and (v) reaction ratio.
10. A single stage axial turbine with axial outlet velocity of 350 m/s is designed on free vortex principle, which expands 30 kg/s of hot gas from inlet total pressure of 0.6 MPa and 1200 K. The temperature drop across the stage is 175 K. The mean blade velocity is 300 m/s at rotational speed 13000 rpm. The axial velocity at nozzle exit is 330 m/s.The turbine is designed with constant annulus area through the stage based on the principle of free vortex concept. If the nozzle loss coefficient is 0.06, then determine the following: (i) blade height (ii) inlet relative Mach number at root rotor radius, and (iii) diameter ratio at outlet.

10. Gas Turbine Combustor

It is difficult to say what is impossible for the dreams of yesterday in the hope of today and the reality of tomorrow.

Robert. H. Goddard, 1904

10.1 Introduction

The combustion system is a very critical component of any gas turbine engine. Most advanced aircraft gas turbine engines usually employ two combustors namely *main burner* and *afterburner*. In both the burners, liquid fuel is being burnt releasing a large amount of heat. Even though, the utility of main burner and afterburner is different, but they work on almost similar principles.

In these two burners, bulk liquid fuel is converted into fine droplets with the help of atomizers which interact with the prevailing complex fluid flow in it. The fuel vaporizes and mixes with the incoming air before the actual chemical reactions take place. On ignition, flame is established in the combustion chamber, raising temperature and thus, making combustion to be self sustained. In order to have a compact combustion chamber, high intensity burning of fuel must be accomplished. As a result, the heat release rate in an aircraft engine combustor (5,000 MW/m^3) is quite high in comparison to that of industrial burner, (500 MW/m^3) in a typical stationary power plant. Hence, it is quite a challenging job for designers to design and develop a gas turbine combustor. Beside these requirements, several desirable features that must also be met by gas turbine combustors for aircraft applications are given below:

1. Completeness of combustion $\Rightarrow$ higher combustion efficiency, η_{com}.
2. Proper temperature distribution at its exit.
3. Lower stagnation pressure loss.
4. Absence of any hot spots.
5. Better flame stability (free from flame blowout, pressure oscillations).
6. Relight ability.
7. Absence of deposits on the surface.
8. Smaller cross-sectional area and length.

Many of these requirements are contradictory to each other. For example, high combustion efficiency is quite difficult to achieve with lower total pressure loss. We know, that high air mass flow rate per unit cross-sectional area is very desirable to have a higher thrust and a compact engine. In other words, the velocity of fluids inside the engine must be high enough to achieve these objectives. Of course, care must be taken such that total pressure losses are not too excessive due to either shock formation, wall friction or viscous mixing. However, higher air velocity inside combustion chamber poses serious problem in stabilising the flame. Let us see now, how it is quite difficult to stabilise a flame within a moving air stream by considering a duct containing stoichiometry fuel/air mixture. On igniting, the flame front will move with certain velocity towards unburnt mixture which is known as the *burning velocity*. If the incoming velocity of mixture is quite higher than that of the burning velocity, then, the flame will be moved out of the tube. Such a phenomena is known as *blowout*. Let us assume that turbulent flame prevails inside the combustion chamber for *stoichiometric hydrocarbon/air mixture*. We know that for stoichiometric hydrocarbon/air mixture used in gas turbine combustor system, representative laminar burning velocity S_L at P = 0.5 MPa, 650 K is around 1.7 m/sec. The empirical relation between turbulent burning velocity, S_t and S_L is given by

$$S_t/S_L = 5 \tag{10.1}$$

Hence, the turbulent flame speed, S_t would be around 8.5 m/sec. But, typical air velocity at the inlet of the combustor will be around 30-60 m/sec. Naturally, the flame can be moved out of combustor easily which is known as *blowout* phenomenon. Hence, it is quite difficult to stabilise a flame inside the combustion chamber. Besides this, question arises, is it really possible to maintain stoichiometric A/F ratio? Let us consider an aircraft flying at M_0 = 0.6 at an altitude of 12 km with compressor pressure ratio, π_c = 20, such that inlet temperature, T_{t_3} is around 650 K. Of course, turbine inlet temperature T_{t_4} is generally limited to 1600 K. Let us strike an energy balance between station (3) and (4) considering one dimensional steady flow in the combustion chamber as shown in Fig. 10.1, that gives

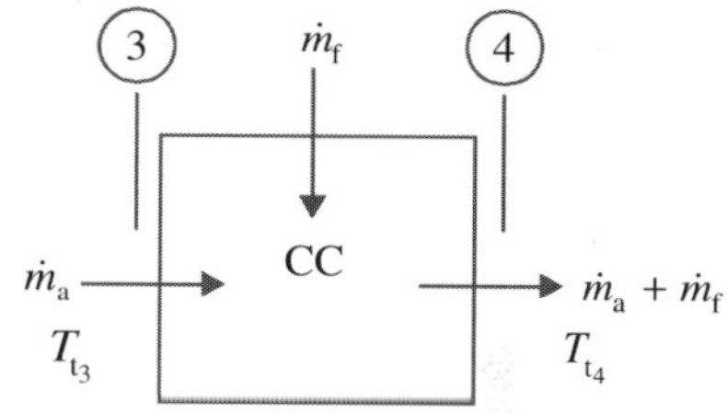

Fig. 10.1. Control volume of a one dimensional combustor.

$$\dot{m}_a C_{pa} T_{t_3} + \eta_b \dot{m}_f \Delta H_C = (\dot{m}_a + \dot{m}_f) C_{pg} T_{t_4} \tag{10.2}$$

By manipulating the above equation, we can have

$$\frac{\dot{m}_f}{\dot{m}_a} = f = \frac{C_{pg}T_{t4} - C_{pa}T_{t3}}{\eta_b \Delta H_C - C_{pg}T_{t_4}} \tag{10.3}$$

where, ΔH_C is heat of combustion = 45,000 kJ/kg for aviation turbine fuel (ATF); η_b is the burner efficiency = 0.94; C_{pa} is representative specific heat of inlet air = 1.005 kJ/kg K; C_{pg} is representative specific heat of hot gas = 1.148 kJ/kg K; T_{t3} is representative inlet temperature = 650 K; T_{t4} is turbine inlet temperature = 1600 K. The fuel/air ratio, f from the above equation can be obtained as

$$f_{act} = \frac{\dot{m}_f}{\dot{m}_a} = 0.0293 \tag{10.4}$$

If we consider the aviation turbine fuel, ATF (JP4) as C_8H_{18} (Octane), then balanced reaction equation between the ATF and air can be represented as

$$C_8H_{18} + 12.5\left(O_2 + \frac{79}{21}N_2\right) \to 8CO_2 + 9H_2O + 12.5\left(\frac{79}{21}\right)N_2 \tag{10.5}$$

The stoichiometric fuel/air ratio, f becomes

$$f_{stoi} = \frac{\dot{m}_f}{\dot{m}_a} = 0.0664 \tag{10.6}$$

The equivalence ratio can be determined as

$$\text{Equivalence Ratio} = \phi = \frac{f_{\text{act}}}{f_{\text{stoi}}} = 0.44 \tag{10.7}$$

Hence, a lean mixture of fuel/air must be maintained in gas turbine main combustor to achieve requisite exit temperature as dictated by the limitation imposed by turbine blade cooling. But, this requirement poses problem of flame stabilisation and ignition. It can be noted from Fig. 10.2 that lean fuel/air mixture of equivalence ratio, 0.4 is outside the flammability limit. Hence, instead of mixing all the air with the fuel initially, a small amount of air known as *primary air* is mixed with the fuel, forming stoichiometric fuel/air mixture in the primary zone of the combustion chamber. As a result, this mixture can be ignited and burnt easily. Then the hot gases from the primary zone are mixed with the remaining air known as *secondary air* in the secondary zone to achieve desired temperature at the exit of main combustor.

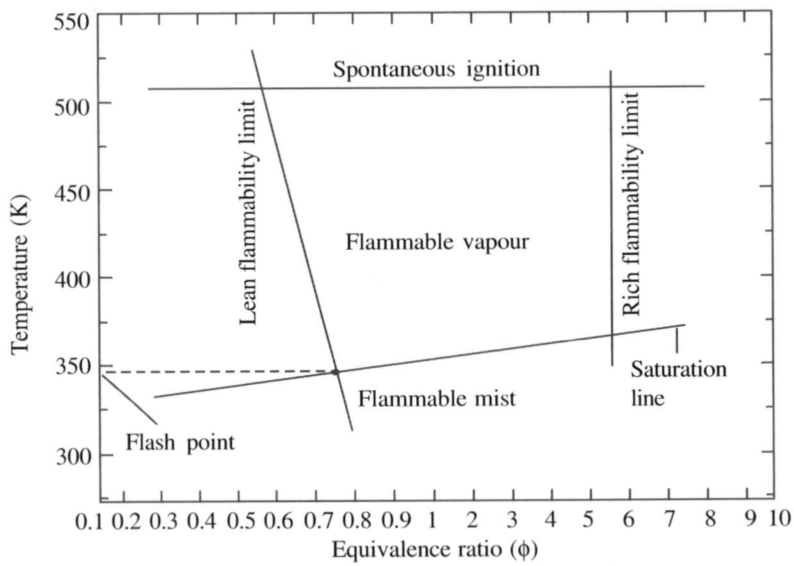

Fig. 10.2. Flammability characteristics of a typical aviation turbine fuel (ATF) [3].

Even though, the air flow in main combustion chamber is divided into two parts namely (1) Primary and (2) Secondary streams. However, air velocity is quite high and does not allow the flame to stabilise inside the combustion chamber. Hence, we must find out other additional ways and means to reduce local air velocity such that the flame can stay inside the combustor.

10.2 Flame Stabilisation

We know that aircraft engine combustor has to function effectively over a wide range of operating conditions from idling to full load. During this period, flame has to be contained within the burner. However, we have learnt from the above discussion that it is quite difficult to stabilise the flame in the combustor as the prevailing air velocity is much higher than the turbulent burning velocity. Hence, there is a need to have a flame stabiliser to contain the flame inside the combustion chamber. The basic idea of flame stabilisation is to create a low velocity zone in the combustor such that flame can be stabilised easily. In other words, the flame can be stabilised only when local burning velocity is matched with local flow velocity. Besides this, it must provide favourable conditions such that exothermic chemical reactions can take place releasing large amount of heat and the flame can be self-sustained easily. In order to achieve this objective, there are mainly two types of flame stabilisers which are extensively used in modern gas turbine engines namely (i) Bluff body, and (ii) Swirler.

10.2.1 Flame Stabilisation by Bluff Body

The flame can be stabilised by placing an object in the flow of combustible mixture as shown in Fig. 10.3. This can be possible due to the fact that flow velocity is reduced in the boundary layer near the solid surface and thus provide ample chances for local flow velocity to match with local burning velocity, necessary for flame stabilisation.

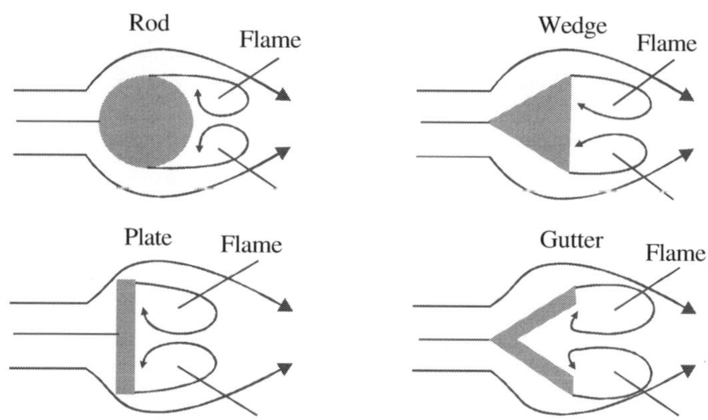

Fig. 10.3. Flame stabilisation by bluff bodies.

Generally, bluff bodies (non-streamlined bodies) are used for this purpose such that flow separates from the bodies creating a recirculating zone downstream of its wake. Because of intense recirculation of flow, there will be high rate of heat and mass transfer between wake and the flow past the wake. If there is combustion in the flow, surrounding the recirculation zone, hot products will penetrate the recirculation zone and can be carried upstream, where they can mix with the fresh combustible mixture and preparing it for ignition. The flow patterns in a typical bluff body is shown in Fig. 10.3. The recirculating flow can be assumed to consist of fully reacted mixture (stirred reactor). This zone not only transfers active species to the combustible mixture but also acts as a thermal reservoir of heat. This transfer of heat, and of active species to combustible mixture together with retardation of local flow velocity in the wall boundary layer of bluff bodies enables the flame to be stabilised at approach velocities far exceeding the local burning velocity.

Let us analyze the flame stabilisation process, considering flow over a *V*-gutter. Due to flow separation over *V*-gutter, recirculating zone is created as shown in Fig. 10.4 which is very much essential for stabilising the flame in the combustor. It must be kept in mind that the length of recirculating zone is very insensitive to the inlet flow condition particularly when the flow is turbulent in nature. Rather, its length is dependent on the shape of the bluff body. Several methods to analyze this flame stabilisation process have been reported in literature. Most of these methods are based on scaling law in which the residence time of combustible fluid element must be greater than the reaction time for the stabilisation of the flame. Considering all the flow quantities with reference to upstream condition of the stabilising device, we can find out the blowoff velocity, V_{bo}, for a given dimension of the stabiliser, d and reaction time, t_{reac} of the combustible mixture. Note that the blowoff velocity, V_{bo}, is defined as the value of upstream velocity of the mixture that causes the flame to blowoff from the flame stabiliser. Of course, if the residence time, t_{res} of the fluid element in the vicinity of hot recirculating zone is smaller than the reaction time, t_{reac} the flame will be moved out of the combustor, causing the burner to blowoff. Thus, the blowoff criteria would become

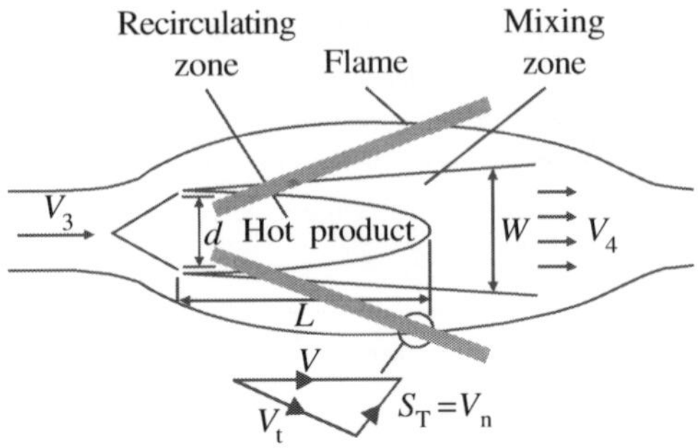

Fig. 10.4. Flame stabilisation by a V-gutter.

$$t_{res} \leq t_{reac} \tag{10.8}$$

However, the residence time, t_{res} can be expressed in terms of d and V_{bo}

$$\frac{d}{V_{bo}} \leq t_{reac} \text{ but, we know that } t_{reac} = \frac{\rho}{\dot{m}_f'''} \tag{10.9}$$

where, $\dot{m}_f'''$ is the mass burning rate of fuel per unit volume. Then, Eq. (10.9) becomes

$$\frac{d}{V_{bo}} \propto \frac{\rho}{\dot{m}_f'''} \propto \frac{\alpha}{S_L^2} \tag{10.10}$$

where, S_L is the laminar burning velocity. We can now derive a dimensionless expression for blowoff velocity known as Peclet number, Pe which is given by

$$Pe = \frac{V_{bo}d}{\alpha} = K\left(\frac{S_L d}{\alpha}\right)^2 \tag{10.11}$$

where, α is the thermal diffusivity of the mixture and K is the proportionality constant. Keep in mind that 2nd order reaction exponent has been incorporated in the above equation through burning velocity term. From an extensive experimental data, it has been found out that the value of K can be taken as 6 for three dimensional stabiliser (sphere) and K appears to be 3 for two dimensional bluff bodies. In practice, V-gutter is the most widely used flame stabiliser, particularly in afterburners. It not only incurs lowest total pressure loss and provides better performance, but also imparts better structural rigidity. It is one of the easier method to stabilise flames in actual afterburners of jet engines.

10.2.2 Swirler

This kind of flame stabilisation device imparts a tangential velocity to the inlet air flow, entering into the primary zone. As a result, the rate of entrainment and rate of mixing air and fuel jet increases providing improved flame stabilisation. Hence, the combustor can operate over a wide range of fuel-air ratio.

A schematic of an axial swirler is shown in Fig. 10.5 which is employed widely in current gas turbine main combustors. Generally, it is mounted in the combustor dome which creates a vortex flow as shown in Fig. 10.5. In the center of the vortex, a low static pressure is created which induces a recirculating (Fig. 10.5). This recirculating flow helps in stabilising the flame. Let us understand the

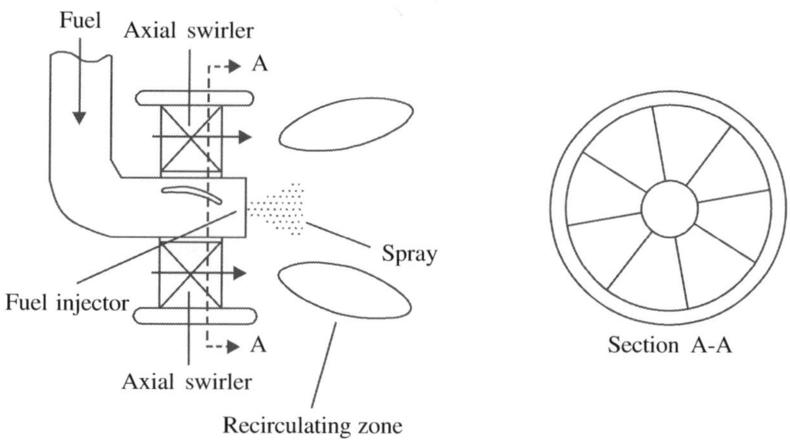

Fig. 10.5. Schematic of a typical axial swirler.

qualitative mechanism of flame stabilisation. Due to the recirculation of flow, an intense mixing zone similar to that of the bluff body flame stabilisers is formed, which not only transfers heat but also transfers active species to the incoming combustible mixture. As a result, this zone acts like a continuous ignition source which ensures good lean flame stability with high combustor efficiency and uniform exit temperature profile. Besides this, the swirling flow is usually supplemented by a series of primary air jets from the liners. The advantage of the swirler is that it produces very less pressure loss in comparison to a bluff body stabiliser. The use of swirler makes the combustor shorter and compact for which it is being used extensively in the main burner of modern gas turbine engines.

10.3 Main Combustor

In the last section, we discussed about the process of combustion that usually takes place in the gas turbine combustor. Let us now initiate a discussion on various components of the main combustor which is essential to provide its basic requirements. The present day gas turbine main combustor has undergone several changes in the past, resulting in a variety of basic combustor configurations. However, we will restrict our discussion to the diffusion flame based main combustor which seems to be the best choice for gas turbine engines till date. The distinguishing feature of this combustor is that fuel is directly injected into the air stream forming diffusion flames with fuel/air ratio closer to stoichiometric value at the flame boundary. As a result, it has an excellent combustion stability characteristic with high turndown ratio (ratio of maximum fuel flow rate to minimum stable fuel flow rate). But the main problem with this type of combustor is the emission levels produced, which are quite high as per modern emission standards.

Let us now look at a cross-sectional view of a typical diffusion flame based main combustor (Fig. 10.6). The outer container of this combustor is known as *casing*. The air after leaving the compressor enters into the inlet diffuser of the main combustor. The purpose of this inlet diffuser is to decelerate the incoming flow from the compressor to requisite velocity which will help in stabilising the flame. Besides this, it must provide the required amount of air to the primary combustion zone as far as possible with stable uniform flow field while recovering maximum possible dynamic pressure head. In fact, the designer has to compromise with the performance of inlet diffuser while attempting to achieve the desired compactness, low pressure loss and high degree of uniform flow.

The incoming air then is divided into the two streams namely (i) Primary and (ii) Secondary (dilution) streams, effected by *snout* as shown in Fig. 10.6. It also helps in streamlining the flow into the combustor *dome* and allows to use a large divergence angle for decelerating the incoming flow, thus, reducing its overall length. After passing through the snout (diffuser), the incoming air enters into the dome (see Fig. 10.6) which has a hemispherical shape. This is generally designed to produce high shear flow and high turbulent flow zone near the fuel injector for promoting atomization of the liquid fuel spray and rapid fuel/air mixing. The fuel injector surrounded by a swirler is placed at the center of this dome. Generally, air enters into the primary zone through the swirler. But some holes are provided in the dome itself to cool its interior surface. These holes are known as flare holes. On ignition of fuel/air mixture by an igniter, flame is established in the primary zone, contained inside the flame tube. The liners in the flame tube are provided with several rows of holes which also permit the introduction of secondary air into the combustion chamber to cool the combustion products as well as the flame tube. The liner is generally designed to withstand stress due to both thermal and pressure loads. In a typical gas turbine engine (see Fig. 10.7) 20% of total air enters into the primary zone through the swirler and dome which is meant to maintain stoichiometric fuel/air mixture. Note that 12% of the primary air is passed through the swirl while 8% of primary air is allowed to pass through the flare holes on the dome. The secondary air is again distributed and used for primary zone (20%), cooling (40%) and dilution (20%) purposes in the main combustor.

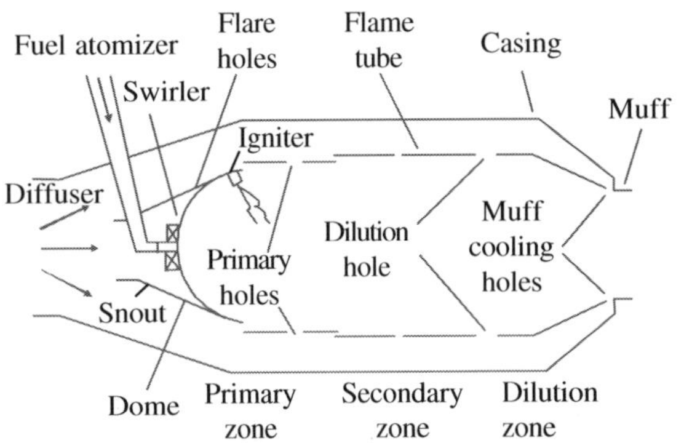

Fig. 10.6. Schematic of a typical gas turbine main combustor.

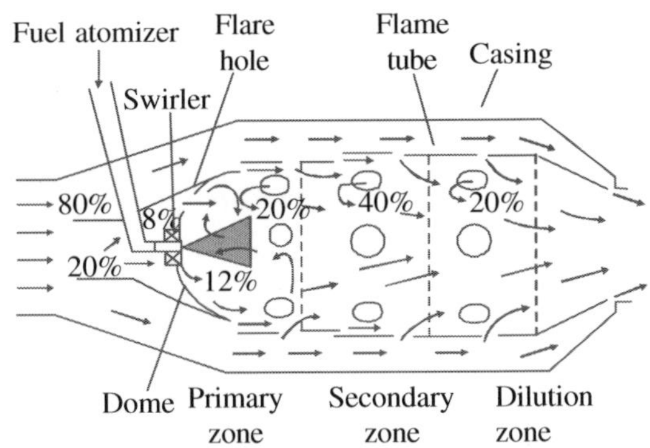

Fig. 10.7. Airflow distribution in a main gas turbine combustor.

10.4 Types of Main Combustor

Based on the geometry of the combustor, it can be broadly classified into three categories namely (a) Can, (b) Annular and (c) Can-annular.

10.4.1 Can Combustor

In this type, individual burners are placed in a circle around the engine axis as shown in Fig. 10.8b. The cutaway view of a typical can combustor is shown in Fig. 10.9 with igniter, injector, swirler and liners, and other requisite components. In other words, each can combustor is a separate entity in itself. This type of combustors were used in older

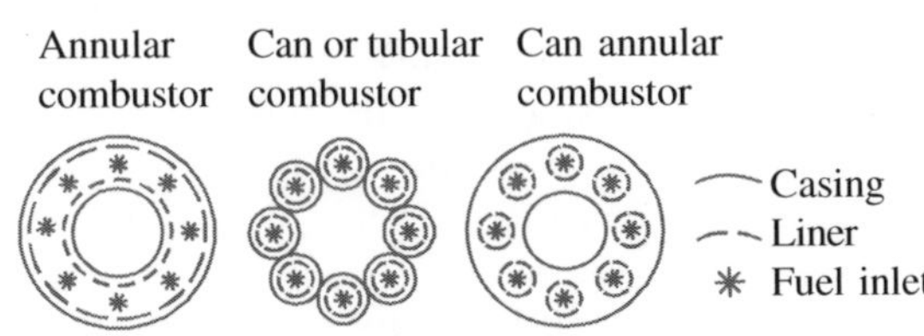

Fig. 10.8. The cross-sections of annular, can and can-annular combustor.

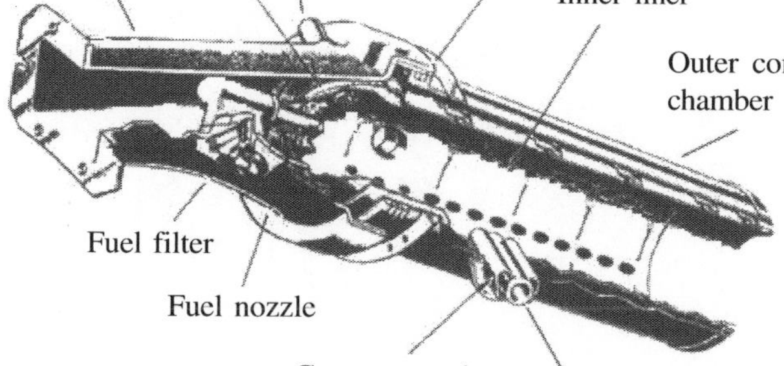

Fig. 10.9. Cut-away view of a can combustor [4].

aircraft engines. Modern aircrafts seldom use can type combustors because they do not use optimally available annulus area for releasing requisite heat input particularly for axial compressor operated engines. Furthermore, it is difficult to achieve uniform temperature distribution at its exit. The pressure drop is around 7% which is quite high. Since it only requires a small fraction of total engine fuel and air flow, production of a number of small can combustors is much easier than developing a single large combustion system. Besides this, replacement of damaged components is much easier and cheaper. In this case, it is quite easy to control fuel-air ratio. This type of combustor may be preferred for aircraft with centrifugal compressor, because the flow in centrifugal compressor is divided into separate streams in the diffuser which can lead to each can combustor. Can combustors find applications in small gas turbine engines such as auxiliary power units (APU), and lower power engines. However, it is not preferred for large size aircraft engines due to its undesirable characteristics in terms of weight, volume, and frontal area.

10.4.2 Annular Combustor

The annular combustor has a single flame tube with a concentric annular casing as shown in Fig. 10.10. The cutaway view of a typical annular combustor is shown in Fig. 10.10 in which full combustor is placed in the annular passage which connects the compressor delivery with the inlet of turbine. Air from the compressor enters the annular diffuser at the entrance of the combustor. Fuel is injected into the inner annular zones by a series of injectors as shown in Fig. 10.10. The liner geometry in this case is quite different which carries a large number of holes for stable combustion, cooling and dilution. This type of combustor geometry is best suited for axial compressor operated engines.

This arrangement makes the fullest use of available volume for combustion. It has lowest pressure loss (around 5%) among all the three types. It is also simple to design and cheaper to produce due to reduction of its linear area. For these reasons, annular combustors are being used in modern engines, for example in Olympus 593, PT 6, JT 15D, etc. However, it also has several disadvantages. Firstly, it is structurally weak due to large diameter and thin wall cylinder. This problem is much severe for larger engines. Secondly, it is quite difficult to confirm fuel-air distribution in spite of using number of fuel jets. The mal-distribution of fuel-air mixture can lead to non-uniform exit temperatures. Thirdly, it is not easy to develop and repair this combustor since it requires a large amount of air and fuel.

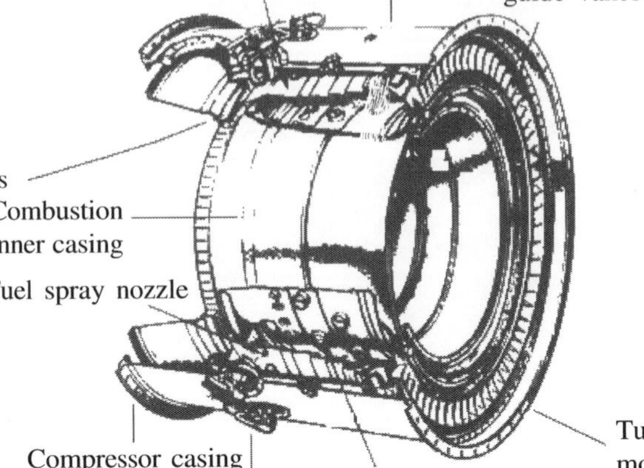

Fig. 10.10. Cut-away view of annular combustor (Rolls-Royce).

10.4.3 Can-Annular Combustor

This is a hybrid of both can and annular combustors. It has individual flame tubes, uniformly spaced in the annular casing as shown in Fig. 10.11. Such a combustor is called can-annular combustor. Fig. 10.11 shows a typical configuration of a can-annular combustor. Each flame tube has individual swirler, injectors, ignites and liners. Compared to can type, the annular space is appreciably better utilised. The fact is that a number of small flame tubes are used which eases the development and production problem as compared to annular combustor. However, there will be a symmetrical flows around the flame tube at its exit. The other disadvantages include expensive repair, thermal expansion of liners and non-uniform outlet temperature.

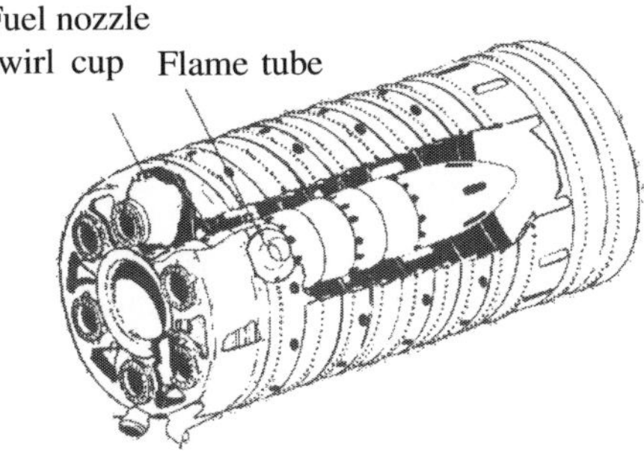

Fig. 10.11. Cut-away view of can-annular combustor (Pratt and Whitney (J57)).

Two important components of combustor which require special attention are (i) Fuel atomizer (injector) and (ii) Igniter. A detailed discussion about various types of atomizers is presented in next section.

10.5 Fuel Atomization

We know that in order to accomplish combustion of the liquid fuel, it must be converted into fine droplets so that evaporation and mixing of liquid fuel with air could be achieved easily within least possible time. This process of converting bulk liquid fuel into array of several smaller droplets is known as *atomization*. As discussed earlier, the main function of a fuel injector (atomizer) is to produce a high ratio of surface to mass in the liquid phase such that high evaporation rate can be accomplished. It must also produce proper fuel/air mixture ratio, which must be distributed uniformly across the primary zone. It should also produce nearly stoichiometric fuel-air mixture in the primary zone for ensuring stable combustion over the whole range of engine operating conditions, starting from idling to full load. Besides this, it must also meter the requisite amount of liquid fuel.

10.5.1 Desirable Characteristics of an Atomizer

An ideal fuel atomizer must have the following characteristics:

1. Better atomization over the whole range of fuel flow rates.
2. Quick response to changes in throttle setting.
3. Free from blockage due to either foreign particles or carbon formation.
4. Free from flow instabilities.
5. Low cost and light weight.
6. Low pressure drop across atomizer.
7. Higher discharge coefficient.
8. Easy to manufacture and installation.
9. Higher turn-down ratio.

10.5.2 Spray Formation Mechanism

We know that during the process of atomization bulk liquid is converted initially into a liquid sheet or jet. Basically this liquid sheet or jet gets disrupted subsequently due to various forces such as internal and external forces and surface tension force. Of course in the absence of any other disruptive forces, the surface tension force tends to pull the liquid in the form of sphere as it has minimum surface energy. However, the viscosity of the liquid affects the atomization process adversely as it opposes any change in the shape and geometry of the liquid. In contrast, the aerodynamic force helps in disrupting the liquid ligaments into droplets. In summary, it can be stated that when the sum of disruptive forces exceeds the overall surface tension force, the liquid is disintegrated into ligaments and subsequently smaller droplets. Note that this disintegration of liquid jet into droplet is governed by a nondimensional number known as *Weber number* that is given by

$$We_j = \frac{\text{Inertial force}}{\text{Surface tension force}} = \frac{\rho V_j^2 d_j}{\sigma} \tag{10.12}$$

where V_j is the velocity, d_j is the diameter of undisturbed liquid jet and σ is the surface tension of the liquid in contact with the surrounding gas. A liquid jet breaks up into droplets only when inertial forces are quite larger in comparison to the surface tension force. Therefore, the Weber number must be greater than unity for droplet formation. The jet Reynolds number that plays an important role in the formation of droplets is defined as

$$Re_j = \frac{\text{Inertial force}}{\text{Viscous force}} = \frac{\rho V_j d_j}{\mu} \tag{10.13}$$

The jet disintegration can be described comprehensively by another nondimensional number known as *Ohnesorge number* which is given by

$$Oh_j = \frac{\mu_l}{\sqrt{\rho \sigma d_j^2}} = \frac{\left[\left(\rho_l/\rho_g\right) We_j\right]^2}{Re_j} \tag{10.14}$$

This nondimensional number is sometimes known as stability number which helps in describing the various mechanism of jet breakup. Fig. 10.13 shows the variation of Ohnesorge number with the jet Reynolds number in which three regions can be identified. Region I is known as the *Rayleigh zone* in which the jet breakup is due to the effects of surface tension forces as the jet Reynolds number is low. The jet structure in this zone is predominantly varicose in which the jet breakup occurs only when jet disturbance wavelength is about 4.6 d_j as predicted by Rayleigh. With further increase in Re_j, the disintegration of jet is initiated by the jet oscillations with respect to the jet axis. Such a kind of mechanism is found to occur in Zone II in which sinuous waves in the jet are observed prior to its breakup into ligaments. This zone marks beginning of the influence of the ambient air and is also known as *wind-induced zone*. At high Reynolds number, liquid jet is disrupted within a short distance from the orifice. In this case, the jet breakup is caused due to the interactions of jet with ambient air combined with aerodynamic drag and flow turbulence. It has been observed that the size of ligaments decreases with increase in relative air speed and resulting in breakup into smaller droplets. As a result, average drop diameters in this zone are found to be much less than jet diameters.

It has been found out from experimental observation that the onset of droplet disintegration due to aerodynamic forces is likely to occur when the droplet Weber number is greater than about 12. The droplet Weber number is given by

$$We_d = \frac{\rho V_R^2 d}{\sigma} \tag{10.15}$$

where V_R is the relative velocity of ambient gas, d is the droplet diameter and σ is the surface tension force of the liquid in contact with the surrounding gas. But at larger Weber number, the droplet breakup is governed by boundary shearing phenomenon related to the Reynolds number. Such kind of phenomena is likely to occur when $\left(We_d/Re_d^{0.5}\right)$ is greater than 0.7: where $Re_d = \dfrac{\rho_g V_R d}{\mu_g}$ Question arises what is the simplest way to generate a fuel spray? If we could force liquid fuel with high pressure through a suitably shaped orifice, a conical spray of fine droplets as shown in Fig. 10.12 is formed. Such kind of atomizer is known as *Pressure Atomizer* (*Injector*). Note, we will be using often two names such as injector and atomizer interchangeable.

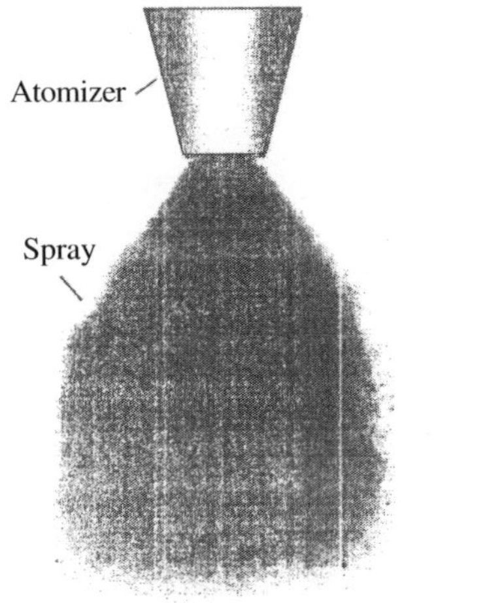

Fig. 10.12. Conical spray of a fuel atomizer.

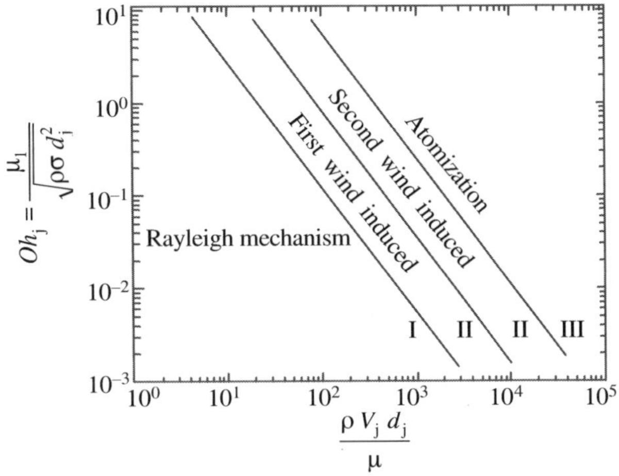

Fig. 10.13. Three stages of liquid fuel disintegration in a simple orifice [5].

We may ask ourselves, how is the liquid converted into droplets? Let us understand the mechanism of droplet formation in a qualitative manner. Liquid is forced through a small hole as shown in Fig. 10.12. Note that pressure energy is used to impart a high velocity to the liquid fuel. As a result, a liquid sheet/jet is created at the exit of orifice which quickly becomes unstable and subsequently breaks up into fine droplets. During droplet formation, the following processes shown in Fig. 10.14 may occur in sequence:

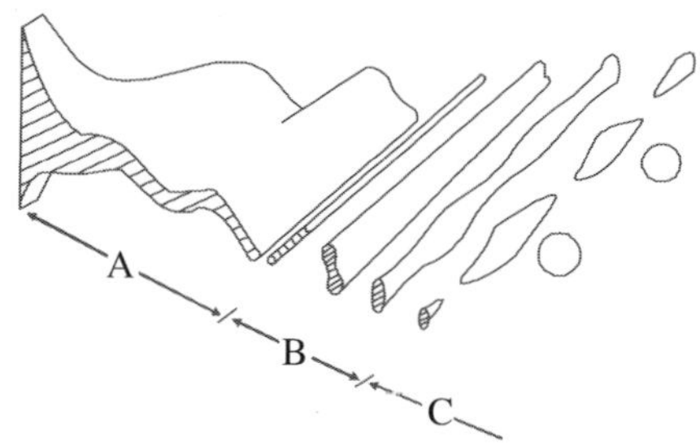

Fig. 10.14. Various processes involved in fuel atomization: (A) Growth of waves in sheet, (B) Formation of ligaments, (C) Break-off of ligaments and droplet formation [5].

- Stretching of fuel into liquid jet/sheet
- Appearance of ripples and protuberances
- Disintegration of liquid jet/sheet by atomizing forces (surface tension, inertia and aerodynamic forces)
- Formation of ligaments or holes in the sheet/jet
- Breaking up of ligaments into droplets leading to spray formation
- Agglomeration or shredding of droplets

10.5.3 Droplet Size Distribution

Most of the atomizers used in combustion applications can generate droplets in size ranging from 5 to 300 μm. In order to describe drop size distribution, one may resort to the statistical means as the spray contains large number of individual droplets. Unfortunately, it has not been possible to derive such a kind of function on theoretical basis. However, in engineering practice, it is essential to define the average droplet diameter such that it can represent the overall characteristics of the entire spray comprising of several droplet diameters. Such kind of droplet diameter can be used in design calculation of combustor. For example, while estimating evaporation rate, ignition time, minimum ignition energy, blowoff limits, etc, such kind of droplet diameter representing the overall characterisation of the spray will be very useful. This also helps in judging the quality of spray. Let us now learn how to estimate average size of droplet in a typical spray by considering an example given below:

Example 10.1 An experimentally determined droplet number and size distribution are shown in Table 10.1. Determine size distribution, cumulative number and volume fraction distributions. Estimate mean diameter (MD), area mean diameter (AMD), and sauter mean diameter (SMD).

Solution We can estimate various quantities from the data given in Table 10.1 for determination of average droplet diameter as given in Table 10.2. The most probable diameter is 46.5 μm corresponding to greatest number constant, $\Delta N_i = 0.244$.

The mean diameter (MD) is 46.98 μm. The other diameters are given below

$$\text{Mean diameter} = \text{MD} = \sum d_i \left(\Delta N_i\right) = 46.98\ \mu\text{m}$$

$$\text{Area mean diameter} = \text{A MD} = \sqrt{\sum d_i^2 \left(\Delta N_i\right)} = 53.2\ \mu\text{m}$$

$$\text{Volume mean diameter} = \text{VMD} = \left(\sum d_i^3 \left(\Delta N_i\right)\right)^{\frac{1}{3}} = 53.2\ \mu\text{m}$$

$$\text{Sauter mean diameter} = \text{SMD} = \frac{\sum d_i^3 (\Delta N_i)}{\sum d_i^2 (\Delta N_i)} = 61.67\ \mu\text{m}$$

Several average droplet sizes can be defined to characterise the spray in simple way. For example, SMD is used extensively in combustion problem as it represents the ratio of volume to surface area of the whole spray. Because, the requisite heat for evaporation of the droplet of spray is provided through the surface of droplet whereas its volume determines the liquid mass to be evaporated.

Table 10.1. Droplet distribution of a typical atomizer

S.No.	Droplet size (microns)	number
1	0-10	103
2	11-25	216
3	26-40	328
4	41-50	405
5	51-60	251
6	61-70	205
7	71-90	123
8	91-140	55
9	141-200	13
10	201-230	0

Table 10.2. Estimated values of terms to determine average diameter of the spray

S.No.	d_i(μm)	ΔN_i	$d_i \Delta N_i$	$d_i^2 \Delta N_i$	$d_i^3 \Delta N_i$	CVF(%)
1	5	0.061	0.305	1.525	7.625	0.0036
2	18	0.122	2.196	39.528	712	0.34
3	33	0.183	6.039	199.3	6576	3.48
4	45.5	0.244	11.102	505.14	22983.9	14.43
5	55.5	0.152	8.436	463.2	25707.76	26.69
6	65.5	0.125	8.188	536.3	35218.6	43.4
7	80.5	0.073	5.876	473.0	38077.9	61.58
8	115.5	0.033	3.881	440.2	50839.7	85.5
9	170.5	0.006	1.023	174.4	29738.8	100
10	215.5	0.0	0.0	0.0	0.0	0
Total		1.0	46.98	2831.9	20977	

10.6 Types of Fuel Atomizer

We know that the fuel atomizers play a very important role in controlling the combustion process in the combustor. According to sources of energy employed for acceleration of liquid fuel, atomizers can be classified into three categories namely (i) Pressure Swirl, (ii) Air-blast, and (iii) Rotary atomizer which have been designed and developed for gas turbine main combustors. We will discuss in detail about these types of atomizers in subsequent section.

10.6.1 Pressure Swirl Atomizers

We have already discussed about simple pressure atomizers where high velocity is imparted to the liquid fuel by applying high pressure to the liquid fuel. As a result, the liquid fuel is forced through small orifice and is disintegrated at high velocity into fine droplets. However, the angle of the spray cone in this atomizer is quite low (5-15°). Thus, it is not very suitable for gas turbine applications. Besides this, it is not very easy to achieve a well-controlled fuel distribution inside the combustion zone due to non-availability of exorbitant high pressure. But, if the liquid fuel can enter into the fuel nozzle tangentially, a wider angled cone is formed and subsequently breaks down into fine droplets. If the tangential component of the fuel is properly chosen, one can easily get a spray cone in the range of 80-100°, which provides a desired fuel distribution in the combustion zone. Such a kind of atomizer is known as *Pressure Swirl Atomizer.*

Let us see, what happens when fuel pressure deviates from the optimum pressure that provides desired degree of atomization of fuel. With increase in fuel pressure, the conical sheet of liquid will break into droplets nearer to the orifice. Then the spray will consist of droplets covering wide range of droplet diameters. If the droplets are too small, then it would not penetrate far enough into the adjacent air stream. On the other hand, if the droplets are too large, then evaporation time will be too much which leads to poor combustion efficiency. Generally, 50-100 μm diameter droplets are preferred in gas turbine combustors. Hence the effective minimum supply of pressure of the fuel would be such that it provides the desired degree of fuel atomization.

In practice, several types of pressure swirl atomizers are used. However, we will restrict our discussion to three types only namely (i) simplex, (ii) spill-return, and (iii) duplex.

(i) Simplex Atomizer

A schematic of a typical simplex pressure swirl atomizer is shown in Fig. 10.15. Note in this figure the fuel enters the vortex chamber tangentially. As it moves along the conical part of this swirl chamber, the tangential velocity of liquid fuel increases with decrease in radius, since the product of tangential velocity, V_θ and local radius inside this conical chamber remains constant, of course, for inviscid flow. Hence, the liquid fuel leaves the final nozzle orifice with a high tangential velocity. The actual cone angle is dependent on the relative magnitude of tangential and axial component of the velocity at exit. A liquid cone is formed at the exit and subsequently is disintegrated, forming fine droplets. It is also dependent on the injection pressure. Several stages in spray development with increase in pressure have been observed as shown in Fig. 10.16, starting from zero pressure to higher pressure which are as follows:

- Liquid fuel dribbles from the orifice.
- Liquid fuel takes a shape of a distorted pencil.
- A smooth hollow cone at exit of the orifice which is contracted at downstream by surface tension force taking a shape of an onion.
- The hollow cone forms a tulip shape.

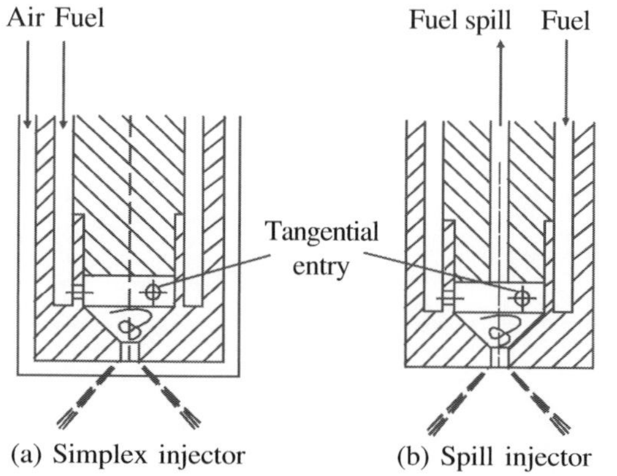

Fig. 10.15. Schematic of (a) simplex and (b) spill pressure swirl atomizers [6].

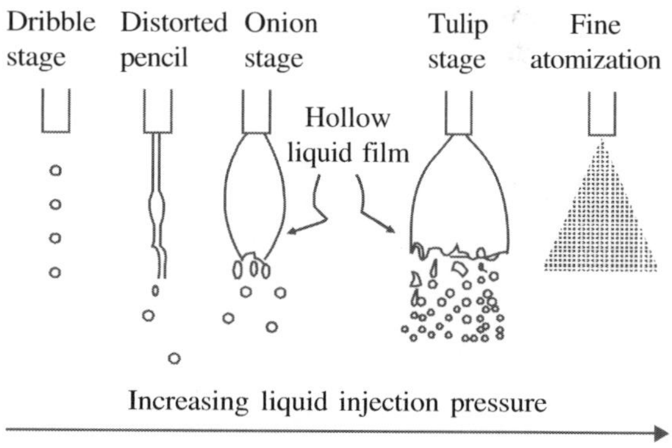

Fig. 10.16. Evolution of spray formation with increase in fuel injection pressure [5].

While designing such kind of atomizers, the tangential entry ports into swirl chamber must be arranged with good symmetry. Besides this, enough entry passage must be provided to avoid asymmetric disturbances in the resulting spray cone. The axis of the orifice must match with the axis of the swirl chamber. If these requirements are not met, the nozzle would not perform satisfactorily.

It can be appreciated from the above discussion that the physical processes involved in the atomization of liquid fuel in a pressure swirl atomizer is quite complex to predict droplet size distribution. Hence, designers rely on experimental data. It has been found out from experiments that the Sauter Mean Diameter (SMD) of a pressure swirl atomizer is dependent on mass flow rate of liquid fuel, the pressure drop across the atomizer, viscosity and surface tension of the liquid fuel. As mentioned earlier, it is very difficult to model such a kind of atomizer to predict droplet size distribution which can provide correct SMD. Hence, in practice semiempirical experimental relationship between SMD and the parameters mentioned above has been obtained by several researchers which can be used for design and development of the atomizer. However, most popular relationship by Lefebvre [5] is given below:

$$SMD = 2.25\,\sigma^{2.25}\,\mu_l^{0.25}\,\dot{m}_l^{0.25}\,\delta P^{-0.5}\rho_a^{-0.25} \tag{10.16}$$

where SMD is in μm and all others are in SI units. It can be noted that SMD is function of surface tension force, σ, viscosity, μ, mass flow rate of liquid fuel, dm_l, pressure difference across the injector, δP and density of ambient air, ρ_a.

The problem with this simple swirl atomizer is that there is no provision for controlling fuel flow rate because fuel flow rate can only be controlled by varying the pressure in the fuel supply line. Since the flow rate through an orifice is proportional to square root of pressure difference across it $\left(\dot{m}_f \propto \sqrt{\delta P}\right)$, for a fuel flow range of 10:1, the supply pressure must be varied in the range of 100 :1. Hence, if the burner is designed to operate optimally at full load, it would not provide desired droplets characteristics at low load. As a result, fuel rich condition may prevail near the orifice, leading to formation of soot (carbon) particles. In order to avoid this problem, a thin air shroud is generally provided as shown in Fig. 10.15 which also helps in disintegrating fuel cone into fine droplets.

In order to obtain good atomization of fuel over a wide range of fuel flow rates, the simplex atomizer is provided with a spill line as shown in Fig. 10.15b. The excess fuel can be spilled off from the swirl chamber. In this case, the flow through the orifice can be reduced by reducing the pressure in the spill line, while keeping higher supply pressure needed for good atomization. As a result, the range of optimum operation can be extended considerably by controlling both supply pressure of fuel line and spill line. But the problem with this atomizer is that undesirable heating and consequent deterioration of desired properties of fuels have been encountered particularly when a large quantity of fuel is recirculated to pump inlet.

(ii) Duplex

In order to extend the operating fuel flow rate range, without deteriorating the characteristics of spray, duplex atomizer as shown in (Fig. 10.17) has been designed and developed. This consists of two fuel manifolds, each supplying fuel to two separate orifices (see Fig. 10.17). The small central orifice is used alone when the flow requirement is quite small. The larger annular orifice is operated for higher flow rate situation. As a result, better characteristics of

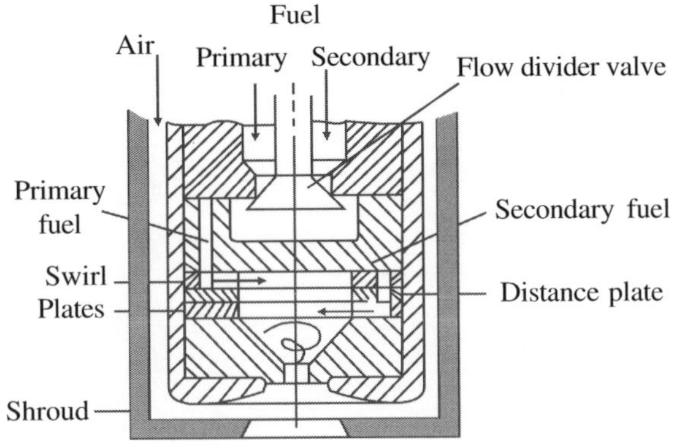

Fig. 10.17. Schematic of duplex pressure swirl atomizers [5].

spray are obtained over a wide range of fuel flow rates. Air is also supplied through the air shroud to prevent the formation and deposition of soot particles on the face of its orifice. Because of its wide range of operation, it is being used profusely in most of the modern large size gas turbine combustors.

10.6.2 Air Blast Atomizer

We have seen that higher pressure is required in case of pressure swirl atomizer to have better spray characteristics as required by gas turbine combustor. Since the velocity of flow in combustor is quite high, it can be used to disintegrate fuel sheet into fine droplets. This concept is used in air blast atomizer in which a high velocity air stream impinges on a relatively low velocity liquid either within the atomizer body or externally. Based on this principle, several air blast atomizers have been devised. In general, we will restrict our discussion to two specific types of air blast atomizers namely (i) Plain-jet air blast atomizer and (ii) Prefilming air blast atomizer, which have recently been used in gas turbine engines.

(i) Plain Jet Atomizer

The schematic of plain-jet atomizer is shown in Fig. 10.18(a). In this atomizer, discrete jets of fuel are impinged into the incoming high velocity air stream, which may be swirling air flow. As a result, the dynamic head of air is used to disintegrate liquid jets into fine droplets.

(ii) Prefilming Atomizer

The schematic of a typical Prefilming injector is shown in Fig. 10.18(b). In this case, the liquid fuel is spread across a prefilming surface to form a thin film. When this film of liquid leaves behind its surface edges, it is disintegrated into droplets due to action of shearing force created by high velocity incoming air stream. In order to achieve good atomization, both sides of liquid film must be exposed to high velocity air stream as shown in Fig. 10.18(b).

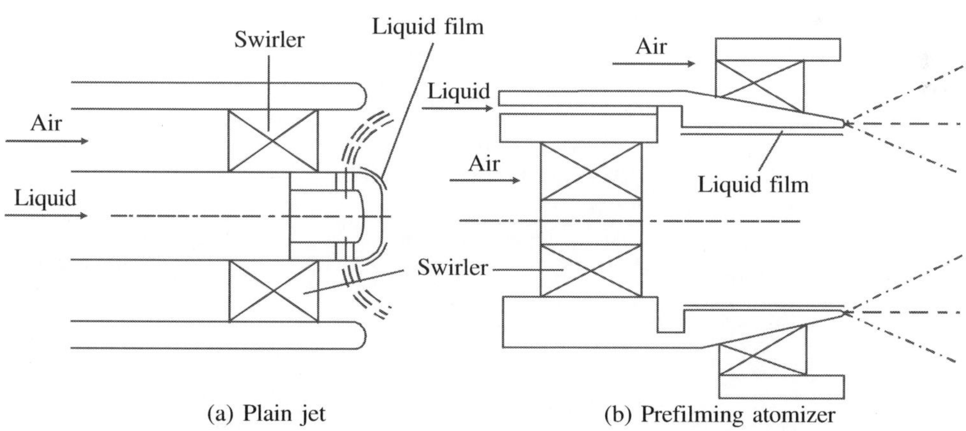

Fig. 10.18. Schematic of air blast atomizers (a) plain jet and (b) prefilming atomizer [6].

The semi-empirical relationship [5] obtained from experimental data for prefilming airblast atomizer that can be used for designing purpose is given below

$$SMD = 3.3\times10^{-3}\left(\frac{\sigma\rho_l\, d_f}{\rho_a^2\, V_a^2}\right)\left(1+\frac{\dot{m}_l}{\dot{m}_a}\right)+13\times10^{-3}\left(\frac{\mu_l^2}{\sigma\,\rho_l}\right)^{0.425} d_f^{0.575}\left(1+\frac{\dot{m}_l}{\dot{m}_a}\right)^2 \qquad (10.17)$$

where, d_f is the diameter of film being formed, ρ_l is the density of liquid fuel. Other symbols are earlier defined.

The advantages and disadvantages of air blast atomizers over swirl atomizers are enumerated below:

Advantages

- Requires low fuel injection pressure except at the start of engine.
- Ensures better mixing of fuel and air.
- Produces finer spray over a wide range of fuel flow rate.
- This reduces soot formation.
- Fuel distribution is determined by air flow thus independent of any changes in fuel flow rate.

Disadvantages

- Narrow stability limit due to intense mixing of fuel and air.
- Formation of larger droplets due to lower aerodynamic force at start-up. But these problems can be solved by using a pilot pressure nozzle.

10.6.3 Rotary Atomizer

In rotary atomizers, mechanical forces are used to produce thin sheets or jets of liquid fuel, which then get disintegrated into small droplets. A schematic of a typical rotary atomizer is shown in Fig. 10.19. Note that rotating cup, discs or bores are employed to impart a centrifugal force to the liquid fuel. When liquid fuel is passed through the inner surface of a rotating cup, the liquid flows toward the edges. Since the cup is rotating, centrifugal motion of the fuel forces the liquid to leave the sharp lip of tapered edge as a thin fuel sheet, which is subsequently disintegrated into small droplets. The ratio of the axial to tangential flow velocity determines the resulting spray cone angle. However, when fuel flow rate is quite low, the surface of rotary device is not completely wetted. Thus, differential droplet sizes are produced. This kind of atomizer is widely employed in industrial gas turbines particularly, which uses high viscous fuel oils. Its application is restricted to small gas turbine engines for propulsion. Most of the time, an air flow is combined with rotary atomizer to assist in disintegrating of fuel sheet into droplets.

In similar way, atomization can be achieved by applying the fuel to the face of spinning disc or by feeding the fuel to a radial bores in disc or in the wall of hollow shaft as shown in Fig. 10.19.

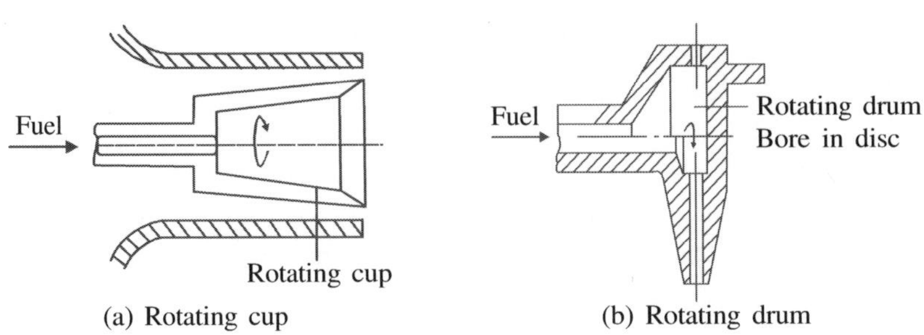

Fig. 10.19. Schematic of rotary atomizers.

In this case a finite number of liquid jets issuing from the bores, rotate in the combustor volume and produces a fairly uniform fuel distribution in the combustion zone. This principle is known as *slinger atomization.* It has been used in small gas turbine engines for *aero propulsion* due to its simplicity and efficacy.

10.7 Igniter

We know that igniter is used to add certain amount of initial energy required to initiate the combustion process in the combustor. This can be accomplished with the help of a spark plug or glow plug. On application of certain high voltage across the electrodes of a spark plug, initial energy is added to fresh fuel-air mixture. As a result, local gas around the region spark plug attains high temperature creating high energy species and ions. These high energy species diffuse into adjacent region and cause chemical reactions to occur liberating heat. This generated heat will make the chemical reactions to proceed to its adjacent area and thus establish a flame. But if the amount of heat energy is less than the heat being diffused away, then the ignition would not be successful. Hence a minimum energy must be supplied so that ignition of fuel-air mixture can be accomplished successfully. This minimum ignition energy is dependent on the type of fuel used, equivalence ratio, initial temperature and pressure. The variation of minimum ignition energy with equivalence ratio is shown in Fig. 10.20 for several hydrocarbons. Interestingly, it can be noted from this figure that the minimum ignition energy does not always occur at stoichiometric mixture ratio as one would expect.

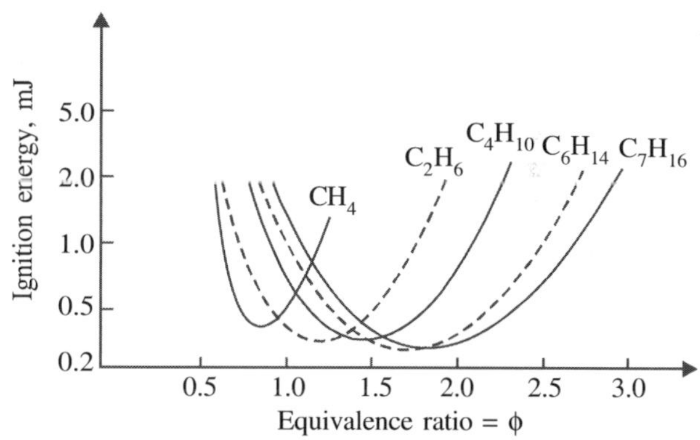

Fig. 10.20. The variation of minimum ignition energies with equivalence ratio at standard temperature and pressure [8].

From both the experimental data and simple theory, it has been established that ignition energy is inversely proportional to square of pressure for hydrocarbon/air systems ($IE \propto P^{-2}$). Hence higher amount of ignition energy will be required to initiate combustion at higher altitude. Thus, igniter for an aircraft gas turbine combustors must be designed to meet two additional requirements namely

(i) It must ignite the combustor for successful re-establishment of flame at higher altitude in case of flame extinguishments.
(ii) It must ignite at idle power with high water content air that may be encountered during heavy rain.

Generally, the high-tension spark plugs can be used to establish combustion at ground level flight. But this kind of spark plug may not be adequate to reestablish a flame at higher altitude due to its low atmospheric pressure. Hence, a spark plug capable of producing much greater energy must be employed to ensure successful ignition at high altitude and other adverse situations.

10.7.1 Surface Discharge Igniter

In recent times, surface discharge igniters are being used profusely in aircraft engines. A schematic of a typical surface discharge igniter is shown in Fig. 10.21. It consists of two electrodes one at the center and other at outer periphery. Both the electrodes are separated by ceramic insulation except near the top region. In the top region, a layer of semiconductor material is used for surface discharge. Generally the electrodes are made of inconel chromel-D, tungsten or any of the various nickel alloys depending on specific service requirement. The insulators are made of an aluminum-oxide ceramic with a diamond like hardness which provide mechanical strength, high voltage insulation and rapid heat conductivity. The sizes and shapes of igniters have not been standardised yet. Therefore, several variations in igniters can be found in different aircraft engines.

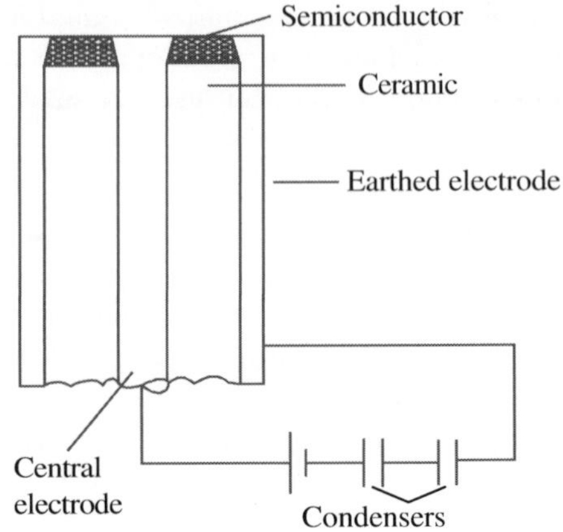

Fig. 10.21. Schematic of typical surface discharge igniter.

When the voltage is applied to the condenser, current flows through the semiconductor at the surface. This semiconductor becomes incandescent and provides an ionised path of low resistance such that energy stored in the capacitor can be discharged. Once ionisation takes place, the main discharge occurs as an intense flash over, thus ignition of fuel/air mixture is accomplished. Generally, about three joules per second ignition energy at spark interval of 60 to 100 times per minute is used in most modern aircraft engines. However, high energy current may cause more rapid erosion of igniter electrodes than encountered in piston engine spark plugs. But it would not pose any problem since igniter in gas turbine engine is operated for a very short duration. Keep in mind that the location of igniter in the combustion chamber is very critical if one wants to have a good performance and long life. It must be placed at outer edge of fuel spray such that ignition can be successful. At the same time, a layer of cooling air must prevail near the igniter, which can protect the igniter being subjected to intense excessive heating. It must be kept in mind that igniter must manage to initiate combustion process in a combustor successfully either for starting condition or at high altitude relight conditions. Another igniter known as *glow plug* is used in some models of gas turbine engines. This consists of mainly a heating element, which makes this plug to glow when current is passed through this coil. This glow plug attains an extremely high temperature at a faster rate which ignites the fuel spray in combustor.

In case of an afterburner, ignition of the fuel spray laden air is achieved in a different manner than that of the main burner. Generally pilot burner is used for ignition of fuel air mixture in afterburner. Generally this pilot burner (torch), consisting of spark igniter and fuel nozzle is placed down stream of the spray bar as shown in figure the fuel-air mixture prevailing in the after burner. This ignition torch burns continuously which ignites the fuel-air mixture prevailing in the after burner. Recently a simple technique of using platinum based igniter is devised which acts as a catalysis to ignite the fuel-air mixture prevailing in the after burner. Besides this, igniters used in the main burner can be employed in the after burner. We will now discuss in detail about an afterburner in next section.

10.8 Afterburner

We can recall from the cycle analysis sections that afterburner are used in an aircraft when the thrust of the engine is to be augmented without increasing the size (frontal area) and weight of the engine exorbitantly. Of course, the fuel consumption and TSFC are increased considerably due to the operation of the afterburner. The afterburner augments the thrust by the release of thermal energy due to burning of fuel in the exhaust stream of the turbine. Keep in mind that there will be enough oxygen available in the gas stream at the inlet of the afterburner to sustain combustion. This availability of requisite oxygen level and higher inlet temperature helps the designer to accomplish the afterburning of liquid fuel in a very simple configuration as compared to the main burner. When the fuel is sprayed into this hot oxygen laden gas stream, combustion takes place releasing large amount of heat which is subsequently expanded in the nozzle resulting in increased exhaust velocity. As a result, there will be enhancement of thrust of the engine. But the heat released in relatively high Mach number flows not only incurs exorbitant pressure losses but also limits the amount of heat being released at extreme conditions. Such kind of phenomenon that limits the heat release is known as *thermal choking*. Hence, provisions are made to use a diffuser between the turbine and afterburner to reduce inlet velocity to help in stabilising the flame and increasing heat release rate in the afterburner.

Generally in modern afterburner, approximately 50% of thrust augmentation can be achieved at the expense of 150% increase in fuel flow rate. As a result, the application of afterburner is restricted to shorter period particularly during take off, climb, maximum burst of speed, etc. Sometimes the afterburner is used for transonic acceleration from Mach number of 0.9 to 1.4 like in the Concord engine. Keep in mind that an engine with afterburner must be equipped with a variable area exhaust nozzle in order to provide optimum performance both under afterburning and non-afterburning operation. The specific requirements of an afterburner along with general features of a main combustor are:

- Large temperature rise across the afterburner.
- *Low dry loss*: Additional pressure losses do occur during cold flow operation due to restriction caused by flame holders, spray bars, liners, etc. This loss is termed as dry loss.
- Wide temperature modulation for better control of thrust augmentation.
- High combustion efficiency with no combustion instabilities.

The configuration of typical afterburner is shown in Fig. 10.22. Note that this after-burner mainly consists of the following components (i) diffuser, (ii) spray bar, (iii) torch igniter (pilot burner), (iv) flame holders, (v) cooling liners, and (vi) variable area exhaust nozzle.

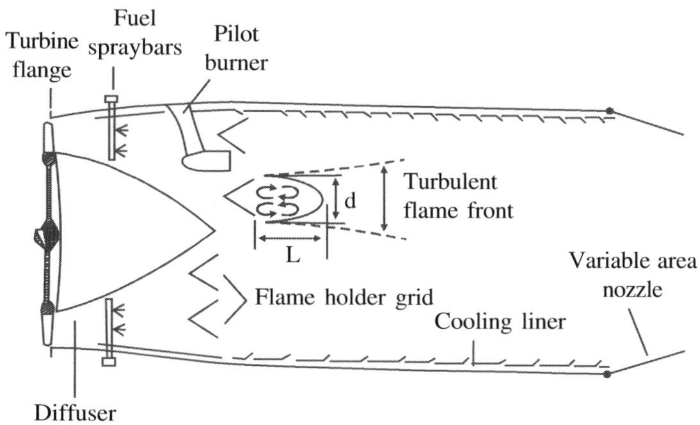

Fig. 10.22. Schematic of a typical afterburner.

The gas from the turbine outlet enters into the diffuser (Fig. 10.23) of the after-burner in which the gas is deswirled and decelerated to a lower velocity. Then the fuel is injected into the hot gas with the help of spray bar (Fig. 10.23) forming fine sprays of liquid fuel. In fact, the combustion process can be initiated by an igniter (pilot burner). The flames formed by the combustion processes are contained inside the afterburner with the help of bluff-body flame stabilisers. Flame holders usually in the form of several concentric rings with V-cross sectional shapes are used in modern afterburner. These flame holders stabilise the flame front by creating a recirculating zone as shown in Fig. 10.23. These recirculating zones enhance the blow of limit off the flame considerably.

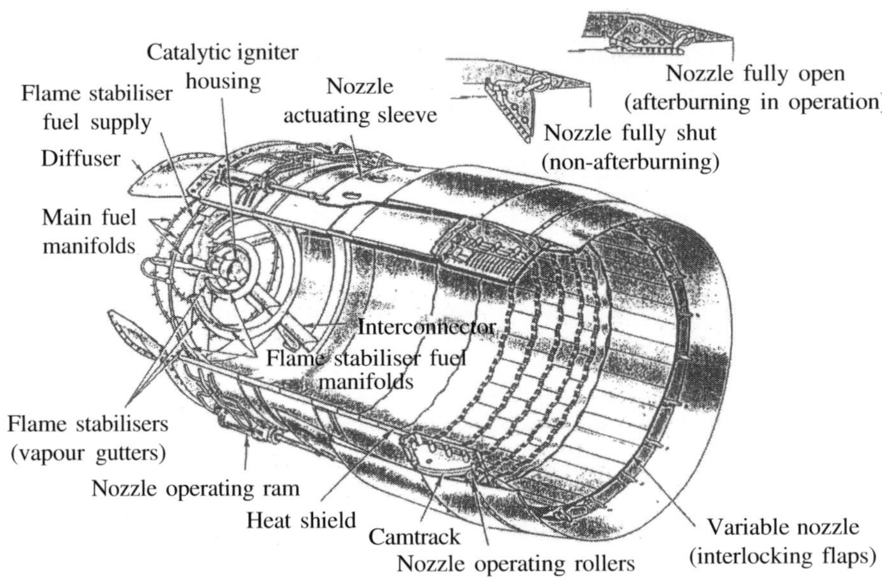

Fig. 10.23. Cutaway picture of an afterburner (Rolls Royce).

Secondary air is allowed to pass through the cooling liners, which helps in stabilising the flame. But its main function is to protect the wall of the afterburner and provide air required for cooling the exhaust nozzle. Besides this, the liners also prevent acoustic combustion instabilities sometimes accompanied with howling/screeching sound.

The performance parameters of an afterburner are similar to that of the main combustor, which will be discussed in detail in the next section.

10.9 Gas Turbine Engine Fuel and its Properties

Generally Aviation Turbine Fuel (ATF) which is mainly composed of refined kerosene is used as a standard fuel for aircraft gas turbine engines. Fuel for any engine is normally chosen on the basis of some important considerations such as fuel availability, its cost, transportability, engine performance requirements and etc. Note that ATF is a mixture of aliphatic (straight chain) and aromatic (ring structure: C_nH_{2n-6}) hydrocarbon compounds which is approximated as C_nH_{2n+2}, where n is varied from 8 to 10. It can be noted that a 'wider-cut' fuel, a under refined kerosene consisting of more number of hydrocarbon compounds were used in the early gas turbine engines. It is conjectured that such kind of fuel may find applications in future as it is cheaper to process such kind of fuel. Based on application, aviation turbine fuels can be classified into (i) military (JP series) and commercial (Jet-A, Jet-A1, Jet-B) grades. Till date total ten fuel types under JP series and three types under commercial grades are being devised as shown in Table 10.3 along with their respective applications. Historically, JP-1 is the first fuel developed for gas turbine engines. It has a low vapour pressure which incurs less fuel loss due to evaporation. But it is quite difficult to ignite this fuel particularly at high altitude. Besides this, it has high viscosity. Due to several reasons it has been outdated by other new jet fuels. JP-3 is a mixture of kerosene, aviation gasoline and diesel oil whose characteristics lies between those of kerosene and aviation gasoline. It is considered to be quite cheap as compared to JP 1 and JP 2. But its nonuniform composition affects the performance of combustor significantly. Subsequently JP-4 was developed to improve its vapour pressure characteristics. It has combined characteristics of kerosene and aviation gasoline. In order to store the aviation fuel, a high flash point low vapour pressure kerosene type fuel is developed which is known as JP-5. It is generally preferred in the Navy as it has best hazard free storage capability. It has potential to be used in high speed vehicles where fuel can be heated safely with aerodynamic heating. JP-6 is developed to meet the demand of higher temperature for supersonic flight. It has lower volatility and higher thermal stability than JP-4. But it has lower freezing point than that of JP-5. JP-7 has lower flash point which is most suitable for hypersonic vehicles. Jet A has similar properties than that of JP-4. Besides this, hydrogen has already been used in B-57 bomber aircraft which is contemplated to be future fuel for aircrafts. Not only it is considered to be greener fuel but it has higher calorific value (116 MJ/kg). However, issues of safety and price slabs will dictate use of hydrogen as a variable aviation fuel in future.

Table 10.3. Types of aviation turbine fuels

Sl Fuel	Droplet size Type	Number Applications and its characteristics
JP-1	Kerosene	Military, first ATF
JP-2	Wide-cut	Military, experimental ATF
JP-3	Wide-cut	Military, around 1/3 kerosene, 2/3 gasoline
JP-4	Wide-cut	Military, very low freezing point, similar to Jet-B
JP-5	Kerosene	Navy, high flash point
JP-6	Wide-cut kerosene	Military, low freezing point
JP-7	Kerosene	Hypersonic engine, low volatility
JP-8	Kerosene	Military, similar to Jet-A1
Jet-A	Kerosene	Commercially use in USA
Jet-A1	Kerosene	Commercial use in other countries and India, lower freezing point than Jet-A
Jet-B	Wide-cut	Commercial use, lower freezing point than Jet-A1

Generally the properties of commercial aviation fuel are characterised by ASTM standard. Some of the important properties of jet fuels pertaining the combustion characteristics such as calorific value, specific gravity, boiling point, flash point and freezing points are provided in Table 10.4. The basic definitions of some of these properties are given below:

Calorific Value

The calorific value of liquid fuel can be determined by using bomb calorimeter. Generally, a stainless steel container (bomb) is surrounded by a large temperature bath. The liquid fuel is burnt in the bomb in presence of oxygen at about 2.5 MPa. The change in temperature of water bath due to burning of a known quantity of fuel is measured using high precision thermometer, which provides the calorific value of the fuel. The calorific values of typical liquid aviation fuels are given in Table 10.4. It can be noted that calorific value of the most of aviation fuels is around 43 MJ/kg.

Table 10.4. Properties of aviation turbine fuels [11, 12]

Property	JP-1	JP-3	JP-4	JP-5	JP-6	JP-7	JP-8	Jet-A	Jet-A1	Jet-B
Aromatic (%V)	—	25	10	19	25	3	18	23	20	—
Naphthenes (%V)	—	—	29	34	—	32	20	29	1.5	—
Paraffins (%V)	—	—	59	45	—	65	60	—	—	—
Olefins (%V)	—	5	2	2	—	5	—	—	5	—
Sulphur (%)	—	—	0.3	0.3	—	—	0.3	0.07	0.25	0.4
H/C ratio (molar)	—	—	2	1.92	—	2.07	1.91	—	—	—
Sp. gravity (max)	0.85	0.78	0.802	0.845	0.84	—	0.71	0.83	0.83	0.802
Initial BP (K)	—	—	333	460	438	453	438	—	450	429
End BP (K)	—	—	514	530	—	523	538	—	553	573
Flash point (K)	316	—	—	336	341	523	538	—	311	—
Freezing point (K)	213	213	223	233	255	223	233	233	226	222
Viscosity (*cp*) at 288 K	—	—	0.7	1.8	—	1.7	1.8	—	—	—
Density (kg/m^3)	—	—	777	810	—	808	795	820	808	801
Cal. value *MJ/kg*	—	—	43.6	43.2	—	43.9	43.2	43.2	42.8	42.8

Specific Gravity

The density of liquid fuel can be expressed in nondimensional form known as specific gravity (SG), which is defined as the ratio of mass density of fuel and mass density of water at the same temperature. Generally, the reference temperature for fuel and water is taken as 288.8 K. It must be noted that fuel having low SG will have a higher APISG and vice versa. The specific gravity of certain liquid fuels is given in Table 10.4. Densities of liquid hydrocarbon fuels are related to their molecular structure and molecular weight by their carbon-hydrogen ratio. As the hydrogen to carbon ratio for paraffin is higher, its specific gravity is higher than aromatic fuels.

Flash Point

It is the minimum temperature at which liquid fuel will produce sufficient vapours to form a flammable mixture with air and can produce a momentary flame (flash) when an external heat source is brought in contact with vapour. It indicates the maximum temperature at which liquid fuel can be stored without any fire hazard. The flash points of some of the aviation fuel is given in Table 10.4. It can be noted that JP-5 has the highest flash point temperature.

Fire Point

In contrast, it is the minimum temperature at which liquid fuel produces sufficient vapours to form a flammable mixture with air that continuously supports combustion establishing a flame instead of just flashing, even after ignition is withdrawn. Note that for the same fuel, fire point is always higher than the flash point.

Smoke Point

It is a measure of the tendency of a liquid fuel to produce carbon particle known as soot. Generally, it is measured by burning fuel in a special wick lamp in which the flame height is increased slowly till it starts producing smoke. The maximum height in terms of millimeter of smokeless flame at which flame starts smoking is termed as smoke point. Hence, higher the smoke point, lower will be the tendency of the fuel to produce soot particles.

Several other properties such as freezing point, pour point (lowest temperature at which fuel becomes fluid), thermal stability lubricity, luminosity number (indicating flame visible radiation), sulfur content, kinematic viscosity, vapour pressure are to be considered during characterisation of aviation fuel. Apart from these, some other secondary characteristics such as microbial growth rate, contamination electrical conduction, coefficient of thermal expansions, corrosiveness and etc are being assessed for characterisation of aviation fuels.

10.10 Performance Parameters of Combustor

The performance of a combustor can be assessed by the following parameters, namely (i) Total pressure loss, (ii) Stability limit, (iii) Exit temperature distribution, and (iv) Combustion efficiency.

10.10.1 Total Pressure Loss

Both main burner and afterburner incur undesirable losses in total pressure. These pressure losses in combustor are mainly caused due to two distinct reasons: (i) frictional loss and turbulent flow mixing and (ii) gas dynamic loss due to combustion (heat addition). Frictional loss are generally contributed by inlet diffuser, burner dome, liner and coolant holes. Besides this pressure losses occur due to the effect of flameholder drag. The turbulence intensity, which is helpful in increasing mixing and combustion, adds frictional losses to the system.

In this subsection, we will carry out a simplified analysis considering the combustor as a constant area duct to predict total pressure losses. This analysis will be more suitable for afterburner. However it can be as well used for the main burner. For this analysis, we will use mass, momentum, energy conservation equations along with equation of states with the following assumptions.

1. One dimensional steady flow.
2. Negligible small fuel mass rate as compared to air mass flow rate.
3. Calorifically perfect gas.
4. No work interaction.
5. No gravitation forces.
6. Ideal gas law.
7. Constant thermo-physical properties.

Let us consider a control volume with heat addition as shown in Fig. 10.1. By invoking the integral conservation equation of mass momentum, energy between station (1) and (2), we can write down following equation.

Mass conservation:
$$\rho_3 V_3 = \rho_4 V_4 \tag{10.18}$$

Momentum conservation:
$$P_3 + \rho_3 V_3^2 = P_4 + \rho_4 V_4 + \frac{1}{2} C_D \rho_3 V_3^2 \tag{10.19}$$

Energy conservation: $$h_3 + \frac{V_3^2}{2} + q = h_4 + \frac{V_4^2}{2} \qquad (10.20)$$

Equation of states: $$P_3 = \rho_3 R T_3; \quad P_4 = \rho_4 R T_4 \qquad (10.21)$$

where, q is the amount of heat added per unit mass (kJ/kg). The flow properties, P_3, ρ_3, T_3 and V_3 can be obtained by solving the above equations for known upstream flow properties and amount of heat additions, q.

By using the basic definition of Mach number, the momentum equation can be rewritten as given below

$$\frac{P_4}{P_3} = \frac{1 + \gamma_3 M_3^2 \left(1 - C_D/2\right)}{1 + \gamma_4 M_4^2} \qquad (10.22)$$

By using equation of state and mass conservation equation, we can derive expression for the static temperature ratio across the combustor as given below

$$\frac{T_4}{T_3} = \left(\frac{P_4}{P_3}\right)^2 \frac{\gamma_4}{\gamma_3} \frac{M_4^2}{M_3^2} \qquad (10.23)$$

Using the above equation and energy equation, we can have an expression for total temperature as

$$\frac{T_{t4}}{T_{t3}} = \frac{T_4}{T_3} \frac{1 + \frac{\gamma_4 - 1}{2} M_4^2}{1 + \frac{\gamma_3 - 1}{2} M_3^2} \qquad (10.24)$$

Combining Eq. 10.22 and Eq. 10.24, we can have

$$\frac{M_4^2 \left(1 + \frac{\gamma_4 - 1}{2} M_4^2\right)}{\left(1 + \gamma_4 M_4^2\right)^2} = \frac{\gamma_3}{\gamma_4} \frac{M_3^2 \left(1 + \frac{\gamma_3 - 1}{2} M_3^2\right)}{\left[1 + \gamma_3 M_3^2 \left(1 - \frac{C_D}{2}\right)\right]^2} \frac{T_{t4}}{T_{t3}} = \$ \qquad (10.25)$$

Note that the above equation for M_4 is expressed in terms of upstream variables and prescribed stagnation temperature ratio, $\frac{T_{t4}}{T_{t3}}$. This can be expressed in the form of quadratic equation as given below

$$\left(\gamma_4 - 2\$ \gamma_4^2 - 1\right) M_4^4 + \left(2 - 4\gamma_4 \$\right) M_4^2 - 2\$ = 0 \qquad (10.26)$$

By considering positive root of the solution for above quadratic equation, we can get an expression for M_4^2 as given below

$$M_4^2 = \frac{2\$}{1 - 2\gamma_4 \$ + \sqrt{1 - 2\left(\gamma_4 + 1\right)\$}} \qquad (10.27)$$

By evaluating the M_4, we can determine the desired total pressure ratio across the burner, π_b by following equations

$$\pi_b = \frac{P_{t4}}{P_{t3}} = \frac{P_4}{P_3} \frac{\left(1 + \frac{\gamma_4 - 1}{2} M_4^2\right)^{\frac{\gamma_4}{(\gamma_4 - 1)}}}{\left(1 + \frac{\gamma_3 - 1}{2} M_3^2\right)^{\frac{\gamma_3}{(\gamma_3 - 1)}}} \tag{10.28}$$

It can be noted that total pressure ratio across the burner, π_b can be always less than unity. For two cases of constant total temperature ratio across the burner, a parametric study is carried out by varying inlet Mach number, M_3 from 0.01 to 0.3 which is shown in Fig. 10.24 along with exit Mach number M_4. It can be noted that total pressure ratio across the burner, π_b increases with inlet Mach number and heat release rate. There is also increase exit Mach number, M_4 which is resulted due to heat release rate and frictional losses. For higher heat release rate, inlet Mach number, M_3 must be restricted to a lower value in order to keep total pressure ratio across the burner, π_b below 0.1 resulting due to increase in gas total temperature and frictional losses. This restriction of low Mach number can lead to stringent design limitations on the requisite burner cross-sectional area, particularly in an afterburner.

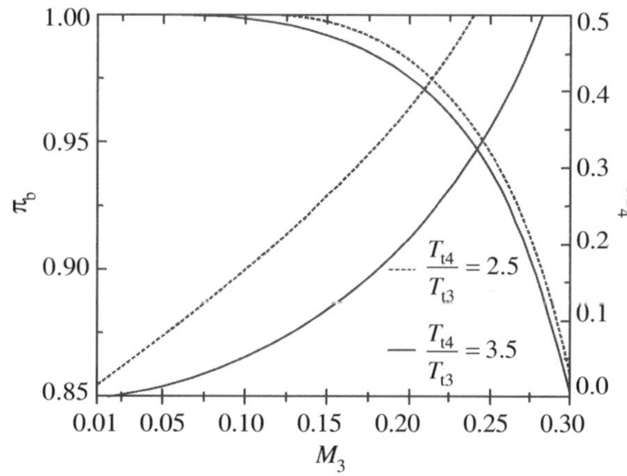

Fig. 10.24. Variation of pressure ratio across combustor, π_b with inlet Mach number, M_3 for $\tau_b = \frac{T_{t4}}{T_{t3}} = 2.5$ and 3.5.

10.10.2 Combustion Stability

We know that there is a range of fuel-air ratios for a particular combustor beyond which the combustion process is unstable. Hence combustion stability becomes a very important performance parameter. Combustion stability is basically conceived as the ability of the combustor to sustain the combustion process by itself in continuous manner over a wide range of fuel-air ratios. Stable combustion process can become unstable if either too lean or rich mixture ratio is used. This happens basically due to the imbalance between heat release rate and heat required to raise the temperature of the incoming fuel and air well enough for self sustenance of combustion. Such a situation may lead to even flame blowout. Of course, instability can occur even below this limit fuel-air mixture. Such kind of instability can lead to the rough running of the combustor, which not only decreases combustion efficiency but also sets up aerodynamic pulsation throughout the combustor. This unstable operation of combustor if allowed for more number of hours may reduce the life of the combustion chamber considerably. Therefore it is important to find out a criterion to judge the combustion instability. The process involved in combustion is quite complex to analyze either analytically or numerically. However, we will restrict our discussion to a simple model known as the *stirred reactor model* to simulate the combustion instability in gas turbine engine described as follows.

10.10.3 Stability of Flame in Gas Turbine Combustor

A stirred reactor is one in which fuel-oxidiser in the entire combustor are assumed to be completely mixed and combustion phenomenon is dominated by reaction processes similar to that is occurred in the premixed flame. At steady state, let $\dot{m}$ be mass flow rate of fuel-air mixture, in which oxidiser is reacted with the fuel consumption rate per unit volume $\dot{m}_f'''$ releasing certain amount of heat. Note this heat release rate $\left(\dot{m}_f''' \cdot V \cdot \Delta H_C\right)$ raises the temperature of product gases from T_3 to T_f as given below:

$$\dot{m}C_p\left(T_f - T_3\right) = \dot{m}_f''' \cdot V \cdot \Delta H_C \tag{10.29}$$

where $\dot{m}_f'''$ is the fuel consumption rate per unit volume, V is the effective volume of combustor and ΔH_C is the heat of combustion and T_f is the flame temperature. Considering a simple single step reaction as $f\,F + oOx \rightarrow pP$, we can write down an expression for $\dot{m}_f'''$ as given below:

$$\dot{m}C_p\left(T_f - T_3\right) = A_f \cdot Y_F^{n1} \cdot Y_O^{n2} \cdot P^n \cdot e^{-\left(\frac{E}{RT_f}\right)} \cdot V \cdot \Delta H_C \tag{10.30}$$

where A_f is the preexponential factor, E is the activation energy of the reaction, P is the chamber pressure, n is the order of reaction which varies from 1.8 to 2.4 for typical hydrocarbon-air mixture, Y is the mass fraction of species (fuel or oxidiser), and R is the universal gas constant. Now let us plot heat loss (LHS) which appears as a straight line against flame temperature, T_f as can be seen in Fig. 10.25. All these straight lines shown in this figure represent rate of heat removal corresponding to different values of $\dot{m}$. In this plot, the RHS representing heat release rate is also shown. At T_{ad}, RHS becomes zero as either fuel or oxidiser attains zero value. Note that the reaction rate varies with fuel-air ratios much like their respective burning velocities.

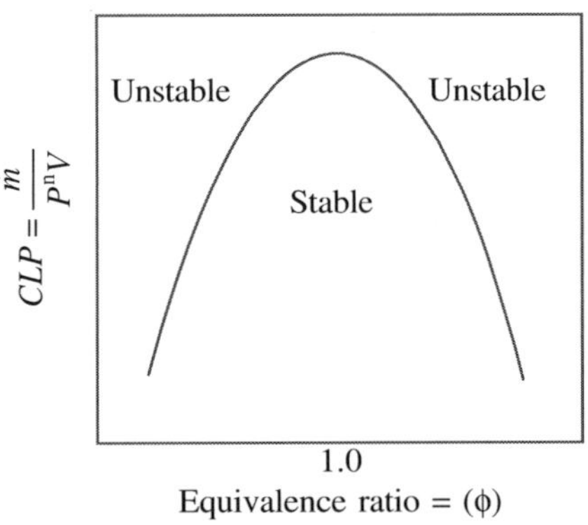

Fig. 10.25. Stable operation of a typical combustor.

The intersection point, B of these two curves corresponds to an operating point. Unfortunately, this point, B is an unstable one. Let us see how it is an unstable point. Suppose by some mechanism one disturbs the system by raising the temperature beyond B, then heat generation is more than the heat carried away by hot gases $\dot{m}\,C_p(T_f - T_o)$.

This causes further increase in temperature and it goes on till point C is reached. At point C, if one raises the gas temperature further, the heat taken away by the outgoing flue gases is more than heat generated, then temperature falls back to point C. If the temperature at C decreases due to certain reason heat release becomes higher than heat loss, then the operating point moves back to the original point, C. Hence, this point C is considered to be stable operating point. However, when engine is operated at point B, if one disturbs the temperature toward lower side, then additional heat is carried away by convection, which becomes more than heat generated resulting in a fall in gas temperature. This leads to moving the operating point away from the original point B. Hence, this point is considered to be an unstable operating point which must be avoided at any cost during the actual operation of the combustor.

We have already learnt from the simple stirred reactor analysis that stable and efficient combustion is affected by fuel-air mixture ratios, pressure and the volume of the combustion zone. Therefore, we can combine these three parameters namely mass flow rate $\dot{m}$, combustion volume, V and pressure, P to form a new parameter known as *combustion loading parameter*, CLP to ascertain the stability of the combustion process which is defined as

$$CLP = \frac{\dot{m}}{P^n V} = f(\phi) \tag{10.31}$$

Note that the denominator of this parameter is based on the rate of the combustion reaction. For most hydrocarbon air system, the pressure index, n in reaction rate expression turns out to be 1.8 for aviation turbine fuel (ATF) as obtained from experimental data. But for high pressure combustion, the value of n becomes around one as the physical mixing processes dictates the combustion process as compared to the chemical reaction rate.

The generalised variation of CLP with equivalence ratio for a typical gas turbine combustor is presented in Fig. 10.26. The equivalence ratio corresponds to the fuel-air ratio in the primary zone. Note that the stable combustion is only possible between left and right branch of the curve. Interestingly, the region of maximum flame stability occurs for the slightly fuel-lean region around $\phi = 1$. In case the mass flow rate increases for same combustor, the fluid velocity will be greater than the burning velocity for same fuel-air mixture resulting in increase in CLP. As a result, the operating condition may fall outside of the stable zone. In case the burner size is quite small, CLP becomes higher such that it can fall beyond stable region. This figure also indicates that there is a greater danger of unstable combustion when the pressure becomes low that occurs at high altitudes. Hence, the combustor must be designed properly to have a stable combustion even at higher altitudes.

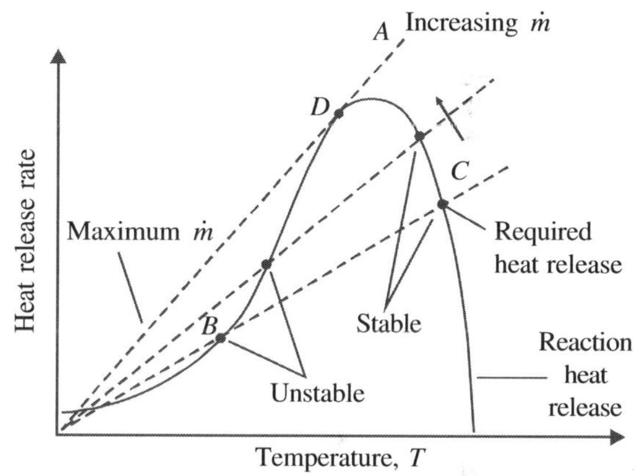

Fig. 10.26. The variation of CLP with equivalence ratio for a typical gas turbine combustor.

10.10.4 Exit Temperature Distribution

We know that the exit temperature distribution is an important combustor performance parameter which affects very much the life of the engine. It can be noted that the exit temperature distribution at burner exit is also affected by any change in operating conditions. Its uniformity gets affected by any reduction in pressure beyond certain threshold pressure. In contrast, more uniform temperature distribution can be achieved for a given burner size by creating better mixing of hot flue gas and cold gases, of course with an increase in total pressure losses. Besides this, if the fuel-air mixture are increased, the exit temperature may become less uniform as more heat is released. On the other hand, if the flow velocity is increased, exit temperature becomes more non-uniform as very less time will be available for mixing.

We know that it is desirable to have a uniform temperature distribution as shown in Fig. 10.27 to avoid uneven stressing of turbine blades due to thermal load. But it is quite difficult to achieve uniform exit temperature distribution in practice. Rather, a normal temperature profile, as shown in Fig. 10.27, can be accepted as an operating profile. But a poor temperature distribution with marked hot spots, shown in Fig. 10.27, can affect seriously the mechanical condition of the turbine stator and rotor blades, although the overall performance of engine is decided by mean temperature.

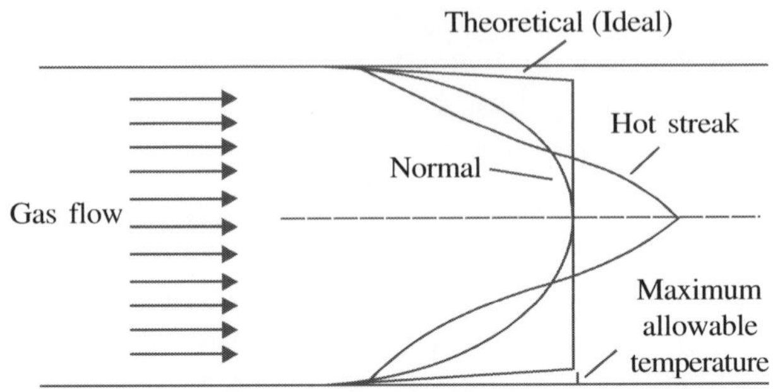

Fig. 10.27. Exit temperature distribution in a typical gas turbine combustor [4].

In actual practice, it is quite difficult to achieve uniform exit temperature. Of course, the combustor exit temperature distribution can be obtained by creating better mixing of hot gas with air dilution at the expense of an increase in total pressure loss. But still then, there will be certain variations in exit temperature profile. Now, question arises how we could assess the quality of the temperature distribution? Several parameters have been devised and reported in the open literature. But we will restrict our discussion to only two parameters known as (i) Temperature Distribution Factor, TDF, and (ii) Profile Factor, PF. The TDF is defined as

$$TDF = \frac{T_{t4,\,max} - T_{t4,\,avg}}{T_{t4,\,avg} - T_{t3,\,avg}} \tag{10.32}$$

where $T_{t,\,max}$ is the maximum measured local exit temperature and $T_{t4,\,av}$ and $T_{t3,\,avg}$ are the average temperature at exit and inlet plane of combustor respectively. Note that TDF is also known as *Pattern Factor* which is commonly used for characterising temperature distribution in the exit of combustor. Note that this pattern factor or TDF is one of the major design requirement that is to be achieved particularly for the first stage stationary aerofoils of the turbine. The TDF for a modern main combustor lies in the range of 0.15-0.3. This TDF can normally be determined by carrying out detailed temperature survey using high temperature thermocouples at the exit plane of the main combustor. If this factor is quite large, then designer has to change hole patterns in the flame tube to achieve the desired exit temperature distribution. But the ultimate deciding factor is whether the turbine with certain temperature distribution has an acceptable life or not.

In order to characterise the circumferentially average exit temperature of the main burner profile factor, PF at each radial location is defined as

$$\mathrm{PF} = \frac{T_{t4,max,avg} - T_{t3,avg}}{T_{t4,av} - T_{t3,avg}} \tag{10.33}$$

where $T_{t,\,max,avg}$ is maximum circumferential average temperature at the exit plane of the main burner. Typical value of PF for a modern main combustor lies in the range of 1.04 to 1.08. Similar to TDF, it is one of the major design requirement that is to achieved particularly for first stage stationary aerofoils of the turbine. Both TDF and PF describe the possible thermal load on the turbine blades which is considered to be an important factor in determining performance of both main burner and turbine. Failure to achieve requisite TDF or PF can shorten the life of turbine which calls for redesign of either turbine or main burner for achieving larger life of engine.

10.10.5 Combustion Efficiency

We know from previous studies that fuel consumption of an engine directly affects aircraft range, payload and operating cost. Hence it is desirable to have higher combustion efficiency (as close to 100% as possible). The combustion efficiency is defined as the ratio of the actual enthalpy rise of the gas stream to the theoretical enthalpy rise for a given fuel/air ratio that is given by:

$$\eta_b = \eta_{comb} = \frac{\Delta h_{actual}}{\Delta h_{ideal}} \approx \frac{\Delta T_{actual}}{\Delta T_{ideal}} \tag{10.34}$$

It is a measure of completeness of the combustion process in the gas turbine combustor. To determine combustion efficiency experimentally, total temperature at inlet and outlet of the combustor and fuel-air ratio must be determined. The ideal (theoretical), ΔT can be easily estimated from simple analysis at a particular fuel-air ratio. It is desirable to have higher combustion efficiency since it has a direct effect on the TSFC of the engine. Normally, combustion efficiency of 99% can easily be obtained in gas turbine engines at design conditions. But combustion efficiency η_{comb} can even fall to as low as 90% at off design conditions particularly when engine is running at an idle condition. Of course, to achieve higher η_{comb}, it is essential to provide adequate residence time and space for fuel and air to vaporise and mix properly so that chemical reaction can be accomplished efficiently. Of course, high velocities which is required to make the combustor more compact pose serious problems for achieving higher efficiency due to poor combustion. As soon as the operating conditions are shifted far away from design conditions, the η_{comb} will start to fall because, fuel spray pattern may change adversely leading to poorer atomization at reduced fuel flow rates. Besides this, the η_{comb} is dependent on inlet air condition at the combustor entry. The inlet conditions can vary a lot with change in altitude. We know that unfavourable conditions such as low inlet temperature, pressure and high air velocity are likely to prevail at higher altitude. Let us look at how typical trend data of η_{comb} obtained for a gas turbine main combustor varies with engine rpm and altitude, as shown in Fig. 10.28. It can be noted that combustion efficiency decreases with altitude. This is so because with reduction in temperature and pressure of air, there will be less intense mixing of fuel and air while the reaction time for complete combustion increases considerably. The residence time would be quite less in comparison to reaction time, which leads to poor combustion. Hence the η_{comb} falls with altitude. It is also evident in this figure that the engine has to operate at high rpm to achieve high values of combustion efficiency. This can only be achieved if chamber pressure can be enhanced at a particular altitude. Hence maximum altitude limitation of a jet engine is dictated mainly by maximum pressure that can be achieved by the compressor and air intake.

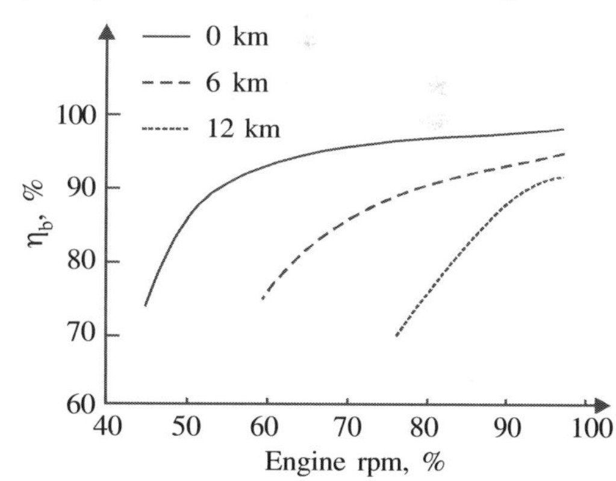

Fig. 10.28. The variation of η_{comb} with equivalence ratio [2].

Besides this, η_{comb} also depends on the fuel/air ratio, f. Of course, it would be better if one can get as close as possible to the η_{comb} vs. equivalence ratio curve as shown in Fig. 10.29. In practice, it is quite difficult to achieve expected η_{comb}. However, a curve similar to B as shown in Fig. 10.29 can be easily achieved in practice with peek η_{comb} value at design conditions accompanied by gradual fall at both lower and higher sides of fuel/air ratio.

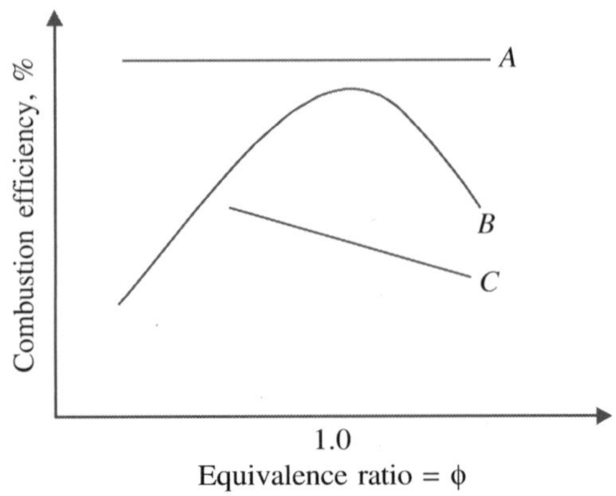

Fig. 10.29. The variation of η_{comb} with equivalence ratio.

10.11 Gas Turbine Emission

The exhaust gases from the gas turbine engine are mainly composed of carbon monoxide (CO), carbon dioxide (CO_2), water vapour (H_2O), oxides of nitrogen (NO_x), oxides of sulfur (SO_x), unburnt hydrocarbons (UHC), nitrogen (N_2), oxygen (O_2) and particulate matters (soot, ash, etc). We know that CO is formed when the combustion of fuel occurs in the absence of sufficient O_2. Besides this, CO_2 may get dissociated to CO during combustion lean fuel mixture. In gas turbine engine, incomplete combustion leading to formation high level of CO and UHC can be caused due to inadequate residence time in the primary zone, improper mixing of fuel and air, quenching of combustion products by cooling air around liner walls. Let us look at the how does the species concentrations vary along the length of combustion in a typical diffusion flame based gas turbine engine as shown in Fig. 10.30. The temperature rises rapidly due to burning of fuel in the primary zone. As a result, there will be rapid formation of CO, CO_2 and H_2O in a relatively short time. As the temperature is quite large, CO_2 can be converted easily to CO in the prevailing high temperature primary zone. But the NO species is formed at a slower rate as its formation is governed by the Zeldovich reactions mechanism which are endothermic in nature with a large value of activation energy. As a result it proceeds only at a very high temperature and it is completed at a very fast rate as shown in Fig. 10.30. This is possible if the product gases are kept at high temperature for sufficient time. The formation of NO_x is enhanced with chamber pressure in conventional combustor. Besides this, NO_x emission increases with the size of droplets at lower equivalence ratio as local high temperature are likely to develop around large droplets.

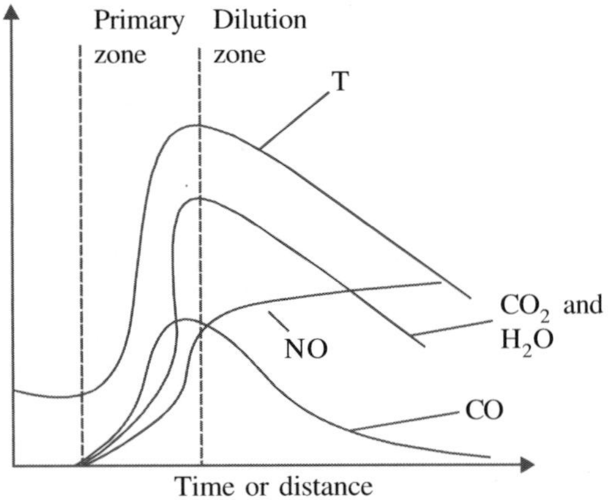

Fig. 10.30. The variation of combustion products along the length of a typical gas turbine engine.

Soot and smoke formation in the combustor is affected by fuel-air distribution, extent of recirculation, temperature and nature of fuel. When burnt gases are recirculated toward the spray zone, soot in the form of carbon particles are formed in the primary zone. Smoke is formed when local regions of the fuel vapour are surrounded by oxygen deficient burnt gases at relatively high temperature. Note that soot and smoke are formed when the combustor is operated at high pressure. This may be due to increase in reaction rate with high pressure which causes the fuel to get burnt near the spray leading to the formation of larger fuel rich zone. Besides this, fuel properties such as viscosity and volatility can influence soot and smoke formation as they affected droplet size, penetration length and evaporation rate significantly. Care must be taken to reduce the soot and smoke formation. Interestingly both soot and smoke can be reduced by enhancing the hydrogen content of the fuel. This problem of smoke can be easily reduced by carrying out combustion in the primary under fuel lean conditions. Generally most of the modern gas turbine engines do not smoke visibly particularly when they are operated at full throttled condition. But when operated at part throttle and idle conditions, combustion in the primary zone is likely to occur at fuel rich condition leading to formation of large amount of CO, UHC, soot and smoke.

Generally, emission of a particular species from the aero gas turbine engines is specified in terms of grams of pollutant of individual species per unit thrust per hour of operation in a standard approach-taxi-taxi-takeoff cycle. However for both aero and stationary gas turbine combustors, emission of a particular species is more often specified in terms of grams of pollutant per kg of fuel burnt. For example, around 30 g NO/kg of fuel is produced for a high-bypass turbofan engine at full takeoff power. Note that samples from the exhaust of the combustor can be used to measure values of major species concentration using standard gas analyzers. For example, NDIR (Non Dispersive Infra Red) gas analyzer is used to measure CO and CO_2 concentrations while Chemiluninsce based gas analyzer is used to measure concentrations of NO_x. Smoke and soot level is specified in terms of smoke number (SN) which is defined as a measure of visibility of the exhaust gas from the combustor. It can be determined by drawing a sample of exhaust gas through a filter paper at set flow rate for certain time and then measuring the change in optical reflectance of the paper due to the deposited soot.

$$SN = 1 - \frac{R_s}{R_p} \tag{10.35}$$

where R_s is the reflectance of the sample and R_p the reflectance of the clean filter paper. Typical value of smoke number for high-bypass turbofan engine at full takeoff power is in the range of 19-20.

10.11.1 Emission Control Methods

We know that emissions from the aero gas turbine engines can affect surroundings of airports significantly. Besides this, upper atmosphere is polluted due to emissions from aero engines. In recent times, its considered that emissions from supersonic aircrafts flying in the lower region of the stratosphere pose potential threats to the Earth's Ozone layer. During taxing and ground manoeuvring, considerable amount of fuel is burnt producing large amounts of CO, CO_2, UHC, NO_x, etc. It has been reported that the respiratory ailments like acute and chronic bronchitis, and emphysema are increasing in alarming rate near airport regions, which are mainly caused due to (SO_2, NO_x) and particulate matter. The photochemical smog is known to be causing eye irritation. The excessive exposure of CO level by human beings for a long duration may cause headache, reduction in mental activity, vomiting, etc. Hence, to reduce emissions from gas turbine combustors several techniques have been devised to control emission of sulfur oxides from combustion systems which can be broadly divided into two categories such as (i) Combustion Modification Method, and (ii) Post-combustion Method.

(i) Combustion Modification

In the first method, combustion processes are to be modified suitably to reduce emission levels without deteriorating the combustion efficiency. For example, any reduction in CO and UHC levels can be achieved only when complete combustion is achieved. Generally CO and UHC levels can be reduced by enhancing mixing of fuel and air in the primary zone with the help of higher level of recirculation. Improved fuel atomization and increased residence time in the primary zone can enhance combustion efficiency while reducing CO and UHC emission level. Besides this, a reduction in cooling air flow rate in the liner surface of the primary zone can lead to reduction in CO and UHC levels. But this method may enhance soot and smoke levels due to the formation of a fuel-rich region.

The formation of NO_x in combustor is mainly influenced by the prevailing temperature in the primary zone. One method to diminish temperature in a stationary gas turbine combustor is the injection of water or steam to extract combustion heat. Water or steam can be injected into primary zone in the format of spray along with fuel injection. Besides this, water can be introduced upstream of combustor liner into air stream. Generally steam is introduced at the compressor discharge. Water injection is considered to provide better results when peak NO_x emission levels are prevailed due to higher operating temperature and pressure in the combustion chamber. A less quantity of water is required as compared to steam to achieve the same level of reduction in NO_x emission level. But all these methods incur 2-3% higher fuel consumption with an increase in the initial cost. It has also been observed that water injection can result in an increase in pressure oscillations in combustion chamber.

Staged combustion is one of the effective method of controlling the emission levels in a gas turbine engine. In this method, fuel is provided in stages such that it is switched from one zone to another. Different quantity of fuel is supplied in stages in selected combination of injectors at different engine operating conditions. For example, in an aircraft engine, certain fuel injectors will not be activated during lightoff or idling conditions. As more power is to be supplied at takeoff condition, it is expected to operate all fuel injectors. It can be noted that at higher power levels, the combustion products from the pilot zone helps in anchoring the flame in the main combustion zone which will be discussed in detail below. At certain adverse off-design operation modes, equivalence ratio in the combustion zone may be enhanced to raise temperature in local regions such that lean blow-out limits can be enhanced in addition to controlling CO and UHC emission levels. However, incomplete combustion is likely to occur in certain local regions, leading to formation of undesirable pollutants. Besides this, staged combustion may yield nonuniform temperature distribution, tending to reduce turbine efficiency. There are several ways by which staged combustion can be accomplished. The most popular method in this category is the rich-lean staging.

A radial staging burner is shown in Fig. 10.31 above in which combustion takes place in two zones such as pilot zone and main zone. Generally the pilot stage combustion takes place during starting, idling, and relight conditions. It is operated at certain fuel-air ratio such that low CO and UHC emission levels are maintained particularly at lower power levels. At normal operating condition, combustion takes place in both pilot and main zones. Normally, fuel lean condition is maintained in the main zone such that it can produce lower levels of NO_x particularly at higher power level. Such a kind of staging method has been implemented in the CFM56-5B aircraft engine of the General Electric Company [10]. In radial staging method, the diameter of combustor increases without affecting overall length of the engine, and the dynamic aspects of the rotating system. However the arrangement of fuel injectors and their tubings become quite complex. Besides this, it lowers the lean blow-out limit due to asymmetric airflow. As a result radial temperature profile gets distorted which require more amount of cooling air in the outer liner. It has been observed that performance characteristics gets deteriorated even at the intermediate power level.

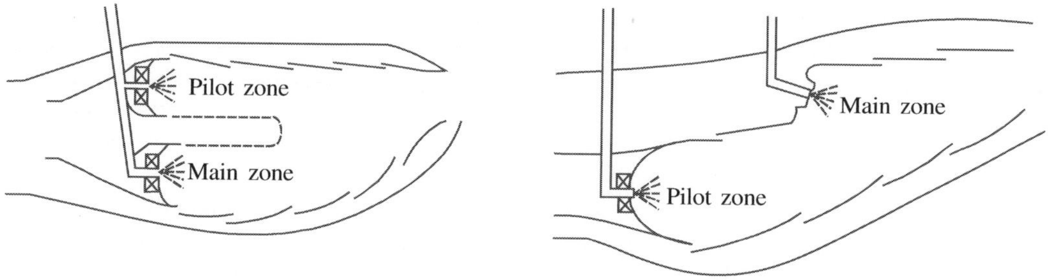

Fig. 10.31. Low emission gas turbine combustor configuration (a) axial staging (b) radial staging [10].

Axial staging combustor overcomes some the problems of axial staging based combustor. A schematic of typical axial staging combustor configuration is shown in Fig. 10.31. It can be noted that both pilot stage and main stage are located in series. In this method, combustion in the pilot zone takes place under fuel rich conditions which subsequently helps in stabilising lean fuel burning in the main zone. Generally pilot stage is optimised to operate in stand alone mode at a lower power level while emitting least amount of CO and UHC. As mentioned earlier, at higher loads and speeds, fuel lean condition is maintained in the main stage which uses high intensity swirler to enhance rapid mixing leading to formation low levels of NO_x due to low residence time. This kind of staging has been incorporated in the V2500-AS engine of Pratt and Whitney company [Koff]. It can be noted that axial staging method provides more reliable and a faster ignition at adverse situation as hot gases from the pilot zone aids in igniting fuel-air mixture in the main stage. Besides this, it ensures complete combustion in the main stage region even at low overall equivalence ratios. A uniform radial temperature profile at the exit of combustor can be easily obtained using available dilution holes as compared to the radial staging method. However, the length of the combustor gets enhanced due to the use of the axial staging of fuel. This length penalty is compensated with improved emission levels and other performance parameters as compared to the axial staging method.

(ii) Postcombustion Method

In this method, pollutants from the exhaust gas streams are being treated separately for each components. In recent times, several methods for each constituent pollutants of exhaust gas are devised which can not be covered in this book. However, few methods will be described briefly. It can be noted that these methods are generally suitable for stationary gas turbine engines.

Recently several post combustion methods have been devised to reduce CO_2 emission level. In these methods, CO_2 gas from the flue gas is absorbed before allowing it into the atmosphere. Recently an amine based CO_2 capturing method has been devised which can be employed in stationary gas turbine engines to capture CO_2 gas from the flue gas. In this method, CO_2 is separated from the exhaust gas by means of absorption using monoethanol amine (MEA). This CO_2 capture system consists of mainly absorber, regenerator and exchanger. The exhaust gas containing CO_2 in addition to mainly N_2, O_2 and H_2O is first cooled to 40-50°C and fed to the absorption tower. In the absorber, the flue gas is mixed with using mono-ethanol amine (MEA) solvent, which captures solvent nearly 90% of the CO_2 in the flue gas as it has more affinity for CO_2 molecules than other compounds. The CO_2 rich amine is fed to the amine stripper, in which the amine is regenerated and fed back to the amine absorption tower. It can be noted that this device must be heated to 120°C with the help of steam in order to break the bonds between the solvent and CO_2. The released CO_2 and steam are cooled for water removal and the CO_2 is compressed for transportation and storage.

Although, reduction in NO_x emission can be achieved by the combustion modification discussed above. However beyond certain limits, they may not meet the requirements of the standards enforced by the new stringent regulation. Hence, several post combustion methods are devised in recent times. The most effective post combustion methods for reducing NO_x emissions from combustion devices can be classified into two categories namely (i) Selective Non-Catalytic Reduction (SNCR) (ii) Selective Catalytic Reduction (SCR).

In Selective Non-Catalytic Reduction (SNCR) method, a non-catalytic additive can be mixed with the NO_x contaminated gases in which chemical reduction of NO to N_2 can be accomplished. Generally, three non-catalytic additives namely ammonic, NH_3, Urea, and Cyanuric acid are being employed in combustion systems. It must be noted that the operating temperature plays a very important role in deciding the effectiveness of NO removal. It has been found out that, such non-catalytic additives work well within a narrow range of temperature (1100-1400 K). It has been observed that the maximum NO removal, as high as 90% can be obtained by this method. Besides this, the effectiveness of this NO removal is dependent on the ratio of NO concentration to the injected additive concentration. Generally, it has been observed that the additives added to NO molar ratio in the range of 0.8-1.5 can provide more than 50% NO reduction. However, N_2O is one of the possible byproducts of NO reduction method, which is formed by the reaction of NO with NH or NCO.

In Selective Catalytic Reduction (SCR) method, a catalyst is used along with other additives like NH_3 to reduce NO to N_2. The catalysts used in this process are mainly metal oxides like V_2O_5, WO_3, and MoO_3. On a support material (e.g. Zeolite or TiO_2), noble metal catalysts (Pt, Pb, etc) have also been used in flue gases, particularly in sulfur free gas turbine engines. For effective reduction of NO, proper temperature window must also be chosen which is of course dependent on the type of catalyst being used. Typical temperature window required for SCR is around 450-750 K as compared to 1100-1400 K for SNCR. The lower temperature range for SCR is limited by the reaction rate, whereas upper temperature is limited by the physical damage to the catalyst through sintering and oxidation of NH_3 to form NO.

We know that particulates are formed during the combustion processes due to oxidation of inorganic materials present in the fuel. It can be noted that burning of certain low grade fuel oil generates particulates which are generally less than 10 micron unless otherwise these particles are coagulated and sintered to form bigger size particles. These particulates are generally termed as *fly ash*. Besides this, soots are formed during the combustion process. These particles can be removed from flue gases either using long bag house filter or an electrostatic precipitator. The particles in the flue gas must be charged to be effectively collected in an electrostatic precipitator. When these charged particle laden flue gases are passed through a strong electric field, they attach themselves to the wall and are subsequently removed by shaking devices. Besides these devices, cyclones and hydrocyclones are also used to remove small particulates from flue gases. This method can be used in stationary gas turbine engines.

10.12 Chapter Summary

The combustor is the heart of the gas turbine engine as it provides the thermal energy which is converted into high kinetic energy gas to produce power and thrust while getting expanded in the turbine and nozzle, respectively. A brief introduction is provided enumerating the requirements of a typical gas turbine combustor particularly for aircraft applications. The technical challenges off designing a compact high powered combustor with low emission level are brought out in a subtle way to inculcate interests among students to appreciate the complexities involved in the gas turbine combustor. The fundamental aspects of combustion pertaining to gas turbine combustor is covered adequately. Various methods of flame stabilisations are discussed briefly emphasising the basic principles of flame stabilisation in high speed flow. Based on the scaling principle, the phenomenological analysis of flame stabilisation is carried out to derive a semi-empirical relationship for non-dimensional blow out velocity in terms of laminar burning velocity, flame holder characteristics length and thermal diffusivity which can be used as a design tool. A typical gas turbine combustor is discussed in detail to identify its main components and their respective functions. The main combustor is broadly divided into three zones namely (i) primary, (ii) cooling and (iii) dilution zone which ensures complete combustion and uniform temperature distribution at its exit. As a result, the flow structure in this combustor becomes quite complex due to presence of cooling and dilution holes. Based on geometry, the gas turbine main combustor is classified into three types namely (i) can (ii) annular and (iii) can-annular which are discussed adequately enumerating their respective advantages and, disadvantages. A section is devoted to the basics of fuel atomization which plays a dominant role in controlling the combustion in the main burner. Besides this, various kinds of atomizers are covered highlighting their respective merits and demerits. The process involved during ignition is discussed briefly. Various kinds of igniters used in both main and afterburner are covered adequately. A brief account of aviation turbine fuel and its properties is provided in a particular section. The performance parameters such as total pressure loss, combustion loading parameter, temperature pattern factor and combustion efficiency employed for characterisation of both main and afterburner are defined and discussed while carrying out phenomenological analyses for certain parameters. As emission is assumed of global importance in recent times, the formation of various chemical species during combustion process are discussed and some of recent methods under both combustion and post combustion categories are included in this chapter which are considered to control emission level in gas turbine engines to meet stringent regulations across the globe.

Nomenclature

A/F	Air-fuel ratio
C_{pa}	Specific heat of air (1.005 kJ/kg K)
C_{pg}	Specific heat of gas (1.148 kJ/kg K)
d	Droplet diameter (μm)
ΔH_c	Heat of combustion (kJ/kg)
E	Activation energy (J/kmol)
f	Fuel-air ratio by mass
IE	Ignition energy (MJ)
$\dot{m}_f'''$	Mass burning rate of fuel per unit volume
M	Mach number
$\dot{m}$	Mass flow rate (kg/s)
N	Order of reaction
Oh_j	Ohnesorge number
Pe	Peclet number (= $V_{bo}d/\alpha$)
PF	Profile factor
q	Heat addition (kJ/kg)
R	Universal gas constant (kJ/kg K)
Re_j	Jet Reynolds number ($\rho V_j d_j/\mu$)
S_L	Laminar burning velocity (m/s)
S_t	Turbulent burning velocity (m/s)
t_{reac}	Reaction time (s)
t_{res}	Residence time (s)

V_{bo}	Blow off velocity
V_j	Jet velocity (m/s)
We_j	Jet Weber number ($\rho V_i^2 d_i/\sigma$)
y_i	Mass fraction of species

Greek Symbols

π	Pressure ratio
α	Thermal diffusivity (m^2/s)
σ	Surface tension
δ_p	Pressure difference across injector
ρ	Density (kg/m^3)
ΔN_i	Number constant
μ	Dynamic viscosity (NS/m^s)
ϕ	Equivalence ratio (f_{act}/f_{stoic})

Subscripts/Superscripts

a	Air
act	Actual
b	Burner
bo	Blow off
c	Compressor
Com	Combustion
g	Gas
j	Jet
l	Liquid fuel
L	Laminar
Reac	Reaction
Res	Residence
Stoic	Stoichiometric
t	Turbulent

Abbreviations

AMD	Area mean diameter (μm)
APU	Auxiliary power units
ATF	Aviation turbine fuel
BP	Boiling point
CC	Combustion chamber
CLP	Combustion loading parameter
CVF	Cumulative volume fraction
SCR	Selective catalytic reduction
SG	Specific gravity (S_{fuel}/S_{water})
SMD	Sauter mean diameter (μm)
SN	Smoke number
TDF	Temperature distribution factor (pattern factor)
VMD	Volume mean diameter (μm)

References and Suggested Readings

1. Hill, P.G. and Peterson, C.R., *Mechanics and Thermodynamics of Propulsion*, 2nd Edition, Addison-Wesley Reading, Massachusetts, 1992.

2. Koff, B.L., *Aircraft Gas Turbine Engine Challenges*, ASME Paper 93-GT-422, New York, 1993.
3. Oates, G.C. (Ed.), *Aerothermodynamics of Aircraft Engine Components*, AIAA Education Series, 1985.
4. Treager, I.E., *Aircraft Gas Turbine Engine Technology*, 3rd Edition, Tata McGraw-Hill, 1996.
5. Lefebvre, A.H., *Atomization and Sprays*, Taylor and Francis, 1989.
6. Meller, A.M., *Design of Modern Turbine Combustors*, Academic Press, London, 1990.
7. Lefebvre, A.H., *Gas Turbine Combustion*, Taylor and Francis, 1998.
8. Lewis, B. and von Elbe, G., *Combustion Flames and Explosion of Gases*, Academic Press, New York, 1987.
9. Saravanamuttoo, H.I.H., Rogers, G.F.C. and Cohen, H., *Gas Turbine Theory*, Pearson Education, 2001.
10. Bahr, D.W., "Technology for Design of High Temperature Rise Combustor", *Journal of Propulsion and Power*, 3(2), 179-186, 1987.
11. Mattingly, J.D., Heiser, W.H. and Daley, D.H., *Aircraft Engine Design*, AIAA Education Series, AIAA Washington, 1987.
12. Mukunda, H.S., *Understanding Aerospace Propulsion*, Interline Publishing, 2004.

Review Questions

1. What are the specific requirements of the main combustor of a gas turbine engine?
2. What is flame stabilisation? What are the types of flame stabilisation? Discuss them along with their respective advantages and disadvantages.
3. What are the methods of bluff-body flame stabilisations? Which is the one preferred in afterburner? Why is it so?
4. What is blowoff velocity? Derive an expression for dimensionless blowoff velocity by stating its assumptions.
5. Draw a typical main combustor of a gas turbine engine. Describe functions of each components of this combustor.
6. What are the types of main combustor in gas turbine engine? Which one is used in modern gas turbine engine? Why is it so?
7. What are types of atomizers used in gas turbine engine?
8. What are the advantages and disadvantages of air blast atomizer over pressure swirl atomizer?
9. What is prefilming air blast atomizer? How is it different from plain jet atomizer?
10. What is Weber number? What is its physical significance?
11. What is Ohnesorge number? How is it different from Weber number?
12. What are the processes involved in liquid atomization?
13. What is Duplex atomizer? How is it different from simplex and spill-return atomizer?
14. What is the basic principle of rotary type atomizer? Draw a diagram and describe its working principles. Enumerate its advantage and disadvantage over pressure swirl atomizer.
15. What is surface discharge igniter? How is it different from spark plug?
16. What is SMD? Why is it used profusely in combustion literature?
17. What are the specific requirements of an afterburner?
18. What is combustion stability? Explain it by a diagram.
19. What is the combustion loading parameter? How does it vary with the equivalence ratio and combustion chamber pressure?
20. What is temperature distribution factor? Why is it important in characterising gas turbine combustor?
21. Define combustion efficiency. How does it vary with altitude in a gas turbine combustor?

Problems

1. Estimate F/A ratio required by a gas turbine at sea level, 5 km, and 10 km for a flight Mach number 0.8 and turbine inlet temperature 1300°C. The compressor is operating at compression ratio of 15 whose efficiency is 85%. The heating value of ATF is 42000 kJ/kg. Take η_{comb} = 95%, C_{pa} = 1005 J/kg K, C_{pg} = 1240 J/kg K.

2. Derive an expression for adiabatic flame temperature. Determine the adiabatic flame temperature when the combustor is operating at a mass air-fuel ratio of 16 for air preheated to 600 K. The fuel enters at 300 K. Take T_0 = 300 K, $MW_F = MW_A = MW_P$ = 29 kg/kmol, $C_{p,f} = C_{p,a} = C_{p,P}$ = 1200 J/kg-K. $h^0_{f,a} = h^0_{f,p} = 0$, $h^0_{f,f}$ = 4.107 J/kg.
3. A supersonic aircraft is flying at Mach number of 2.5 in an atmosphere whose temperature is 235 K. The pressure ratio across compressor is 10. The air enters into combustor (constant area duct) at Mach number of 2.0 and leaves at temperature of 2200 K. Estimate the Mach number of gas leaving the burner assuming negligible friction. Estimate the stagnation pressure loss.
4. The drop size distribution obtained from an experiment is given in Table P1. Determine CVF, MPD (most probable diameter), mean diameter, SMD, VMD, AMD.
5. Compute SMD for an ATF spray from a pressure swirl atomizer of 5 mm orifice diameter. The upstream pressure of atomizer is 6 bar. The discharge coefficient of orifice is 0.91. The pressure and temperature of the combustor is 5 bar and 1200 K respectively. The density, viscosity and surface tension force of the ATF are 780 *kg/m*^{3}, 0.0013 *N.s/m*^{3} and 0.03 N/m respectively.
6. Show the variation of SMD with $\dot{m}_a/\dot{m}_l$ for a fuel spray obtained from a prefilming air-blast atomizer with film diameter of 30 mm over a range of air flow to fuel flow ratios from 1 to 8. The air is at temperature of 500 K and 5 bar. The average air velocity is varied from 25 to 100 m/s in the interval of 25 m/s. Use the fuel properties as given in previous question.

Table P1. Distribution of droplet size

S. No.	Droplet size (microns)	Numbers
1	0-10	73
2	11-30	112
3	31-50	175
4	51-60	205
5	61-65	314
6	66-70	421
7	71-80	301
8	81-100	103
9	100-130	59
10	131-150	13

7. A gas turbine engine is flying at an altitude of 15 km with flight Mach number of 1.5. The pressure ratio across the air intake is 0.95 and its isentropic efficiency is 0.98. The compressor is operated with an polytropic efficiency of 0.85 at pressure ratio of 12. The main burner has an inlet Mach number of 0.1 and drag coefficient, C_D of 1.7. If maximum exit temperature of burner is 1600 K, determine the total pressure ratio across the main burner, assuming it to be constant area burner. Determine the burner stagnation temperature ratio that will just cause choking at its exit.

11. Air Intake

We engineers of course know that machine building, through widely extended practical experimenting, has solved problems, with utmost ease, which baffled scientific investigations for years. But this "cut and try method", as engineers ironically term it, is often extremely costly; and one of the most important questions of all technical activity, that of efficiency, should lead us not to underestimate the results of scientific technical work.

Aurel Bolesav Stodola 1903

Till now we have discussed about various components of gas turbine engine, which are independent of any aeronautical applications. In other words, the analysis of compressors, turbines and combustors so far carried out in previous chapters can be applied to any gas turbine engine, used either for power plant or aircraft propulsion. The aero engines differ only in their emphasis on compactness and weight. Besides this, the aero engines always have two additional components namely (i) Air intake and (ii) Nozzle, which are integral part of a propulsion unit in a flight vehicle. In this chapter, we will mainly discuss about air intake both for subsonic and supersonic flight conditions. In the next chapter, we will discuss about the salient features of nozzles.

The main function of air intake is to deliver the required air mass flow rate to the compressor with minimum possible stagnation pressure loss and with smoothest possible velocity distribution as any loss in stagnation pressure would lead to deterioration in overall performance of engine. Note that any distortion in velocity profile at compressor inlet can severely affect the performance of the compressor and sometimes may lead to compressor surging. These requirements must be met adequately over a wide range of operating flight conditions, namely take off, landing, taxing, combat operation, etc to mention few. Besides this, the off design performance of air intake must also meet these requirements to a large extent. Another engineering difficulty encountered while designing a good air intake is restricted availability of space in the aircraft. Besides these restrictions, the air intake must be integrated easily with rest of the engine, which has to also provide lowest possible drag. The air intake must have lowest possible aerodynamic noise. Hence, the design of a high performance intake is a challenging task for designers.

Based on flight conditions the air intake can be classified into two categories namely (i) subsonic air intake, and (ii) supersonic air intake. Note that deceleration is common to both types of intake. The subsonic intake is basically a device in which deceleration of air velocity takes place converting kinetic energy of the flow into static pressure rise. Whereas, in case of supersonic intake, the supersonic flow is decelerated first to sonic flow and subsequently is decelerated further from sonic flow to subsonic velocity in a subsonic diffuser.

11.1 Subsonic Intake

We know that the velocity at the compressor inlet must be lower than flight speed of the aircraft particularly at design mass flow rate. As a result, air has to undergo deceleration in the intake, which can be achieved by proving the necessary change in cross sectional area of air intake. Of course, we can arrive at this area variation by using the relation of area and pressure for quasi-one-dimensional isentropic flow as discussed earlier, which is given below

$$\frac{dA}{A} = \frac{1}{\gamma}\left(\frac{1}{M^2} - 1\right)\frac{dP}{\rho} \tag{11.1}$$

where, M is the Mach number, A is the cross sectional area, P is the pressure, ρ is the density of air. From the above equation, it can be easily seen that for diffusion of a subsonic flow, the cross-sectional area of the intake must increase in direction of the flow. In other words, a subsonic flow can be decelerated either (i) by causing it to flow through a diverging passage or (ii) by causing the stream lines of the flow to diverge by providing proper shape to the cowl at entrance of the intake. Based on these concepts, subsonic intake can be classified further into (i) Internal compression intake and (ii) External compression intake, which are discussed below.

11.1.1 Internal Compression Subsonic Intake

We have seen that the air can be diffused by allowing it to pass through a divergent duct. However, one may be tempted to use a larger divergence angle to have a shorter intake due to space limitation. Let us see what happens when we use high divergence angle duct for intake by considering experimental observations. Keep in mind that viscous effects on the solid surface of the air intake are present unlike in inviscid analysis.

When the intake area does not increase very rapidly in the flow direction as shown in Fig. 11.1(a) the boundary layer growth adheres to its surface in a well behaved manner. As discussed earlier, there will be an increase in the static pressure along the flow direction. This adverse pressure gradient retards the slow moving fluid near the solid surface, leading to thickening of the boundary layer. However, if this pressure gradient is not too large, there would not be any flow separation and the flow will behave in the desired manner providing better pressure recovery. However, with high rate of area increase, the flow is likely to separate from the solid surface which may be characterised by unsteady zones as shown in Fig. 11.1(b). In this case, the shear forces would not manage to overcome the pressure forces resulting in local flow separation at certain point as shown in Fig. 11.1(c). Under these conditions, the separation zone is quite unstable leading to unsteady flow patterns. As a result, there will be irreversible mixing of slow moving fluid with fast moving fluid of the main stream, causing a loss of stagnation pressure. With further increase in divergence, although flow separation attains a steady pattern, there will be enormous increase in stagnation pressure loss. On increasing the divergence angle further, the flow will separate completely from the wall of the intake and will behave like a jet as shown in Fig. 11.1(d). Under this condition, there would not be any rise of static pressure, because all the kinetic energy is being utilised for supporting the separation of

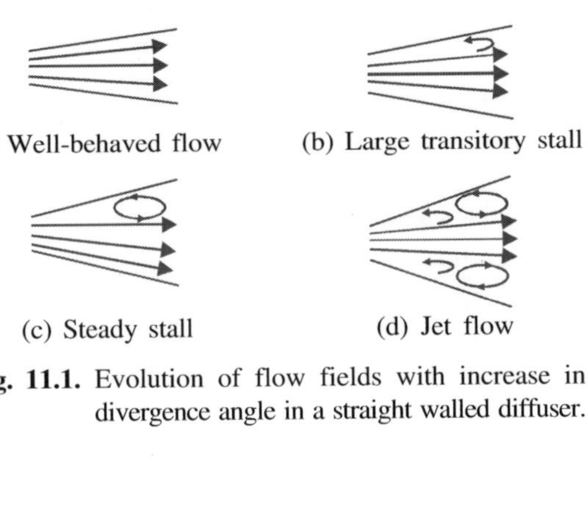

(a) Well-behaved flow (b) Large transitory stall (c) Steady stall (d) Jet flow

Fig. 11.1. Evolution of flow fields with increase in divergence angle in a straight walled diffuser.

boundary layer. From the above discussion, it is apparent that it is not practical to use high divergence duct as intake. But if the divergence of the duct is made very small to minimise flow separation, the pressure losses due to skin friction would be quite high due to increase in the length of the intake. Hence an optimum divergence angle must be used to minimise overall stagnation pressure loss.

Question arises how one can find out the divergence angle for an intake. One may be tempted to use boundary layer analysis to predict intake stall. Unfortunately, we cannot really employ boundary layer theory to predict the separation of boundary layer in an intake. Hence, the designer has to rely on experimental data only. For this purpose, let us discuss experimental studies of two-dimensional straight walled intake [2]. Figure 11.2 shows variation of limiting divergence angle with L/W_1 ratio for two dimensional diffuser. It certainly indicates that there will be appreciable flow separation, as the divergence angle increases by a certain value for a particular L/W_1 ratio. In other words, the limiting divergence angle depends on diffuser length-to-width ratio. For a given L/W_1 ratio, if the divergence angle is too large, the rate of increase diffuser area is quite high to cause flow separation. On the other hand, for given divergence angle, if the L/W_1 ratio is too large, flow separation is likely to occur. Kline [2] reported data of divergence angle, θ for two-dimensional straight walled diffuser. By curve fitting these data, two empirical relations between θ vs L/W_1 are obtained as given below

$$\ln \theta_{min} = 3.28 - 0.46 \ln (L/W_1) - 0.031 [\ln(L/W_1)]^2 \quad (11.2)$$

$$\ln \theta_{max} = 3.39 - 0.38 \ln (L/W_1) - 0.020 [(\ln(L/W_1)]^2 \quad (11.3)$$

where, $\ln \theta_{min}$ is the minimum divergence angle and $\ln \theta_{max}$ is the maximum divergence angle of the diffuser. For example, if $L/W_1 = 4.0$, flow separation is likely to take place between 13 and 17° for two-dimensional straight walled diffuser.

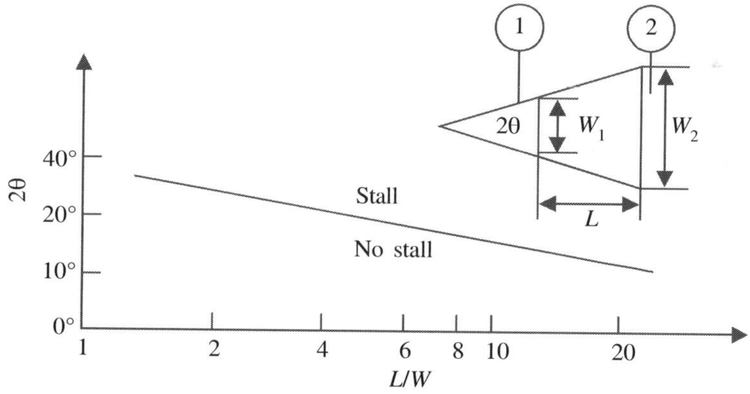

Fig. 11.2. Stall limits in 2D straight walled diffuser [2].

Generally, the pressure coefficient (C_{PR}) is used to indicate the extent of stagnation pressure loss in the intake. An expression for C_{PR} can be derived by using Bernoulli's equation considering one-dimensional flow in the diffuser as

$$C_{PR} = \frac{P_2 - P_1}{\frac{1}{2}\rho V_1^2} = 1 - \left(\frac{V_2}{V_1}\right)^2 = 1 - \left(\frac{A_1}{A_2}\right)^2 \quad (11.4)$$

where A_1 and A_2 are inlet and exit area of the intake respectively. This will provide the ideal pressure recovery as boundary layer is present during flow in actual intake. This C_{PR} will provide theoretical maximum value of pressure recovery in the intake which can be used to ascertain the effectiveness of actual intake. However, if the flow is separated in the diffuser then there will be variation of static pressure along the inlet area. Hence, pressure cannot be considered to be constant across the whole cross-sectional area for which the C_{PR} must be defined appropriately. The more generalised definition of C_{PR} is given below.

$$C_{PR} = \frac{\frac{1}{A_2}\int_{A_2} PdA - \frac{1}{A_1}\int_{A_1} PdA}{\frac{1}{A_1}\int_{A_1} \frac{1}{2}\rho V^2 dA} \tag{11.5}$$

Let us now see the typical measured values of C_{PR} in two-dimensional intake as shown in Fig.11.3 for given length-to-width ratio and several divergence angles. It can be observed that the measured C_{PR} is quite low even for the case of no flow separation. However, this difference in C_{PR} between measured and ideal case increases appreciably with appearance of stall resulting in large stagnation pressure loss in the intake.

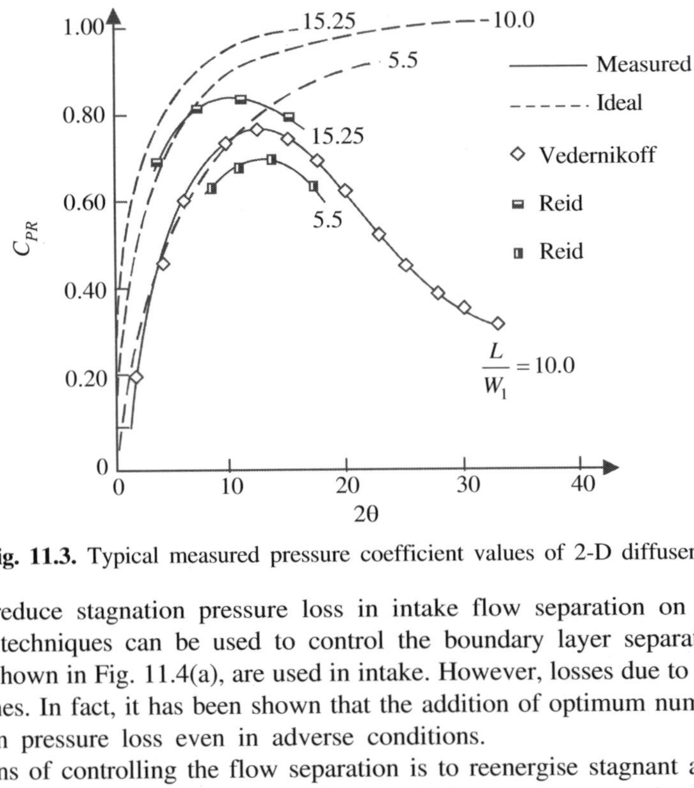

Fig. 11.3. Typical measured pressure coefficient values of 2-D diffuser [2].

In order to reduce stagnation pressure loss in intake flow separation on its surface must be avoided. Several techniques can be used to control the boundary layer separation. In some cases, splitter vanes, as shown in Fig. 11.4(a), are used in intake. However, losses due to skin friction increase due to splitter vanes. In fact, it has been shown that the addition of optimum number of splitter plates reduces stagnation pressure loss even in adverse conditions.

Another means of controlling the flow separation is to reenergise stagnant air in boundary layer by using vortex generators. The air flow path in a typical vortex generator is shown in Fig.11.4(b). It can be noted from this figure that trailing vortex energises the fluid in the boundary layer due to mixing of fluid resulting in a flattened velocity profile. Hence, vortex generator makes it possible

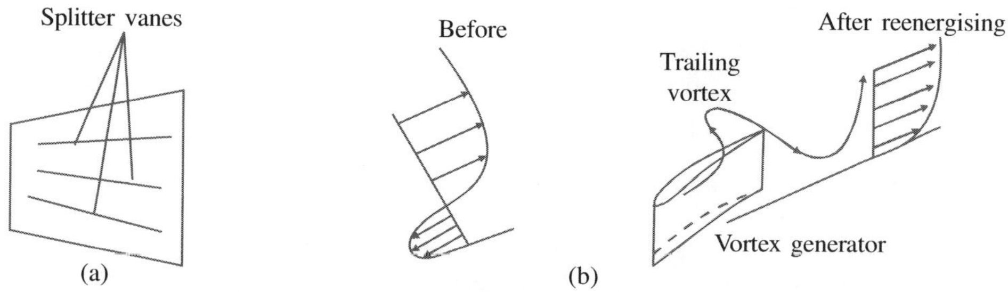

Fig. 11.4. Methods of reducing flow separation (a) splitter plate (b) vortex generator.

to use a wide angle intake with lower total pressure loss. Of course, there will be losses in total pressure due to presence of vortices. But frictional losses due to flow separation are in general much higher than the vortex losses for well-designed vortex generators. Thus, we can use intake with shorter length, which will not only reduce engine weight but also facilitate smooth engine installation.

Let us now see the cross-sectional view of a typical subsonic intake with its geometric parameters as shown in Fig.11.5 where, D_i is the inlet diameter of intake, D_e is exit diameter of intake, D_c is the diameter of center body at the exit, L is its length, D_{max} is the maximum diameter of intake. The diffuser can accomplish compression of air flow over a wide range of subsonic flight condition. The design of lip geometry is very crucial for successful external compression of air in a subsonic air intake which is discussed in the next subsection.

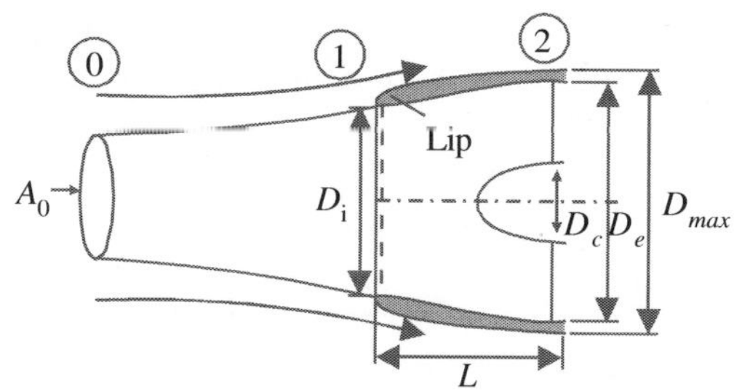

Fig. 11.5. Schematic of a typical subsonic intake.

11.1.2 External Compression Subsonic Air Intake

In view of difficulties in compression in an internal air intake, one can go for an external compression subsonic diffuser. A typical streamline pattern for large external deceleration is shown in Fig. 11.6(a). The shape of the lip causes the streamlines to diverge as they approach the inlet, the flow is decelerated, giving rise to increase in static pressure. This compression process is almost isentropic since most of the diffusion takes place without any physical boundary. Hence all relationship derived for isentropic flow can be applied to external compression diffuser. However, there are limits to the

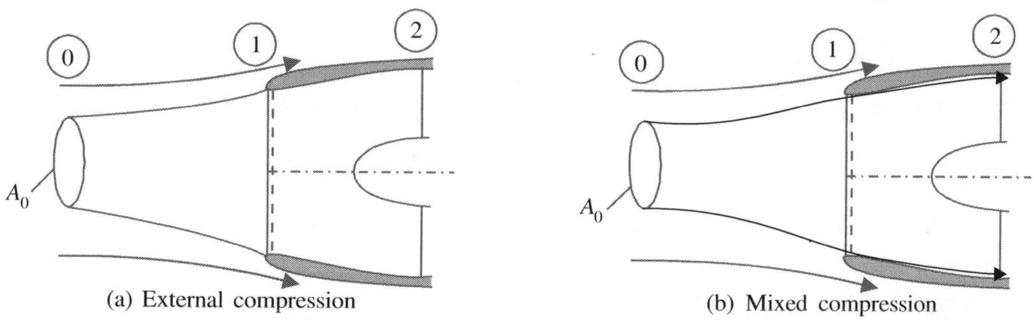

Fig. 11.6. Flow streamlines of (a) an external and (b) a mixed compression air intakes.

amount of free stream diffusion, because there is a great possibility of flow passing over external surface of the inlet leading edge which gets separated from the surface as shown in Fig. 11.6(a). Thus, this flow separation can lead to higher external drag which will eventually reduce the effect of positive thrust generated by the engine. For high subsonic speed region, partial supersonic flow can occur around outer surface of the rounded inlet lip that can result in a normal shock formation, which may cause boundary layer to separate and incurs exorbitant external drag.

It can be concluded from the above discussion that neither internal nor external diffuser can serve the complete purpose. Rather a combination of both external and internal compression types may be the best choice for designing an optimum air intake.

11.1.3 Mixed Compression Subsonic Intake

Figure 11.6(b) shows the schematic of a mixed compression subsonic intake, used in actual practice. Generally a small portion of total compression across the intake is carried out externally. It must be kept in mind that this external, compression intake may function well only when it is operated under design conditions. However, the intake has to operate from static condition to cruise flight speed with high efficiency at all altitude and operational flight speeds. Let us see what happens to the flow pattern in the intake during its operation at different flight conditions.

11.1.3.1 Airflow Pattern

We know that the operating conditions of an intake can be varied not only by the flight speed but also by the mass flow as demanded by the engine. Let us see the streamline patterns of the flow as shown in Fig. 11.7 for three typical subsonic conditions. Figure 11.7(a) shows the streamline pattern of intake at design level cruise flight speed. It can be seen that the streamlines diverges even before entering into intake duct without the aid of duct surface and hence ensuring isentropic external compression of air. Of course, the streamlines are further diverged inside intake duct due to increase in area and thus, decelerating the flow further to the desired value. The combined diffusion processes (external and internal compression) are described in the T-s diagram in 11.8.

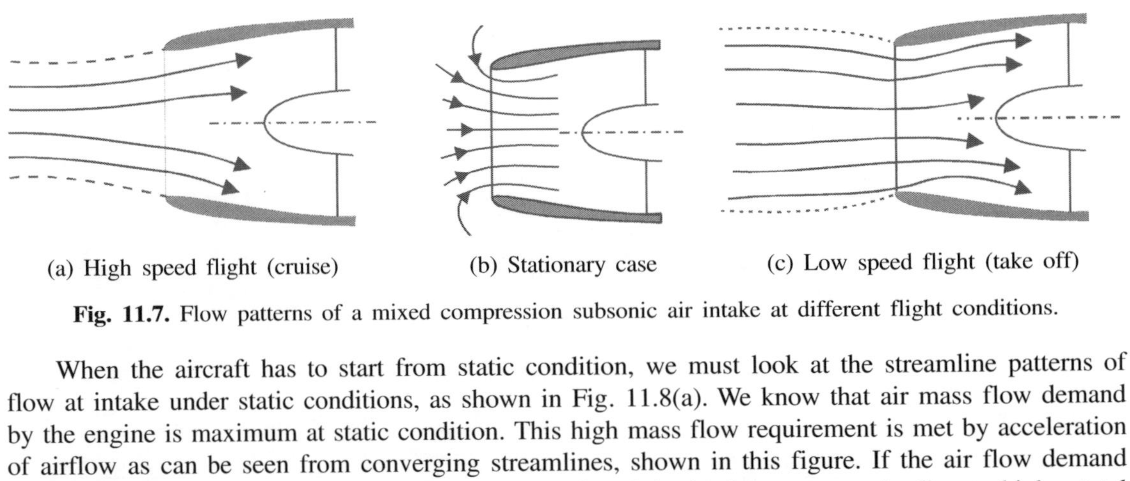

(a) High speed flight (cruise) (b) Stationary case (c) Low speed flight (take off)

Fig. 11.7. Flow patterns of a mixed compression subsonic air intake at different flight conditions.

When the aircraft has to start from static condition, we must look at the streamline patterns of flow at intake under static conditions, as shown in Fig. 11.8(a). We know that air mass flow demand by the engine is maximum at static condition. This high mass flow requirement is met by acceleration of airflow as can be seen from converging streamlines, shown in this figure. If the air flow demand is too high, then severe break down of flow at the lip of the intake can occur leading to higher total pressure loss. To overcome this problem, some aircraft manufacturers had used spring-loaded doors in the intake duct in earlier days. When the pressure loss in intake duct is quite high the differential

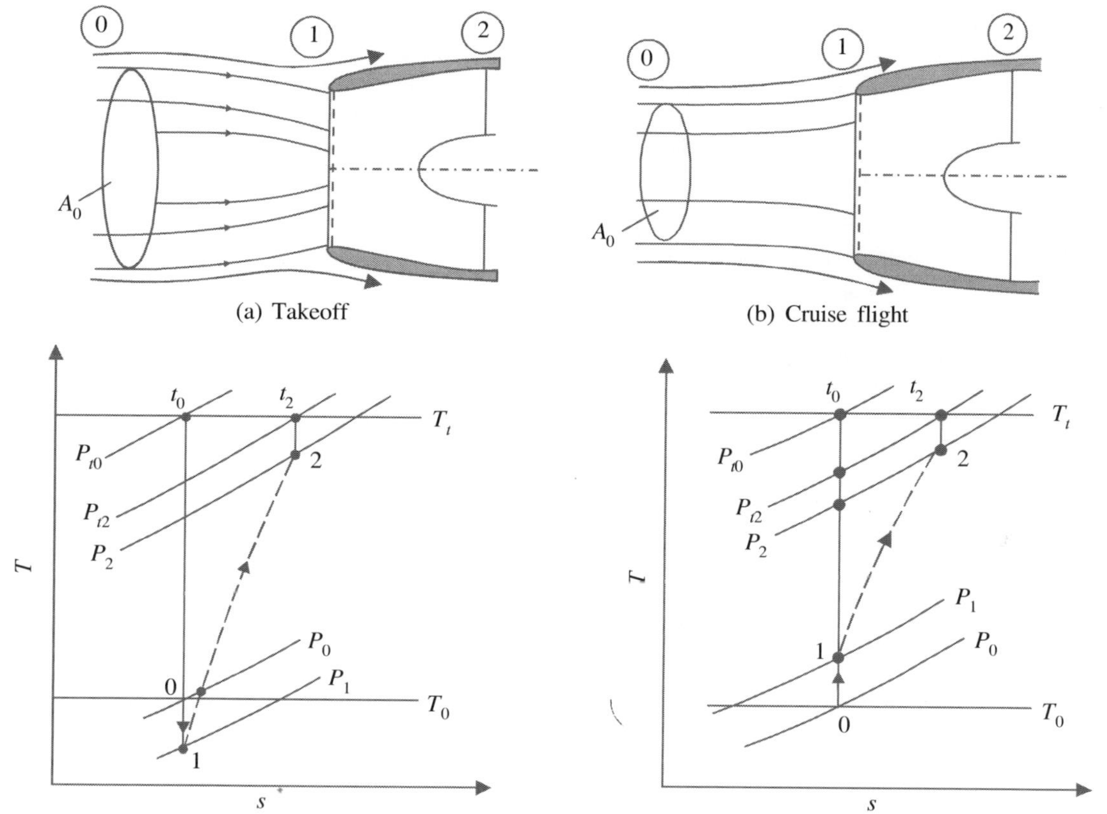

Fig. 11.8. Flow patterns of air intake at (a) takeoff and (b) cruise flight.

pressure between the duct and ambient, causes the door to open automatically to supply the requisite airflow rate to the engine. This kind of arrangement is not in use in recent times due to the inherent noise problems. Of course, during low speed and high thrust operation corresponding to climb/acceleration, the flow pattern will be distorted. The flow has to be less accelerated in the external region followed by deceleration in its duct similar to takeoff condition. When the aircraft is moved at low flight speed, the flow pattern changes as shown in Fig. 11.7(c). It can be observed that flow in the external portion of inlet is accelerated isentropically to supply the deficit amount of air flow at the inlet of the compressor. As the flow is subsonic, flow is decelerated in the internal portion of air intake. The process of compression under both takeoff and cruise flight has been described on T-s plot (see Fig. 11.8). Note that at cruise flight the flow gets compressed and decelerated both externally and internally as described in Fig. 11.8b unlike at take off condition. Before leaving this section, let us see what are the vulnerable zones in intake where flow separation is likely to occur as shown in Fig. 11.9. Since the flow is accelerated over the outer surface around the lip in zone (1) followed by subsequent deceleration, the flow is likely to separate in this zone. Similar situation of acceleration of flow

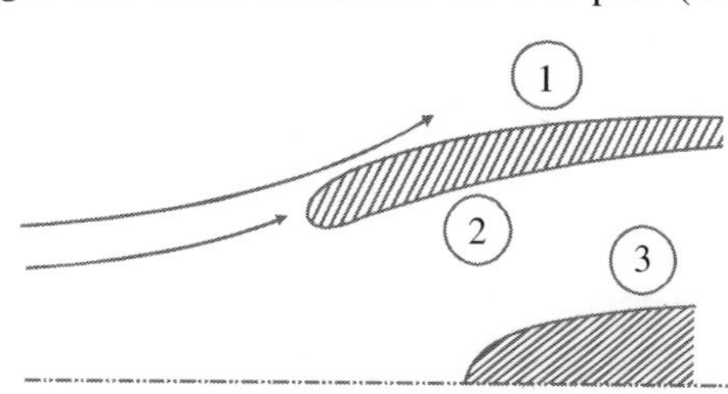

Fig. 11.9. Potential regions of boundary layer separation in a subsonic intake.

followed by deceleration under adverse pressure gradient arises in zone (3) around center body resulting in flow separation. The separation of flow on the internal surface of zone (2) may also occur as it is dependent on operating conditions and geometry. Hence, proper care must be taken during design phase to avoid flow separation. Keep in mind that all three zones may be subjected to adverse pressure gradient leading to flow separation at high angles of attack, which must be avoided at any cost.

11.2 Supersonic Intake

We know that supersonic intake decelerates the incoming supersonic flow to an approximately sonic speed followed by subsonic deceleration of flow to requisite velocity. For decelerating the supersonic flow, it has to pass through at least one shock wave. But the deceleration of flow through a shock is always accompanied by stagnation pressure loss. In fact, the intensity of this total pressure loss is dependent on the strength of the shock. Hence, multiple shocks of low intensities are employed to keep the pressure loss to a minimum possible value. When multiple shocks are used, it is not easy to control all of them for which care must be taken during design. But it incurs an intolerably large increase in external drag. Both these features are in conflict with each other. Therefore, the designer should strive to obtain the best compromise between maximum pressure recovery and minimum external drag. Besides this, the supersonic intake with multiple shocks poses serious operational problem, which will be discussed in subsequent sections.

Based on the type of compression, the supersonic intake can be classified under three categories namely (i) External compression, (ii) Internal compression, and (iii) Mixed (External and Internal) compression.

11.2.1 External Compression Supersonic Intake

As the name indicates, the external compression intakes are those in which shocks can be formed ahead of intake duct. Several types of external compression intakes can be found in literature. But we will restrict our discussion to only four representative external compression supersonic intakes namely, (i) Pitot, (ii) Single cone, (iii) Double cone and (iv) Isentropic (Oswatitsch) cone intake.

11.2.1.1 Pitot Intake

A simple subsonic intake, used for supersonic flight condition is the *pitot air intake*. A schematic of a typical pitot intake is shown in Fig. 11.10. Let us understand how does the pitot intake work in supersonic flow. When the flight speed exceeds the speed of sound, a shock wave builds up in front of the intake running almost perpendicular to the incoming flow. As a result, the flow is decelerated instantaneously from supersonic to subsonic speed. There will be a rise of static pressure and density across the shock wave. Hence, this kind of deceleration of

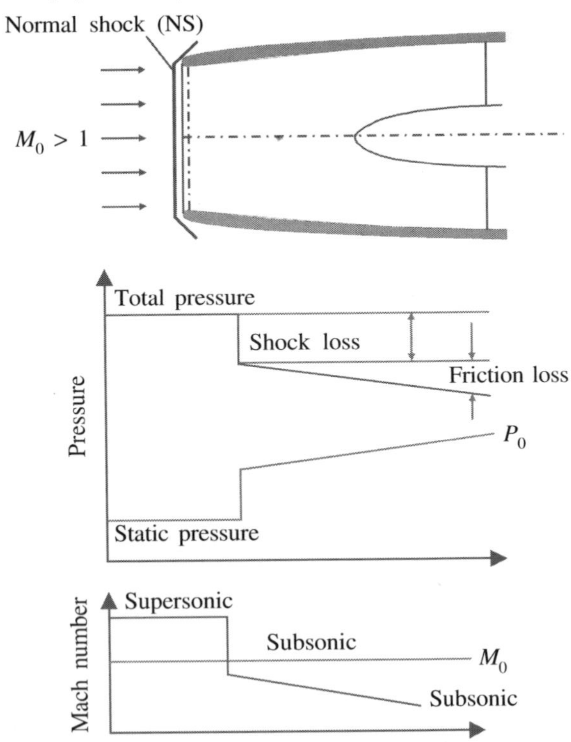

Fig. 11.10. A schematic of a typical Pitot intake and variation of P, P_t, M_0 along with its length.

flow is also known as normal shock compression air intake. Keep in mind that the formation of shock is exploited in air intake to compress the air, in contrast to usual notion of avoiding the formation of shocks at any cost. But in supersonic air intake, the formation of low strength shocks is desirable to achieve deceleration of flow which incurs minimum total pressure loss. But the flow while passing through a normal shock always suffers a higher loss in total pressure. This loss becomes exorbitant with increasing flight Mach number. Therefore, the upper limit of pitot intake is restricted to approximately a flight Mach number of 1.5. When the flow enters into a diverging duct, it is decelerated to a designed velocity. Of course, it incurs additional skin friction losses. To reduce skin friction losses, shorter intake duct is preferred. Both total and static pressure and Mach number change along the intake are shown in Fig.11.10 which clearly illustrates the deceleration of flow accompanied with both shock and friction losses. Note that this type of intake has been used in several aircrafts for example on Mirage III, V and F1 as well as on Fighter F-104.

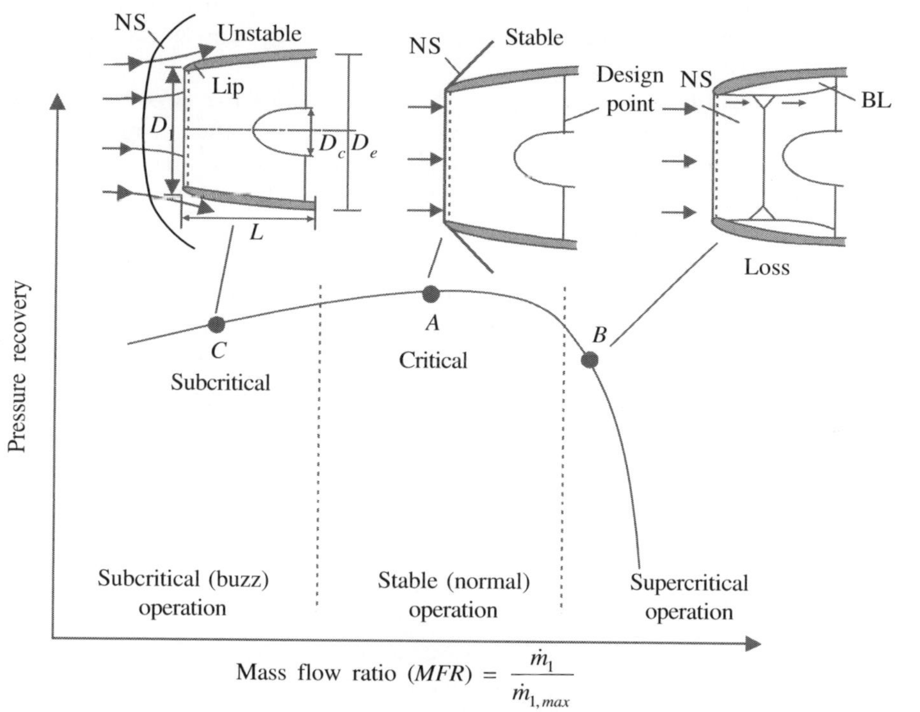

Fig. 11.11. Characteristic intake behaviour of a pilot intake.

Its operational behaviour: The normal shock in the pitot intake can occur adjacent to its lip only when it is operated at design conditions. This is known as critical operation, shown in Fig. 11.11 as point A. In this case, the mass flow captured by area A_0 can easily pass through inlet cross-section A, completely and the capture area ratio A_0/A, becomes unity. Under this condition, the maximum possible mass flow rate $\dot{m}_{1,max}$ for certain flight speed is being reached. Hence, in order to ascertain its effect over the whole operational range, we must define a term known as the *Mass Flow Ratio*, MFR as follows

$$MFR = \frac{\dot{m}_1}{\dot{m}_{1,max}} = \frac{\text{mass flow rate actually ingested}}{\text{maximum swallowed mass flow rate}} \tag{11.6}$$

This *MFR* states what is the mass flow rate, $\dot{m}_1$, actually ingested into intake as per engine demand, at particular flight condition in comparison to the maximum possible mass flow rate, $\dot{m}_{1,max}$ that can be swallowed by the engine.

However, in actual practice, the aircraft has to fly over the whole range of flight conditions. When the aircraft moves with a higher speed, the capture area for the required engine mass flow rate must be small. But the flow being supersonic the actual capture area, A_0 becomes quite higher than requisite capture area to deliver required mass flow rate to the engine. As a result, there will be accumulation of mass in the intake duct, giving rise to increase of the static pressure. Due to which, the static pressure ratio across the shock increases resulting in an increase in propagating speed of the shock relative to the fluid. Hence, the shock moves away from the lip of the intake. Since the flow will be subsonic downstream of shock wave, the streamline will diverge allowing only air mass flow to enter into the intake duct as required by the engine. This phenomenon is known as *spillage*. Such an operation is called *subcritical* operation (see point *C* in Fig. 11.11).

Now if the mass flow ratio decreases further, then the shock location moves outward further. Due to the position of the shock far in front of the inlet, the mass flow decreases more strongly than in keeping with the engine requirement. If the mass flow requirement of the engine increases suddenly for a short time, the shock will jump into the inlet duct (point *B* in Fig. 11.11). This mass flow is too large as compared to engine requirement. As a result, the shock is squeezed out again and the process starts afresh. This phenomenon may lead to periodic pressure fluctuations, which may initiate unwanted surging process in the compressor. This phenomenon is known as *buzz*. It also produces intense noise, which must be avoided at any cost. Sometimes this intense aerodynamic instability may result in extinguishing flames in the combustion chamber. It has not been thoroughly understood till date. Boundary layer bleed can delay the onset of buzz. In some case, the use of a lengthy non-divergent subsonic passage may have beneficial effect because of establishment of healthy boundary layer.

Example 11.1 A subsonic air intake with A_2/A_1 of 4.0, is used in aircraft, flying at an altitude of 15 km, with M_0 of 1.5 forms a normal shock at its inlet. If the isentropic efficiency is 0.85, determine (a) Mach number (b) Static pressure and temperature at its exit.

Given P_0 = 12.11 kPa, T_0 = 216.7 K, M_0 = 1.0, $\frac{A_2}{A_1}$ = 4.0, η_{id} = 0.85.

Solution A normal shock is standing at the inlet of a subsonic air intake.

To find M_2, P_2 and T_2.
From normal shock table, we can get Mach number downstream of normal shock, M_1 = 0.7011 given in Appendix B2. We need to know the Mach number at the exit of intake, M_2, which can be obtained by area-Mach number relationship for 1D isentropic flow as given below

$$\frac{A_2}{A^*} = \frac{1}{M_2}\left[\frac{2}{\gamma-1}\left(1+\frac{\gamma-1}{2}M_2^2\right)\right]^{\frac{\gamma+1}{2(\gamma-1)}}$$

But we need to know the value of $\frac{A_2}{A^*}$ as given below

$$\frac{A_2}{A^*} = \frac{A_2}{A_1}\times\frac{A_1}{A^*} = \frac{A_2}{A_1}\times\frac{1}{M_1^2}\left[\frac{2}{\gamma-1}\left(1+\frac{\gamma-1}{2}M_1^2\right)\right]^{\frac{\gamma+1}{2(\gamma-1)}} = 4\times 1.0928 = 4.37$$

For known value of $\frac{A_2}{A^*}$, we can get value of $M_2 = 0.134$ from the isentropic table given in Appendix B1. From the definition of isentropic efficiency, η_{id}, we can have

$$\eta_{id} = \frac{\left(\frac{P_{t2}}{P_0}\right)^{\frac{\gamma-1}{\gamma}} - 1}{\frac{(\gamma-1)}{2} M_0^2}$$

Using the above relationship, we can evaluate easily $\frac{P_{t2}}{P_0}$ and P_{t2} as given below

$$\frac{P_{t2}}{P_0} = 3.11 \Rightarrow P_{t2} = \frac{P_{t2}}{P_0} \times P_0 = 37.66 \text{ kPa}$$

From isentropic relationship, we can get the value of P_2 for $M_2 = 0.134$ as given below

$$P_2 = \frac{P_2}{P_{t2}} \times P_{t2} = \frac{1}{\left[1 + \frac{\gamma-1}{2} M_2^2\right]^{\frac{\gamma}{\gamma-1}}} \times P_{t2} = 0.987 \times 37.66 = 37.17 \text{ kPa}$$

Similarly the static temperature, T_2 at the intake exit is found to be 430 K noting that $T_{t2} = T_{t0} = 432.1$ K.

Note that this problem can be solved easily using tables for both normal shock and isentropic flow given in the Appendix. However, the isentropic relationship is used sometime deliberately to demonstrate that this problem can be solved using relationships derived in this book.

11.2.1.2 Single Cone Intake

We have already seen that the pitot air intake incurs exorbitant high total pressure loss when the flight Mach number exceeds around 1.5. For example, there will be 25% of total pressure loss at flight Mach number of 2.0. Hence, we will have to decelerate the approaching flow by using oblique shocks. For this purpose, a single conical center body is employed in the pitot intake with a sharp lip as shown in Fig. 11.12 which forces the fluid to form an oblique shock followed by a normal shock. Such kind of intake is known as *single cone intake*. Although such shock is not isentropic, but stagnation

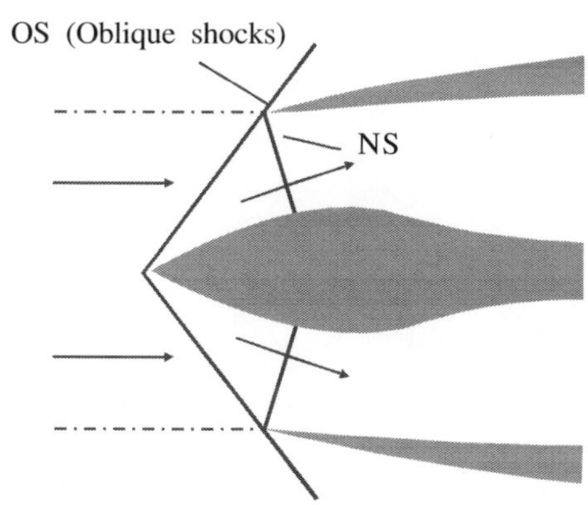

Fig. 11.12. Schematic of a single cone supersonic intake.

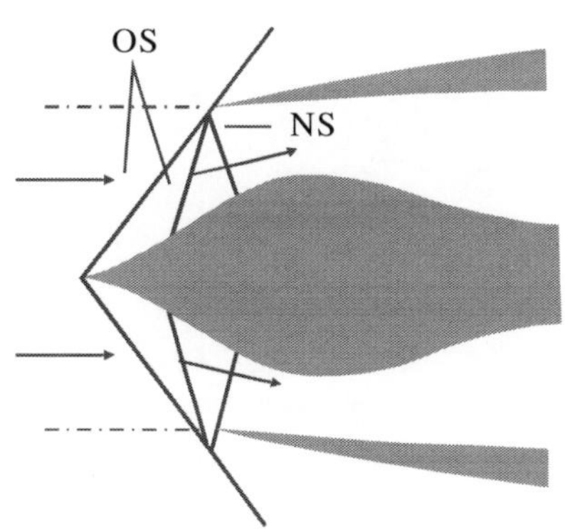

Fig. 11.13. Schematic of a double cone supersonic intake.

pressure loss in this intake is less than that of pitot intake at the same flight velocity. This pressure loss is dependent on the flight Mach number and geometry of the center body (cone). By selecting suitable intake geometry and cone angle, the shock can be made to fall at the cowl lips and the normal shock needed to decelerate the flow to subsonic condition can also be located at the lip. Of course, usual subsonic diffusion takes place inside its duct. Since shock and boundary layer (in this external compression process) may interact in a complex manner, resulting in more pressure losses. The designer must attempt to locate the shocks at the lower portion of lip where boundary layer is likely to be absent.

11.2.1.3 Double Cone Intake

When the flight Mach number becomes high, one has to go for double cone intake, a straightforward extension of the single cone intake in order to have lower total pressure loss. A schematic of a double cone intake is shown in Fig. 11.13 in which two oblique shocks are formed, the first one by the tip of conical center body and second one by the change in cone angle followed by a weak normal shock. It is not that easy to make both the oblique shocks to fall on the cowl lip, which is required to have minimum total pressure losses. Of course, with good design and proper fabrication, two conical shocks can meet at the cowl lips at design flow conditions from which normal shock can emanate. As a result, the combined effect of the two conical shocks as compared to that of a single cone intake will be quite less and thus its performance is improved for the same flight speed. The pressure recovery for two and three cone intakes are shown in Fig. 11.15. It must be noted that the pressure recovery improves with increasing number of oblique shocks for a particular flight Mach number.

11.2.1.4 Isentropic (Oswatitsch) Intake

We have seen in the previous section that by increasing number of oblique shocks, we could decrease total pressure loss for a particular flight Mach number. Then one can imagine that if infinite numbers of weak oblique shocks can be employed for decelerating the air, then the compression will be isentropic. Thus, at least theoretically, one can achieve zero total pressure loss. This noble concept was shown for the first time, theoretically by Oswatitsch, a German inventor (Fig. 11.14). Intake based on this principle is known as *Oswatitsch (isentropic) intake*. The pressure ratio across the external compression intake are shown in Fig. 11.15 with n number of oblique shocks of uniform strength and

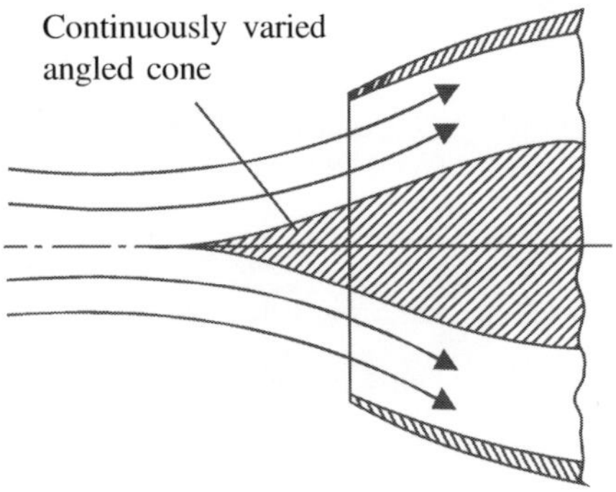

Fig. 11.14. Schematic of isentropic supersonic intake [4].

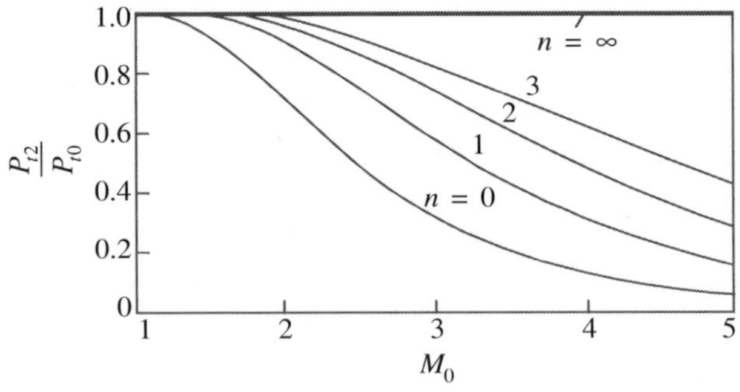

Fig. 11.15. The variation of stagnation pressure ratio with flight Mach number for supersonic intake with several oblique shocks [4].

one normal shock. It can be noted that with increasing number of oblique shocks, the pressure losses decreases and attains zero value theoretically when infinite (large) number of oblique shocks are formed. However it is quite difficult and cumbersome to achieve this performance in actual air intake under flight conditions.

Figure 11.14 indicates that for infinite number of oblique shocks, the theoretical pressure recovery factor is unity over at the flight Mach number which is desirable. Unfortunately, the efficacy of this type of intake has not been practically demonstrated till date due to several practical difficulties. It is quite difficult to design and develop a continuous varied cone that the cowl could produce isentropic compression waves. Furthermore, the boundary layer over the curved surface of isentropic cone unlike that along a plane surface would be subjected in high-pressure gradient, which may cause the flow to separate. In addition, flow has to be turned through a large angle particularly before reaching the sonic velocity for which a large cowl is needed and thus giving rise to higher external drag. Besides these difficulties, the cone would only be of the correct contour at one particular flight Mach number. The performance of isentropic intake is greatly affected even when there is small change in operating conditions. For these reasons, the isentropic intake has not been used till now in aero engines.

11.2.1.5 Two dimensional Ramp Intake

We know that flow downstream of a conical shock in an axisymmetric intake is not uniform. In fact the streamlines downstream of the shock for a cone are curved as shown in Fig. 11.16 as compared to supersonic flow over a wedge. This curvature of streamlines result in further diffusion but with non-uniform Mach number. Hence, its pressure loss will be lower in comparison to two-dimensional intake. Although, an axisymmetric intake can have a high-pressure recovery, but it incurs high-pressure loss when the flow is not symmetrical to the axis of the cone. This kind of situation arises during sideslip and turning flight manoeuvres leading to severe distortion of the flow. As a result, the compressor may be forced to operate around a dangerous surge limit. In contrast, the two-dimensional intake is less sensitive to unsymmetrical flows even if it incurs higher total pressure loss. Thus, instead of cone, wedge is used for forming oblique shocks, in certain fighter aircrafts.

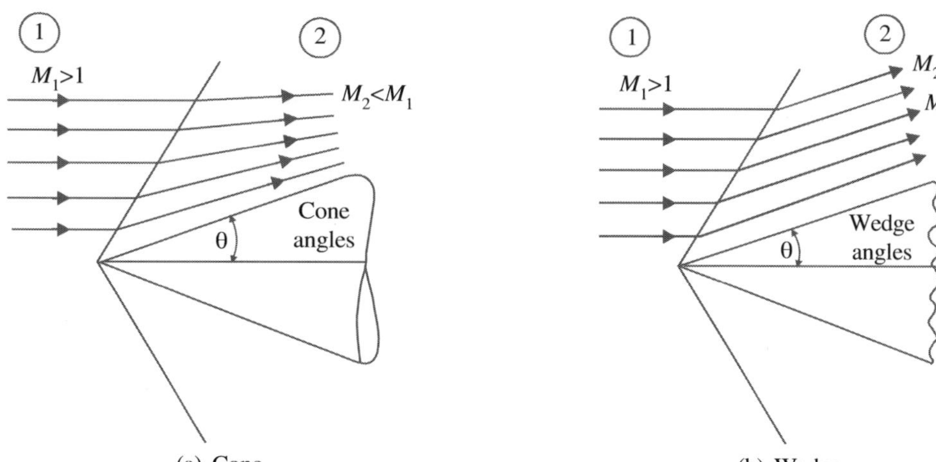

(a) Cone (b) Wedge

Fig. 11.16. Comparison of supersonic flow at cone and wedge.

In the two-dimensional intake, streamlines immediately run parallel after the stock has passed as shown in Fig. 11.17. Several modern combat aircrafts such as the F-14, F-15, MIG-25 and Tornado are using two-dimensional intakes. These types of intakes can also be found in other civilian supersonic aircraft like the famous Concorde, Tu-14, to name few.

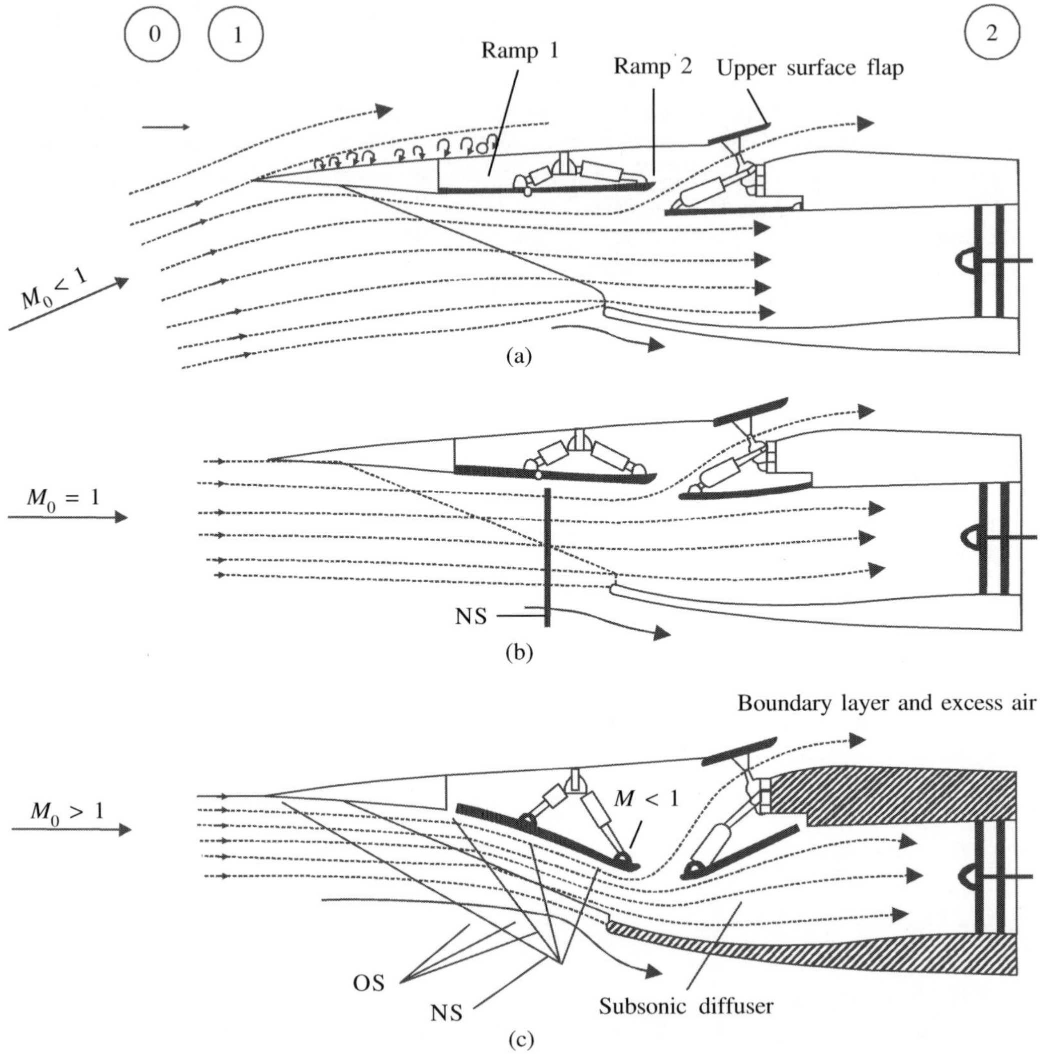

Fig. 11.17. The ramp position and flow of a typical 2D intake at various flight conditions.

Besides these reasons, the analysis of two-dimensional shock waves can be carried out much easily for design purpose. Therefore, we will consider only two-dimensional air intake for illustration purpose. Generally, fixed geometry intake can be used without any problem up to flight Mach number of 1.5, because adjustment mechanisms are not required. Besides this, the flow in the fixed geometry intake becomes quite unstable during acceleration from low flight Mach numbers up to design Mach numbers or during sudden throttling of the engine. The normal shock in the narrowed cross-section is squeezed out of inlet within few milliseconds. As a result, there will be high-pressure losses, which cause the compressor to surge instantaneously. In order to reduce this so called constant area intake problem, several methods such as perforated walls, inlets with bypass apertures as well as half-open inlets are used. To overcome these difficulties, provision must be made to vary the position of the center body. By this means, the shock fronts are adapted properly to achieve better performance for a particular flight Mach number. Such kind of variable intake has been used in F-14. The ramp position and corresponding flow pattern is shown in Fig. 11.17 for various flight conditions (a) subsonic speed and high angle of a attack (typical manoeuver cone) (b) transonic flow, and (c) supersonic flow. An open-loop control system has been used in practical air intake due to its simplicity, operational safety and low cost. This kind of air intake is preferred up to flight Mach number of 2.5. Interestingly, the position of the terminating normal shock does not create any problem because the normal shock moves only a very small area and has negligible effect on pressure recovery and flow distortion.

Operational Behaviour of External Compression Axisymmetric Air Intake

Let us turn our attention to the operational behaviour of external compression axisymmetric air intake as shown in Fig. 11.18. We know that, the oblique shock in the external compression axisymmetric intake can occur adjacent to its lip as shown in Fig. 11.18(b) with an emanating normal shock from its lip, only when it is operated at design conditions and this is known as critical operation. In this case, the mass flow captured by area can easily pass through inlet cross-section with very little spillage during actual operation. Under this condition, the maximum possible mass flow rate $\dot{m}_{1,max}$ for certain flight speed is being reached. As mentioned earlier the *Mass Flow Ratio*, *MFR* can be employed to ascertain its effect over the whole operational range. Recall that the *MFR* indicates what is the mass flow rate, $\dot{m}_1$, actually ingested into intake as per engine demand, at particular flight condition in comparison to the maximum possible mass flow rate, $\dot{m}_{1,max}$ that can be swallowed by the engine.

Note that in actual practice, the aircraft has to fly over a wide range of flight conditions. When the aircraft moves with a higher speed, the capture area for required engine mass flow rate must be small. But the flow being supersonic the actual capture area becomes quite higher than requisite capture area to deliver required mass flow rate to the engine. As mentioned earlier, there will be accumulation of mass in the intake duct, giving rise to increase of the static pressure. Due to which, the static pressure ratio across the shock increases resulting in an increase in propagating speed of the shock relative to the fluid. Hence, the oblique shock moves away from the lip of the intake and

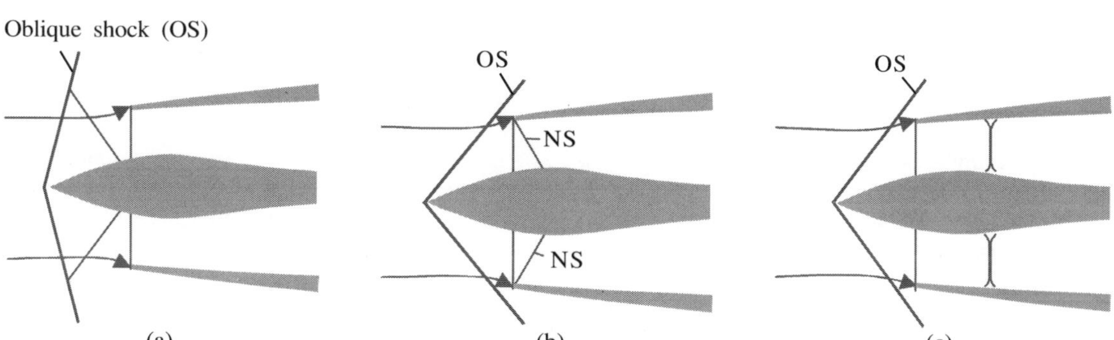

Fig. 11.18. Three operational modes of external compression axisymmetric air intake: (a) Subcritical (b) Critical, and (c) Supercritical.

another normal shock is formed upstream of the intake as shown in Fig. 11.18(a). It results in an increase in spillage causing a higher external drag. Such an operation is called *subcritical* operation during which the position of shock keeps changing leading to rapid oscillations resulting in onset of *buzz*. As discussed earlier, due to the position of the shock far in front of the inlet, the mass flow decreases more strongly than in keeping with the engine requirement. If the mass flow requirement of the engine increases suddenly for a short time, the shock will jump into the inlet duct. This mass flow is too large as compared to the engine requirement. As a result, the shock changes its position again and the process starts afresh. As it results in violent fluctuations in internal flow field of air intake, it affects the overall performance of entire gas turbine engine. It may lead to compressor surging and combustion instabilities and intense noise which must be avoided at any cost. Sometimes this intense aerodynamic instability may result in extinguishing flames in the combustion chamber. Although this phenomena of buzz has not been understood fully. It has been known that it can be brought under control by controlling downstream conditions of the air intake. It has also been found that the onset of buzz can be delayed considerably by bleeding air from the boundary layer over the center body. Another way is to provide longer subsonic passage just downstream of intake which can result in the occurrence of healthy boundary layer growth over the actual divergent portion of subsonic passage of the air intake.

When the aircraft moves at a lower speed than that of the design speed, the capture area for the required engine mass flow rate becomes higher. But the flow being supersonic, the actual capture area becomes smaller than the requisite capture area to deliver required mass flow rate to the engine. As a result, normal shocks are pushed into divergent passage of the air intake as shown in Fig. 11.18(c) and this operating condition is known as the supercritical condition. The position of normal shock is dependent on the back pressure downstream of air intake.

We know that the formation of shock waves is essential for the deceleration of supersonic to subsonic flow in supersonic air intake. However care must be taken to minimise the pressure losses due to shock and boundary layer interaction which is of course dependent on flight Mach number. In order to operate supersonic intake optimally, variable geometry can be employed to avoid flow instability over entire flight envelope. However, too much of variation in inlet geometry may lead to deterioration in its performance and must be avoided.

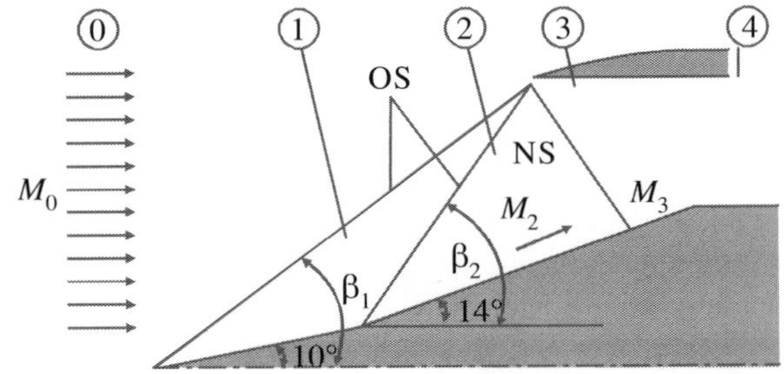

Fig. 11.19. Schematic of shocks formation in a 2d external compression supersonic intake with two ramp angles.

Example 11.2 A two-dimensional external compression supersonic air intake with two ramp angles of 10° and 14° respectively produces two oblique shock waves followed by a normal shock wave when it is operated at an altitude of 15 km with flight Mach number of 2.5. If the stagnation pressure loss is 5% in the subsonic portion of air intake, determine (i) Angle of oblique shock, and (ii) Overall total pressure ratio.

Given P_0 = 12.11 kPa, T_0 = 216.7 K, M_0 = 2.5, θ_1 = 10°, θ_1 = 14°, total pressure loss across subsonic inlet = 5%.

Solution Two oblique shocks are formed at the inlet of this air intake followed by normal shock as shown in Fig. 11.19.

To find β_1, β_2 and overall total pressure loss.

For $M_0 = 2.5$ and $\theta_1 = 10°$ we can obtain oblique shock angle $\beta_1 = 32°$ using $\theta - \beta - M$ curve as given in chapter 2.

The normal component of M_0 becomes

$$M_{n0} = M_0 \sin \beta_1 = 2.5 \sin 32° = 1.32$$

From the normal shock table for $M_{n0} = 1.32$, we can get

$$M_{n1} = 0.776; \frac{P_{t1}}{P_{t0}} = 0.976; M_1 = \frac{M_{n1}}{\sin(\beta_1 - \theta_1)} = \frac{0.776}{\sin 22°} = 2.07$$

For $M_0 = 2.07$ and $\theta_1 = 14°$ we can obtain oblique shock angle $\beta_2 = 43°$ from $\theta - \beta - M$ curve. The flow Mach number perpediculate to the 2nd oblique shock can be determined as

$$M_{n1} = M_1 \sin \beta_2 = 2.07 \sin 43° = 1.41$$

Then the supersonic flow undergoes a normal shock as shown in Fig. 11.19. From the normal shock table for $M_{n1} = 1.41$, we can get

$$M_{n2} = 0.736; \frac{P_{t2}}{P_{t1}} = 0.955; M_2 \frac{M_{n2}}{\sin(\beta_2 - \theta_2)} = \frac{0.736}{\sin 29°} = 1.518$$

Finally, there is a normal shock at exit as shown in Fig. 11.19. From the normal shock table, for $M_2 = 1.518$, we can get

$$M_3 = 0.694; \frac{P_{t3}}{P_{t2}} = 0.923$$

Then, the total pressure loses across this air intake can be estimated as

$$\frac{P_{t4}}{P_{t0}} = \frac{P_{t4}}{P_{t3}} \times \frac{P_{t3}}{P_{t2}} \times \frac{P_{t2}}{P_{t1}} \times \frac{P_{t1}}{P_{t0}} = 0.95 \times 0.923 \times 0.955 \times 0.976 = 0.817$$

Note that pressure loss incurred in two ramp external compression air intake is lower as compared to a single ramp air intake at same design condition. But it is quite to control the position of shocks during off-design operation. As a result, it may cause exorbitant pressure loss and other operational problems particularly when it is operated under either subcritical condition.

11.2.2 Internal Compression Supersonic Air Intake

Another way of achieving isentropic compression at least theoretically is the reverse convergent-divergent (de Laval) nozzle. Because, we know from simple one dimensional analysis that a Laval nozzle, operated in a reverse mode will decelerate the supersonic flow to subsonic velocity with a consequent rise in static pressure. Figure 11.20 illustrates the schematic of an internal compression intake which is similar to an inverted Laval nozzle. Note that a very sharp cowl lip is provided to avoid shock formation. Theoretically, the supersonic flow can be decelerated in the convergent portion at design condition. But if the flow velocity at the throat is still supersonic, then on entering into divergent portion, the air will accelerate, which must be avoided. But, the flow into the compressor has to be subsonic. Hence, a normal shock must be formed at the throat such that the strength of the shock will be quite small. Because, the total pressure loss is dependent on the strength of the shock. The operation of this intake incurring least pressure loss is known as critical operation which is most desirable.

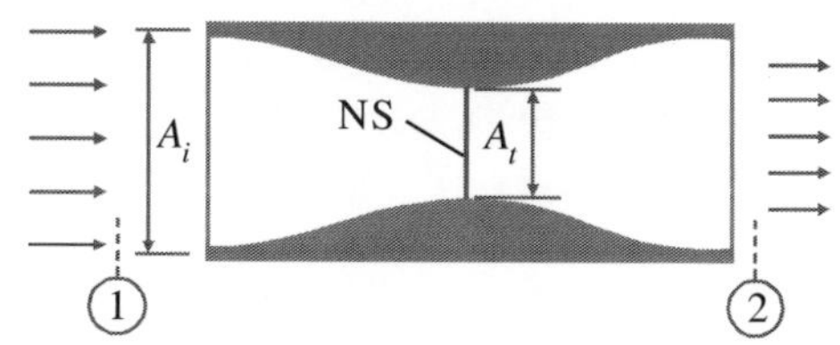

Fig. 11.20. Schematic of an internal compression supersonic intake.

In subcritical operation, the normal shock is located upstream of the throat (convergent portion) as shown in Fig. 11.21. Since this operation is quite unstable in nature, situation may arise when the normal shock may be expelled out of this intake (see Fig. 11.22). Then it will act like a pitot intake. Under this condition, total pressure loss will be exorbitantly high and flow spills over the cowl, producing a high drag and aircraft control problem. In contrast, the shock is located in divergent portion in supercritical operation which is quite stable. For this reason, the supersonic intake are designed to operate at supercritical condition. This is done to maintain a safe margin of stability in the event of sudden change in the inlet flow.

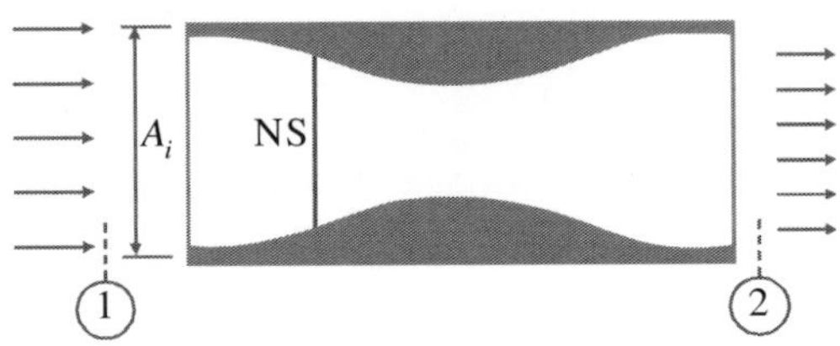

Fig. 11.21. Subcritical operation of an isentropic CD intake.

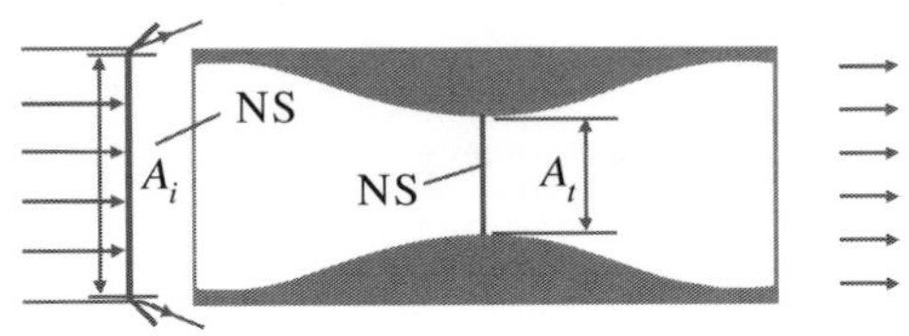

Fig. 11.22. The operation of a CD supersonic intake with expelled normal shock.

Till now, we have not considered the boundary growth into account while discussing about performance of intake. The flow is likely to separate from the inner surface of the intake incurring high total pressure loss. Hence, its wall contour must be carefully designed and constructed to avoid excessive pressure loss. Even for a well wall contoured intake, it is not possible to have isentropic deceleration due to formation of boundary layer which leads to shock formation, incurring high pressure loss. Even by considering the boundary layer effects, it is not easy to establish the desired flow at design condition. Such a problem is known as *starting problem* which is discussed in the next section.

11.2.3 The Starting Problem

We will discuss this starting problem of de Laval intake by considering the flow to be one dimensional and isentropic except losses due to shock formation. Although, it looks to be quite simplistic model for this kind of intake, however, it represents essential features associated with starting problem phenomenon.

Let us consider a convergent-divergent intake for a design Mach number, M_0 of 1.8 as shown in Fig. 11.23. From isentropic flow relationship, the area ratio, A_i/A^* (A_i = intake area, A^* = throat area), would be 1.65. For design conditions, the throat area, A_t, is the area A^* as the Mach number at the throat is unity. When the intake receives the air at M_0 = 1.8, then the capture area, A_0 would be equal to A_i. The deceleration of air would be taking place isentropically up to the throat where the Mach number is equal to unity. Then, the air is decelerated further to desired subsonic velocity while passing through the divergent duct, as shown in Fig. 11.20.

Now, we will see how does an intake behave under off design conditions. Suppose an aircraft takes off from the ground with increasing forward speed to achieve its design flight speed. In order to simplify the analysis, let us isolate intake behaviour from the rest of the engine. In other words, we can assume that whatever the air mass flow reaches the intake exit can always be swallowed by the engine. Hence, the mass flow rate is limited only by the chocking of flow at its throat, A_t.

We can see a subsonic operation of this intake as shown in Fig. 11.23(a). In this case the flow is accelerated both in external region and convergent portion of the intake, but the throat is not yet

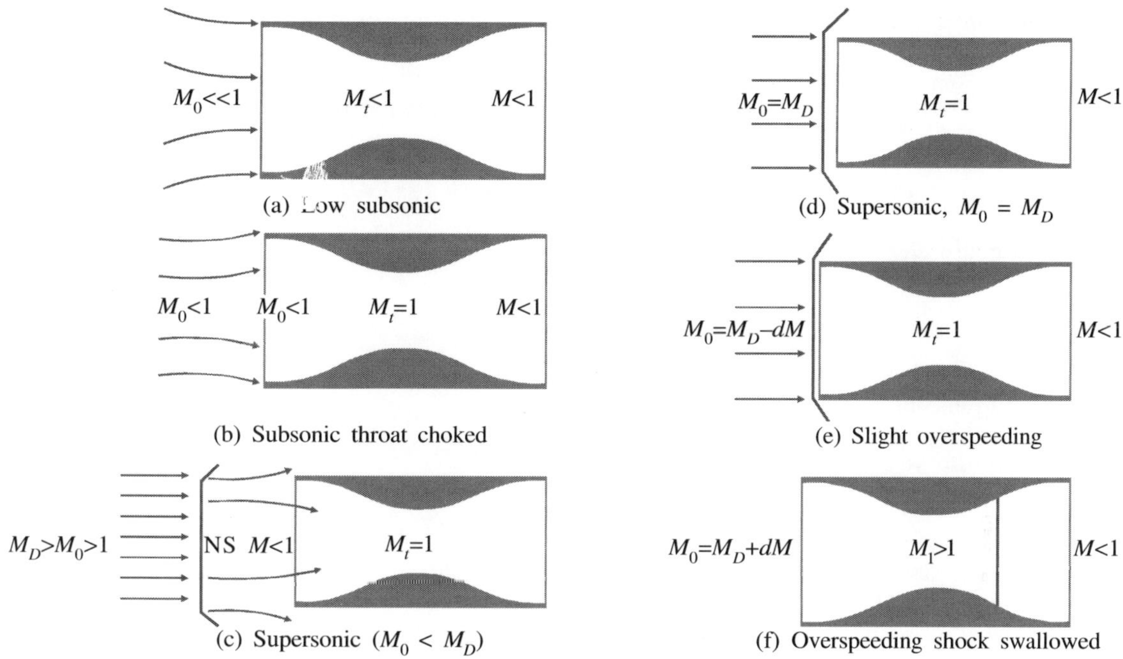

Fig. 11.23. Successive steps in the operation of CD supersonic intake.

chocked ($M_0 < 1$). Then the air decelerates in the divergent portion of the air intake. Of course, the upstream area A_0 is determined by downstream flow condition of the intake. In this case, A_0/A^* would be as shown in Fig. 11.24.

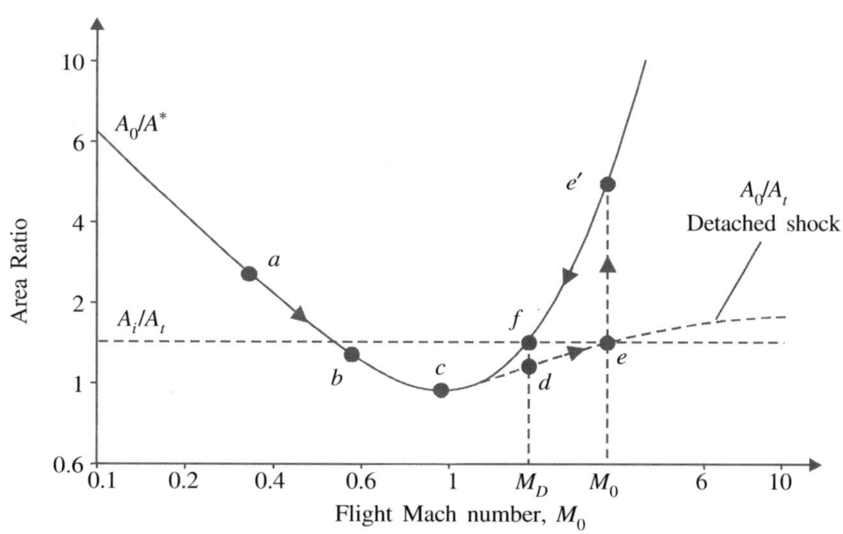

Fig. 11.24. Acceleration and overspeeding of 1-D supersonic CD intake.

Let us consider next situation (Fig. 11.23(b)), when the flight Mach number is still subsonic, but the throat is chocked ($M_t = 1$). Of course, M_0 for this chocked condition is dictated by the area ratio, $\frac{A_i}{A_t}$ of the intake. Keep in mind that the inlet mass flow rate is limited by chocking condition at A_t. Since, the flow is assumed to be isentropic, the A_t is equal to A^*. Then, for a particular $\frac{A_i}{A_t}$, the upstream capture area, A_0 can be estimated by

$$\frac{A_0}{A^*} = \frac{A_0}{A_t} = \frac{1}{M_0}\left[\frac{2}{(\gamma+1)}\left(1+\frac{(\gamma-1)}{2}M_0^2\right)\right]^{\frac{\gamma+1}{2(\gamma-1)}} \tag{11.7}$$

For high subsonic flow, (see Fig. 11.24),we have

$$\frac{A_0}{A^*} = \frac{A_0}{A_t} < \frac{A_i}{A_t} \tag{11.8}$$

As the capture area A_o is less then inlet area A_i for this condition, there will be spillage around the inlet. Then if M_o increases further, while keeping M_t fixed, there will be external deceleration of air from A_0 to A_i so that M_o would be unity as shown in Fig.11.23(c). Since the flow is at sonic condition, the stream lines cannot bend. Therefore, a normal shock will occur ahead of the intake (see Fig.11.23(c)). The stream lines of subsonic flow behind the shock can easily be diverged to provide enough mass flow rate to the choked throat, of course, for a fixed A_i/A_t ratio. Let us understand the mechanism of this shock formation.

Suppose, there is no shock wave, when the intake air attains firstly sonic/supersonic velocity. The flow will definitely enter into inlet area of the convergent-divergent duct without any deviation. But for this case, the allowable capture area is limited by chocking condition at A_t, $\left(\frac{A_0}{A_t} < \frac{A_i}{A_t}\right)$. As a result, the mass of air entering into the CD duct is accumulated in the convergent portion of the intake and result in a rise in static pressure in intake. This pressure rise will continue to increase rapidly till a shock is formed before the intake lip whose position is decided by amount of spillage required. Of course, as M_o increases further, the downstream subsonic Mach number behind the shock wave decreases and less-spillage is required. Hence the shock is formed closer to inlet lips (see Fig. 11.23(d)).

After a shock is formed in front of the intake, the flow entering the inlet cannot be considered to be isentropic. Because, there will be pressure losses due to presence of shock wave. Hence, when the aircraft reaches the design flight speed, then the throat would not allow even the designed mass flow to pass through it. Let us understand why is it so? We know that choked mass flow through a given A_t is given by

$$\frac{\dot{m}}{A^*} = \frac{\dot{m}}{A_t} = \frac{P_t\sqrt{\gamma}}{\sqrt{(RT_t)}}\left(\frac{2}{\gamma+1}\right)^{\frac{\gamma+1}{2(\gamma-1)}} \tag{11.9}$$

It can be seen from this expression that the chocked mass flow rate is proportional to P_t. But total pressure P_{t2} downstream of shock is much lower corresponding to actual flight Mach number without the presence of any normal shock. Considering a normal shock to be formed in front of the cowl lip, the downstream Mach number, M_2 (see Fig. 11.24) can be expressed in term of upstream Mach number M_0 by the normal shock relation as

$$M_2 = \sqrt{\frac{\frac{2}{(\gamma-1)} + M_0^2}{\frac{2\gamma}{(\gamma-1)} M_0^2 - 1}} \tag{11.10}$$

We can combine the above expression with Eq. 11.8 to get A_0/A_t for this case with flight Mach number M_0 is plotted in Fig. 11.24. It can be seen in this figure that it is quite different than isentropic case. For a particular A_t, inlet area A_i, is quite large and thus, spillage has to occur even beyond the M_D. Such a problem is known as *starting problem* of intake.

Question arises, how could we overcome this problem? Keep in mind that this problem arises due to the formation of normal shock in front of intake. Then, if this shock can be avoided or swallowed inside, then, this starting problem may not arise. One method is to increase the flight Mach number, M_0 beyond M_D to the point where the downstream Mach number of the shock is just that corresponding to A_i/A_t = 1.69. At this Mach number, the intake is capable of ingesting the incident mass flow rate without spillage, when the shock is just on its lip as shown in Fig. 11.23(e). With a little increase in flight speed (Fig. 11.24), the shock will enter into converging portion of the intake. As the shock is not very stable in this zone, it will move into divergent portion and attains a stable position as shown in Fig. 11.23(e). Now the flow through the throat is isentropic. Thus, from the isentropic relations, we can find out that A_0/A^* is greater than the actual A_0/A_t (see Fig. 11.24). Thus, the supersonic flow will decelerate from A_i to A_t and then it is again accelerated in the divergent part of this intake. Then, one has to reduce the flight Mach number to original design Mach number, as shown in Fig. 11.24. Such a method of overcoming starting problem of an air intake is known as *overspeeding*.

However, overspeeding method of overcoming the starting problem may not be possible, particularly when the flight Mach number is quite high. Another method is also used in which the shock can be swallowed by varying the throat area at a constant flight speed. The mechanism of this method can be easily illustrated by considering the case in which a starting shock is in front of the intake as shown in Fig. 11.25. The actual area ratio can be decreased by increasing momentarily the

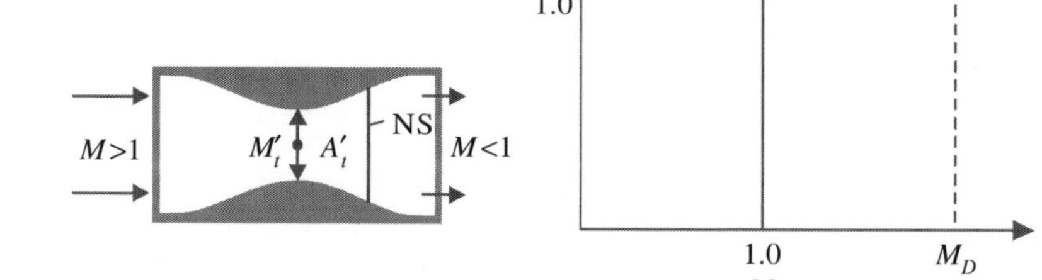

Fig. 11.25. Shock swallowing method by throat area variation.

throat area such that this new intake can manage to swallow all the flow behind the shock. Now the flow will be isentropic in front section of intake and throat Mach number M_t' will be greater than unity. Hence, a relatively strong shock will be formed downstream of the throat. Of course, the complete isentropic flow can be established throughout the intake by changing back to the original throat area. Thus the operating point can move from the point (e) to (d) through (f') as shown in Fig. 11.25. But this variation of throat area would be quite difficult in practice from mechanical point of view particularly for axisymmetric flow.

Another method of overcoming this starting problem is to use porous convergent wall. When the shock is formed in front of the intake, high static pressure downstream of the shock causes a large amount of air to flow through the porous walls which swallows the shock into the air intake. However, this method is not very popular. Because, the total porosity required to swallow the shock is even greater than the throat area, leading to large mass flow loss, which incurs excessive drag. Another reason is the non-availability of cheaper light weight porous materials for this purpose.

Note that when the normal shock is swallowed, the air intake is said to be operated under supercritical condition. However, if the shock is detached from the lip of an air intake leading to the spillage of incoming flow this operating condition is termed as subcritical. Recall that subcritical operation is occurred due to chocking of flow at the throat of intake. A de Laval air intake is said to be operated at critical condition when the normal shock falls on its lip itself.

11.3 Mixed Compression Intake

The features of both internal and external compression can be combined to design a new type of air intake which is known as Mixed compression air intake. A schematic of a typical mixed compression intake is illustrated in Fig. 11.26. As can be seen in this figure, a cone/wedge is employed to form initial oblique shock which lies on the lip of the intake, particularly at design flight Mach number. In contrast to external compression intake, in the case of mixed compression air intake, the duct after lip is converging one. In this intake, the supersonic flow prevails even after external oblique shocks. Then the flow is to decelerate further, while passing through the converging portion of this intake. Therefore, it is very suitable even for high flight Mach number aircraft, beyond 2.5 without using large intake diameter, in comparison to external compression intake. If the flow is still supersonic, then it will accelerate to form a normal shock in the diverging portion of intake, downstream the normal shock, the flow will be subsonic and hence will decelerate in remaining divergent portion to desired subsonic velocity. Generally, in normal operation, the normal shock should be held downstream of the throat to have a stable operation. However if it moves into upstream of the throat, then it can be easily stabilised with the help of center body. Of course, position of center body must be varied accurately. Thus, the problem of stability is not very severe like internal compression intake. Hence, it is preferred for high flight Mach number applications.

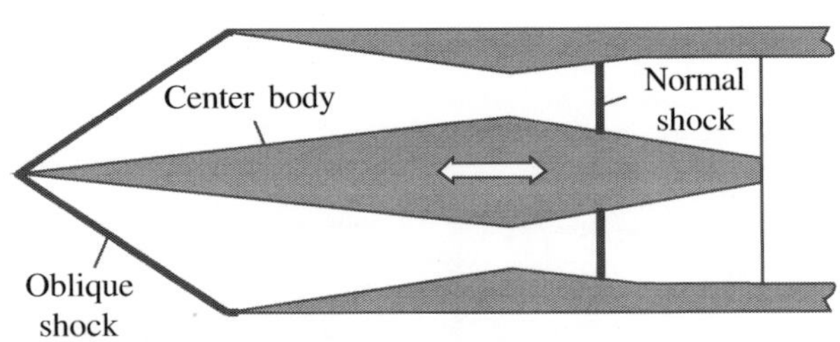

Fig. 11.26. A schematic of a typical mixed (internal and external) compression intake.

Let us see a practical example of mixed compression intake. An axisymmetric mixed compression intake is shown in Fig. 11.27 which is used in the Lockheed SR-71(USA) long range reconnaissance aircraft. Note, it has a center body which has combination of straight and curved profiles to produce number of low intensity oblique shocks, under supersonic condition. Three incoming external oblique shocks are clustered inside the intake cowl behind the sharp lips. Here, they are reflected as single oblique shock. Since more number of oblique shocks of low intensity are formed, the overall total pressure loss is quite low. Then the flow is decelerated in diverging duct subsonically. Beside this, it has a provision to bleed stagnation fluid on the boundary layer to avoid the flow separations over the center body. This can operate efficiently at extreme altitude even beyond flight Mach number of 3.0.

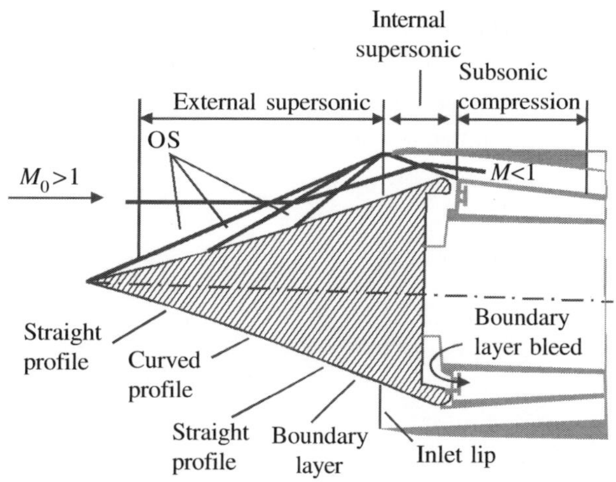

Fig. 11.27. Lockheed SR-71 (USA): Mixed compression intake.

11.4 Performance Criteria of Air Intake

We have already touched on briefly about the performance parameters of the intake while dealing with gas turbine cycle analysis which are revisited again for completeness. One may find several parameters in literature to describe the relative performance of air intake. However, we will restrict our discussion mainly to three parameters which are often employed to characterise its performance. These parameters are:

1. Isentropic Efficiency, η_{id}
2. Pressure Recovery Factor, PRF
3. Kinetic Energy Efficiency.

11.4.1 Isentropic Efficiency

The process of diffusion is described in T-s diagram in Fig. 11.8 for subsonic intake. Recall that the definition of isentropic efficiency, η_{id} is given by

$$\eta_{id} = \frac{T_{t_{2s}} - T_0}{T_{t_2} - T_0} = \frac{\dfrac{T_{t_{2s}}}{T_0} - 1}{\dfrac{T_{t_2}}{T_0} - 1} = \frac{\left(\dfrac{P_{t_2}}{P_0}\right)^{\frac{\gamma-1}{\gamma}} - 1}{\dfrac{(\gamma-1)}{2} M_0^2} \tag{11.11}$$

11.4.2 Pressure Recovery Factor, PRF

This is defined as the ratio of stagnation pressure at compressor face, P_{t_2} to the value which have resulted if the compression would have taken place isentropically in the air intake from ambient pressure

$$\text{PRF} = \frac{P_{t_2}}{P_{t_0}} \tag{11.12}$$

This is used widely as a performance parameter because it is simple to comprehend and use. Note that both the isentropic efficiency and PRF can be easily related as given below

$$\eta_{id} = \frac{(\text{PRF})^{\frac{\gamma-1}{\gamma}}\left[1+\frac{\gamma-1}{2}M_0^2\right]-1}{\frac{\gamma-1}{2}M_0^2} \tag{11.13}$$

Since, we know that

$$\frac{P_{t_2}}{P_0} = \frac{P_{t_2}}{P_{t_0}}\frac{P_{t_0}}{P_0} = \text{PRF}\left(1+\frac{\gamma-1}{2}M_0^2\right)^{\frac{\gamma}{\gamma-1}} \tag{11.14}$$

11.4.3 Kinetic Energy Efficiency

We know that the main purpose of propulsive device is to provide a high jet velocity. Hence, another criteria must be defined based on kinetic energy available for further expansion after undergoing diffusion process in the air intake that is given by

$$\eta_{KE} = \frac{h_{t_2} - h_e}{h_{t_2} - h_o} \simeq \frac{T_{t2} - T_e}{T_{t2} - T_0} \tag{11.15}$$

This criterion can be very useful for supersonic and hypersonic flight. This can be easily related to η_{id} and PRF as follows

$$\eta_{kd} = \frac{\eta_{id}}{(\text{PRF})^{\frac{\gamma-1}{\gamma}}} \tag{11.16}$$

11.5 Chapter Summary

This chapter presents the fundamental characteristics of air intake whose main function is to decelerate ingested air flow with an increase in its pressure and supply in sufficient quantity with adequate smooth profile to the inlet of the compressor. Although both subsonic and supersonic air intakes are used profusely in aircrafts, but the understanding of flow features and their associated complexities are yet to be understood fully. It is believed that modern tools like computational fluid dynamics will be helpful in resolving unsolved problems of air intakes in future. The processes involved during the operation of both external and internal subsonic air intakes are discussed with the help of T-s process diagram and streamlines. It is concluded that although external compression is free from losses, but it can only provide marginal compression of air in an aircraft engine. Hence mixed subsonic compression is being used in the actual aircraft engines. The losses in a subsonic air intake are caused due to mainly friction resulting from boundary layer and its separation, and turbulence. Both external and internal compression supersonic air intakes are discussed adequately in this chapter. Four types of external compression intake such as (i) Pitot, (ii) Single cone, (iii) Double cone, and (iv) Isentropic (Oswatitsch) air intakes are covered highlighting their respective operational limitations. The supersonic flow characteristics over both two dimensional (wedge) and axisymmetric (cone) ramps are delineated to demonstrate the differences between them. In order to make acquainted students about the applications of two dimensional ramp in actual aircraft engine, flow features over an external supersonic intake during its operating range are covered briefly. Besides this, shock relations are used to design two dimensional external air intake which are illustrated through examples. This kind of air intake has advantages of simple in design and manufacturing, shorter length and better off-design performance. It may incur excessive external drag which can be avoided by using adjustable ramp angle as shocks position and its length can be controlled easily. Isentropic compression of supersonic flow can be achieved at least ideally by using convergent and divergent (CD) air intake. This air intake may undergo through various operating conditions such as critical, supercritical and subcritical during its operation in actual aircraft. It incurs quite less external drag as compared to external compression air intake particularly at a higher flight Mach number. The starting of supersonic air intake are demonstrated considering an example of CD air intake. The mixed compression air intake is discussed briefly as it has several advantages over both external and internal compression air intake. The performance parameters of air intake such as isentropic efficiency, kinetic energy efficiency and pressure recovery factor (PRF) are discussed which are generally used for its characterisation. In the next chapter, we will discuss about the basic flow features of exhaust nozzle.

Nomenclature

A Cross sectional area
A Throat area
A_i Intake area
A_t Throat area
A_o Capture area
C_{PR} Pressure coefficient
D_c Center body diameter
D_e Exit diameter
D_i Inlet diameter
D_{max} Maximum diameter
L Diffuser length
M Mach number
M_t Throat Mach number
$\dot{m}$ Mass flow rate
P Pressure
P_t Total pressure
θ Divergence angle
θ_{max} Maximum divergence
θ_{min} Minimum divergence angle
S Entropy
T Temperature
W Width ratio

Greek Symbols

ρ Density
η_{id} Isentropic efficiency
η_{KE} Kinetic energy efficiency

Subscripts/Superscripts

c Center body
e Exit
i Isentropic
KE Kinetic energy
min Minimum

max Maximum
P Pressure
t Total

Abbreviations
CD Convergent divergent
MFR Mass flow rate
NS Normal shock
OS Oblique shocks
PRF Pressure recovery factor

References and Suggested Readings

1. Seddon, J. and Goldsmith, E.L., *Intake Aerodynamics*, AIAA Education Series, AIAA New York, 1985.
2. Kline, S.J., "On the Nature of Stall", *Journal of Basic Engineering*, Vol. 81, Series D, No. 3, 1959, pp. 305-320.
3. Hill, P.G. and Peterson, *Mechanics and Thermodynamics of Propulsion*, Addison-Wesley Longman Inc., 1992.
4. Oswatitsch, K., *Pressure Recovery for Missiles with Reaction Propulsion at High Supersonic Speed*, NACA, TM-1140, 1947.
5. Mattingly, J.D., *Elements of Gas Turbine Propulsion*, McGraw-Hill International, 1996.
6. Anderson, J.D., Jr., *Fundamentals of Aerodynamics*, 4th Edition, McGraw-Hill, New York, 2005.

Review Questions

1. Draw schematic of a typical subsonic air intake and describe its operating principle.
2. What are the advantages and disadvantages of internal compression air intake over external compression air intake?
3. Why does the shape of the intake lip affect the performance of the subsonic air intake?
4. What is the critical, subcritical and supercritical operation of the subsonic air intake?
5. Show schematically the variation of total enthalpy, pressure and temperature across a subsonic air intake. How are these properties comparable with their respective static quantities?
6. What is a starting problem in a CD air intake? Can you suggest methods to overcome this problem?
7. What is the critical, subcritical and supercritical operation of the supersonic air intake? Explain it in details.
8. Why is the two dimensional wedge preferred over cone for external compression supersonic air intake?
9. What is Oswatitsch air intake? Why is it not being used in practice?
10. What is buzz in an air intake? Why is it occurred at all? How to overcome it?
11. Why a normal shock would not attain a stable position in the convergent portion of the CD air intake? Explain this phenomena for both choked and unchoked flow condition.
12. Which kind of shock is preferred while designing a supersonic air intake? Why is it so?
13. Why is variable geometry air intake preferred in supersonic aircraft?
14. What is ram recovery point? What do you mean by isentropic efficiency? Will these be considered for designing of an air intake?

Problems

1. An air intake is to be designed operate at Mach number of 0.8 while consuming 50 kg/s of air. The freestream air at pressure of 10 kPa and 228 K enters through air intake with an inlet area of 3.2 m^2 and leaves its exit with Mach number of 0.3. If the isentropic efficiency of the air intake is 0.82, determine (i) Static pressure ratio across the intake (ii) PRF, and (iii) Kinetic energy efficiency.

2. An aircraft is flying at an altitude of 12 km with flight Mach number of 2.5. The exit velocity from this air intake turns out to be 220 m/s. If the kinetic energy efficiency of the air intake is 0.94, determine exit stagnation pressure and Mach number.
3. Air enters into an air intake with area ratio of 1.8 when the aircraft is flying at Mach number of 0.4. If the isentropic efficiency of the air intake is 0.82, determine (i) air intake inlet Mach number (ii) PRF, and (iii) Maximum possible engine intake Mach number.
4. Air with freestream Mach number of 0.7 enters into a two dimensional air intake with area ratio of 1.7. The air intake is 100 cm long and 80 cm diameter with inlet cross sectional area of 0.5 m^2. The total pressure and temperature at inlet of the air intake are 80 kPa and 270 K. If the Fanning friction factor is 0.02, determine (i) exit total pressure, and (ii) PRF.
5. A wedge with an angle of 28° is used in the supersonic air intake flying at Mach number of 2.5. Assuming a shockfree flow inside the air intake, its internal recovery factor is 0.9. What is the overall recovery factor? What is the inlet Mach number at the intake inlet?
6. An aircraft is flying at Mach number of 2.2 with a rectangular supersonic air intake with a 2D planar ramp. It consumes around 50 kg/s during its operation forming an oblique shock at its cowl lip followed by shockfree air intake. The pressure recovery factor for internal compression of air intake is 0.95, determine (i) pressure recovery factor (ii) isentropic efficiency, and (iii) inlet area for maximum deflection angle.
7. A planar 2D wedge with angle of 32° is resulted in forming an oblique shock followed by a normal shock when the aircraft is flying at flight Mach number of 2.3 at altitude of 20 km. If the internal recovery factor is 0.96, determine (i) Mach number behind the normal shock (ii) Angle of oblique shock (iii) Total pressure at exit of intake, and (iv) Minimum flight Mach number for which shock remains attached to its cowl.
8. A two-dimensional supersonic air intake with two ramp angles of 12 and 22° respectively produces two oblique shock waves followed by a normal shock wave when it is operated at flight Mach number of 2.7. If the internal recovery factor is 0.93, determine (i) Angle of oblique shock, and (ii) Overall total pressure ratio.
9. A mixed compression supersonic air intake with a 2D spike with cone angle of 26 and 52° relative to its axis is flying at flight Mach number of 2.8. During its operation, two oblique shocks are falling at lip of the cowl. Then air is decelerated further in the convergent inlet portion with a throat area. Assuming $\pi_r = 1$, determine the π_d inlet is already started by swallowing the normal shock. Estimate the required A_n/A_1. If the inlet of the air intake has a shock in the front of spike, determine π_d.
10. A turbojet engine powered aircraft is flying at altitude of 15 km with the flight speed of 2.5. The internal compression air intake is designed for flight Mach number of 3.0. Determine (i) Maximum pressure recovery for started and unstarted condition, and (ii) What will be the ratio of unstarted inlet mass flow rate to the started inlet mass flow rate?
11. An internal compression air intake with fixed capture area of 1.0 m^2 is designed for flight speed of 2.5. The Mach number at the throat is estimated to be 1.1. Determine (i) inlet area (ii) ratio of inlet and throat area, and (iii) can this inlet be started for Mach number of 3.0. If not, what will be the change in throat area required for this?

12. Exhaust Nozzle

Supersonic airplanes have carried men at more than 2,000 miles per hour and there are reasons to believe that this speed will be doubled by 1960 or so.

Igor Sikorsky, 14 January 1958

The main purpose of an exhaust nozzle in an aircraft engine is to expand the high pressure hot gas to a higher jet velocity for producing the requisite thrust. Recall that the nozzle contributes all the thrust in case of the turbojet engine. Whereas, in turboprop engine, majority of gas expansion occurs across the turbine, and exhaust nozzle contribution is a very small portion of the total thrust. As stated earlier, the nozzle also serves to straighten the swirl flow from a turbine, to have a higher thrust. In order to have a large value of specific thrust, the kinetic energy of exhaust gas must be large enough to produce a higher exhaust velocity. Keep in mind that the pressure ratio across the nozzle controls the expansion process. We can achieve maximum thrust for a given engine only when exit pressure P_e is same as ambient pressure, P_0. In other words, maximum thrust can be achieved only when exhaust nozzle is fully expanded.

While designing an exhaust nozzle, the designer has to keep in mind the types of aircraft engine applications. For example, nozzle in jet engine with afterburner, used to augment the thrust must have a provision to vary the exit area of the nozzle to perform optimally during both afterburning and normal operations. When the aircraft is used as a fighter plane, then both noise level and infrared reduction must be reduced at nozzle exit such that the enemy's radar cannot detect it. In some modern aircrafts, thrust reversing and thrust vectoring are required, which can be accomplished by the exhaust nozzle. Hence, the designer has to consider applications such that required features can be incorporated in the exhaust nozzle. A modern aircraft must be designed to fulfill the following objectives:

(i) Efficient expansion of hot exhaust gases with minimum total pressure losses.
(ii) Efficient thrust reversing and vectoring capabilities.
(iii) Low installation drag.
(iv) Low noise, radar, thermal and infrared (IR) signature.
(v) Low weight.
(vi) Low manufacturing cost.
(vii) Longer life and reliability.

Generally two types of exhaust nozzles are used in modern jet engines: (i) Convergent nozzle and (ii) Convergent-Divergent nozzle.

12.1 Convergent Nozzle

The schematic of a simple convergent exhaust nozzle employed in jet engines is shown in Fig. 12.1. It is designed to maintain a constant internal pressure and still produce sonic velocities at the nozzle exit, resulting in an increase in its velocity in the flow direction. Keep in mind that at nozzle exit, the gas velocity cannot exceed the speed of sound. This convergent nozzle can operate under two different conditions:

(i) When its exit pressure matches with the prevailing ambient pressure
(ii) When its exit pressure is greater than equal to the critical pressure resulting in choking of nozzle.

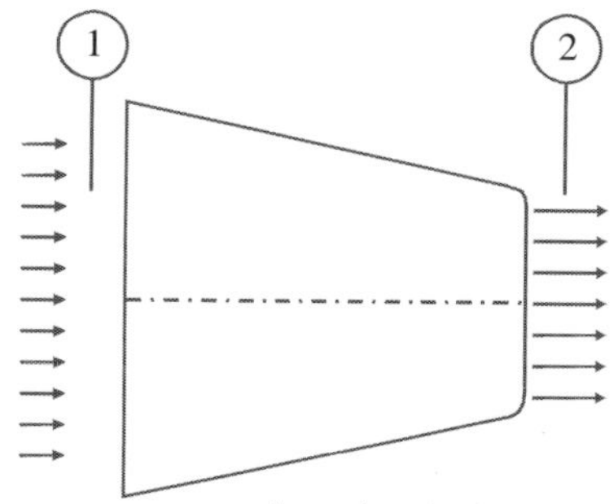

Fig. 12.1. Schematic of a simple convergent nozzle.

Generally, the convergent nozzle is preferred when the nozzle pressure ratio is less than around three. Because, it has been noted from experimental data that the thrust produced by a convergent nozzle is comparable to that of convergent-divergent (CD) nozzle at this pressure ratio. Besides this, we can easily accomplish variation in exit nozzle area and thus, effecting thrust reversal and noise suppression in a convergent nozzle. Because of above reasons, the convergent nozzle has been incorporated in modern aircrafts. For designing of both convergent and convergent-divergent (CD) nozzles, we can use the area-velocity relationship obtained from quasi-one-dimensional analysis for isentropic flow, given as

$$\frac{dA}{A} = \frac{dV}{V}(M^2 - 1) \tag{12.1}$$

where M is the Mach number, V is the flow velocity and A is the cross-sectional area. Although this expression is valid for isentropic flow, but it can be used for design calculation.

12.2 Flow through a Convergent Nozzle

We can look at the flow field through a convergent nozzle by invoking quasi one-dimensional analysis for isentropic flow in the exhaust nozzle that has already been discussed in Chapter 2. Let us consider an ideal convergent nozzle as shown in Fig. 12.2 with a large area ratio, $\frac{A_1}{A_2}\left(=\frac{A_1}{A^*}\rightarrow\infty\right)$, indicating that nozzle is supplied with gas from a large reservoir at total pressure P_t and temperature, T_t respectively. Now, we will see what happens to the flow through the nozzle when backpressure, P_B outside the nozzle is varied independently. Of course, if the backpressure is equal to total pressure, P_t, certainly there would not be any flow in the nozzle (see curve '*a*' in Fig. 12.2). But when P_B is slightly reduced below P_t, then there will be flow inside the nozzle which is represented by curve '*b*' and '*c*' in Fig. 12.2. Since the flow is subsonic in the nozzle, pressure decreases in x direction with increase in mass flow rate as shown in Fig. 12.2(b). If the nozzle pressure is allowed to reduce further, at certain P_B (see curve *d*), the flow will reach the sonic velocity at the exit plane of the exhaust nozzle. Now if we further decrease the backpressure, there would not be any increase in velocity at its exit plane of the nozzle. Question arises why is it so? Let us understand the physical

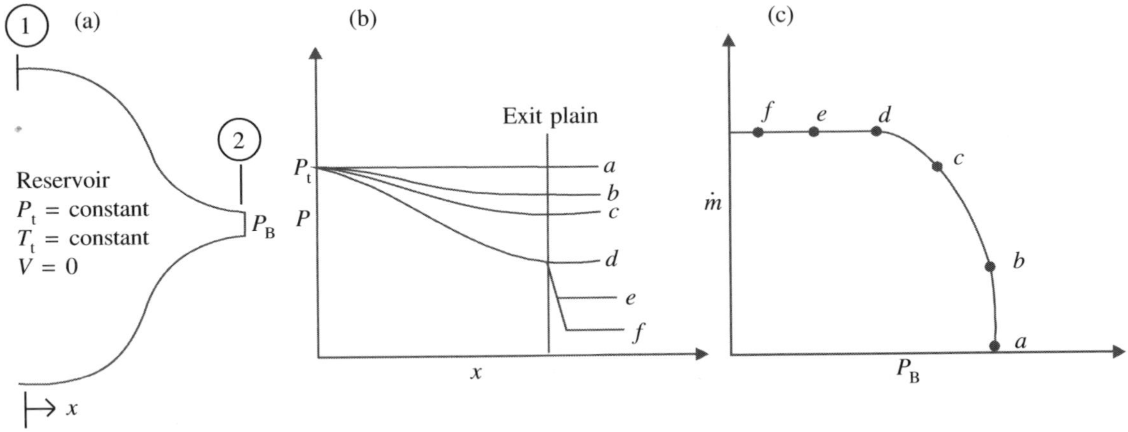

Fig. 12.2. Flow characteristics of a typical convergent nozzle: (a) convergent nozzle, (b) variation of P along nozzle and (c) mass flow rate with P_B.

mechanism that is responsible for this situation. Since the flow has already attained a sonic speed, any disturbances due to change in backpressure cannot reach the reservoir in order to adjust the incoming fluid flow. This is so because the absolute wave velocity of the disturbance becomes zero when the exit velocity attains a sonic value. For all backpressure below d, the nozzle is said to be choked. In other words, the flow in the nozzle cannot adjust to any further change in backpressure. But the exit flow pressure does eventually decrease to the backpressure outside the nozzle by creating expansion waves (see 'e', 'f' curves of Fig. 12.2). It must be borne in mind that there will be a difference between the nozzle exit pressure and backpressure P_B over certain range of backpressure. As the flow has not expanded to the back (ambient) pressure within nozzle, then expansion is known as underexpansion (supercritical).

It can be noted that when the flow is subsonic the velocity increases in a convergent duct. In contrast, at supersonic flow condition, there will be an increase in velocity while passing through a divergent duct. The design of a simple, convergent nozzle is relatively easy from both the standpoint of gas dynamics and mechanical engineering. Generally half angle of convergence around 15° can be easily used in nozzle design. As a result, much shorter nozzle can be designed as compared to air intake for same pressure ratio. Of course, in practice, a variable area nozzle is being used and it becomes mechanically complex to design. A series of pie-shaped segments is often used to construct a variable area nozzle. A variable area convergent nozzle, used in F404 turbojet with afterburner is shown in Fig. 12.3. Note that smaller exhaust area in Fig. 12.3(a) corresponds to cruise operation without afterburner. A larger area shown in Fig. 12.3(b) is required when afterburner is operational.

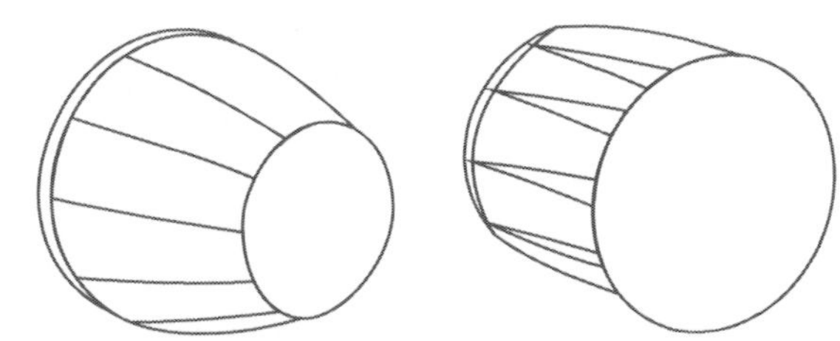

Fig. 12.3. A schematic of variable area exhaust nozzle: (a) smaller exit area and (b) larger exit area.

Example 12.1 A fixed exhaust nozzle with an exit area of 0.15 m^2 and isentropic efficiency of 0.95 is expanded to atmospheric pressure (0.1 MPa) in a turbojet engine for producing thrust. If this nozzle is operated with total inlet pressure of 350 kPa and total inlet temperature of 1000 K, determine the mass flow rate passing through this nozzle. Assume $\gamma_g = 1.33$.

Given $A_9 = 0.15\ \text{m}^2$, $P_{t5} = 350$ kPa, $P_0 = 100$ kPa, $T_{t5} = 1000$ K, $\eta_{in} = 0.95$, $\gamma_g = 1.333$.

Solution We need to determine whether the nozzle is choked under this condition. Let us invoke relation for pressure ratio at critical condition ($M_9 = 1$) as given below.

$$\frac{P^*}{P_{t9}} = \left[1 - \frac{\gamma_g - 1}{(\gamma_g + 1)\eta_{in}}\right]^{\frac{\gamma_g}{\gamma_g - 1}} = \left[1 - \frac{0.333}{(1.333 + 1)0.95}\right]^{4.0} = 0.52$$

As $P_{t5} = P_{t9} = 350$ kPa, we can evaluate, $P^* = 0.52 \times 350 = 182$ kPa. Note that P_0 is less than P^*, hence flow is choked ($M_9 = 1.0$) and thus P_9 is equal to P^*. We need to evaluate mass flow rate, $\dot{m}_9$ at the nozzle exit as

$$\dot{m}_5 = \dot{m}_9 = \rho_9 V_9 A_9$$

We can estimate ρ_9 by using equation of state and V_9 by determining the speed of sound, a_9. For this purpose, we will have first evaluate T_9 by using isentropic relation as given below

$$T_9 = \frac{T_{t5}}{1 + \frac{\gamma_g - 1}{2} M_9^2} = \frac{1000}{1 + \frac{0.333}{2}} = 857.3\ \text{K}$$

By using ideal gas law, we can determine density at its exit as

$$\rho_9 = \frac{P_9}{RT_9} = \frac{182 \times 10^3}{287 \times 857.3} = 0.74\ \text{kg/m}^3$$

For an ideal gas, velocity at nozzle exit, V_9 is determined as

$$V_9 = a_9 = \sqrt{\gamma_g R T_9} = \sqrt{1.333 \times 287 \times 857.3} = 572.7\ \text{m/s}$$

By using the values of $\rho_9 V_9$ and A_9, we can estimate mass flow rate at nozzle exit as

$$\dot{m}_5 = \dot{m}_9 = \rho_9 V_9 A_9 = 0.74 \times 572.7 \times 0.15 = 63.57\ \text{kg/s}$$

Note that mass flow rate through the exhaust nozzle remains same when P_9 is less than P_0 as nozzle is choked.

12.3 Convergent Divergent (CD) Nozzle

We know that when an aircraft will fly at a high flight Mach number (supersonic), then the pressure ratio across the nozzle becomes greater than its critical value. As a result, the maximum gas velocity attained is equal to the speed of sound and under such situation a convergent divergent (CD) nozzle must be employed. Recall from Eq. (12.1) that in the convergent section, the gas velocities can not be increased above the speed of sound. Hence, the convergent-divergent nozzle is to be employed when one has to achieve higher exhaust velocity beyond sonic speed. This nozzle is basically a convergent duct followed by a divergent duct as shown in Fig. 12.4. Note that there is a minimum cross-sectional area existing between convergent and divergent portion, which is known as *throat* as discussed earlier. The gas attains a sonic speed at its throat under supersonic flow condition. Recall that this is known as de *Laval* or simply *Laval* nozzle which is employed for high performance supersonic aircraft.

12.3.1 Flow through a Convergent Divergent (CD) Nozzle

We can investigate flow through a CD nozzle in the similar way that has been carried out for convergent nozzle in the previous section. The flow characteristics through a CD (Laval) nozzle is shown in Fig. 12.4. When the back pressure (P_B) is reduced below total pressure (reservoir), P_t, then the pressure decreases in the divergent portion as shown by a curve '*a*' in Fig. 12.4. If we decrease the backpressure further, at certain value of P_B sonic flow will be attained at the throat section, as shown by curve '*b*'. Now, the sonic flow at the throat decreases to subsonic flow in the divergent section. The mass flow rate through the nozzle attains a maximum value. Recall, this condition at the throat is called as choked condition. That means, the mass flow cannot be changed by any further changing of P_B for a constant P_t and T_t.

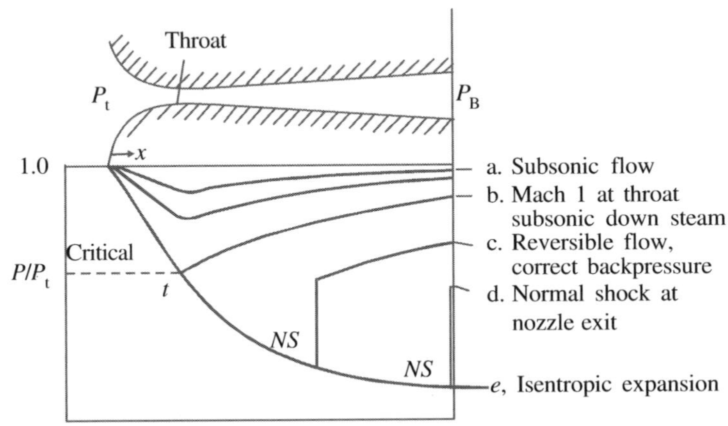

Fig. 12.4. Flow through convergent-divergent nozzle.

Now if the backpressure P_B decreases further, say to P_{Be}, then the flow cannot attain isentropic conditions since the throat is choked. Rather, the pressure will decrease further from the throat until a normal shock is formed in the divergent portion of the CD nozzle as shown in Fig. 12.4 (curve '*c*'). After the formation of this normal shock, the flow will get decelerated in the divergent duct to the subsonic flow at its exit. With further reduction in P_B, the normal shock will move downstream and at certain P_{Bd}, it will stand exactly at nozzle exit plane as shown in Fig. 12.4. If now, P_B is lowered further, oblique shocks are formed at the nozzle exit (see Fig. 12.5a). This situation is known as *overexpansion* because the exit pressure has already expanded below the backpressure. In such a situation, the patterns of reflected shocks, expansion waves are formed as shown in Fig. 12.5a to adjust the flow to the prevailing backpressure. Of course, if the backpressure matches with the correct one for which nozzle is designed, then isentropic expansion will take place as shown in Fig. 12.4. In this case, the flow will be supersonic throughout the divergent portion of the nozzle. This is known as *complete expansion*. Further lowering of backpressure will lead to the formation of expansion wave as can be seen in Fig. 12.5(b). The expansion waves are reflected from the jet boundary layer forming compression waves followed by expansion waves. This situation is known as underexpansion and the nozzle is said to be *underexpanded.* A small amount of *underexpansion* would not affect much the performance of aircraft engine as compared to overexpansion because, the flow is more likely to be separated inside the nozzle, when it is overexpanded and thus, reducing the overall performance of the aircraft.

12.3.1.1 Variable Geometry CD Nozzle

We know that most supersonic and fighter aircrafts use afterburner to achieve supersonic flight and

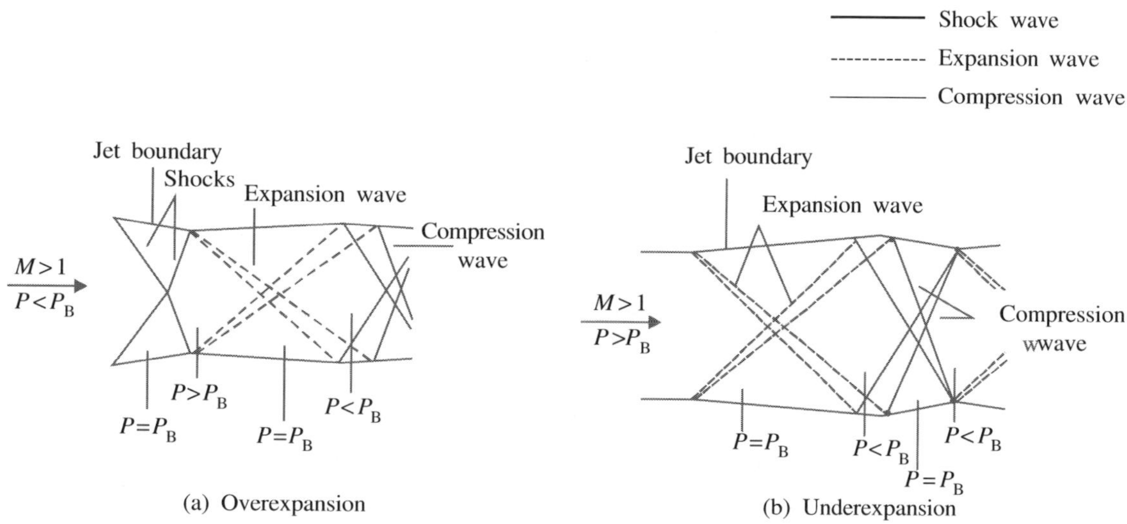

Fig. 12.5. Flow pattern at downstream of nozzle exit under (a) overexpansion and (b) underexpansion.

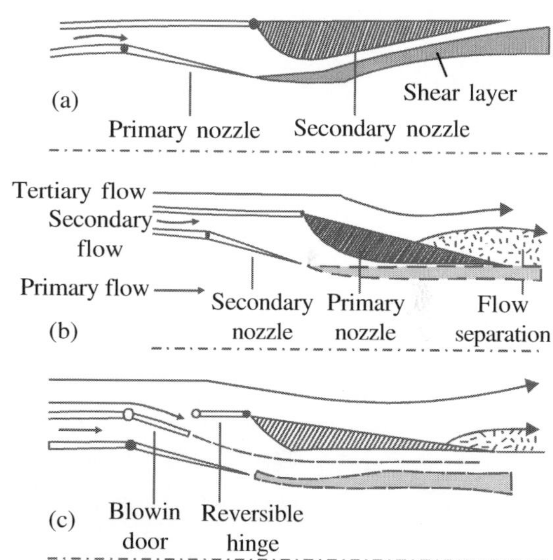

Fig. 12.6. Schematic of ejector type variable geometry CD nozzle for (a) supersonic flight, (b) takeoff and (c) subsonic transition flight [2].

manoeuvrability. During supersonic cruise, idling, landing there is no need to use afterburner. Hence, minimum two-position variation of throat area of CD nozzle must be provided to have a successful supersonic flight. In earlier supersonic aircraft, ejector nozzle has been used. Two versions of the ejector nozzles are shown in Fig. 12.6. In either ejector nozzles, a secondary flow is exploited to provide an aerodynamically varied expansion ratio. This secondary air generally bled either from the intake or from the compressor is allowed to pass through the outer surface of the nozzle. By controlling the amount of this secondary air, we can vary the effective nozzle area available for engine gas flow and thus, expansion ratio across the nozzle. In order to avoid flow separation in the divergent portion of the nozzle, it may be contracted as shown in this figure while operating at low flight Mach number. During takeoff, the flight Mach number is generally quite low. But the nozzle may require high airflow rate, which may not be available from the intake alone. Hence, a blow-in-door is provided before the nozzle throat as shown in Fig. 12.6(c) to supply tertiary air. This tertiary air helps in avoiding the flow separation. Keep in mind that the blow-in-doors open only when the pressure difference across the door is favourable. At high flight speed, the pressure of the secondary air will be definitely above the ambient pressure and thus makes the blow-in-door to get closed. Then the nozzle operates like a two stream ejector as shown in Fig. 12.6(c). However, in the modern high

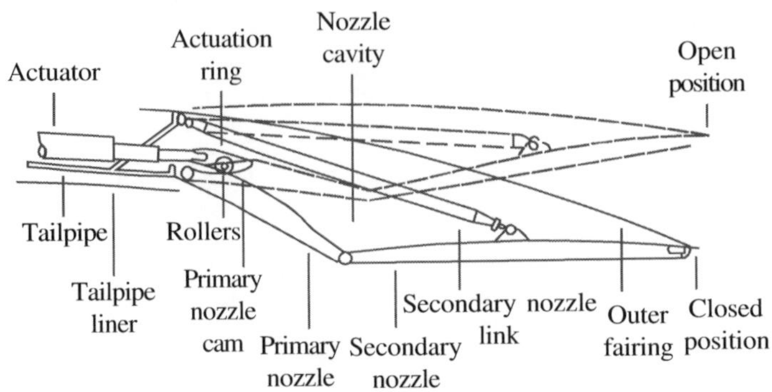

Fig. 12.7. Schematic of modern variable geometry CD nozzle [2].

performance turbofan engines with afterburner, a more simpler CD nozzle is used as shown in Fig. 12.7, which is devoid of secondary air as in ejector nozzle. Note this kind of variable geometry has already used in F100 engine.

Example 12.2 A fixed CD nozzle with throat area of 0.2 m^2 is expanded fully to atmospheric pressure (50 kPa) in a turbojet engine for producing thrust. This nozzle is operated with total inlet pressure of 350 kPa and total inlet temperature of 1000 K and inlet Mach number of 0.5. If the nozzle allows maximum mass flow rate, determine (i) exit Mach number (ii) exit mass flow rate passing through this nozzle and (iii) exit area. Assume γ_g = 1.333.

Given A_8 = 0.2 m^2, P_{t7} = 350 kPa = P_{t9}, M_7 = 0.5 P_{atm} = 50 kPa, T_{t7} = 1000 K.

Solution Assuming the flow to be isentropic, the expression for exit Mach number, M_9 can be expressed as

$$M_9 = \sqrt{\frac{2}{\gamma_g - 1}\left[\left(\frac{P_{t9}}{P_9}\right)^{\frac{\gamma_g - 1}{\gamma_g}} - 1\right]} = \sqrt{\frac{2}{0.333}\left[\left(\frac{350}{50}\right)^{0.25} - 1\right]} = 1.94$$

We know that nozzle exit velocity can be obtained from M_9 as

$$V_9 = M_9\sqrt{\gamma_g RT_9}$$

We need to evaluate T_9 by using following isentropic relation as

$$T_9 = T_{t9}\left(\frac{P_9}{P_{t9}}\right)^{\frac{\gamma_g - 1}{\gamma_g}} = 1000 \times \left(\frac{50}{350}\right)^{0.25} = 614.8 \text{ K}$$

Let us evaluate V_9 as

$$V_9 = M_9\sqrt{\gamma_g RT_9} = 1.94 \times \sqrt{1.333 \times 287 \times 614.8} = 940.9 \text{ m/s}$$

For maximum flow to occur, the throat of the nozzle must be chocked. Let us evaluate chocked mass flow rate thorough its throat as

$$\dot{m} = \rho^* A^* V^* = \rho_8^* A_8^* \sqrt{\gamma_g RT_8^*}$$

The nozzle exit area can be determined as

$$A_9 = \frac{\dot{m}_9}{\rho_9 V_9} = \frac{88.2}{\dfrac{50 \times 10^3}{287 \times 614.8} \times 940.9} = 0.33 \text{ m}^2$$

But, the density at throat can be estimated as

$$\rho_8^* = \left(\frac{2}{\gamma_g + 1}\right)^{\frac{1}{\gamma_g - 1}} \frac{P_{t9}}{RT_{t9}} = \left(\frac{2}{1.333 + 1}\right)^{\frac{1}{1.333 - 1}} \frac{350 \times 10^3}{287 \times 1000} = 0.77 \text{ kg/m}^3$$

Then, T^* can be determined as

$$T_8^* = \left(\frac{2}{\gamma_g + 1}\right) T_{t9} = 857.3 \text{ K}$$

Then, the velocity at the throat which is equal to speed of sound can be estimated as

$$V_8^* = a^* = \sqrt{\gamma_g R T_8^*} = \sqrt{1.333 \times 287 \times 857.3} = 572.7 \text{ m/s}$$

The mass flow rate through the nozzle is determined as

$$\dot{m}_9 = \dot{m} = \rho_8^* \, A_8^* \, V_8^* = 0.77 \times 0.2 \times 527.7 = 88.2 \text{ kg/s}$$

12.3.2 Thrust Reversing

We know that engine produces thrust only in a certain direction by expanding hot gases through the nozzle. But the direction of the thrust can be reversed at the time of need, if proper provisions are made in nozzle while designing it. Of course, thrust reversing is very much required in the high-powered aircraft during its landing, which enhances effectiveness of the existing brakes. Besides this, the usual wheel brakes may not be very effective on wet or icy runways. And, the maneuverability of the combat aircraft can be enhanced considerably by thrust reversing or vectoring for which special provisions are to be made in the exhaust nozzle. In practice, two basic type thrust vectors are used: (a) Cascade reverser and (b) Clamshell reverser.

The clamshell type is based on the principle of mechanical blockage of exhaust flow which can be accomplished simply by placing an obstruction in the flow path of the exhaust nozzle. The schematic of a typical clamshell type is shown in Fig. 12.8. It can be observed that the exhaust jet is turned in the reverse direction of its flow. As a result, reverse thrust is obtained which retards the aircraft.

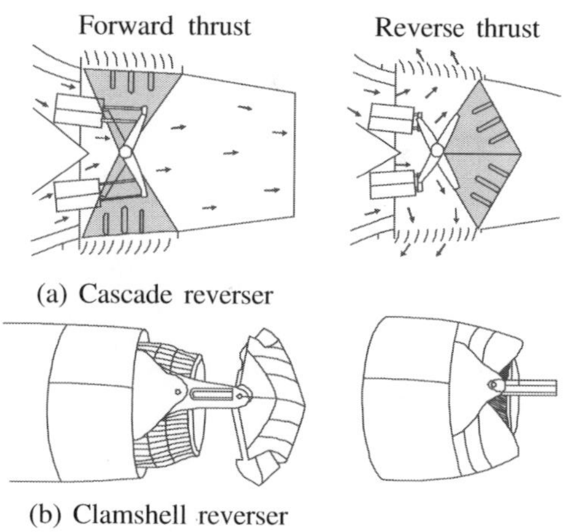

Fig. 12.8. A schematic of two types of (a) cascade (b) clamshell thrust reverser [3].

But the cascade blocker type thrust reverser is based on aerodynamic blockage. In this case the gases are turned in the reverse direction by means of rows of airfoils (cascades), as shown in Fig. 12.8(a), and thus, provides the reverse thrust. Note that these cascades are generally blocked during normal forward thrust operation. During reverse thrust, the doors used to cover cascades, are moved forward as can be seen in Fig. 12.8(b) so that they can now block the exhaust gas stream. As a result, the gas can exit through the cascade and are deflected in the opposite direction providing deceleration to the aircraft. It must be kept in mind that both types should provide an increase in effective nozzle throat area during the deployment of thrust reverse while preventing aerodynamic stall in the compressor blades.

12.4 Performance of an Exhaust Nozzle

The performance of an exhaust nozzle can be characterised by several performance parameters, reported in the literature. However, we will restrict our discussion to three parameters, which are used profusely by the designer. These parameters are:

(i) Isentropic nozzle efficiency, η_{in}
(ii) Discharge coefficient, C_d
(iii) Thrust coefficient, C_T

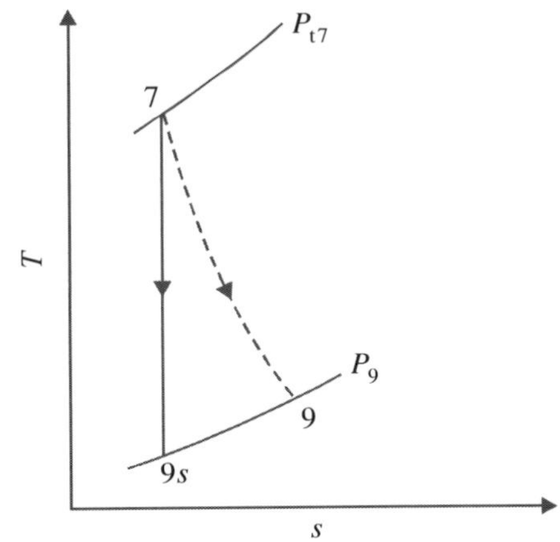

Fig. 12.9. The expansion processes in a T-s diagram.

12.4.1 Isentropic Efficiency

The expansion process in a nozzle is shown in T-s diagram in Fig. 12.9. We can recall from the cycle analysis that it is defined as ratio of actual to isentropic enthalpy drop across the nozzle. Using the station numbers of cycle analysis.

$$\eta_{in} = \frac{T_{t7} - T_9}{T_{t7} - T_{9s}} \tag{12.2}$$

where, C_d is assumed to be constant and the nozzle is fully expanded.

12.4.2 Discharge Coefficient

We know that actual effective flow area of the nozzle is always slightly less than geometrical area due to presence of boundary layer. Hence, we will have to define a discharge coefficient C_d as

$$C_d = \frac{\text{Actual mass flow rate through the throat}}{\text{Ideal mass flow rate through the throat}} = \frac{\dot{m}_{8a}}{m_8} = \frac{\rho_8 V_8 A_{8a}}{\rho_8 V_8 A_8} = \frac{A_{8a}}{A_8} \tag{12.3}$$

Note that discharge coefficient, C_d is dependent on Reynolds number of the flow which eventually is dependent on the pressure ratio P_{t7}/P_9 across the nozzle. If we can assume to have same density at the throat for both ideal and actual condition, then C_d is dependent on the throat area. Hence, the variation of C_d with the pressure ratio P_{t7}/P_9 for a typical convergent nozzle is shown in Fig. 12.10a. Note that C_d increases with P_{t7}/P_9 till the nozzle is choked. At this condition, the discharge coefficient attains a maximum value. Interestingly this maximum value, $C_{d,\,max}$ is dependent on the angle of the nozzle. Now let us see how does the C_d varies with pressure ratio, P_{t7}/P_9 for both convergent and CD nozzle. The variation of C_d vs P_{t7}/P_9 for a typical CD nozzle is shown in Fig. 12.10b. Interestingly, it can be observed from this figure that the C_d does not vary with P_{t7}/P_9 due to venturi effects during initial subsonic flow. With increase in P_{t7}/P_9 the C_d decreases due to increase in losses and attains a minimum value when the nozzle is choked.

12.4.3 Thrust Coefficient

We know that in actual nozzle, the flow streamlines would not be parallel to axis of the nozzle. Rather, there will be velocity vector angularity in the flow. Hence, the actual thrust will be less than the ideal one as the velocity component in the nozzle axis only contributes to the actual thrust. In addition, there will be a reduction in thrust due to decrease in velocity magnitude caused by the presence of boundary layers over its surface. Besides this the flow may have non-uniformities due

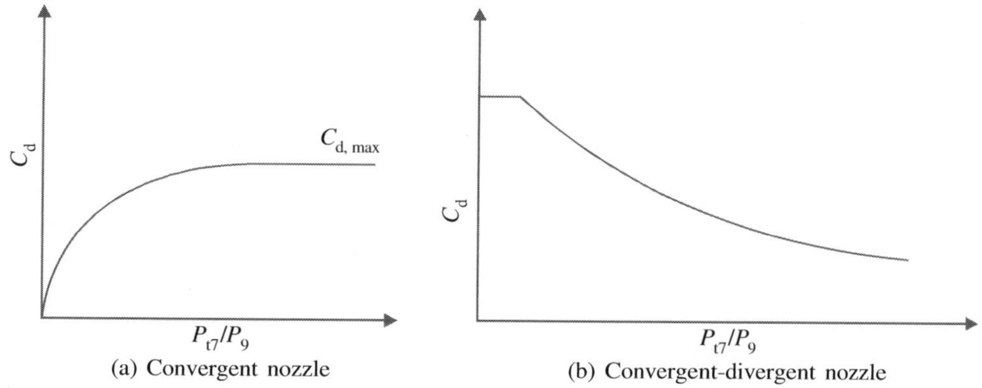

Fig. 12.10. Variation of C_d vs pressure ratio, P_{t7}/P_9 across (a) a convergent nozzle and (b) a CD nozzle.

to swirl in the turbine exhaust gas. Due to all above stated reasons, the actual thrust will be different than ideal one. Hence, let us define a term known as thrust coefficient, C_T that is given by

$$C_T = \frac{\text{Actual thrust produced by exhaust nozzle}}{\text{Ideal thrust produced by exhaust nozzle}} \tag{12.4}$$

By carrying out analysis for isentropic quasi one-dimensional steady flow, we can derive an expression for C_T that is given by

$$C_T = \sqrt{\frac{2\gamma_g^2}{\gamma_g - 1}\cdot\left(\frac{2}{\gamma_g - 1}\right)^{\frac{\gamma_g+1}{\gamma_g-1}}\left[1-\left(\frac{P_9}{P_{t7}}\right)^{\frac{\gamma_g-1}{\gamma_g}}\right]} + \frac{P_9 - P_0}{P_{t7}}\left(\frac{A_9}{A_8}\right) \tag{12.5}$$

It can be noted from this expression that C_T is dependent on the nozzle geometry and the properties of fluid, apart from pressure ratios, P_{t7}/P_9, P_{t7}/P_0. Hence let us plot the variation of C_T against A_9/A_8 in Fig. 12.11 for various P_{t7}/P_0. It is interesting to note that C_T attains unity when the nozzle is completely expanded ($P_9/P_{t7} = P_0/P_{t7}$). But under both underexpansion ($P_9/P_{t7} > P_0P_{t7}$) and overexpansion ($P_9/P_{t7} < P_0P_{t7}$), C_T is lower than unity.

Recall that the overexpansion in the nozzle must be avoided to achieve higher performance in the nozzle, because strong oblique shocks are formed inside the divergent portion of nozzle as shown in Fig. 12.12, when it is overexpanded. If the strength of the oblique shock is high, then the boundary layer is likely to separate resulting exorbitant losses in the thrust. Hence, the area ratio A_9/A_8 for a CD nozzle must be properly selected so that the flow in the nozzle should not get separated when it is overexpanded.

Generally, the nozzle pressure ratio in the range of 3 to 5 is used for subsonic flight speed of turbofan and turbojet engines. While using a convergent nozzle for subsonic flight, C_T is in the range of 1 to 3% when the nozzle is fully expanded. But if the pressure ratio across the nozzle is too high, then one should go for CD nozzle. For example, for aircraft with flight Mach number of 2, the pressure ratio across the nozzle would be around 12. But if one will use convergent nozzle, under this condition, then the actual thrust will be 9% less than the ideal one. However, a CD nozzle will provide larger thrust gain even at supersonic flight.

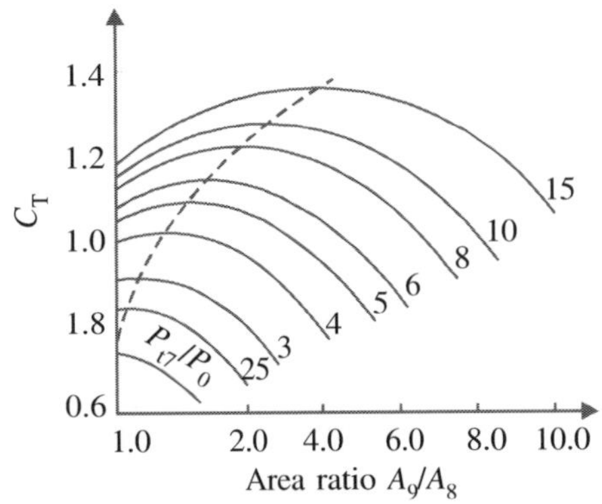

Fig. 12.11. The variation of C_T against A_9/A_8 [4].

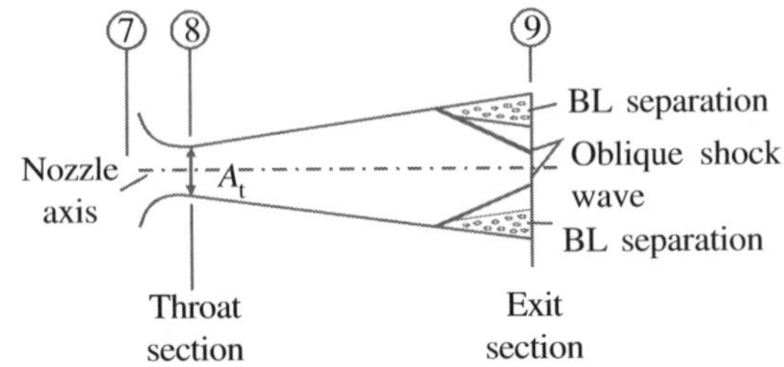

Fig. 12.12. Effect of shock-boundary layer interaction in CD nozzle under overexpanded condition.

12.5 Chapter Summary

The exhaust nozzle is an important component in which high temperature and pressure gas undergoes expansion and produces thrust in jet engines. Both convergent and convergent-divergent (CD) nozzle are employed in the aircraft engines. However, either fixed or variable exit area convergent nozzle is preferred in aircraft engines. Quasi-one dimensional steady inviscid flow analysis is used to analyze nozzle flow. A brief discussion is provided about the boundary layer over its internal surface which invariable incurs losses in the thrust. The converging nozzle can be subjected to two types of operating conditions namely (i) unchocked, and (ii) choked. In contrast, five different conditions are likely to occur in CD nozzle. Choking of flow is one of the major problem being encountered during operation of nozzle particularly in fixed area nozzle which limits its operating range. Beside this, formation of shocks, expansion waves either inside or outside of the nozzle geometry, and boundary layer separation can cause exorbitant losses leading to decrease in design thrust. Generally variable area nozzle can be used to tide over this problems particularly in case of convergent nozzle. The interaction of boundary layer and shock waves can enhance flow losses in the nozzle. The concept of thrust reversing/vectoring which can be used in fighter aircrafts to change its direction are discussed briefly. The performance parameters of exhaust nozzle such as isentropic efficiency, discharge coefficient and thrust coefficient are introduced which are generally used for its characterisation.

Nomenclature

A	Area
A_t	Throat area
A_e	Exit area
C_d	Discharge coefficient
$C_{d,\,max}$	Maximum discharge coefficient
C_T	Thrust coefficient
M	Mach number
$\dot{m}$	Mass flow meter
$\dot{m}_a$	Actual mass flow rate
P_e	Exit pressure
P_o	Ambient pressure
P_t	Total pressure
S	Entropy

T	Temperature
T_t	Total temperature
V	Velocity

Greek Symbols

ρ	Density
η_{in}	Isentropic nozzle efficiency

Subscripts/Superscripts

a	Actual
B	Back
D	Discharge
e	exit
in	Isentropic
max	Maximum
o	Ambient
T	Thrust

Abbreviation

CD	Convergent divergent nozzle

References

1. Hill, P.G. and Peterson, C., *Mechanics and Thermodynamics of Propulsion*, Addison-Wesley, 1999.
2. Oates, G.C. (Ed.), *Aircraft Propulsion Systems Technology and Design*, AIAA Education Series, 1989.
3. Treager, I.E., *Aircraft Gas Turbine Engine Technology*, 3rd Edition, Tata McGraw-Hill, 1996.
4. Mattingly, J.D., *Elements of Gas Turbine Propulsion*, McGraw-Hill International, 1996.

Review Questions

1. What are the type of exhaust nozzle used in an aircraft engine?
2. Show schematically the variation of total enthalpy, pressure and temperature across a convergent. How are these comparable with their respective static quantities?
3. What is back pressure control? Explain it.
4. What is thrust reverting? Describe various methods being used in practical system.
5. When is the CD nozzle preferred? Can you suggest methods to have a variable throat area CD nozzle?
6. What is underexpansion in a CD nozzle? When is it likely to occur?
7. Describe the flow processes involved during the overexpansion of a CD nozzle.
8. Why is the variable-geometry nozzle preferred in an aircraft engine with an after-burner?
9. Is it possible to have an arrangement of converging variable-geometry nozzles to produce the flow similar to a CD nozzle? Describe its operating principle.

Problems

1. A convergent nozzle with length-to-diameter ratio of 3 and area ratio of 0.3 is operated with hot gas at total pressure of 0.7 MPa and total temperature of 800 K from the turbine. If it is operated under ideal and choked condition, estimate static pressure, temperature and Mach number at its exit. Determine the isentropic efficiency of the nozzle.

2. In the fan stream of a turbofan engine, a convergent nozzle is used to expand the cold air at total pressure and temperature of 150 kPa and 330 K respectively. This nozzle has an exit area of 0.56 m^2. If the isentropic efficiency of the nozzle is 0.98, estimate (i) exit Mach number, and (ii) mass flow rate. If the inlet total temperature is increased by 1.5 times, what will be the increase in mass flow rate?
3. The gas at total temperature of 1000 K and pressure of 0.6 MPa enters into a CD exhaust nozzle with area ratio (A_e/A_{th}) of 10. A normal shock wave stands in the divergent portion of nozzle where the Mach number is 2.5. Determine (i) mach number (ii) static pressure, and (iii) static temperature at the exit of nozzle.
4. A convergent nozzle is used in turbofan engine to expand the cold air at total pressure and temperature of 120 kPa and 330 K respectively. This nozzle has an exit area of 0.6 m^2. If the isentropic efficiency of the nozzle is 0.95, estimate (i) exit Mach number, and (ii) mass flow rate. If the inlet total temperature is increased by 1.5 times, and pressure is increases to 200 kPa what will be the new mass flow rate?
5. A convergent-divergent nozzle has an area ratio of 2 is designed to expand the hot at total pressure and temperature of 0.15 MPa and 1330 K respectively. What is the location of the normal shock wave and exit Mach number if the back pressure becomes 0.1 MPa?
6. An afterburning turbojet engine is operated at flight Mach number of 2.5 at an altitude of 12 km. The hot gas from the afterburner at total temperature of 2000 K and pressure of 0.4 MPa enters into the convergent-divergent exhaust nozzle with throat area of 0.33 m^2 and leaves with exit Mach number of 2.0. The total fuel flow rate consumed by the engine is 4 kg/s. If the isentropic efficiency of the nozzle is 0.98, determine (i) total mass flow rate (ii) Thrust, and (iii) TSFC.
7. The hot gas at total pressure and temperature of 0.8 MPa and 2100 K respectively enters into a convergent nozzle. This with exit area of 0.25 m^2 is designed to operate at an exit Mach number of 1.3 with an efficiency of 0.97 when exit pressure matches to ambient pressure of 100 kPa and 300 K. Determine (i) exit Mach number (ii) mass flow rate when the nozzle is operated under off-design condition (80 kPa and 300 K). If the total inlet temperature is decreased to 1500 K, estimate change in mass flow rate.
8. A convergent-divergent nozzle is designed to expand the hot at total pressure and temperature of 0.5 MPa and 1330 K respectively. This nozzle has an exit area of 0.6 m^2. If the isentropic efficiency of the nozzle is 0.95, estimate (i) exit Mach number, and (ii) mass flow rate, and A_e/A_{th} when the nozzle is operated under on-design condition (50 kPa and 226 K).
9. An ideal turbojet engine flying at flight Mach number of 2.0 at an altitude of 15 km consumes 110 kg/s of air. The pressure ratio across the compressor is 14. The variable nozzle area can vary its exit pressure from 50 to 200 kPa. If the total temperature at inlet of nozzle is 1200 K, determine the variation in nozzle exit area, exit velocity and thrust with variation in exit pressure. Plot these results with exit pressure and comment on your result. Assume the heating value of the fuel to be 43,000 kJ/kg.

Appendix A

Table A1. Physical constants

Universal gas constant	R_u = 8.314 J/(g mol k) = 1.987 cal/(g mol = 82.05 cm^3 atm/(g mol k)
Standard acceleration due to gravity	g = 1 kg m/(Ns^2) = 9.80665 m/s^2
Stefen-Boltzmann constant	$\sigma = 5.6697 \times 10^{-8}$ W/(m^2 K^4)
Boltzmann constant	k_B = Boltzmann constant = 1.38054×10^{-16} erg/(molecule K)
Avogadro's number	$N_A = 6.02252 \times 10^{23}$ molecules/(g mol)
Planck's constant	$h = 6.256 \times 10^{-27}$ erg s
Speed of light	$c = 2.997925 \times 10^{8}$ m/s (in vacuum)
Standard atmosphere pressure	P_{atm} =101 325 Pa = 1 atm

Table A2. Conversion factors

Energy	1 J	= $9.47817.10^{-4}$Btu
		= $2.3885.10^{-4}$ kcal
Energy rate	1 W	= 3.41214 Btu/hr
Force	1 N	= 0.224809 Ib_f
Heat flux	1 W/m^2	= 0.3171 Btu/($hr.ft^2$)
Kinematic viscosity and diffusivities	1 m^2/s	= $3.875.10^4$ ft^2/hr
Length	1m	= 39.370 in
		= 3.2808 ft
Mass	1 kg	= 2.2046 Ib
Mass density	1 kg/m^3	0.062428 $1b/ft^3$
Mass flow rate	1 kg/s	= 7936.6 Ib/hr
Pressure	Pa	1 = 1 N/m^2
		= 0.020885 4 $1b_f/ft^2$
		= $1.4504.10^{-4}$ $1b_f/in^2$
		= $4.015.10^{-3}$ inches of water
	1.10^5 N/m	= 1 bar
Specific heat	1 J/kg.K	= $2.3886.10^{-4}$Btu/(1b°F)
Temperature	K	= (5/9)°R
		= (5/9) (°F + 459.67)
		= °C + 273.15
Time	3600S	1 hr

Appendix B

Table B1. Isentropic flow table (γ = 1.4)

M	T_t/T	P_t/P	ρ_t/ρ	A/A*	MFP
0.1	1.002000	1.007018	1.005008	5.821829	1.998998
0.2	1.008000	1.028281	1.02012	2.963520	3.986079
0.3	1.018000	1.06443	1.045609	2.035065	5.949679
0.4	1.032000	1.116552	1.08193	1.590140	7.878913
0.5	1.050000	1.186213	1.129726	1.339844	9.763859
0.6	1.072000	1.275504	1.189836	1.188200	11.59578
0.7	1.098000	1.387101	1.263298	1.094373	13.36728
0.8	1.128000	1.52434	1.351365	1.038230	15.07237
0.9	1.162000	1.691303	1.45551	1.008863	16.7065
1.0	1.200000	1.892929	1.577441	1.000000	18.26651
1.1	1.242000	2.135136	1.719111	1.007925	19.7505
1.2	1.288000	2.424965	1.882737	1.030440	21.15775
1.3	1.338000	2.770744	2.07081	1.066305	22.48855
1.4	1.392000	3.182272	2.286115	1.114926	23.74404
1.5	1.450000	3.671031	2.531745	1.176167	24.92605
1.6	1.512000	4.250414	2.811121	1.250235	26.03696
1.7	1.578000	4.935993	3.128005	1.337606	27.07956
1.8	1.648000	5.745796	3.486527	1.438982	28.05692
1.9	1.722000	6.700636	3.891194	1.555257	28.97231
2.0	1.800000	7.824449	4.346916	1.687500	29.82908
2.1	1.882000	9.144683	4.859024	1.836944	30.63061
2.2	1.968000	10.69271	5.433288	2.004975	31.38024
2.3	2.058000	12.50428	6.075939	2.193131	32.08125
2.4	2.152000	14.62002	6.793689	2.403100	32.7368
2.5	2.250000	17.08594	7.59375	2.636719	33.34993
2.6	2.352000	19.95403	8.483856	2.895975	33.92351
2.7	2.458000	23.28287	9.472284	3.183011	34.46029
2.8	2.568000	27.1383	10.56787	3.500123	34.96284
2.9	2.682000	31.59408	11.78004	3.849768	35.43356
3.0	2.800000	36.73272	13.11883	4.234568	35.87471
3.1	2.922000	42.64624	14.59488	4.657311	36.28839
3.2	3.048000	49.43703	16.2195	5.120958	36.67656

Contd.

Table B1. Continued

M	T_t/T	P_t/P	ρ_t/ρ	A/A*	MFP
3.3	3.178000	57.21878	18.00465	5.628647	37.04104
3.4	3.312000	66.11746	19.96300	6.183699	37.38349
3.5	3.45000	76.27230	22.10791	6.789621	37.70549
3.6	3.592000	87.83693	24.45349	7.450111	38.00848
3.7	3.738000	100.9805	27.01458	8.169066	38.29378
3.8	3.888000	115.8889	29.80682	8.950585	38.56263
3.9	4.042000	132.7661	32.84663	9.798973	38.81616
4.0	4.200000	151.8352	36.15124	10.71875	39.05544

Table B2. Isentropic flow table (γ = 1.333)

M	T_t/T	P_t/P	ρ_t/ρ	A/A*	MFP
0.1	1.00165	1.006667	1.005008	5.866473	1.948722
0.2	1.00660	1.026867	1.020134	2.984731	3.887850
0.3	1.01485	1.061210	1.045682	2.047994	5.808022
0.4	1.02640	1.110733	1.082163	1.598603	7.700335
0.5	1.04125	1.176934	1.130309	1.345408	9.556535
0.6	1.05940	1.261825	1.191076	1.191701	11.369180
0.7	1.08085	1.367996	1.265667	1.096361	13.131770
0.8	1.10560	1.498695	1.355549	1.039139	14.838800
0.9	1.13365	1.657933	1.462473	1.009100	16.485840
1.0	1.16500	1.850604	1.588502	1.000000	18.069450
1.1	1.19965	2.082638	1.736038	1.008192	19.587250
1.2	1.23760	2.361167	1.907859	1.031584	21.037740
1.3	1.27885	2.694731	2.107152	1.069083	22.420300
1.4	1.32340	3.093514	2.33755	1.120283	23.735060
1.5	1.37125	3.569617	2.603185	1.185281	24.982790
1.6	1.42240	4.137367	2.908722	1.264569	26.164780
1.7	1.47685	4.813677	3.259422	1.358968	27.282780
1.8	1.53460	5.618456	3.661186	1.469590	28.338890
1.9	1.59565	6.575067	4.12062	1.597817	29.335450
2.0	1.66000	7.710851	4.645091	1.745290	30.274980
2.1	1.72765	9.057716	5.242796	1.913907	31.160140
2.2	1.79860	10.652800	5.922827	2.105829	31.993620
2.3	1.87285	12.539190	6.695248	2.323484	32.778140
2.4	1.95040	14.766800	7.571165	2.569583	33.516400
2.5	2.03125	17.393210	8.562810	2.847130	34.211040
2.6	2.11540	20.484740	9.683623	3.15944	34.864630
2.7	2.20285	24.117550	10.94834	3.510155	35.479640
2.8	2.29360	28.378900	12.37308	3.903265	36.058460
2.9	2.38765	33.368470	13.97545	4.343125	36.603330
3.0	2.48500	39.199920	15.77461	4.834484	37.116410
3.1	2.58565	46.002460	17.79145	5.382498	37.599730
3.2	2.68960	53.922700	20.04859	5.992764	38.055210

Contd.

Table B2. Continued

M	T_t/T	P_t/P	ρ_t/ρ	A/A*	MFP
3.3	2.79685	63.126560	22.57059	6.671339	38.484630
3.4	2.90740	73.801420	25.38399	7.424767	38.889690
3.5	3.02125	86.158430	28.51748	8.260109	39.271970
3.6	3.13840	100.435000	32.00197	9.184969	39.632940
3.7	3.25885	116.897400	35.87076	10.20752	39.973990
3.8	3.38260	135.844000	40.15964	11.33655	40.296400
3.9	3.50965	157.608000	44.90705	12.58147	40.601370
4.0	3.64000	182.561300	50.15419	13.95235	40.890010

Table B3. Normal shock properties for $\gamma = 1.4$

M_1	M_2	P_2/P_1	ρ_2/ρ_1	P_{t2}/P_{t1}	T_2/T_1	s_2-s_1
1	1	1	1	1	1	0
1.1	0.91177	1.245	1.169082	0.998928	1.064938	0.147349
1.2	0.84217	1.513333	1.341615	0.992798	1.127994	0.927027
1.3	0.785957	1.805	1.515695	0.979374	1.190873	2.635738
1.4	0.739709	2.12	1.689655	0.958194	1.254694	5.372076
1.5	0.701089	2.458333	1.862069	0.929787	1.320216	9.134332
1.6	0.668437	2.82	2.031746	0.8952	1.387969	13.87009
1.7	0.640544	3.205	2.197719	0.855721	1.45833	19.50258
1.8	0.616501	3.613333	2.359223	0.812684	1.531578	25.94507
1.9	0.595616	4.045	2.515679	0.767357	1.607916	33.10892
2	0.57735	4.5	2.666667	0.720874	1.6875	40.90801
2.1	0.561277	4.978333	2.811902	0.674203	1.77045	49.26107
2.2	0.547056	5.48	2.95122	0.628136	1.85686	58.09292
2.3	0.534411	6.005	3.084548	0.583295	1.946801	67.33484
2.4	0.523118	6.553333	3.211896	0.540144	2.040332	76.92463
2.5	0.512989	7.125	3.333333	0.499015	2.1375	86.8064
2.6	0.503871	7.72	3.44898	0.460123	2.238343	96.93014
2.7	0.495634	8.338333	3.558991	0.42359	2.342892	107.2514
2.8	0.488167	8.98	3.663551	0.389464	2.451173	117.7306
2.9	0.48138	9.645	3.762864	0.357733	2.563207	128.3329
3	0.475191	10.33333	3.857143	0.328344	2.679012	139.0276
3.1	0.469534	11.045	3.946612	0.301211	2.798603	149.7874
3.2	0.464349	11.78	4.031496	0.276229	2.921992	160.5887
3.3	0.459586	12.53833	4.11202	0.253276	3.049191	171.4107
3.4	0.4552	13.32	4.188406	0.232226	3.180208	182.235
3.5	0.451154	14.125	4.26087	0.212948	3.315051	193.046
3.6	0.447413	14.95333	4.329621	0.195312	3.453728	203.8297
3.7	0.443948	15.805	4.394864	0.179194	3.596244	214.5743
3.8	0.440732	16.68	4.45679	0.16447	3.742604	225.2695
3.9	0.437742	17.57833	4.515586	0.151027	3.892813	235.9063
4	0.434959	18.5	4.571429	0.138756	4.046875	246.4771

Table B4. Normal shock properties for γ = 1.333

M_1	M_2	P_2/P_1	ρ_2/ρ_1	P_{t2}/P_{t1}	T_2/T_1	s_2-s_1
1	1	1	1	1	1	0
1.1	0.911378	1.239742	1.175051	0.998912	1.055054	0.135695
1.2	0.840904	1.502318	1.355527	0.992645	1.108291	0.920104
1.3	0.783604	1.787725	1.539547	0.97882	1.161202	2.668293
1.4	0.736191	2.095966	1.725404	0.956865	1.214768	5.495886
1.5	0.696397	2.427039	1.911577	0.927246	1.269653	9.414983
1.6	0.662601	2.780944	2.096738	0.891011	1.326319	14.38354
1.7	0.633609	3.157682	2.279751	0.849487	1.3851	20.33198
1.8	0.608523	3.557253	2.459664	0.804087	1.446236	27.17807
1.9	0.586655	3.979657	2.635697	0.756175	1.509907	34.83537
2	0.567466	4.424893	2.807229	0.706992	1.576249	43.21805
2.1	0.550528	4.892961	2.973779	0.657607	1.645368	52.2437
2.2	0.535497	5.383863	3.134994	0.608903	1.717344	61.83473
2.3	0.522097	5.897597	3.290627	0.561579	1.792241	71.91917
2.4	0.510097	6.434163	3.440525	0.51616	1.870111	82.43091
2.5	0.499309	6.993562	3.584615	0.47302	1.950994	93.30964
2.6	0.489575	7.575794	3.722889	0.4324	2.034923	104.5007
2.7	0.480762	8.180858	3.855392	0.394436	2.121927	115.9547
2.8	0.472758	8.808755	3.982211	0.359174	2.212026	127.6274
2.9	0.465467	9.459485	4.10347	0.326596	2.30524	139.4788
3	0.458807	10.13305	4.219316	0.296632	2.401585	151.4735
3.1	0.452709	10.82944	4.329917	0.269176	2.501074	163.5798
3.2	0.447111	11.54867	4.435455	0.244097	2.603717	175.7694
3.3	0.44196	12.29073	4.536121	0.221252	2.709524	188.0173
3.4	0.437211	13.05562	4.632111	0.200487	2.818504	200.3013
3.5	0.432822	13.84335	4.723624	0.181647	2.930662	212.6017
3.6	0.42876	14.65391	4.810859	0.164578	3.046006	224.9013
3.7	0.424993	15.4873	4.894012	0.149132	3.16454	237.1848
3.8	0.421492	16.34352	4.973275	0.135168	3.286269	249.4387
3.9	0.418235	17.22258	5.048837	0.122553	3.411197	261.6515
4	0.415198	18.12446	5.120879	0.11116	3.539327	273.8127

Table B5. Rayleigh flow properties for γ = 1.4

M	P/P^*	T/T^*	ρ/ρ^*	Pt/Pt^*	Tt/Tt^*
0.1	2.366864	0.05602	42.25	1.259146	0.046777
0.2	2.272727	0.206612	11	1.234596	0.173554
0.3	2.131439	0.408873	5.212963	1.198549	0.34686
0.4	1.960784	0.615148	3.1875	1.156577	0.529027
0.5	1.777778	0.790123	2.25	1.114053	0.691358
0.6	1.595745	0.916704	1.740741	1.075253	0.818923
0.7	1.423488	0.992895	1.433673	1.043104	0.908499
0.8	1.265823	1.025477	1.234375	1.019343	0.963948
0.9	1.124649	1.024516	1.097737	1.004856	0.992073
1	1	1	1	1	1
1.1	0.890869	0.960313	0.927686	1.004858	0.993924
1.2	0.795756	0.911848	0.872685	1.019415	0.978717
1.3	0.713012	0.859174	0.829882	1.04366	0.957979
1.4	0.641026	0.805391	0.795918	1.077652	0.934254
1.5	0.578313	0.752504	0.768519	1.121545	0.909276
1.6	0.52356	0.701735	0.746094	1.175611	0.884186
1.7	0.475624	0.653771	0.727509	1.240235	0.859709
1.8	0.433526	0.608941	0.711934	1.315925	0.836279
1.9	0.396432	0.567342	0.698753	1.4033	0.814136
2	0.363636	0.528926	0.6875	1.503096	0.793388
2.1	0.334541	0.493558	0.677816	1.616159	0.774064
2.2	0.308642	0.461058	0.669421	1.743446	0.756135
2.3	0.28551	0.43122	0.662098	1.88602	0.739543
2.4	0.264784	0.403836	0.655671	2.045055	0.724213
2.5	0.246154	0.378698	0.65	2.221831	0.710059
2.6	0.229358	0.35561	0.64497	2.417741	0.696995
2.7	0.214171	0.334387	0.640489	2.634285	0.684935
2.8	0.200401	0.314858	0.63648	2.87308	0.673796
2.9	0.187882	0.296869	0.632878	3.135853	0.663502
3	0.176471	0.280277	0.62963	3.424452	0.653979
3.1	0.166044	0.264954	0.626691	3.740844	0.645162
3.2	0.156495	0.250783	0.624023	4.087118	0.636989
3.3	0.147729	0.237661	0.621595	4.465489	0.629405
3.4	0.139665	0.225492	0.619377	4.878303	0.622359
3.5	0.132231	0.214193	0.617347	5.328035	0.615805
3.6	0.125366	0.203686	0.615484	5.817298	0.609701
3.7	0.119012	0.193904	0.613769	6.348844	0.60401
3.8	0.113122	0.184783	0.612188	6.925566	0.598698
3.9	0.107652	0.176269	0.610728	7.550504	0.593732
4	0.102564	0.16831	0.609375	8.226849	0.589086

Table B6. Rayleigh flow properties for $\gamma = 1.333$

M	P/P^*	T/T^*	ρ/ρ^*	Pt/Pt^*	Tt/Tt^*
0.1	2.299418	0.052873	43.48927	1.250806	0.04546
0.2	2.212305	0.195772	11.30043	1.227569	0.169154
0.3	2.080915	0.389718	5.339533	1.193279	0.33949
0.4	1.921174	0.590546	3.253219	1.153089	0.520288
0.5	1.748593	0.764394	2.287554	1.112057	0.683198
0.6	1.575602	0.893708	1.762995	1.074316	0.812699
0.7	1.410668	0.975092	1.446702	1.042788	0.904659
0.8	1.258643	1.013877	1.241416	1.019301	0.962182
0.9	1.121648	1.019057	1.100673	1.00487	0.991634
1	1	1	1	1	1
1.1	0.89296	0.964826	0.925513	1.004922	0.993523
1.2	0.799259	0.919894	0.86886	1.019766	0.977219
1.3	0.717431	0.869855	0.824771	1.044677	0.954862
1.4	0.646002	0.817944	0.789787	1.079872	0.929157
1.5	0.583594	0.76631	0.761564	1.125691	0.901977
1.6	0.528968	0.716307	0.738466	1.182606	0.874571
1.7	0.481037	0.668737	0.719322	1.251244	0.847746
1.8	0.438861	0.62402	0.70328	1.332387	0.821993
1.9	0.401634	0.582329	0.689703	1.426978	0.797591
2	0.368671	0.543673	0.678112	1.536129	0.774676
2.1	0.339388	0.507962	0.668136	1.661122	0.753288
2.2	0.31329	0.475049	0.65949	1.803419	0.73341
2.3	0.289956	0.444754	0.651947	1.964664	0.714985
2.4	0.269028	0.416887	0.645327	2.146697	0.697937
2.5	0.250201	0.391254	0.639485	2.351558	0.682176
2.6	0.233215	0.36767	0.634304	2.581502	0.667613
2.7	0.217845	0.345956	0.629689	2.839006	0.654154
2.8	0.203899	0.325948	0.625558	3.126785	0.641712
2.9	0.191214	0.307493	0.621848	3.447803	0.630203
3	0.179645	0.290452	0.618503	3.805288	0.619548
3.1	0.16907	0.274698	0.615476	4.202747	0.609675
3.2	0.159379	0.260115	0.612728	4.64398	0.600518
3.3	0.150481	0.246598	0.610226	5.1331	0.592016
3.4	0.142292	0.234055	0.607942	5.674546	0.584113
3.5	0.13474	0.222399	0.605851	6.273102	0.576757
3.6	0.127764	0.211553	0.603932	6.933917	0.569904
3.7	0.121306	0.201449	0.602166	7.662524	0.563512
3.8	0.115317	0.192023	0.600537	8.464858	0.557542
3.9	0.109754	0.183219	0.599033	9.347276	0.55196
4	0.104578	0.174985	0.597639	10.31658	0.546735

Table B7. Fanno flow properties for $\gamma = 1.4$

M	P/P^*	T/T^*	ρ/ρ^*	Pt/Pt^*	$4fL^*/D$
0.1	10.94351	1.197605	9.137833	5.821829	66.92156
0.2	5.455447	1.190476	4.582576	2.96352	14.53327
0.3	3.619057	1.178782	3.070167	2.035065	5.299253
0.4	2.695819	1.162791	2.318405	1.59014	2.308493
0.5	2.13809	1.142857	1.870829	1.339844	1.06906
0.6	1.763364	1.119403	1.575272	1.1882	0.490822
0.7	1.493452	1.092896	1.366509	1.094373	0.208139
0.8	1.289277	1.06383	1.21192	1.03823	0.07229
0.9	1.129133	1.032702	1.093377	1.008863	0.014512
1	1	1	1	1	0
1.1	0.893588	0.966184	0.924863	1.007925	0.009935
1.2	0.804362	0.931677	0.863348	1.03044	0.033638
1.3	0.728483	0.896861	0.812258	1.066305	0.064832
1.4	0.663198	0.862069	0.769309	1.114926	0.099738
1.5	0.606478	0.827586	0.732828	1.176167	0.13605
1.6	0.556794	0.793651	0.701561	1.250235	0.172357
1.7	0.512966	0.760456	0.67455	1.337606	0.207803
1.8	0.474067	0.728155	0.651052	1.438982	0.241886
1.9	0.43936	0.696864	0.630482	1.555257	0.274328
2	0.408248	0.666667	0.612372	1.6875	0.304997
2.1	0.380243	0.63762	0.596348	1.836944	0.333851
2.2	0.35494	0.609756	0.582102	2.004975	0.36091
2.3	0.332002	0.58309	0.569383	2.193131	0.38623
2.4	0.311142	0.557621	0.557981	2.4031	0.409889
2.5	0.292119	0.533333	0.547723	2.636719	0.431977
2.6	0.274725	0.510204	0.538462	2.895975	0.452588
2.7	0.258783	0.488202	0.530074	3.183011	0.471819
2.8	0.244138	0.46729	0.522455	3.500123	0.489765
2.9	0.230655	0.447427	0.515514	3.849768	0.506516
3	0.218218	0.428571	0.509175	4.234568	0.522159
3.1	0.206723	0.410678	0.503371	4.657311	0.536777
3.2	0.19608	0.393701	0.498043	5.120958	0.550444
3.3	0.186209	0.377596	0.493142	5.628647	0.563232
3.4	0.177038	0.362319	0.488625	6.183699	0.575207
3.5	0.168505	0.347826	0.484452	6.789621	0.586429
3.6	0.160554	0.334076	0.48059	7.450111	0.596955
3.7	0.153133	0.321027	0.47701	8.169066	0.606836
3.8	0.146199	0.308642	0.473684	8.950585	0.616119
3.9	0.13971	0.296883	0.47059	9.798973	0.624849
4	0.133631	0.285714	0.467707	10.71875	0.633065

Table B8. Fanno flow properties for $\gamma = 1.333$

M	P/P^*	T/T^*	ρ/ρ^*	Pt/Pt^*	$4fL^*/D$
0.1	10.78462	1.163081	9.272461	5.866473	70.53457
0.2	5.379037	1.157361	4.647672	2.984731	15.35358
0.3	3.571419	1.147953	3.11112	2.047994	5.613986
0.4	2.663451	1.135035	2.34658	1.598603	2.453087
0.5	2.115512	1.118848	1.890795	1.345408	1.139696
0.6	1.74776	1.099679	1.589336	1.191701	0.525
0.7	1.48314	1.077855	1.37601	1.096361	0.223389
0.8	1.28314	1.053726	1.217716	1.039139	0.077852
0.9	1.12637	1.027654	1.096059	1.0091	0.015682
1	1	1	1	1	0
1.1	0.895866	0.971117	0.922511	1.008192	0.010808
1.2	0.808521	0.941338	0.858907	1.031584	0.036711
1.3	0.734192	0.910975	0.805941	1.069083	0.070977
1.4	0.670177	0.880308	0.761298	1.120283	0.109524
1.5	0.614488	0.84959	0.723276	1.185281	0.149836
1.6	0.56563	0.819038	0.690602	1.264569	0.190354
1.7	0.522451	0.788841	0.662302	1.358968	0.230119
1.8	0.484053	0.759155	0.63762	1.46959	0.268549
1.9	0.449718	0.73011	0.61596	1.597817	0.305312
2	0.41887	0.701807	0.596844	1.74529	0.340233
2.1	0.391035	0.674326	0.57989	1.913907	0.373243
2.2	0.365825	0.647726	0.564783	2.105829	0.40434
2.3	0.342913	0.622047	0.551265	2.323484	0.433565
2.4	0.322025	0.597313	0.539123	2.569583	0.460987
2.5	0.302929	0.573538	0.528176	2.84713	0.486691
2.6	0.285426	0.550723	0.518275	3.15944	0.510768
2.7	0.269344	0.52886	0.509291	3.510155	0.533316
2.8	0.254534	0.507935	0.501116	3.903265	0.55443
2.9	0.240868	0.487927	0.493656	4.343125	0.574204
3	0.228233	0.468813	0.486832	4.834484	0.592729
3.1	0.216529	0.450564	0.480574	5.382498	0.610092
3.2	0.205669	0.43315	0.474822	5.992764	0.626373
3.3	0.195575	0.41654	0.469524	6.671339	0.641648
3.4	0.186179	0.400702	0.464633	7.424767	0.655989
3.5	0.17742	0.385602	0.460111	8.260109	0.669462
3.6	0.169241	0.371208	0.45592	9.184969	0.682128
3.7	0.161595	0.357488	0.45203	10.20752	0.694045
3.8	0.154438	0.34441	0.448414	11.33655	0.705266
3.9	0.147729	0.331942	0.445045	12.58147	0.715838
4	0.141433	0.320055	0.441904	13.95235	0.725808

Appendix C

Table C1. International standard atmosphere; Note: Density of air at sea level = 1.225 kg/m^3

$z[m]$	$p[kPa]$	$T[K]$	$\rho/\rho o$	$a[m/s]$
0	101.325	288.2	1.0000	340.3
500	95.460	284.9	0.9529	338.4
1000	89.880	281.7	0.9075	336.4
1500	84.560	278.4	0.8638	334.5
2000	79.500	275.2	0.8217	332.5
2500	74.690	271.9	0.7812	330.6
3000	70.120	268.7	0.7423	328.6
3500	65.780	265.4	0.7048	326.6
4000	61.660	262.2	0.6689	324.6
4500	57.750	258.9	0.6343	322.6
5000	54.050	255.7	0.6012	320.5
5500	50.540	252.4	0.5694	318.5
6000	47.220	249.2	0.5389	316.5
6500	44.080	245.9	0.5096	314.4
7000	41.110	242.7	0.4817	312.3
7500	38.300	239.5	0.4945	310.2
8000	35.650	236.3	0.4292	308.1
8500	33.150	233.1	0.4047	306.0
9000	30.800	229.9	0.3813	303.8
9500	28.580	226.7	0.3589	301.7
10000	26.500	223.3	0.3376	299.5
10500	24.540	220.3	0.3172	297.4
11000	22.700	216.8	0.2978	295.2
11500	20.980	216.6	0.2755	295.1
12000	19.400	216.6	0.2546	295.1
12500	17.930	216.6	0.2354	295.1
13000	16.580	216.6	0.2176	295.1
13500	15.330	216.6	0.2012	295.1
14000	14.170	216.6	0.186	295.1
14500	13.100	216.6	0.172	295.1
15000	12.110	216.6	0.159	295.1
15500	11.200	216.6	0.147	295.1

Contd.

Table C1. Continued

$z[m]$	$p[kPa]$	$T[K]$	$\rho/\rho o$	$a[m/s]$
16000	10.350	216.6	0.1359	295.1
16500	9.572	216.6	0.1256	295.1
17000	8.850	216.6	0.1162	295.1
17500	8.182	216.6	0.1074	295.1
18000	7.565	216.6	0.0993	295.1
18500	6.995	216.6	0.09182	295.1
19000	6.467	216.6	0.08489	295.1
19500	5.980	216.6	0.0785	295.1
20000	5.529	216.6	0.07258	295.1

Appendix D

Table D1. Selected properties of air at 1 Atm

T(K)	ρ(kg/m^3)	Cp(kJ/kg-k)	$\mu \cdot 10^7$(N-s/m^2)	$\nu \cdot 10^4$(m^2/s)	$\kappa \cdot 10^3$(w/m.K)	$\alpha \cdot 10^6$(m^2/s)	Pr
100	3.5562	1.032	71.1	2.00	9.34	2.54	0.786
150	2.3364	1.012	103.4	4.426	13.8	5.84	0.758
200	1.7458	1.007	132.5	7.590	18.1	10.3	0.737
250	1.3947	1.006	159.6	11.44	22.3	15.9	0.720
300	1.1614	1.007	184.6	15.89	26.3	22.5	0.707
350	0.9950	1.009	208.2	20.92	30.0	29.9	0.700
400	0.8711	1.014	230.1	26.41	33.8	38.3	0.690
450	0.7740	1.021	250.7	32.39	37.3	47.2	0.686
500	0.6964	1.030	270.1	38.79	40.7	56.7	0.684
550	0.6329	1.040	288.4	45.57	43.9	66.7	0.683
600	0.5804	1.051	305.8	52.69	46.9	76.9	0.685
650	0.5356	1.063	322.5	60.21	49.7	87.3	0.690
700	0.4975	1.075	338.8	68.10	52.4	98.0	0.695
750	0.4643	1.087	354.6	76.37	54.9	109	0.702
800	0.4354	1.099	369.8	84.93	57.3	120	0.709
850	0.4097	1.110	384.3	93.80	59.6	131	0.716
900	0.3868	1.121	398.1	102.9	62.0	143	0.720
950	0.3666	1.131	411.3	112.2	64.3	155	0.723
1000	0.3482	1.141	424.4	121.9	66.7	168	0.726
1100	0.3166	1.159	449.0	141.8	71.5	195	0.728
1200	0.2902	1.175	473.0	162.9	76.3	224	0.728
1300	0.2679	1.189	496.0	185.1	82	238	0.719
1400	0.2488	1.207	530	213	91	303	0.703
1500	0.2322	1.230	557	240	100	350	0.685
1600	0.2177	1.248	584	268	106	390	0.688

Reference

1. Incorpera, F.P. and De Witt, D.P., "Fundamental of Heat and Mass Transfer", 3rd Edition, 1990.

Table D2. Selected properties of hydrocarbon fuels: Heat of formation, and higher and lower heating values all at 298.15 K and 1 atm; boiling points and latent heat of vaporisation at 1 atm; liquid density

Fuel	Fuel	Mw (kg/kmol)	$\bar{h}_f^o$ (kJ/kmol)	HHV (kJ/kg)	LHV (kJ/kg)	Boiling pt.(°C)	h_{fg} (kJ/kg)	ρ_{liq}^* (kg/m³)
CH_4	Methane	16.043	– 74.831	55.528	50.016	– 164	509	300
C_2H_2	Acetylene	26.038	226.748	49.923	48.225	– 84	-	-
C_2H_4	Ethane	28.054	52.283	50.313	47.161	– 103.7	-	-
C_2H_6	Ethane	30.069	– 84.667	51.901	47.489	– 88.6	488	370
C_3H_6	Propene	42.080	20.414	48.936	45.784	– 47.4	437	514
C_3H_8	Propane	44.096	– 103.847	50.368	46.357	– 42.1	425	500
C_4H_8	1-Butene	56.107	1.172	48.471	45.319	– 63	391	595
C_4H_{10}	n-Butane	58.123	– 124.733	49.546	45.742	– 0.5	386	579
C_5H^{10}	1-Pentene	70.134	– 20.920	48.152	45.000	30	358	641
C_5H_{12}	n-Pentane	72.150	– 146.440	49.032	45.355	36.1	358	626
C_6H_6	Benzene	78.113	82.927	42.277	40.579	80.1	393	879
C_6H_{12}	1-Hexene	84.161	– 41.673	47.955	44.803	63.4	335	673
C_6H_{14}	n-Hexane	86.177	– 167.193	48.696	45.05	69	335	659
C_7H_{14}	1-Heptene	98.188	– 62.132	47.817	44.665	93.6	-	-
C_7H_{16}	n-Heptane	100.203	– 187.820	48.456	44.926	98.4	316	684
C_8H_{16}	1-Octene	112.214	– 82.97	47.712	44.560	121.3	-	-
C_8H_{18}	n-Octane	114.230	– 208.447	48.275	44.791	125.7	300	703
C_9H_{18}	1-Nonene	126.241	– 103.512	47.631	44.478	-	-	-
C_9H_{20}	n-Nonane	128.257	– 229.032	48.134	44.686	150.8	295	718
$C_{10}H_{20}$	1-Decene	140.268	– 124.139	47.565	44.413	170.6	-	-
$C_{10}H_{22}$	n-Decane	142.284	– 249.659	48.020	44.602	174.1	277	730
$C_{11}H_{22}$	1-Undecene	154.295	– 144.766	47.512	44.360	-	-	-
$C_{11}H_{24}$	n-Undecane	156.311	– 270.286	47.926	44.532	195.9	265	740
$C_{12}H_{24}$	1-Dodecene	168.322	– 165.352	47.468	44.316	213.4	-	-
$C_{12}H_{26}$	n-Dodecane	170.337	– 292.162	47.841	44.467	216.3	256	749

Index